THE GRANDIN PAPERS

Over 50 Years of Research Articles on Animal Behavior and Welfare That Improved the Livestock Industry

All marketing and publishing rights guaranteed to and reserved by:

FUTURE HORIZONS
(817) 277-0727
(817) 277-2270 (fax)
E-mail: info@fhautism.com
www.fhautism.com

ISBN: 9781957984292

Printed in Canada

THE GRANDIN PAPERS

Over 50 Years of Research Articles on Animal Behavior and Welfare That Improved the Livestock Industry

DR. TEMPLE GRANDIN

FUTURE HORIZONS

CONTENTS

FOREWORD

Looking Back on 50 Years of Research on Farm Animal Behavior, Handling Facility Design, and Welfare of Cattle and Other Livestock

By Temple Grandin
Distinguished Professor of Animal Science, Colorado State University

This book is a compilation of some of my most important scientific papers with an emphasis on livestock facility design, the development of animal welfare auditing programs, and behavior of cattle and horses during handling. They have been placed in chronological order. I chose the papers and book chapters that would most likely be of interest to the general public, policymakers, and people interested in farm animal welfare and behavior. This anthology will show how my thinking has evolved over a fifty-year career. When I started my work in the early 1970s, cattle handling on many ranches, feedlots, and slaughter plants was often rough and sometimes cruel. I wanted to change this, and at the time, I thought I could solve all the problems with improved equipment. This is a common mistake that is made by many young engineers and policymakers. They think that either a technology or a new method will solve all the problems.

Over the years, I learned that well-designed equipment will make livestock handling easier, but it does not replace management. Good animal welfare requires both well-designed equipment and a commitment by top management to maintain high standards. Today handling of cattle and other farm animals has improved, but it is still necessary to constantly remind people about the basics. Some examples of the basics are having nonslip flooring, not yelling at animals, and using behavioral principles of handling instead of electric prods. I often get asked, "What is the most important thing that you did to improve animal welfare?" In the first twenty-five years of my career, I focused on equipment design and then I switched my emphasis to welfare assessment. This switch occurred after I was hired in 1999 by McDonald's Corporation, Wendy's International, and Burger King to assess animal welfare at slaughter plants. When these large corporate buyers enforced standards, big positive changes occurred. In this single year, I saw huge improvements. The plant managers had to repair broken equipment, add nonslip floors, employ behavioral principles of handling, and make simple changes in facilities, such as adding a light to a dark chute entrance. Cattle and other animals do not like to enter dark places.

Prior to this, many large beef plants already used equipment I had designed. Unfortunately, some people tore it up and wrecked it. Some of my major papers describe the development of a simple, objective animal welfare assessment method for slaughter plants. The scoring system was very effective because it was outcome-based, objective, and easy to understand. I like to use traffic rules as an analogy. Enforcement of a few simple rules will greatly improve public safety. To be effective, the most important rules have to be chosen for enforcement, including drunken driving, speeding, stopping violations, seatbelts, and texting. These rules are not vague. Speeding and drunken driving are measured with a device. Vague guidelines, such as "handle cattle properly," do not work, because one person's definition of proper is different from another person's. Numerical outcome measures of the consequences of poor handling are easy to access. Papers are included on scoring the effectiveness of stunning, slips and falls, electric prod use, and vocalization.

Throughout my career, I have also done basic research on animal behavior during handling. Some of the areas of research are how animals make choices, visual perception of rotated objects, and training of animals to cooperate with veterinary procedures. This greatly reduces fear and stress. My first graduate student's paper on cattle temperament and weight gain is also included.

The last section contains my more recent papers. I have included research on animal perception and sustainability. I want people who are interested in animal welfare to learn from my work.

Many times throughout my career, I get asked "Why do I still work for the meat industry, instead of being an activist against it?" Cattle handling was atrocious in the 1970s, but the cattle in the dry Arizona feedlots, where I started my career, had good living conditions. All the feedlots had shades and the pens remained dry. At this early point in my career, if I had been exposed to muddy, filthy pens that lacked shade, it is likely that I could have gone down the activist road.

I visualized a future where cattle handling could be improved, because there were a few people who handled cattle with compassion. Two of the cattle producers who made a big positive impression on me were Bill and Penny Porter at Singing Valley Ranch, a kind stockperson who taught me how to gently run a squeeze chute, and a wonderful dairy manager. These good managers convinced me that it was possible to raise animals that had a decent life. I dedicated my work to making the rest of the industry improve.

Temple Grandin

ONE

Observations of Cattle Behavior Applied to the Design of Cattle Handling Facilities

T. Grandin (1980) Observations of Cattle Behavior Applied to the Design of Cattle Handling Facilities, *Applied Animal Ethology*, Vol. 6, No. 1, pp. 19-31.

This was my first published scientific paper. It contains my hand drawn drawings of designs for cattle-handling facilities. It also contains my first flight zone diagram, which has slowly improved and evolved over the years. I was greatly inspired by the sheep handling work of Ron Kilgore in New Zealand. Jack Albright from Purdue University encouraged me to write this paper. This shows the importance of mentors in helping students get started in research.

ABSTRACT

Field observations were conducted while cattle were being handled in abattoirs, auction markets, yards on ranches, dipping vats and restraining chutes. Mature cattle and calves of many different breeds were observed under commercial conditions. A review of the literature and the observations indicated that cattle can be most efficiently handled in yards and races which have long narrow diagonal pens on a 60° angle. In yards designed by the author, cattle which are waiting to be sorted are held in a 3–3.5 m-wide curved race with an inside radius of 7.5–11 m. From the curved race, the animals can either be sorted into the diagonal pens, or they can be directed to the squeeze chute, dipping vat, or restraining chute at the abattoir. The handler works from a catwalk which is located along the inner radius of the race. This facilitates the movement of the animals because they will tend to circle around the handler in order to maintain visual contact. The curved holding race terminates in a round crowding pen which leads to a curved single file race. Cattle have 360° panoramic vision and poor depth perception. Sharp contrasts of light and dark should be avoided. Single file races, forcing pens, and other areas where cattle are crowded should have high solid fences. This prevents the animals from observing people, vehicles, and other distracting objects outside the facility.

INTRODUCTION

Field observations and a review of the literature were conducted to determine the most efficient and humane designs for yards and races for handling cattle. The Livestock Conservation Institute (1974) estimates that the cattle industry loses $22 million annually from bruises in the U.S.A. The discussion. will be limited to beef cattle-handling facilities where there is little opportunity for learned behavior. At an abattoir or auction market the animals pass through the facility only once. In cattle feeding operations each animal will be handled in the facility once or twice during its lifetime. On ranches, cattle are handled only once or twice a year.

MATERIALS STUDIED

Field observations were conducted while cattle were being handled in abattoirs, auction markets, yards on ranches, dipping vats and restraining chutes. Both *Bos taurus* and *Bos indicus* were observed, along with many crossbreds. The observations were conducted throughout the Southwestern, Midwestern and Northwestern United States, from 1973 to 1978, in over 100 different cattle operations. The author actually worked with the regular employees in order to gain a more complete understanding.

LITERATURE REVIEW AND OBSERVATIONS

Effect of noise on handling

Observations indicated that equipment should be designed to minimize loud noises. The incidence of balking was lower in hydraulically actuated restraining chutes when the motor was located off to one side instead of on top of the restrainer (Grandin, 1975). Webb (1966) reported that cattle would move away from a compressed air jet or siren. When air cylinders are used to operate gates the exhausts should be muffled. Cattle are more sensitive to high frequency noise than humans. The auditory sensitivity of cattle is greatest at 8000 hz and the human ear is most sensitive at 1000–3000 hz (Ames, 1974). Falconer and Hetzel (1964) found that exploding firecrackers caused visible fright in sheep. Cattle will move more easily if handlers are quiet and refrain from yelling. All welded construction and rubber stops on gates will reduce noise in steel facilities.

Effect of odor on handling

Cattle have been observed refusing to enter, or balking at, the entrance to either a stunning pen or a restraining chute when there was blood on the floor. The animal would stop and sniff the blood (Grandin, 1975). People who work in abattoirs have reported that the cattle will often refuse to enter the abattoir when the wind is blowing abattoir odors towards them.

Visual perception and handling

Kilgour (1972) stated that cattle are visual animals that are easily motivated by fear. They have 360° panoramic vision with a binocular visual field of 25–50° depending on the breed (Prince, 1970; McFarlane, 1976a). Cattle have poor depth perception. They can detect movement behind themselves without turning their heads.

Cattle are able to distinguish colors (Albright, 1969). They can distinguish all of the colors from gray except dark blue (Hebel and Sambraus, 1976; Sambraus, 1978). Observations by the author indicated that cattle will often balk and refuse to cross a shadow or drain grate. The illumination should be even and there should be no sudden discontinuity in the floor level or texture (Lynch and Alexander, 1973). Slatted type sunshades which cast a zebra stripe shadow pattern can cause balking. For most efficient handling, yards and races should be painted one uniform color.

Observations also revealed that cattle will often refuse to enter buildings or pass under overhead catwalks. At one abattoir the animals would readily enter the building at night, but they often refused to enter during the daytime. At night the inside of the building was brightly illuminated compared to the stockyards. Extending the single file races out into the stockyards reduced balking. Cattle will enter a truck, building or other dimly lit enclosure more readily if they are lined up in a single file race. Lighting in indoor facilities should be diffused and bare bulbs should be covered. Loading ramps should be directed so that the animals do not have to look directly into the sun. Il luminating the interior of trucks at night will reduce balking during loading.

Curved races with solid sides

In areas where animals are crowded, they should not be able to see either through or over the fences (Rider et al., 1974; Grandin, 1977a; McFarlane, 1976b). Crowding-pen gates should also be solid, otherwise the animals will look out through the Crowding-pen gate instead of facing the entrance to the single file race (Rider et al., 1974). Cattle will stand more quietly and move more easily through a single file race with solid sides. Sheep moved more rapidly through a single file race with solid sides (Hutson and Hitchcock, 1978).

Observations at an abattoir indicated that cattle balked frequently in a crowding pen built from bars which enabled the animals to see people outside the pen. The cattle entered the single file race from the crowding pen with less hesitation after the fences were covered. Experience indicated that straight single file races were inefficient because the animals tend to balk and back up to the end of the race. Curved races prevent the animals from seeing the restrainer, stunning pen, or an auction ring surrounded by people, until they are almost in it. For mature cows the inside radius of a curved single file race should be 5 m (McFarlane, 1976b,c). For smaller cattle a 3.54 m inside radius will work (Grandin, 1978a).

Cattle can be handled more easily if they maintain visual contact with the handler (Williams, 1978). They will turn and face a person and circle around him when he enters the pen. Cattle can be driven more efficiently when the handler orients himself at an angle from the animal's shoulder (Fig.1) instead of standing directly behind it (Williams, 1978). A curved race should have a catwalk for the handler along the inner radius (Fig.2). The catwalk should be alongside the fence and NOT overhead. A handler working from the catwalk is forced to stand at the correct angle from the animal's shoulder (Fig. 1) (Grandin, 1978b). In the curved race in Fig. 2 the cattle can see the handler but outside distractions are blocked.

Zebra could be moved with less excitement with a helicopter if the machine moved slowly, which enabled the animals to maintain visual contact (Oelofse, 1970). Cheviot sheep would twist their heads when they entered a strange pen to maintain visual contact with the handler and his dog (Whateley et al., 1974). Both cattle and sheep maintain visual contact with their herd-mates (Crofton, 1958; Strickland, 1978b).

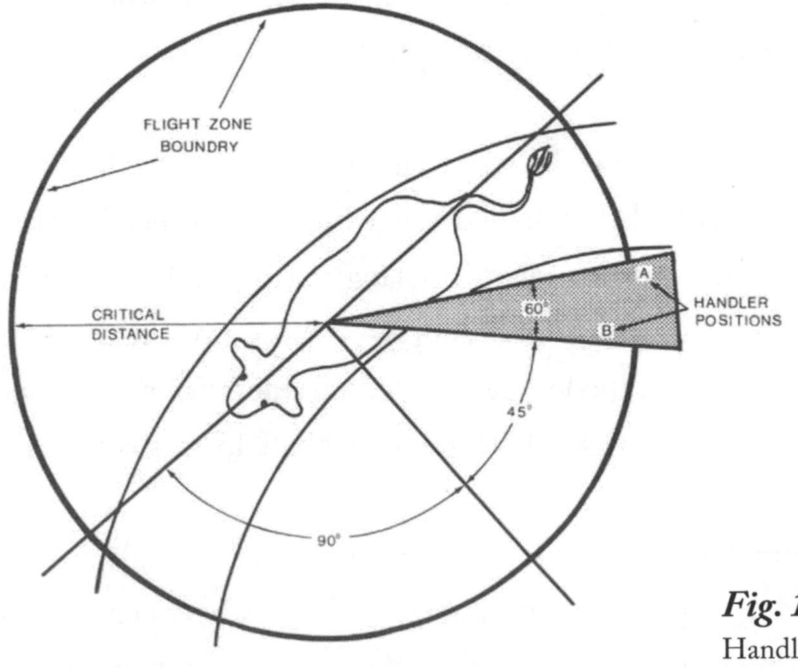

Fig. 1. Handler positions for driving cattle.

Fig. 1.
Handler positions for driving cattle.

Fig. 2.
Wide curved holding race.

Following the leader

A well-designed handling facility takes advantage of the animal's tendency to follow the leader (Ewbank, 1961; Hafez et al., 1969). The next animal in line should be able to observe the animal in front of it moving down the race. The sides of the race should be solid, but sliding gates, one-way gates, and the tailgate on the restraining chute should be constructed so that the animals can see through them. Observations indicate that this will facilitate handling.

The single file race must be long enough to maintain a smooth flow of cattle. Observations in many abattoirs revealed that an abattoir which slaughters 30 cattle or less per hour should have a 9 m-long race. An abattoir slaughtering 100 cattle per hour would need 30 lineal meters. This insures a steady supply of cattle even if the handlers in the yards have difficulties. For most ranches, cattle-feeding operations and auctions a 9–15 m-long race is recommended. A single file race longer than 15 mis is not usually needed for these operations.

Following behavior can impede cattle movement if animals in adjacent races observe each other moving in the opposite direction. Fences between races where cattle move in the opposite direction should be solid. If the cattle movement is in the same direction the fence between two adjacent races should be constructed so that the animals can see through it.

Flight distance

The flight distance is that radius of surrounding area within which intrusion provokes a flight reaction (Fraser, 1974; McFarlane, 1976c). Williams (1978) and Strickland (1978a) have done research on flight distance. The flight distance during handling is usually 1.5–7.6 m for beef cattle raised in a feeding operation and up to 30 m on mountain ranges. Brahman cattle have a larger flight distance than most English breeds (Grandin, 1978b).

The critical distance for most efficient cattle movement is on the boundary of the flight zone (Fig.1). Bulls maintained a fixed distance between themselves, and a moving mechanical trolley and they turned and ran past the trolley when cornered (Kilgour, 1971). Penetrating the flight zone will cause the animal to move away, and retreating from the flight zone will cause the animal to stop moving. Deep penetration of the flight zone will cause the animal to either run away or turn and run back past the handler. If an animal starts to turn back the handler should retreat instead of rushing up closer (Williams, 1978). To make an animal move through a curved race he should move from Position A to Position B (Fig.1). As soon as the animal has moved forward as far as possible the handler should retreat back to Position A. A common cause of excitement and rearing up in a single file race is handlers leaning over the fences over the animals.

Loading and unloading ramps

There is a difference in the design of ramps which will be used solely for unloading trucks and ramps which are used for both loading and unloading.

When cattle are exiting from a confined area they will move more readily if they have a broad clear path to freedom. Ramps in abattoirs which are used for *unloading* only are 2.35–3 m wide. This type of ramp is hazardous for loading.

Ramps for both loading and unloading should be narrow enough to force the animals into single file. Either a curve or a 15° bend is recommended to prevent the cattle from seeing the truck until they are part way up the ramp. The curve must not be too sharp otherwise the cattle are likely to balk because the ramp will appear as a dead end. Observations indicated that the most efficient ramps had solid fences and an inside radius of 3.5–5 m. The inside width should be 70–75 cm for mature cattle. The ramp should have stairsteps with a 10cm rise and a 30 cm tread width (Grandin, 1978c).

Single animals alarmed

Cattle are less likely to become alarmed when they are together as a group and touching each other (Ewbank, 1968). A lone animal will often attempt to jump a fence to rejoin its herd-mates. A tame cow placed in a race with a wild cow will help keep her calm (Swan, 1975; Elings, 1977). In auctions, a single file race should be built to handle cattle which are sold as individuals. At one auction market, many calves were injured because each individual animal had to be separated from the group before entering the ring.

Pen shape

The Apache on the San Carlos Reservation in Arizona and ranchers in the Northwestern U.S.A. have used round yards for gathering range cattle for the last 100 years. Circular yards have also been described by Daly (1970), Ward (1958), and McFarlane (1976b,c). Since a round yard has no corners for the cattle to bunch up in, they tend to circle the fence and stay together in a cohesive mob. A handler can then direct groups of cattle out of the round yard as they circle about him.

Whereas round yards are efficient for gathering and handling large groups of cattle, long narrow pens are more efficient for sorting and holding groups of cattle prior to transport or slaughter. Many large cattle feeding operations in the U.S.A. sort and handle cattle in long narrow pens constructed on a 60° angle. McFarlane (1976b,c) has conducted extensive research on the benefits of long narrow diagonal pens.

RESULTS

Knowledge gained from the observations and literature review was utilized to design new handling systems which would be more efficient, humane and labor-saving. A system was designed by the author for handling cattle through a dipping vat at the rate of 600 per hour with only three people (Figs. 3 and 4). The entrance and exit from the facility are set on a 60° angle to eliminate comers. Prior to dipping, the cattle are held in a wide curved race (Fig. 2).

This race may be 3–3.5 m wide with an inside radius of 7.5–11 m. The handler on the catwalk can easily direct groups of 15–20 cattle to the round crowding pen (Figs. 2 and 3). A 3.5m gate in the round crowding pen has a ratchet latch which latches automatically as the gate is advanced. The cattle then enter a curved single file race which leads to the dip vat (Fig. 3).

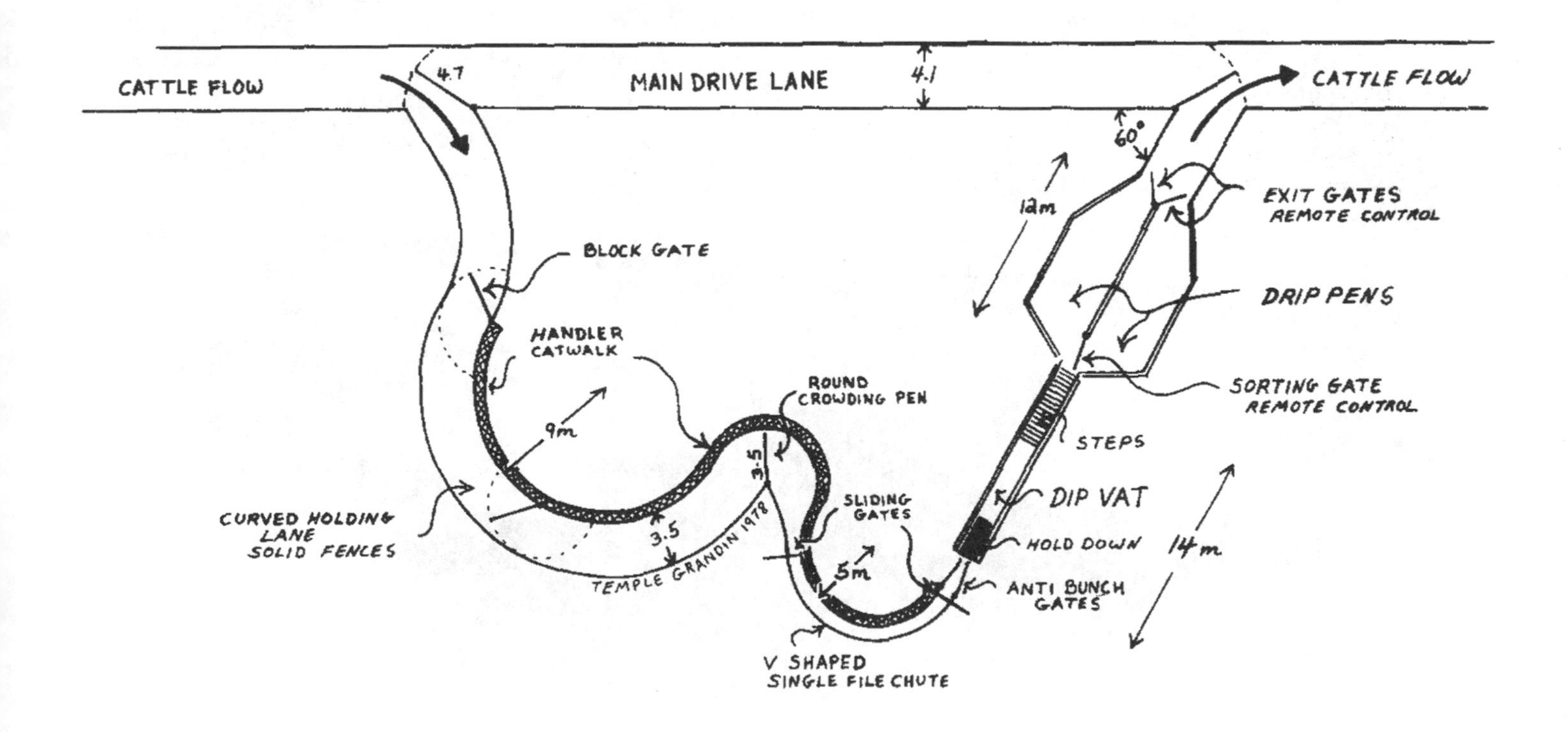

Fig. 3.
Handling system for dipping cattle with curved races.

Dip vat entrance

Figure 5 illustrates a dipping vat entrance which was designed to reduce problems with balky cattle and prevent the animals from jumping on each other. An adjustable hold-down rack forces the animals to immerse their heads. Observations indicated that 95% of the cattle were fully immersed and they did not have to be pushed under the water with a dipping stick.

The cleated ramp in Fig. 5 is on a relatively gradual, 25° angle and it appears to continue on into the water. When an animal steps out over the water it falls in due to the steep drop-off hidden under the water (Fig. 5). The ramp must have a nonskid surface because the animal will usually attempt to back out if it slips. The entrance also has two ant bunch gates (Figs. 3 and 5) which can be adjusted so that only one animal can enter the vat at a time.

Fig. 4.
Aerial view of dipping vat system with curved races.

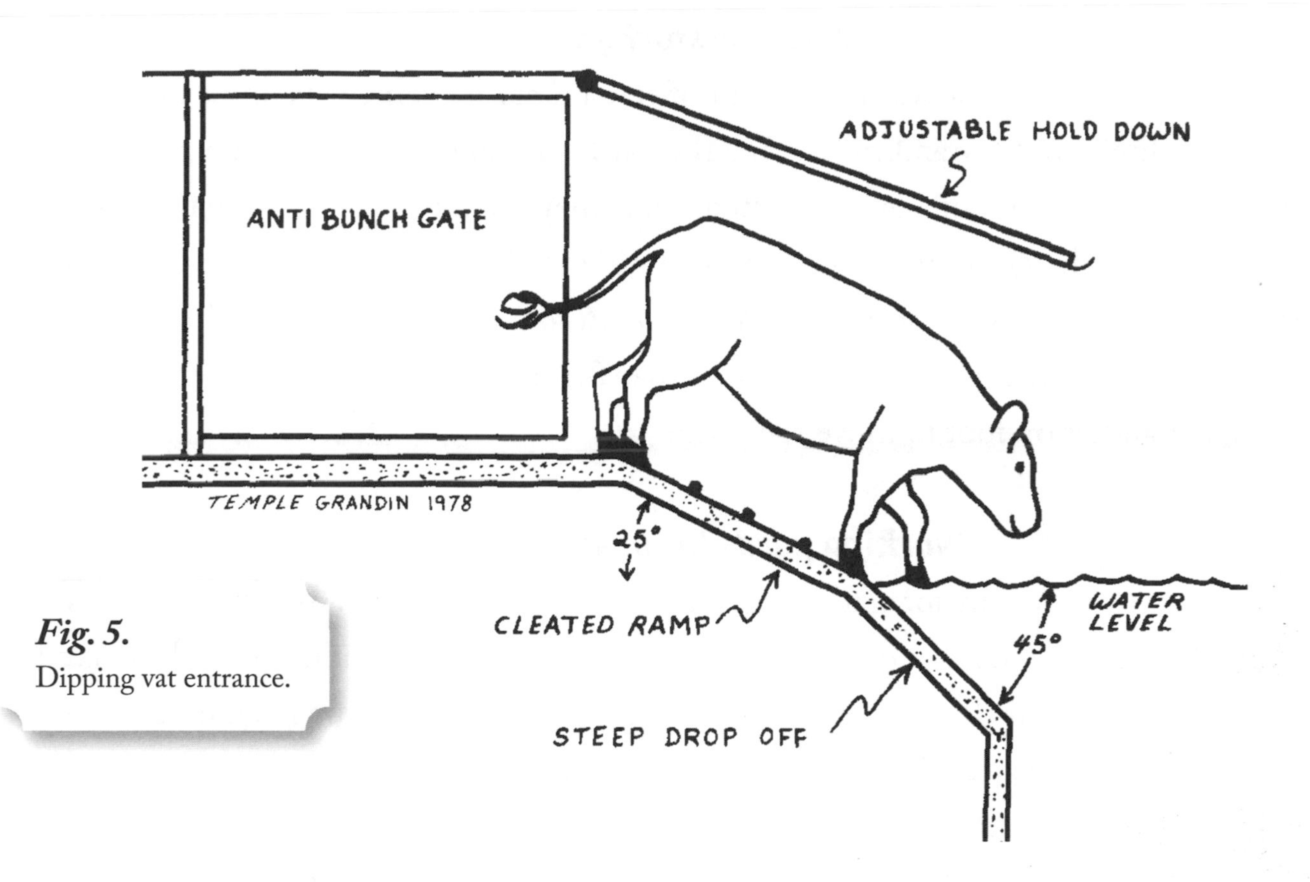

Fig. 5.
Dipping vat entrance.

Drip pens

The handler can release the cattle from the drip pens by opening the exit gates by remote control. The cattle have been observed to leave the drip pens 90% of the time with no assistance from the handler. An advantage of remote-controlled gates is that the handler can open the exit gates without penetrating the animal's flight zone. Cattle confined in a small pen will often become agitated when a person walks up to the fence. In drip pens with manually opened gates the cattle have been observed ramming fences and jumping back in the dip vat. The drip pen exit gates and the dividing fence between the two drip pens is solid to prevent cattle which are confined on one side from pushing on the gate and attempting to follow cattle which have been released.

Abattoir stockyard

Figure 6 is an abattoir stockyard where all of the animal movement is one way and 90° corners are eliminated (Grandin, 1977b). The cattle are unloaded on a wide ramp with stairsteps. The pens are built on a 60° angle, and they are 22 m long. The gates on the diagonal pens are longer than the drive race is wide to eliminate corners. On a 3m wide drive race, 3.5m gates are recommended and 4.1m gates are recommended on a 3.5m drive race. The drive race should not be wider than 3.5 m otherwise it will be very difficult for a single handler on foot to move the cattle.

Working yards for pasture cattle

Figure 7 is a yard system for gathering cattle from pasture for transport, sorting and handling. Most of the actual working and sorting of the cattle is conducted in the curved races and diagonal pens. The large pens labelled "gathering" and "holding" are used to hold cattle either before or after the actual handling or sorting operation. From the gathering pen, a handler can direct groups of 20–40 animals to the curved race which is labelled "sorting reservoir" (Fig. 7). The cattle will circle around the handler in the round gathering pen and follow the fence into the funnel which leads to the wide, curved race.

The curved race labelled "sorting reservoir" serves two functions (Figs. 2 and 7). It can be used to hold cattle which are waiting to go to the squeeze chute or loading ramp, or it can be used as a reservoir to hold cattle which are being sorted back into the diagonal pens. Sorting cattle back into the diagonal pens is efficient because the animals have a strong tendency to move back in the same direction they originally came from. Three people gathered one hundred cow and calf pairs and sorted the calves from their mothers in 45 min in this system.

During most cattle sorting operations, one of the sort-groups contains the majority of the animals. Passthrough gates are provided to allow cattle in the largest sort-group to enter the pen labelled "holding" (Fig. 7). The holding pen is shaped so that a forcing yard is created at both ends. This makes it possible to drive cattle which are in the holding pen back into the diagonal pen.

Grandin, T., 1977a. Processing feedlot cattle. In: I. Dyer and C. O'Mary (Editors), *The Feedlot.* Lea and Febiger, Philadelphia, pp. 213-232.

Grandin, T., 1977b. Cattle handling systems for meat-works. *Meat Research Newsletter*, CSIRO, Division of Food Research, Brisbane, Queensland, Australia, 77: 110.

Grandin, T., 1978a. Design of lairage, yard and race systems for handling cattle in abattoirs, auctions, ranches, restraining chutes and dipping vats. 1st World Congr. Ethology Applied to Zootechnics, Madrid, Spain, pp. 37-52.

Grandin, T., 1978b. Observations of the spatial relationships between people and cattle during handling. Proc. Western Sect., *Am. Soc. Anim. Sci.*, 29: 76-79.

Grandin, T., 1978c. Transportation from the animal's point of view. *Am. Soc. Agric. Eng.*, Technical Paper No. 786013, St. Joseph, Michigan.

Hafez, E.S.E., Schein, M.W. and Ewbank, R., 1969. The behaviour of cattle. In: E.S.E. Hafez (Editor), *The Behaviour of Domestic Animals*, 2nd edn., Williams and Wilkins, Baltimore, Maryland.

Hebel, R. and Sambraus, H.H., 1976. Are domestic mammals colorblind? *Ber. Miinch. Tierarztl. Wochenschr.*, 89: 321-325.

Hutson, G.D. and Hitchcock, D.K., 1978. The movement of sheep around corners. *Appl. Anim. Ethol.*, 4: 349-355.

Kilgour, R., 1971. Animal handling in works. Proc. 13th Meat Industry Conf., *Meat Industry Res. Inst.*, Hamilton, New Zealand, pp. 9-12.

Kilgour, R., 1972. Animal behaviour concepts and the veterinarian. In: J.L. Albright (Editor), *Animal Behavior.* School of Veterinary Science, Purdue University, Lafayette, pp. 73-78.

Livestock Conservation Institute, 1974. Livestock safety is a $61,000,000 word, 1100 Jorie Blvd., Oak Brook, Illinois.

Lynch, J.J. and Alexander, G., 1973. *The Pastoral Industries of Australia.* Sydney University Press, Sydney, Australia, pp. 371-400.

McFarlane, L, 1976a. A practical approach to animal behavior. In: M.E. Ensminger (Editor), *Beef Cattle Science Handbook.* Agri services Foundation, Clovis, California, 13: 420-426.

McFarlane, I., 1976b. Rationale in the design of housing and handling facilities. In: M.E. Ensminger (Editor), *Beef Cattle Science Handbook.* Agri-services Foundation, Clovis, California, 13: 223-227.

McFarlane, I., 1976c. Personal communication.

Oelofse, J., 1970. Plastic for game catching. *Oryx*, 10: 306-308.

Prince, J.H., 1970. The eye and vision. In: M.J. Swenson (Editor), *Dukes Physiology of Domestic Animals.* Cornell, New York, pp. 11351159.

Rider, A., Butchbaker, A.F. and Harp, S., 1974. Beef working, sorting, and loading facilities. *Am. Soc. Agric. Eng.*, Technical Paper No. 744523, St. Joseph, Michigan.

Sambraus, H.H., 1978. Personal communication.

Strickland, W.R., 1978a. Knowledge of animal behavior can reduce livestock handling and management problems. Proc. 62nd Annu. Meet. Livestock Conservation Institute, 1100 Jorie Blvd., Oak Brook, Illinois, pp. 8386.

Strickland, W.R., 1978b. Personal communication.

Swan, R., 1975. About Al facilities. *New Mexico Stockman*, February, pp. 24-25.

Ward, F., 1958. *Cowboy at Work.* Hasting House Publishers, New York, pp. 19-105.

Webb, T.F., 1966. Feasibility tests of selected stimuli and devices to drive livestock. Agric. Res. Serv., U.S. Dept. Agric. Agric. Res. Puhl., No. 5211.

Whateley, J., Kilgour, R. and Dalton, D.C., 1974. Behaviour of hill country sheep breeds during farming routines. *Proc. N.Z. Soc. Anim. Prod.*, 34: 28-36.

Williams, C., 1978. Livestock consultant, personal communication.

TWO

Bruises on Southwestern Feedlot Cattle

T. Grandin (1981) Bruises on Southwestern Feedlot Cattle,
Journal of Animal Science, Vol. 53, Supplement a:213.

This is an abstract of my first research study that I did at the Swift plant in Tolleson, Arizona. Even though this study is over forty years old, it is still relevant today. There were three major findings: bruises increase due to rough handling, overloaded trucks, and the method of producer payment. My abstract uses the term discountable bruises. A discountable bruise results in a deduction from the payment to the producer. Bruises have to be trimmed out of the carcass and cannot be used for human consumption. When cattle are sold based on live weight, the plant absorbs the loss. When the cattle are sold based on carcass weight, the producer's pay is discounted. Holding people financially accountable for bruises greatly reduced them and improved handling and trucking practices.

1523 head of 454kg. fed feedlot cattle were surveyed for discountable bruises in a large southwestern slaughter plant.

- The cattle were crossbreds with 1/8 to 1/2 Brahman breeding.
- In each group of animals 25% to 50% had either tipped or complete horns.
- The cattle were transported to the slaughter plant from six different feedlots.
- The distance transported was 30 to 240 km.

Overall, the cattle had 10.5% discountable bruises, and 5% of them had discountable loin bruises.

- Out of the 5% that had loin bruises, 2.3% had loin bruises that extended all the way through the carcass.

  Location of the bruises:
  - 45% loin
  - 23% back and withers
  - 2% rump
  - 15% flank
  - 11% rib
  - 4% on the shoulder

Cattle sold by live weights had 14% discountable bruises and cattle sold on a carcass basis had 8% discountable bruises.

The producer gets the bruises deducted from his payment when the cattle are sold on a carcass basis.

13 truckloads of cattle were hauled from one feedlot in a 14.6m Wilson double deck fat cattle trailer which unloaded through a rear door.

- Loads consisting of 50 to 51 head had 10.2% discountable bruises overall, and 4.3% discountable loin bruises.
- Loads consisting of 48 to 49 head had 5.3% discountable bruises overall, and 1.3% discountable loin bruises.

Level of significance = < .01

Rough handling at the feedlot of origin was a major cause of bruises. A feedlot where rough handling occurred during weighing and loading was compared to a feedlot which had careful, quiet handling. **The rough feedlot had 15.5% discountable bruises and the careful, quiet feedlot had 8.35% discountable bruises.**

THREE

Double Rail Restrainer for Livestock Handling

T. Grandin (1988) Double Rail Restrainer for Livestock Handling,
Journal of Agricultural Engineering, Research, Vol. 41, pp. 327-338.

This paper covers the initial design of a conveyorized restrainer system I developed for large calves. The center track restrainer that I developed for large beef slaughter plants is based on this paper. It is described in detail in later papers. This system is now in use in many large beef plants in the U.S., Canada, Mexico, Australia, and other countries. All the drawings in this paper were hand-drawn by the author. In the early part of my career, I was primarily focused on improving equipment. Later in my career, I became more focused on the importance of good management and training employees.

A double rail restrainer conveyor has been operating for two years in a commercial calf slaughter plant. It is operating at a production rate of 300 small calves or 150 large veal calves/h. The calves straddle a moving double rail which supports them under the brisket and belly. The moving rail is formed from metal segments which are attached to an endless chain. Adjustable sides on each side of the conveyor can be rapidly positioned to accommodate different sized animals while the slaughter line is running.

This system has many advantages compared to existing V restrainer conveyor systems. Some of the advantages are less expensive to construct, animals enter the restrainer more easily, easier captive bolt stunning, and it is adjustable for a wide variety of animal sizes. It can handle calves weighing 23–226 kg and all sizes of sheep. Systems are under development for adult cattle and automatic stunning of pigs. A double rail for handling cattle for veterinary procedures is also being developed.

INTRODUCTION

There is a need for more efficient and humane systems for rapidly handling large numbers of livestock in slaughter plants and large cattle fattening operations. The V restrainer conveyor which is presently used in many slaughter plants was a major innovation.[1,2] This system consists of two obliquely angled conveyors which form a V. The animals ride in the V with their feet protruding through a space at the bottom. However, there have been some problems with this equipment. Cattle and calves often balk and refuse to enter the V restrainer conveyor. Large 200 kg veal calves had difficulty entering the restrainer and small baby calves crossed their legs and fell through the opening between the two conveyors.[3,4]

Pigs and sheep can be handled efficiently and humanely in a V restrainer conveyor. In lambs the V restrainer may damage meat quality by causing petechial haemorrhages.[5,6]

Petechial haemorrhages are small pinpoint haemorrhages in the fat and connective tissue. During electrical stunning V restrainer conveyors sometimes increase the incidence of petechial haemorrhages because the animal's skin and muscles are stretched when muscle spasms occur against the sides of the conveyor.

On large US cattle fattening operations, thousands of cattle have to be individually restrained for vaccinations and other veterinary procedures. Cattle are caught in a hydraulic squeeze restrainer with a stanchion head bail. They are sometimes injured when they strike the head bail.[7] These restrainers are also inefficient because in some operations 1–3% of the animals escape before the veterinary procedures are completed.[7] Two to four percent of the animals are bruised during handling in these devices.[8] Stressing an animal during handling can lower weight gains.[9,10]

2. PREVIOUS RESEARCH

Researchers at the University of Connecticut developed a laboratory prototype double rail conveyor restrainer system.[4,11] Calves and sheep straddled two moving double rails. The animal was supported under the belly and brisket. Their research demonstrated that calves and sheep would ride quietly on the double rail with a minimum of stress. The prototype was a major innovation, but many components needed to be invented to create a system which would work in a commercial slaughter plant.

Some of the components which needed to be developed were: a device at the entrance to the double rail conveyor which would reliably position a calf's legs on each side of the moving conveyor, a system for rapid adjustment for a wide variety of calf and sheep sizes and compatibility with existing shackling systems.

3. A NEW TYPE OF CALF AND SHEEP SYSTEM

A double rail restrainer system was constructed and installed in a commercial calf slaughter plant (Figs 1 and 2). The plant has a maximum production rate of 300 baby calves, or 150 large formula fed calves. Calf weights varied from 23 to 225 kg. Both large and small calves were often mixed together, so it was essential to have a system for rapid size adjustment. The plant also slaughtered sheep.

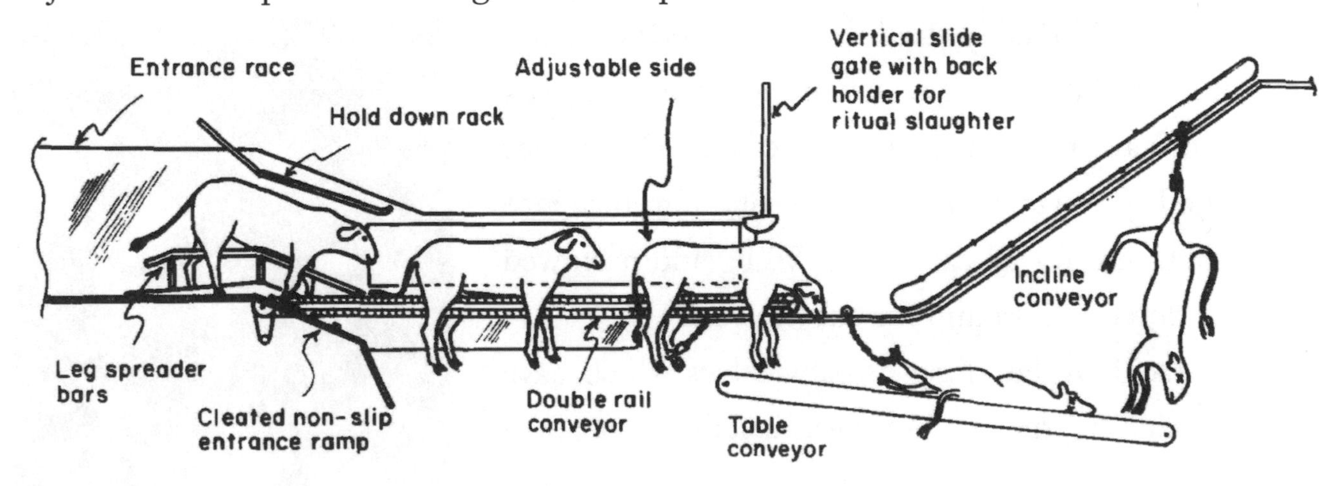

Fig. 1.
Side view of the double rail restrainer system for calves. This system has many advantages compared to the V restrainer conveyor

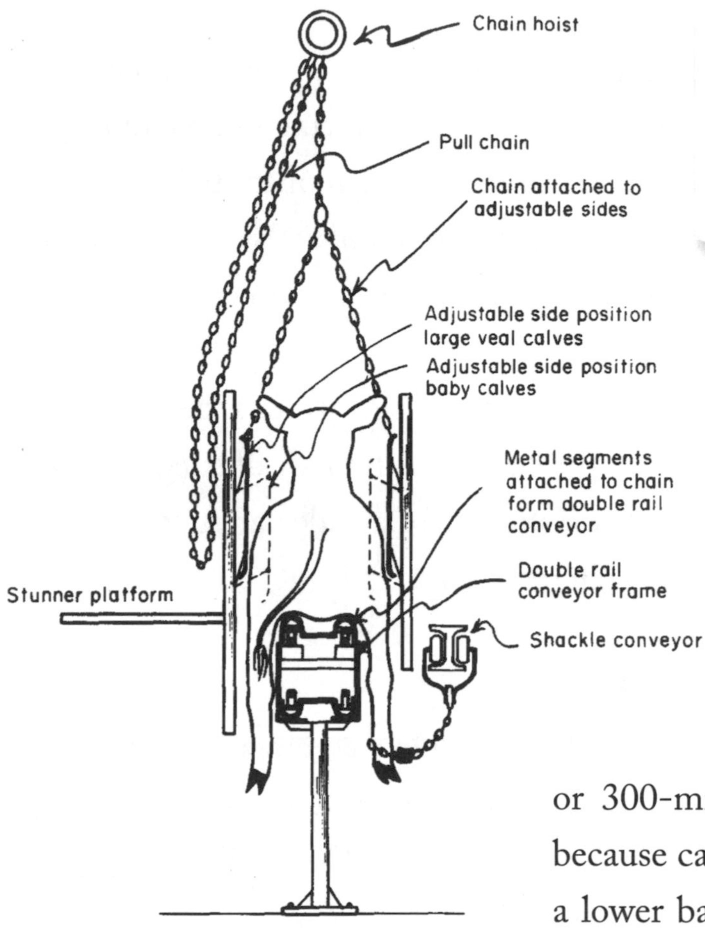

Fig. 2.
Cross-section of the double rail restrainer. Adjustable sides on pivots enable rapid adjustment for a wide variety of animal sizes. A chain hoist is used to adjust the width between the sides

Calves entering the restrainer straddle a 150 mm wide stationary leg spreader bar in the restrainer entrance (Fig. 3). The bar is 480 mm above the entrance race floor and the surface of the cleated ramp. The height of the leg spreader bar was determined by trying bars at different heights. A 200- or 300-mm high leg spreader bar worked poorly because calves stepped over it to one side. For sheep a lower bar is used. A 1–5 m long hold down rack prevents the calves from jumping up.

Calves will readily straddle the bar and walk into the restrainer. Baby calves which are too young to walk unassisted, have to be manually placed in the entrance and pushed. The width between the solid entrance race walls is 500 mm for all calves over 90 kg and narrowed with bars down to 330 mm for baby calves.

To facilitate handling of baby calves which have difficulty walking, the restrainer was installed in a pit to

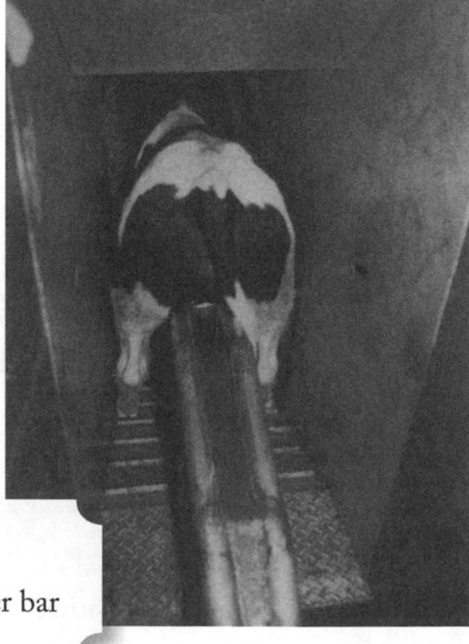

Fig. 3.
The calves' legs are positioned on each side of the conveyor by a leg spreader bar

Fig. 4.
The double rail conveyor is constructed from metal segments attached to a chain

eliminate ramps. If the system is going to be used with bigger calves or sheep, it can be installed above the floor with a ramp leading to the entrance.

The moving double rail conveyor is 5–48 m long. The animals straddle a conveyor constructed from stainless steel segments attached to an endless chain (Fig. 4). The top of the conveyor is even with the floor of the entrance race (Fig. 1). Each one-piece segment is bent to form a double rail configuration (Fig. 5). The total conveyor width of the stationary frame is 215 mm. The moving segments themselves are 190 mm. A 76 x 76 mm space in the middle of each segment accommodates the animal's brisket. The moving conveyor in the prototype constructed by Giger et al.[4] was 170 mm wide. The width of the conveyor was increased to 190 mm to accommodate the two chains the metal segments were attached to. The goal was to create a system which could be constructed from standard components available to the meat industry.

Adjustable sides were mounted above the moving conveyor to adjust the restrainer for different sized animals (Fig. 2). They lightly press against the top portion of the animal's body to hold it upright on the conveyor. The

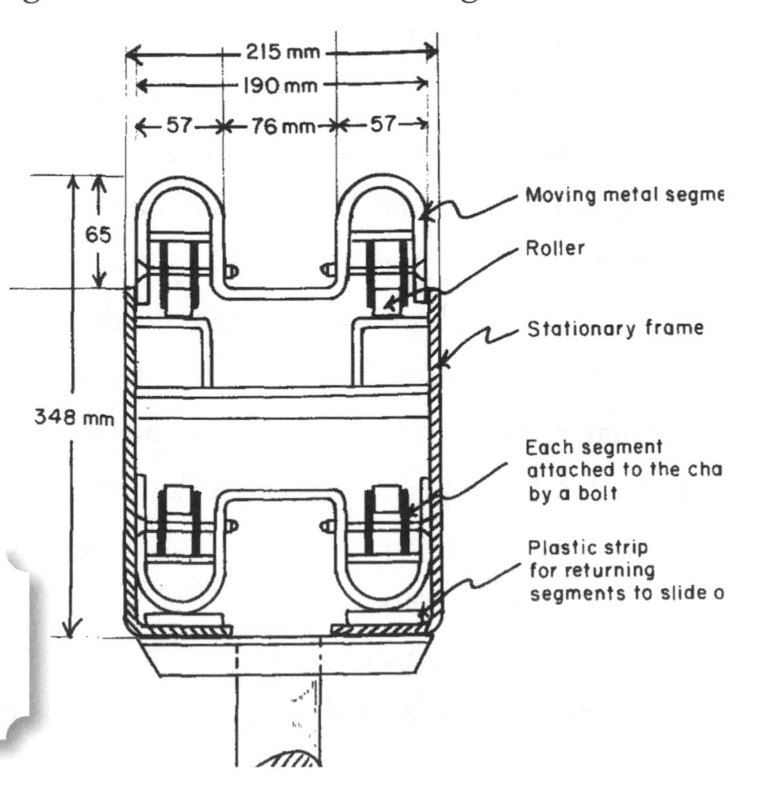

Fig. 5.
Cross-section of double rail conveyor showing the dimensions for. calves and sheep

Fig. 6.
Stunning with a captive bolt is easier in the double rail system because the operator can stand close to the animal. The operator can quickly adjust the restrainer for different sized calves by pulling the chain

sides are on pivots, and they can be rapidly adjusted by pulling a chain hoist. For calves and sheep, the space between the adjustable sides can be varied from 510 mm for large formula fed calves to 250 mm for baby calves. The double rail conveyor does not require adjustment.

The adjustable sides are positioned so that the bottom of the adjustable side is slightly above the top of the double rail. This provides room for the animal's leg joints. The space between the bottom of the adjustable side and the top of the double rail is 50 mm when the sides are spaced 250 mm apart and 127 mm when the sides are spaced 457 mm apart.

The design of the system enables the stunner operator to stand closer to the animal compared to a V restrainer (Fig. 6). After stunning, calves and sheep are shackled while they are held in the restrainer. The shackle conveyor runs along one side of the restrainer in the same manner as the shackle system for a cattle V restrainer conveyor. Since the shackle conveyor could be placed closer to the animal's leg, noose type shackles which loop tightly around the leg could be used.

Ritual slaughter is conducted one day each week on formula fed calves. The double rail restrainer is stopped when each calf reached the end. The configuration of the system is similar to the cattle V restrainer conveyor described in Grandin.[12] A vertical sliding gate with a U-shaped back holder descends against the animal's back (Fig. 1). The head is held by a person for the throat cut. After the throat cut, the animal is ejected onto a stainless-steel slat conveyor table (Fig. 1). The table dimensions are 5–48 m x 1–2 m. The table conveyor is continually washed by a water spray to prevent blood cross contamination between animals. A shorter table could be used for stunning only.

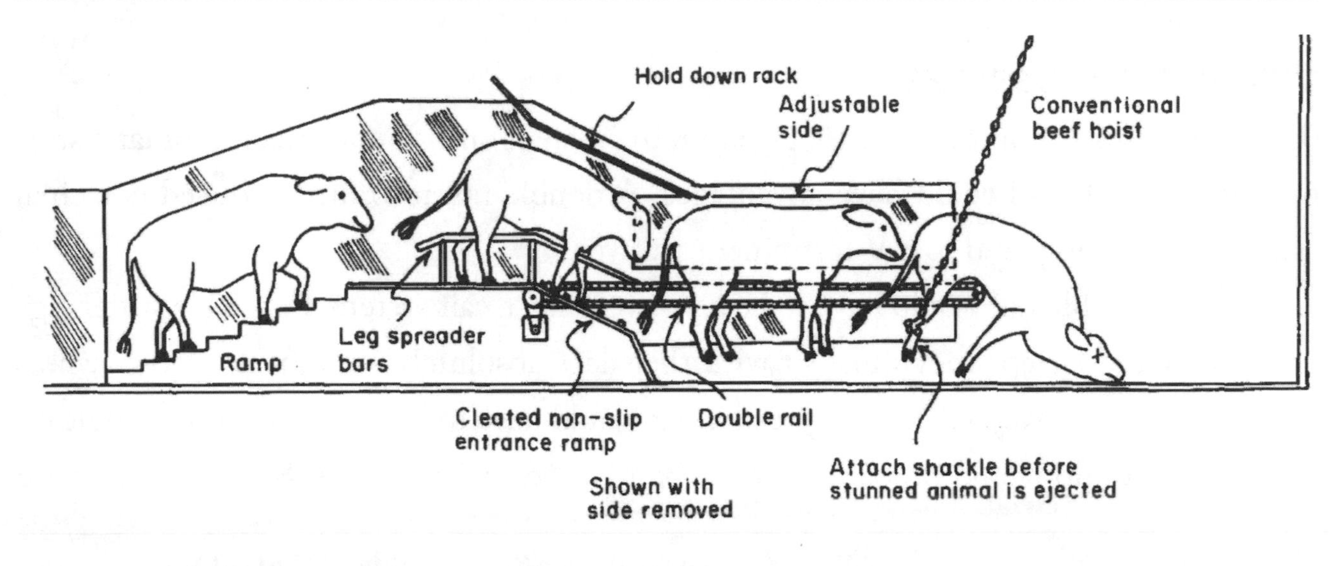

Fig. 7.
The double rail could be used to replace the stunning pen in smaller plants. The existing hoist could still be used, and a minimum of structural modification would be required

4. LARGE CATTLE SYSTEM

A double rail system for large cattle slaughter is under development. Fig. 7 illustrates a large cattle double rail system which could be used to replace an existing stunning box without major plant modifications. It would be suitable for plants under 100 cattle/h and it does not require the large ramp and shackling system which is described in Grandin. This system could utilize the existing hoist, and stunned cattle would be shackled just before they are ejected off the restrainer onto the floor. The double rail could also be used with the standard ramp and shackle system configurations which are used with V restrainers in large American beef slaughter plants.[2]

Measurements for the outermost width of the double rail conveyor were obtained by placing two pipes underneath the armpits of different sized cattle. The animals were restrained in a head bail. The pipes were held parallel to the animal's spine and raised up under the animal's body. A system to handle 275–700 kg cattle would require a double rail 300 mm wide with a 100 x 100 mm space between the rails. The leg positioner bar would have to be 558 mm above the floor, and the width between the adjustable sides would also have to be increased.

S. PIG STUNNING SYSTEM

Fig. 8 illustrates a proposed electric pig stunning system on a double rail. Automatic stunning on a double rail could have advantages. A double rail restrainer reduced petechial haemorrhages compared to a V restrainer conveyor.[13]

A pig double rail would have to be narrower than a calf system. Pigs have stiffer legs than calves and sheep. They must sit with their legs absolutely straight up and down as they ride on the conveyor. The leg spreader bar would have to be lowered and the angle on the entrance ramp would probably have to be reduced to 10 degrees. Since the entrance

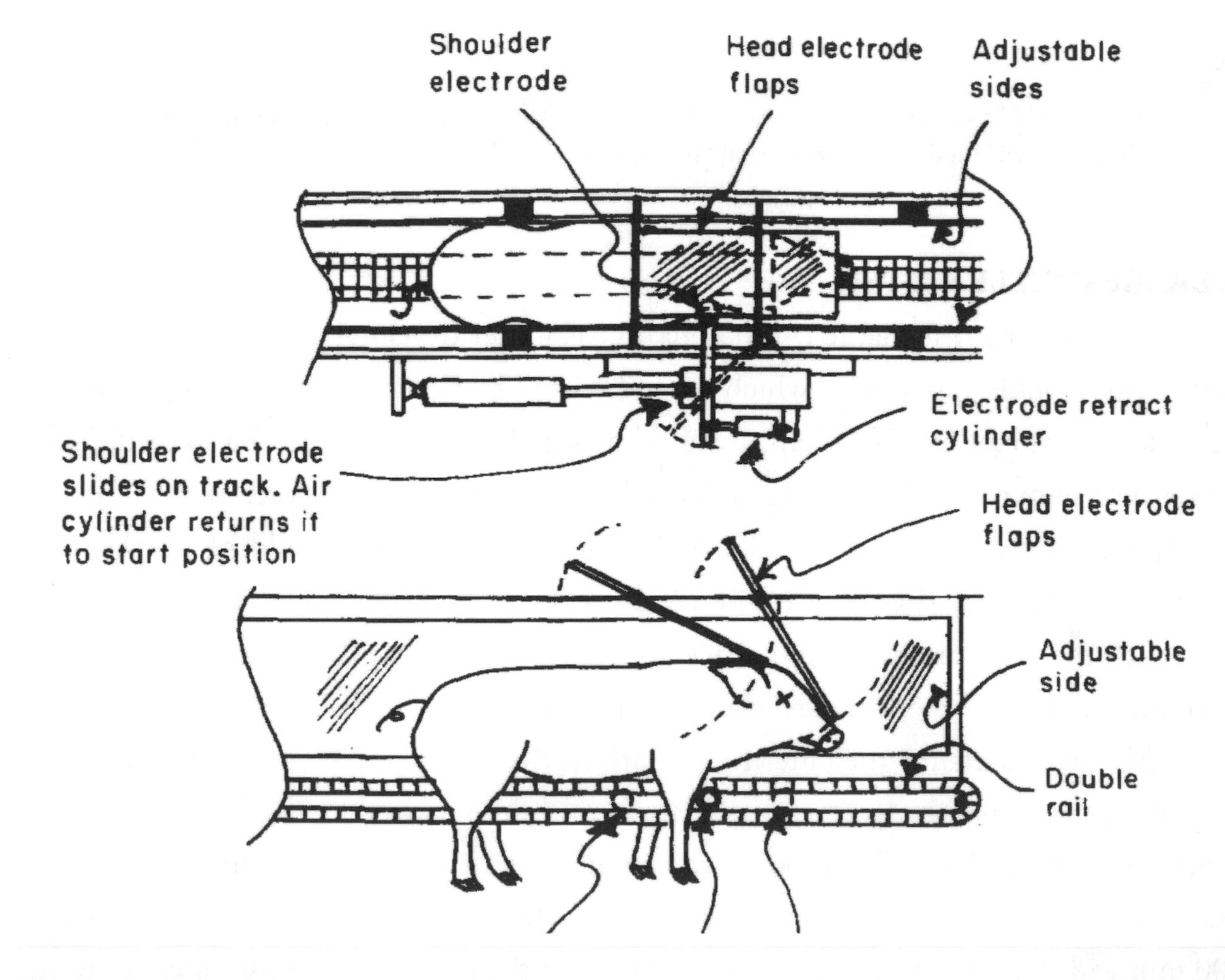

Fig. 8.
A simple automatic stunner for use with the double rail would have swinging flap head electrodes and a shoulder electrode

ramp on a V restrainer for pigs is at a shallower angle than the entrance ramps used for cattle, it is likely the same principle would apply to the double rail.

Two double rail conveyors placed end to end would enable the pigs to be automatically stunned without stopping the restrainer. One conveyor would run faster than the other, to create a separation between the animals. Creating a separation between the animals enables the automatic stunner to reset itself for the next animal while the conveyors are continually moving. This feature enables higher throughputs.

When two conveyors are placed end to end, it is recommended to redesign the segments to eliminate the gaps which open on the ends. Commercially, V restrainer conveyors are available with overlapping segments to eliminate pinch points which occur when the segments close as they turn around the end pulley. Practical experience has shown that this design reduces pinching when an animal transfers from one conveyor to the other. If this is not done, it may be possible to transfer the pigs between the two conveyors by installing rollers. A powered roller at the junction between the two conveyors would help transfer the animal from one conveyor to the next.

The stunner would pass electric current from head to shoulder (Fig. 8). Two to three flaps hinged at the top would contact the pig's head as it rides on the conveyor. The head flap electrodes were invented by Raymond Cooper at the Queen's University in Belfast and Wilson Swilley, Persia, Iowa, USA. The flaps would '1BE constructed from heavy insulating plastic with a flat metal electrode on each one of them. To prevent blood splash (large haemorrhages in the muscle or around the joints), the metal plate on the electrode must be bordered with plastic. This will prevent it from shocking the pig as it passes over the shoulders and back. The flaps are spaced so that as the pig's head passes from one flap to the next, the flow of electricity is continuous. A spring or air cylinder must be attached to the flaps to maintain firm contact with the pig's head. Both the first head flap and the shoulder electrode must be in firm contact with the pig before the electricity is turned on. Current flow could be initiated by the shoulder electrode striking a limit switch.

The shoulder electrode contacts the pig's shoulder. It slides on a track and is pushed forward by the forward movement of the animal. At the end of the stunning cycle, a small air cylinder retracts the electrode. It then returns to the start position for the next pig. The shoulder electrode is positioned slightly below the adjustable side. It must be positioned so that the pig's leg does not bend excessively when the electrode is pushed forward. This stunner should be used with a New Zealand type constant current power supply to reduce blood splash and petechial haemorrhages.[14] Excessive current and current surges increase petechial haemorrhages. A minimum of 1·25 A at 300 V 50 Hz must be used for stunning pigs to insure instantaneous unconsciousness.[15] The head to shoulder configuration has the potential to reduce petechial haemorrhages. Head to front leg electrodes caused fewer haemorrhages than head to back electrodes.[6] In a head to foot system stunning current must not pass through the back feet. In head to shoulder systems, it is easier to ensure proper electrode placement.

6. VETERINARY SYSTEM

A veterinary version of the double rail could be used in large cattle fattening operations (Fig. 9). In the US hundreds of cattle have to be handled every week when they arrive at the feedlots. V restrainer conveyors have not been used in feedlots due to high cost. Cattle weights range from 100 to 450 kg. An experimental prototype double rail conveyor with a width of 260 mm was constructed. The adjustable sides had a maximum spacing of 760 mm. Veterinary procedures such as vaccinations, applying pesticides, dehorning, foot trimming, giving medicines, ear tagging, etc., could be conducted while the animals

Fig. 9.
A veterinary version of the double rail conveyor

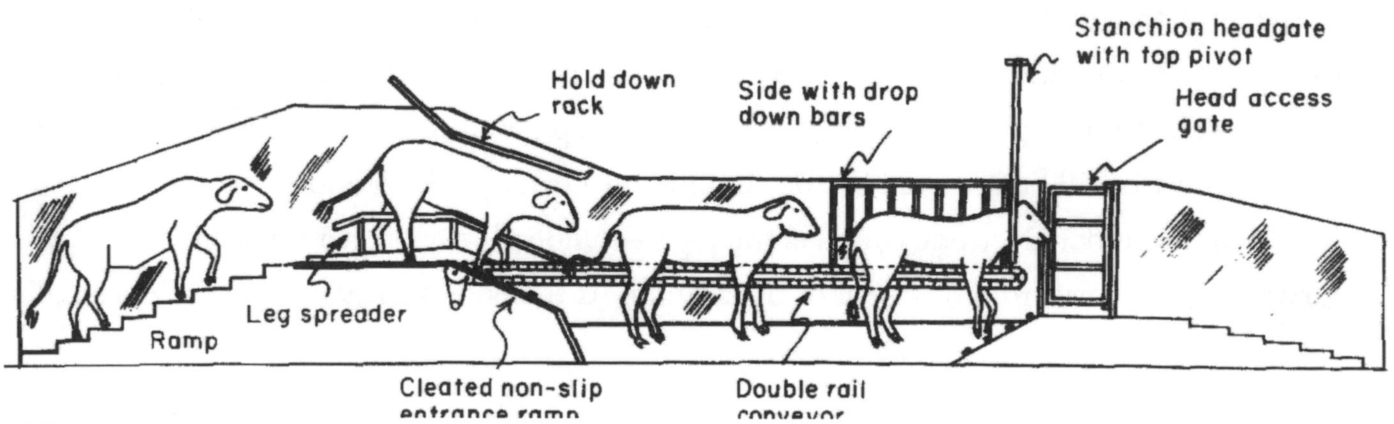

are held on the conveyor. After treatment, the animal would walk off down an exit ramp. For operations requiring head restraint a stanchion head bail could be installed. Possible benefits of a conveyorized system would be reduced injuries and labour reduction.

It may also be possible to develop a veterinary double rail for sheep, to replace the rubber belt V restrainer conveyors which are currently used for this purpose.

7. RESULTS AND DISCUSSION

The double rail system has been operating in a commercial calf slaughter plant for two years. It can easily maintain the plant's top production line speeds of 300 baby calves, or 150 large formula fed veal calves or sheep/h. The existing floor stunning and shackle hoist system had been removed and the plant was completely dependent on the double rail system. Replacement of the shackle hoist with the double rail restrainer has greatly reduced accidents. For an eighteen-month period prior to installation of the restrainer there were five lost time accidents. Three were very serious and three workers were absent for more than three weeks. Injuries were caused by kicking calves or a shackle trolley falling on a person's head. A total of 126 working days were lost. For an eighteen-month period after installation of the restrainer there was only one lost time accident with two days of worker absence. A person's hand was bruised while attaching the shackle chain.

The double rail has several advantages compared to a V restrainer conveyor system. Stunning with a heavy pneumatic stunner was easier and more accurate because the operator could stand closer to the restrainer. The distance is 70–180 mm in the double rail system and 400–430 mm in a V restrainer. A heavy pneumatic stun gun could be easily used without a balancer. Use of the same gun without a balancer is extremely difficult in a V restrainer. The author stunned animals with a pneumatic gun in both the double rail and V conveyor systems. Stunning calves in the double rail system for 3 h was done with minimum exertion by the author. The author had insufficient arm and wrist strength to operate the same brand of pneumatic stunner for 3 h in a V restrainer conveyor.

The calves entered the restrainer more easily compared to a V restrainer. Balking and refusing to enter the restrainer was reduced because the animals were less likely to feel the moving conveyor contact their legs. The animals could walk into the restrainer more easily because their legs were in the natural position. In a V restrainer the small gap between the two conveyors forces the animal to walk in with its legs pinched together.

Another advantage was that calves seldom reared up or attempted escape from the double rail restrainer. Over 250 wild, pasture reared calves of 136–226 kg and over 2400 formula fed veal calves weighing approximately 160 kg were observed. An animal was never observed rearing up over the sides or putting its leg over the stationary side of the restrainer. Approximately 5 to 10% of the 136–226 kg calves placed one front leg on the top of the conveyor. This appeared to cause no discomfort and an animal with its leg in this position was unable to raise its belly or brisket off the top of the conveyor. Over 99% of the animals had both rear legs in the correct position for shackling.

In a V restrainer conveyor cattle will often rear up and are difficult to stun. Head movement in the double rail is reduced due to the relatively narrow solid sided passage formed by the adjustable sides. The solid sides restrict the animal's vision and prevent it from seeing activity outside the restrainer. Visual restriction reduces stress in poultry.[16] Similar results have also been obtained with cattle.[17]

Shackling in the double rail was also easier. The shackle conveyor rail can be placed approximately 150 mm closer to the animal's legs. The closer position of the shackle conveyor rail made it possible to use noose type shackles instead of shackles with an open hook fastener. Noose type shackles are less likely to come off if the chain remains loose. The use of noose type shackles made it possible to eliminate the cumbersome shackle chain tensioning devices that are used in V restrainer systems for cattle. Initially several types of tensioning devices were installed, but they have all been removed. The double rail was also less expensive to construct because it contains only one powered conveyor instead of two.

Entering calves must not be able to see bright light under the restrainer. The restrainer room should be illuminated with bright diffuse ceiling lamps. Sources of bright light under the restrainer must be eliminated.

Calves of all weights ranging from 23 to 225 kg rode quietly on the conveyor. Ample space for the animal's leg joints between the bottom of the adjustable sides and the top of the double rail conveyor was very important. The adjustable sides must not press on the leg joints. Pressure on the leg joints causes calves to struggle and vocalize. When the correct leg joint clearance was determined, calves sat quietly on the conveyor without struggling or vocalizing for 15–30 min.

When large number of baby calves are handled, a single file race cannot be used. In the double rail installation, large calves would enter the double rail restrainer without the aid of a single file race. This is another indicator that animals enter easily. In plants which do not handle baby calves a single file race should be constructed. Designs for races can be found in Grandin.[18-20] The adjustable sides could be moved within 10–15 s to accommodate different sized calves in mixed weight groups.

Cattle weighing 180–410 kg readily entered the experimental veterinary prototype and rode on the conveyor. The animals entered easily despite the fact that a makeshift arrangement of portable livestock panels and a portable loading ramp was used to direct them into the restrainer. These animals were wild, and they had a flight distance of approximately 8 m. The experimental prototype was located outdoors and cattle on the conveyor could see cattle in pens through the end of the restrainer. Some animals struggled in an attempt to escape. Placement of a solid barrier to prevent vision through the exit point of the restrainer immediately caused the animals to calm down and ride quietly.

A multiple flap head electrode has been tested for two weeks on a commercial pig slaughter line. Blood splashing and petechial haemorrhage levels were similar to the best commercially available manual and automatic stunning equipment. Maintaining firm contact with the head during stunning is essential to prevent blood splash. The flaps must not be allowed to bounce and make or break contact while the current is applied. Planning is in progress to install a large cattle slaughter system and a pig double rail in commercial operations.

8. CONCLUSIONS

The double rail restrainer is a superior system for handling calves of varying sizes compared to the V restrainer conveyor. It also worked well with sheep. Initial testing indicates that the system will also work for larger cattle. Animal welfare will be improved because it is easier for the operator to reach the animal's head for proper placement of a captive bolt stunner. A double rail can replace a V restrainer conveyor with minimum modification of existing shackling or race systems. Installation of double rail restrainers on commercial pig and adult cattle slaughter lines will be required to fully evaluate the system for these animals.

ACKNOWLEDGMENTS

Funding for installation of the double rail restrainer in a commercial slaughter plant was from the Council for Livestock Protection, New York, USA, and Utica Veal Co., Marcy, New York, USA. Double rail equipment fabrication by Clayton H. Landis, Souderton, Pennsylvania, USA. Layout and design Grandin Livestock Handling Systems, Inc.

REFERENCES

1. Regensburger, R. W. Hog stunning pen. U.S. Patent No. 2 185 949, 1940
2. Grandin, T. System for handling cattle in large slaughter plants. ASAE 1983, Paper No. 83406
3. Lambooy, E. Automatic electrical stunning of veal calves in a V-type restrainer. Proceedings 32nd European Meeting of Meat Research Workers, Ghent, Belgium, 1986, Paper 2: 2, pp. 77-80
4. Giger, W.; Prince, R. P.; Westervelt, R. G.; Kinsman, D. M. Equipment for low stress animal slaughter. Transactions of the ASAE 1977, 20: 571-578
5. Thornton, R. N.; Blackmore, D. K.; Jolly, R. D.; Harris, R. E.; Marsden, N. A. Petechial hemorrhages in carcasses of slaughtered lambs. *New Zealand Veterinary Journal* 1979, 27: 181-189
6. Gilbert, K. V.; Devine, C. E. Effect of stunning method on petechial hemorrhages and on blood pressure of lambs. *Meat Science* 1982, 7: 197-207
7. Grandin, T. Good cattle restraining equipment is essential. *Veterinary Medicine and Small Animal Clinician* 1980, 75: 1291-1296
8. Brown, H.; Elliston, N. G.; McAskill, J. W.; Tonkinson, L. V. The effect of restraining fat cattle prior to slaughter on the incidence and severity of injuries resulting in carcass bruises. *Proceedings Western Section of Animal Science 1981*, 32: 363365
9. Grandin, T. Reduce stress of handling to improve productivity of livestock. *Veterinary Medicine* 1984, 79: 827-831
10. Grandin, T. Handling stress. *Recueil de Medecine Veterinaire 1988* (In press)
11. Westervelt, R. G.; Kinsman, D. M.; Prince, R. P.; Giger, W. Physiological stress measurement during slaughter of calves and Jambs. *Journal of Animal Science* 1976, 42: 831-837
12. Grandin, T. Problems with Kosher slaughter. *International Journal for the Study of Animal Problems* 1980, 6: 375-390
13. Lambooy, E. Research Institute for Animal Husbandry, Zeist, The Netherlands, 1987
14. Devine, C. E.; Gilbert, K. V.; Ellery, S. Electrical stunning of lambs: The effect of stunning parameters and drugs affecting blood flow and behaviour on petechial haemorrhage incidence. *Meat Science* 1983, 9: 247-256

15. Hoenderken, R. Electrical and carbon dioxide stunning of pigs for slaughter. In: *Stunning of Animals for Slaughter* (Eikelenboom, G., ed.). Martinus Nijhoff, 1983, pp. 59-63

16. Douglas, A. C.; Darre, M. D.; Kinsman, D. M. Sight restriction as a means of reducing stress during slaughter. In: *Proceedings of the 30th European Meeting of Meat Research Workers*, Bristol, England, 1984, pp. 10-11

17. Kinsman, D. M. University of Connecticut, Storrs, Connecticut.

18. Grandin, T. Designing meat packing plant handling facilities for cattle and hogs. *Transactions of the ASAE* 1979, 22: 912-917

19. Grandin, T. Pig behavior studies applied to slaughter plant design. *Applied Animal Ethology* 1982, 9: 141-151

20. Grandin, T. Race system for cattle slaughter plants with a 1.5 m radius curve. *Applied Animal Behaviour Science* 1984, 13: 295-299

FOUR

Behavior of Slaughter Plant and Auction Employees Towards Animals

T. Grandin (1988) Behavior of Slaughter Plant and Auction Employees Towards Animals, *Anthrozoos*, Vol. 1, No. 4, pp. 205-213.

This was my first study where I described three different ways that people handling large numbers of animals behave. At this point in my career, I had learned that designing better equipment would not solve all the problems that would be detrimental to animal welfare. People who handle large numbers of animals can become desensitized, and handling practices need to be constantly monitored.

ABSTRACT

Abuses of animals at auctions and slaughter plants occur often. Commonly observed abuses include the dragging of crippled animals, hitting and excessive prodding of animals. In both auctions and slaughter plants, employees are under pressure to maintain a steady flow of animals to the auction ring or slaughter lines. In both types of facilities large numbers of animals must be moved rapidly. The purpose of this study was to determine the behavior of people handling livestock in these types of facilities. Observations of the behavior of slaughter plant managers were also made to gain a better understanding of how management behavior affects employee behavior.

SURVEY A - LIVESTOCK MARKETS

In 1984, an investigator was hired to make unannounced visits on sale day at 51 livestock markets in 11 southeastern states. His itinerary was objectively predetermined by a person who had no knowledge of the conditions in the markets. Ten percent of the markets on the published lists were visited in each state.

Twenty-one percent of the surveyed markets had excellent handling and 32% had either rough handling or acts of cruelty (Table 1). When the condition of the market facilities was evaluated, it was found that 35% had excellent, well-maintained facilities and 28% had dirty, broken-down, or poorly designed facilities (Table 2). Markets that had good facilities tended to have a lower incidence of rough handling (Table 3). The quality

TABLE 1: Handling Ratings for Southeastern Livestock Markets

CATEGORY	DESCRIPTION	%
Excellent Handling	Animals were moved quietly with a minimum of prodding. Care was taken to avoid slamming gates on animals, and they were never kicked or hit with solid objects	21%
Acceptable Handling	Handling practices did not fall into the excellent or one of the not acceptable categories	47%
Not Acceptable Rough Handling	Many animals were handled roughly by more than one person, and management did not attempt to stop the abuse. A rough handling rating was given if any one of the following abuses was observed as a routine practice: constant prodding with an electric prod when the animals had no space to move, slamming gates on animals, over-crowding and causing animals to pile up, hitting animals with sticks or other objects, and constant whipping of animals with whips	20%
Not Acceptable Cruelty	Animals were dragged, thrown, or picked up by the tail or ears. This rating was also given if the majority of the employees handled most animals roughly and appeared to have no regard for them	12%

Source: Grandin 1985.

TABLE 2: Facilities Ratings for Southeastern Livestock Markets

CATEGORY	DESCRIPTION	%
Excellent	All pens and chutes were clean and well maintained with a minimum of sharp protrusions that could injure animals. Facilities also had to have adequate lighting to be placed in this category. A market with a good pen layout design was also placed in this category	35%
Acceptable	The majority of the pens and chutes were well maintained and clean. A market with a few broken boards or muddy pens was placed in this category	37%
Dirty or needed major repairs	Many of the pens had broken fences or gates and there was a need for major repairs. A market was also placed in this category if it was littered with trash, or chutes showed no evidence of being cleaned out on a regular basis	22%
Design unsatisfactory	This rating was given if a design defect caused a serious handling problem that increased the amount of rough handling and was likely to cause injuries to animals	6%

Source: Grandin 1985.

TABLE 3: Relationship Between Handling and Quality of Facilities

	HANDLING	
FACILITIES	EXCELLENT/ACCEPTABLE	NOT ACCEPTABLE/CRUELTY
Excellent/ Acceptable	28 (76%)	9 (24%)
Not Acceptable	7 (50%)	7 (50%)

(x2 = 3.11; p = 0.08)

of the facilities had little effect on the incidences of overt cruelty nor on the incidence of rough handling due to poor management. Thirty-three percent had no water troughs or feeding facilities.

The size of the market was not related to handling practices, but markets that specialized in one species had a tendency to have better handling. Both the "excellent handling" markets as well as the "not acceptable handling" markets preferred battery-operated electric prods. This indicates that the important factor in handling is how a driving aid is used rather than what is used a good handler will often tap an animal with the prod instead of shocking it. There was a tendency for handling to be more abusive when electric prods connected to an overhead wire were used. This type of prod will give a less localized shock compared to a battery prod.

SURVEY B - SLAUGHTER PLANT EMPLOYEE BEHAVIOR

Twenty-five federally inspected U.S. and Canadian slaughter plants were visited. More than two days were spent in each plant observing the behavior of employees who killed and handled the livestock. The visits were made by the author between 1975 and 1987. Plants were rated as belonging to one of three categories:

1. Acts of deliberate cruelty occurring on a regular basis,
2. Rough handling occurring as a routine practice, and
3. Good to excellent employee behavior.

The condition of plant equipment and facilities was not included in this rating even if they contributed to handling problems. This was strictly an employee behavior survey. Of the 25 plants, 8 (32%) were in category 1, 3 (12%) were in category 2, and 14 (56%) were in category 3. Twelve plants were surveyed before 1982 and 13 after 1982. The incidence of cruelty and abuse dropped from 67% during' 19751982 to 23% in 1982-1987.

FACTORS INFLUENCING THE ABUSE AND CRUELTY INCIDENCE

Operations identified as having humane handling had a manager who enforced a strict code of conduct. If an employee abused an animal, he was either fired or transferred away from the animals. Slaughter plants that had cruelty problems tended to have lax management in the livestock department. In four instances, slaughter plants markedly improved their handling rating after hiring a new manager. In one instance, handling got worse. Management enforcement of a strict code of conduct had a greater influence on employee behavior than the regional location of the plant or employee cultural background.

The size of the slaughter plant or livestock market was not found to be related to the incidence of bad employee behavior resulting in rough handling or cruelty. However, poorly maintained or poorly designed facilities correlated with an increased incidence of rough handling and livestock accidents. Gentle handling is impossible if animals constantly balk, fall down on slick floors, or become jammed in chutes. Facilities should be well lighted and kept clean. It is also easier to encourage good employee attitudes in pleasant surroundings.

Good facilities, however, do not guarantee good handling. The two worst incidents of deliberate animal cruelty witnessed occurred in slaughter plants that had new, well-designed facilities. One man took pleasure in shooting the eyes out of cattle before he killed them. In the other plant, a man stabbed a meat hook deep into a live hog's shoulder and dragged it like a hay bale. One of these plants had lax management and never disciplined employees for cruelty, while the other only gave a reprimand for stabbing the hog. In neither of these cases did management punish employees severely for cruelty.

Personal observations indicate that the incidence of rough handling tends to be lower in Midwestern and more northern areas with an estimated incidence of rough handling for all types of livestock operations at 10%–15%. In the southern U.S. rough handling appears to be higher, probably due to a more widespread "macho" attitude. Kellert (1978, 1980) has also observed regional differences in attitudes toward animals. In Europe there was less interest in animal welfare in southern countries (Curtis and Guither, 1983). There appears to be a correlation between climate and handling. In Australia, there is a greater concern for animals in the cooler, southern parts. In the tropical north handling is more often very rough according to personal observations and discussions. There is a trend for slaughter plant managers in Scandinavian countries and in Canada to be more concerned about humane handling than are U.S. managers. Slaughter plants in Holland and Sweden are very civilized. The employees are concerned about animal welfare and the management is concerned about the welfare of the employees. By contrast, the slaughterhouses are dreadful in Mexico. There is an impression that societies that treat people humanely also tend to treat animals humanely.

In the surveyed slaughter plants, approximately 4% of the employees directly involved with livestock committed acts of deliberate cruelty. These people appeared to enjoy watching an animal suffer. If a plant or feedlot has a cruelty problem, usually only one or two people are involved in the worst incidents. On the other hand, rough handling tends to become widespread in poorly managed operations. In some poorly managed plants and auctions over half the employees engaged in rough treatment of animals.

ABUSE AND CRUELTY BY CHILDREN

A disturbing finding in the livestock market survey was that half the markets rated as having cruel handling allowed young children to abuse animals. At one sale, a young seven- to eight-year-old boy continually hit feeder pigs on the nose in the auction ring. None of the adults sitting around the ring made any attempt to stop it. At another sale teenage boys appeared to enjoy hitting cattle with boards. Three or four different instances of children tormenting calves with an electric prod have been observed this year. Last summer my students observed similar abusive behavior by children in an auction in Texas. Children who enjoy abusing animals may be more likely to engage in cruelty or aggressive behavior as adults (Leyton 1987 and Feldhous et al., 1987). Leyton reports that serial killers Albert DeSalvo (The Boston Strangler) and Edmund Kemper, who cooked part of his victim in a macaroni casserole, both tortured cats when they were children. One slaughter plant employee handled cattle roughly and teased them; he stated that when he was a child, he was forced to kill his pet steer. He said, "I could never love another beef again." His behavior was similar to the taunting of animals prior to sacrifice in the ancient bear cults. Serpell (1986) states that taunting the victim helps distance the killers emotionally from an animal they had tenderly reared. This employee, however, never committed an extreme act of cruelty such as the ones described previously.

FACTORS THAT CORRELATE WITH IMPROVED ANIMAL HANDLING

The two major factors that motivate managers to improve handling in slaughter plants are legal sanctions and economic incentives. Most of the handling improvements observed during 1982–1987 can be attributed to legal sanctions and economic incentives. In 1978, the jurisdiction of the Humane Slaughter Act was extended to handling. In later years a greater percentage of the meat was sold on a carcass basis instead of on a liveweight basis. Ownership changes hands only after slaughter in a carcass-based system, and the seller has to pay for bruises.

Sellers exert strong pressure on plant management to improve handling and to take steps to improve handling at their feedlots. Cattle sold on a liveweight basis had almost twice as many discountable bruises (Grandin, 1981). Rough handling doubles the amount of bruising.

Within the last five years many pork plants have also started to export to Japan. The Japanese reject poor-quality pork. Reducing stress and excitement in the stunning chute will improve pork quality (Grandin, 1986). When a plant starts exporting to Japan, management usually takes immediate steps to improve handling, because they see the Japanese grader rejecting over 50% of their pork. Six of the plants visited during 1982–1987 exported pork to Japan. These plants all had good handling. Four non exporting hog plants were also surveyed. Three of these plants had either rough handling or incidents of cruelty.

Personal observations indicate that severe rough handling, abuse, and neglect on farms, ranches, markets, and feedlots have remained at a steady 10% – 15% of operations for the last ten years over the entire United States. They have not shown the improvement that has occurred in slaughter plants. Even though rough handling causes great economic losses it continues, because the market is segmented. The attitude of some market people is, "I don't care if they get shipping fever—that's the feedlot's problem." The abuses will continue unless there is a direct economic incentive or animal welfare pressure leads to legal sanctions. Approximately 25% of all operations have truly excellent handling.

PSYCHOLOGY OF SLAUGHTER PLANT MANAGERS

Michael Lesy (1987), in his book *The Forbidden Zone*, describes a dreadful plant with a wisecracking, foulmouthed manager. The manager tells endless dirty jokes. This type of manager is in a minority. The most common management psychology is simply denial of the reality of killing. Managers will use words such as "dispatching" and "processing" to avoid this reality.

Over the years the author has made many observations of the behavior of slaughter plant managers. In large plants with corporate offices in a distant city, management tends to deny the reality of killing. The few times they visit the plant they tend to avoid the kill area. Even managers who have their offices on the plant grounds sometimes have this attitude. One manager told the author that he would not expand the stockyard because he did not want to see it from his office window. He wanted his plant to look like a "food factory."

The meat packaging room, coolers, and the dressing line where the carcass is cut up are often much better designed and maintained than are the stockyards and kill chute. Several plants had stockyards that were falling apart and neglected while the rest of the plant was new and modern. Management attitudes are further reflected by the fact that livestock handling and kill chute jobs are often the lowest paid jobs on the line.

This attitude makes no sense economically. Bruises cost the meat industry $46 million per year, and meat quality problems due to animal stress cost even more (Livestock Conservation Institute, 1983; Grandin, 1986). Improving handling can increase Japanese acceptance of pork loins by 10%–25% (Grandin, 1986). Well-designed chutes and stockyards and good handling practices will reduce bruises and stress related meat quality problems (Grandin, 1982, 1982a, b, 1981, 1980a; Kilgour, 1971). The actual cost of livestock accounts for at least half the operating costs of a slaughter plant.

The author has also worked with managers and engineers who really care about animals but who also avoid visiting the kill area because it upsets them. Managers who care about livestock often raise their own livestock or have had previous experience working with animals. These managers will enforce a strict code of conduct. They appear to be motivated by a genuine caring for the animals. One engineering manager who raised cattle made unauthorized expenditures to improve the humaneness of slaughter equipment. A feedlot owner, newly in the slaughter business, built his new office looking out toward the plant stockyards. From his office he could watch and ensure that employees did not abuse the animals. He seldom went inside the slaughter area. Managers promoted from the livestock buying department are usually more concerned about animal treatment than managers promoted from other departments. A manager's background affects his attitudes. One of the best managed and most humane slaughter plants in the United States is owned and run by a Mennonite family. Hard work and good values transformed a small business into a company with almost $400 million in annual sales. The managers have an attitude of humaneness toward both animals and employees and have state-of-the-art equipment throughout the plant. They are proud of their operation, which is one of the few plants that still conduct public tours. While the big corporations attempt to cover up what they are doing, this company is proud of its excellent operation. Another plant with excellent humane handling had many Mormons in high management positions.

PSYCHOLOGY OF SLAUGHTER PLANT EMPLOYEES

Herzog and McGee (1983), in a study of college student attitudes to slaughter, found that when college students first visited a slaughter plant, the killing of the animals bothered them more than gutting the carcass. Owens et al. (1981) did a study on the psychology of euthanasia technicians who kill surplus dogs and cats at the animal shelters. They found that the technicians often felt guilty, but they felt that they were performing a needed service. One technician commented, "I would rather euthanize the animals myself than leave it to someone who does not know what they are doing." Slaughter plant employees have made similar statements. Several said that they took the kill chute job to prevent sadistic people from doing it.

Some technicians adopt the mechanical attitude suggested by Owens' study (Owens et al., 1981): "permeating most responses was the theme of protecting oneself from the full impact of the act by isolating one's feelings from the act. Some accomplished this by talking about euthanasia formally or intellectually." The meat industry also has euphemisms for killing, such as 'dispatching." Animal shelter personnel have euphemisms for killing, such as PTS, "put them to sleep." (Arkow, 1985)

The people who actually do the killing in slaughter plants have three different approaches to their jobs. These are the mechanical approach, the sadistic approach, and the sacred ritual approach. These approaches usually are observed only in the people who actually do the killing or who drive the animal up the chute.

Mechanical approach

The mechanical attitude is most common. The person doing the killing approaches his job as if he was stapling boxes moving along a conveyer belt. He has no emotions about his act. Most people who have the mechanical attitude kill the animals efficiently and painlessly. Employees who use the mechanical approach will chitchat about the weather and gossip while they kill hundreds of animals per day. The animals have become a commodity. A consulting engineer who raised cattle and designed a system for electrocuting hogs stated, "Hogs are just a commodity but I can't stand to watch cattle killed." He used the mechanical approach for hogs only.

Some slaughter plant employees who have humanely killed animals for many years act as if the animals were inanimate objects. They do not talk to animals, call them names, or get angry at them. A person who has fully accepted the mechanical attitude no longer has any emotions about the job. Serpell (1986) states that people who kill animals regularly become progressively desensitized. The first few killings are upsetting, but then the person becomes habituated, and the killing act becomes a reflex without emotion. Slaughter plant employees often comment that they were upset when they first started their jobs.

Sadistic approach

The second approach is sadistic. The person starts to enjoy killing and will sometimes do extremely cruel things and torment the animals on purpose. A typical comment from a person with sadistic tendencies is, "They are just animals, and it does not really hurt them." "It is going to die in five minutes, so it does not matter how I treat it." The above statements are examples of devaluation of the subject according to social psychology terminology. By devaluing the animal, the person justifies in his mind the cruel things he does to it.

This concept was graphically illustrated by a series of experiments with people during the 1960s and early 70s. The first experiments were conducted by Milgram (1963) and Elms and Milgram (1966). Human subjects were instructed by an experimenter to give progressively bigger electrical shocks to another subject when he made mistakes on a learning task. The highest shock level was labeled 450 volts, "Danger Severe Shock." Sixty-five percent of normal American males obeyed the experimenter's orders and administered the highest shock levels.

The entire shocking procedure was fake, but the subject believed he was giving real shocks.

Subjects who obeyed and administered the highest shock level tended to devalue the other subject. A typical comment was, "The good scientist deserved to be followed while the stupid, excitable learner deserved to be given a lesson." (Elms and Milgram, 1966)

A similar study was designed by Zimbardo (1972), who placed college students in a simulated prison. Half the students were "guards" and half were "prisoners." One-third of the student guards treated the prisoners in a sadistic manner. Zimbardo concluded that normal people can be turned into sadists.

Fromm (1973) notes that two thirds of the student guards did not turn into sadists and questions how they might be distinguished from the other third who did. In the

Milgram experiments, many of the obedient subjects had emotional conflict and were nervous and upset when they pushed the shock buttons. On the other hand, some of the obedient subjects were calm and deliberate. Fromm suggests that the subjects who did not experience conflict may turn into sadists.

Sacred ritual approach

The third approach is to make the act of killing a sacred ritual. Many different societies have slaughter ceremonies. The American Indians showed respect for the deer and elk they ate. As a sign of respect, the bones from these animals were not thrown to the dogs (Frazer, 1963). Serpell (1986) also describes slaughter ceremonies practiced by the ancient Greeks, Egyptians, Phoenicians, Babylonians, Hebrews, and Romans. Judaism attaches great seriousness to the act of taking life. One reason for the many laws detailing the precise manner in which animals are killed for food is to maintain controls on the act itself (Grandin, 1980b). A ritual serves to place controls on the act of killing and prevents it from getting out of control.

The *shochet* (ritual slaughterman) must be moral; otherwise, he would be degraded by his work (Lesy, 1987). Grunwald (1955) stated that the person performing *shechitah* (slaughter) should think about the act of taking an animal's life:

> A man may kill an animal, but he should always remember that the animal is a living creature and that taking life from the animal involves responsibility (Levinger, 1979).

Islam has similar controls. The slaughterman must have a clear mind. "The act of slaughter (*Al Dhabh*) starts by pronouncing the name of Allah, the Creator (this is symbolic to take his permission and in order to make the slaughterman accountable and responsible and give compassion and mercy to the animal during this act)." (Katme, 1986)

The builders of highspeed automated pig killing equipment in Holland appear to have similar feelings. The Machinefabriek, G. NIHJUIS B.V., in Winterswijk, Holland, named their most highly automated equipment "Walhalla." In Nordic mythology,

Walhalla is the paradise for warriors who died gloriously in battle (Katme, 1986). Richard Selzer (1987), a surgeon, after a visit to a slaughter plant describes his view of an ideally designed slaughterhouse that definitely falls in the sacred ritual category. He describes an atrium built from columns with carved cattle heads, a labyrinthine, serpentine loading ramp, and workers reciting prayers.

Promoting a humane attitude toward animals is extremely important. These words were written by a blind girl when she visited a slaughter plant and reached over the side of the chute and touched an animal:

> The Stairway to Heaven is dedicated to those people who desire to learn the meaning of life and not to fear death. We, through respect for these animals, can come to respect our fellow man as well (Tester, 1974).

Signs with the above message have been placed over the kill chute in some plants to help improve employee attitudes.

Slaughter rituals usually occur among the people who actually perform animal killing. When animals are killed by hunting dogs or by traps there is no slaughter ritual (Serpell, 1986). The blame for the animal's death is shifted to the dogs. Burkert (cited by Serpell, 1986) states that sacrificial customs are elaborate exercises in blame shifting. The priests are directly responsible for the animal's death, but theirs is a sacred duty and therefore forgivable. The gods are blamed instead, because they demanded the sacrifice.

Rituals also serve a beneficial function by placing controls on the act of killing, and they also help prevent the devaluation and detachment that leads to the mechanical approach or to sadism. For over 12 years the author has designed and operated the equipment used to kill animals in commercial slaughter plants. To prevent herself from degenerating into mechanical "box stapling" she uses the sacred ritual approach. A ritual can be simple and still be effective in controlling behavior and promoting respect for animals. The act of killing is controlled by an act of submission similar to a submissive wolf exposing its throat to a dominant wolf. The author's own ritual is to face the plant and bow her head down when first approaching it. She has also written "Stairway to Heaven" or "Valhalla" on some of the drawings for new systems. The braces and supports on one slaughtering system were designed utilizing the Greek Golden Mean and a mathematical sequence which determines the behavior of many things in nature. Humans do not really know what happens after death. A ritual act of submission before one kills an animal acknowledges the unknown that haunts all people.

The ritual also serves a very practical function of controlling bad behavior. The author has observed Kosher slaughter in 13 different U.S. slaughter plants with a total of over 20 days observation time. Even though plant employees sometimes abused the animals, a *shochet* was never observed taunting, teasing, or deliberately abusing an animal. This observation illustrates the power of the ritual to control behavior. Some Kosher plants have cruel, dangerous methods for restraining the animal, which would have a tendency to encourage cruel behavior. Sixty one percent of the Kosher plants engaged in cruel, dangerous live hoisting, and 23% had employees who abused animals. The *shochets* never engaged in abuse even though their working environment was often worse than that of most non-kosher plants. A total of 19 different *shochets* were observed actually killing animals.

PLANT CASE STUDY

The mechanical approach is the most common. In one large beef slaughter plant the author observed ten regular employees driving livestock, stunning, shackling, and bleeding cattle every week for three years. Seven employees utilized the mechanical approach, two were sadistic, and one worked hard to treat the animals with kindness. The plant manager cared deeply about humane treatment of animals, but he was unable to fire the sadistic employees because his superiors wished to avoid union problems. The people engaged in shackling, in hoisting stunned animals, and in bleeding all had the mechanical attitude. The two sadists worked as a stunner operator and as a cattle driver. This pattern appeared in other plants. Shacklers and bleeders of stunned animals seldom engaged in cruelty. The stunned animal is either clinically dead or at least appears dead when it reaches their stations. Shacklers who hoisted fully conscious animals for ritual slaughter were often observed abusing animals. People do not torment or act sadistically toward dead animals or animals that appear dead.

CONCLUSION

It is important to rotate the employees who do the killing, bleeding, shackling, and driving. Nobody should kill animals all the time. Several plant managers and supervisors state that rotation helps prevent employees from becoming sadistic. The author has worked many full shifts driving livestock and operating the kill chute at slaughter plants. Rotation every few hours between the kill chute and driving cattle up the chute made it easier to maintain a humane attitude. It is also easier to maintain a good attitude in plants with a slower line speed. At 1,000 hogs per hour, it is almost impossible to handle the hogs properly. The constant pressure to keep up with the line leads to abuse. Maintaining respect for animals is much harder at 1,000 hogs per hour compared to 500 hogs per hour. Rotation of employees is even more essential on a high-speed line. One of the worst aspects of a high-speed line is the noise and confusion. Designing equipment to reduce noise would reduce stress on employees and animals.

The three types of approaches (Mechanical, Sadistic, and Sacred Ritual) have been repeatedly observed in over 150 slaughter plants. These three categories only apply to the people who actually do the killing and people who work in the kill-chute area. For managers the most common attitude is simply denial of the reality of killing. Some good managers who really care about animals often become upset when they have to watch the kill chute, but they express their caring by enforcing a strict code of employee conduct and spending money on good equipment. The paradox is that it is difficult to care about animals but be involved in killing them.

REFERENCES

Arkow, P.1985. The Humane Society and the Human Animal Bond: Reflections on the Broken Bond. *Veterinary Clinics of North America* 15:455-466.

Curtis, S. E., and H. D. Guither. 1983. Animal Welfare: An International Perspective. *Beef Cattle Science Handbook*, ed. F. W. Baker 19:1187-1191.

Davidson, H. R. R. 1972. *Gods and Myths in Northern Europe*. London: Penguin Books.

Elms, A. C., and S. Milgram. 1966. Personality Characteristics Associated with Obedience and Disobedience Towards Authoritative Command. *Journal of Experimental Research into Personality* 1: 282-289.

Felthous, A. R., and S. R. Kellert. 1987. Childhood Cruelty to Animals and Later Aggression Against People: A Review. *American Journal of Psychiatry* 144: 710-717.

Frazer, J. G. 1963. *The Golden Bough*, vol.1 (abridged ed.). New York: The Macmillan Co.

Fromm, E. 1973. *The Anatomy of Human Destructiveness*. New York: Holt, Rinehart and Winston.

Grandin T. 1986. Good Pig Handling Improves Pork Quality. European Meeting of Meat Research Workers, Ghent, Belgium.

Grandin, T. 1985a. Treatment of Livestock in Southeast U.S. Markets. *Proceedings Livestock Conservation Institute*, 1424. Commissioned by Humane Information Services.

Grandin T. 1985b. Livestock Handling Needs Improvement. *Animal Nutrition and Health* Aug., 69.

Grandin T. 1982a. Welfare Requirements of Handling Facilities. *Commission of European Communities Seminar on Housing and Welfare*, Aberdeen, Scotland, July 28-30, 1983.

Grandin T. 1982b. Pig Behavior Studies Applied to Slaughter Plant Design. *Applied Animal Ethology* 9:141-151.

Grandin, T. 1981. Bruises on Southwestern Feedlot Cattle. Paper presented at 73rd Annual Meeting American Society of Animal Science, July 26-29.

Grandin, T. 1980a. Observations of Cattle Behavior Applied to the Design of Cattle Handling Facilities. *Applied Animal Ethology* 6:19-31.

Grandin T. 1980b. Problems with Kosher Slaughter. *International Journal for the Study of Animal Problems* 1(6): 375-390.

Grunwald, J.J. 1955. *The Schochet and Schechita in Rabbinic Literature*. New York: Feldheim Publishers.

Herzog, H. A., and S. McGee. 1983. Psychological Aspects of Slaughter Reactions of College Students to Killing and Butchering Cattle and Hogs. *International Journal for the Study of Animal Problems* 4(2): 124-132.

Katme, A. M. 1986. An Up-to-Date Assessment of the Muslim Method of Slaughter. In *Humane Slaughter of Animals for Food*, 37-46. U.K.: Universities Federation for Animal Welfare.

Kellert, S. 1980. American Attitudes Toward, and Knowledge of Animals: An Update. *International Journal for the Study of Animal Problems* 1(2):87-119.

Kellert, S. 1978. Policy Implications of a National Study of American Attitudes and Behavioral Relations to Animals. U.S. Fish and Wildlife Service, U.S. Dept. of Interior, Stock No.024101 004827. Washington, D.C.: U.S. Govt. Printing Office.

Kilgour, R. 1991. Animal Handling in Works, Pertinent Behavior Studies, 912. 13th Meat Industry Res. Conf., Hamilton, New Zealand.

Lesy, M. 1987. *The Forbidden Zone*. New York: Farrar, Straus and Girout.

Levinger, I. M. 1979. Jewish Attitude towards Slaughter. *Animal Regulation Studies* 2:103-109.

Leyton, E. 1987. *Compulsive Killers: The Story of Modern Multiple Murder*. New York: New York Univ. Press.

Livestock Conservation Institute. 1983. *Only You Can Stop Bruising, a $46 Million Annual Drain on the Livestock Industry*. Madison, WI.

Milgram, S. 1963. Behavioral Study of Obedience. *Journal of Abnormal and Social Psychology* 2(1):1926.

Owens, C. E., R. Davis, and B. H. Smith. 1981. The Psychology of Euthanizing Animals; the Emotional Components. *International Journal for the Study of Animal Problems* 2(1): 19-26.

Selzer, R. 1987. *Taking the World in for Repairs*. New York: William Morrow and Co.

Serpell, J. 1986. *In the Company of Animals*. New York: Basil Blackwell.

Shoshan, A. 1971. *Animals in Jewish Literature*. Shoshanim, Rehovat, Israel.

Tester, G. 1974. Cited in Temple Grandin, 1976. The Stairway to Heaven. *National Humane Review* Jan. 1976.

Zimbardo, P.1972. Pathology of Imprisonment. *Trans-Action* 9 (Apr.): 48

FIVE

Voluntary Acceptance of Restraint by Sheep

T. Grandin (1989) Voluntary Acceptance of Restraint by Sheep,
Applied Animal Behavior Science, Vol, 23, pp. 257-261.

This short communication clearly shows that animals will voluntarily enter a restraint device when they are given tasty feed rewards. At this time many people thought this was nearly impossible. The sheep in this research had previously been in a study which proved that a widely advertised electro immobilization device was not humane. The company that made the device advertised that it was a low stress way to restrain animals. It was definitely not humane. When given a choice, the sheep preferred the tilting restraint device.

ABSTRACT

Four Suffolk ewes which had no previous experience with aversive restraining methods were mixed with twelve ewes which had experienced varying amounts of electro-immobilization and restraint in a squeeze tilt table. With successive voluntary passes, previously restrained ewes become more and more willing to be voluntarily restrained in a tilt squeeze table for a grain reward. Initially four ewes which had never experienced electro-immobilization entered the tilt table first. Six of the previously restrained ewes voluntarily entered the tilt table and were squeezed and tilted 6 times in a row without missing a pass. Less time was required to move each ewe during the last 5 passes compared to the first 5 passes.

INTRODUCTION

Livestock producers sometimes have difficulty in moving animals through races, especially if the animals have had an aversive experience previously in the same handling facility. Sheep become increasingly reluctant to enter a race if they have had previous aversive restraint experiences (Rushen, 1986; Hutson, 1985; Grandin et al., 1986).

The purpose of this experiment was to determine if naive sheep and sheep with previous aversive experience could easily be trained to voluntarily enter and be repeatedly restrained in a squeeze tilt table. Training livestock to voluntarily accept restraint would be useful for handling and treatment of animals used for research and breeding.

METHOD

Sixteen mature Suffolk ewes were used. The sheep had been subjected to a 2-way choice test the previous day with a choice between electro-immobilization and a squeeze tilt table (Grandin et al., 1986). Four of the ewes were the decoys used in the choice test and they had never experienced either form of restraint. The other twelve ewes had experienced both electro-immobilization and restraint in the squeeze tilt table (Livestock Systems, Inc., Sidney, Nebraska). The ewes had adlib hay and were not fasted prior to handling.

All sixteen sheep were driven into the crowd pen of the same sheep handling system described in Grandin et al. It consisted of an 8m long, curved race with a tilt table at one end and a crowd pen which could hold sixteen sheep at the other end. Guillotine gates were located at the front and the rear of the tilt table. No hock boards or gates were used to prevent sheep from backing out of the race after voluntary entry. After release from the tilt table the sheep entered a holding pen which led back to the crowd pen. All sheep movements from the crowd pen into the race and squeeze tilt table were voluntary.

A single experimenter used small handfuls of grain to entice the sheep to enter the race and tilt table. A grain reward was used because previous research indicated that feed rewards facilitated movement (Hutson, 1985). Ewes were allowed to eat a small handful (-25 g) of palatable grain while they were in the tilt table and another handful after release. Animals that refused to enter the race and move toward the tilt table were not forced through. After all the volunteers had gone through the tilt table, they were returned to the crowd pen for the next pass.

A total of 10 passes was made on the same day. A new pass was started immediately if no more sheep volunteered to enter the tilt table within 60 s. There was a 2h break in between the 5th and 6th passes. On pass 1, each volunteer was clamped upright in the tilt table for 15 s. On passes 2 and 3, each volunteer was clamped and tilted 70°) for 15 s and on pass 4, they were clamped and tilted 45°) for 30 s. On passes 510 the volunteers were clamped and tilted to a horizontal position for 30 s.

RESULTS

The four-decoy sheep which had no previous experience with either form of restraint entered the tilt table on the 1st pass along with two previously restrained sheep (Table 1). With successive passes, more and more of the previously restrained sheep voluntarily entered the tilt table. On the 9th pass, one decoy sheep refused to enter the tilt table, but additional previously restrained sheep entered. One decoy sheep repeatedly jumped up on the tail gate of the tilt squeeze chute attempting to get in. Two decoys and two previously restrained sheep refused to leave the tilt table after being released on 2 or more passes. Sheep which voluntarily entered the tilt table quietly ate feed while held in the tilted position (Fig. 1). Six (50%) of the previously restrained sheep completed all 6 fully tilted passes and never missed a pass after they started entering the tilt table. Two previously restrained animals entered for the first time on the 7th and 9th passes and completed all remaining passes.

TABLE 1: Number of Sheep Voluntarily Entering a Tilt Squeeze Table during Ten Repeated Passes on the Same Day

	4 sheep with no restraint experience	12 sheep with immobilization and tilt table experience	TOTAL
Pass 1 (Squeeze only)	4	2	6
Pass 2 (Tilt 70°)	3	4	7
Pass 3 (Tilt 45°)	4	4	8
Pass 4 (Tilt 45°)	4	6	10
Pass 5 (Tilt horiz.)	4	8	12
Pass 6 (Tilt horiz.)	4	8	12
Pass 7 (Tilt horiz.)	4	9	13
Pass 8 (Tilt horiz.)	4	9	13
Pass 9 (Tilt horiz.)	3	10	13
Pass 10 (Tilt horiz.)	3	10	13

Fig. 1.
Ewe being rewarded with grain after voluntary entering the tilt table

One previously restrained animal entered the tilt table on the 7th pass and never returned, and another never entered. Another previously restrained animal that entered on the 1st pass skipped the 8th and 10th passes.

As the passes progressed, less time was required to handle each animal. During passes 15, a total of 43 sheep entered the tilt table with a mean time of 147 s per animal. During passes 610, a total of 64 sheep entered the tilt table and the mean time per animal was reduced to 127 s. These times include recycling the sheep back into the crowd pen and the time required to restrain and release each animal.

DISCUSSION

During the retraining passes, previously restrained sheep became progressively more willing to pass through the race and be restrained in the tilt table. These same sheep in the choice test had quickly learned to avoid the immobilizer side. They had been previously restrained in the tilt table 34 times and had experienced immobilization only once or twice. During the choice experiment the tilt table was clearly aversive, but in the retraining experiment the tilt table had lost most of its aversiveness.

Hutson (1985) reported that during the training phase when no restraint was applied, less time was required to push up the sheep after each successive pass. Clamping sheep in a sheep handler in an upright position increased pushup time with each successive pass (Hutson, 1985). In the author's experiment less time was required to handle each animal during the last 5 passes (all animals horizontal tilt) compared to the first 5 passes with less severe restraint. A possible explanation for the difference between Hutson's (1985) results and the author's results may be because of the comfort of the restraint devices. The tilt table used by the author had no sharp edges or protrusions pressing against the sheep.

The number of times the animals were immobilized had no effect on voluntary entry. The ewe which never entered the tilt table and the ewe which entered only once and never returned, experienced electro-immobilization only once. Both these animals remained in the crowd pen. Six animals previously experienced electro-immobilization twice and six experienced it only once. The first previously restrained animal that entered the tilt table had been immobilized twice.

Fourteen out of sixteen ewes returned for one or more additional passes. Three out of four decoys and six out of twelve previously restrained sheep entered the tilt table for all 6 passes where the table was tilted to the horizontal position. The tilt table appeared no longer to be aversive to these sheep.

The tilt table may have been aversive in the choice test because the table may have been associated with the electro immobilizer, even though the sheep had learned to avoid immobilization. This may be similar to the stress response and lost weight gain that occurs in pigs when they are approached by a person who occasionally shocked them. Even though the pigs had learned to avoid the shock by avoiding the person, they were still stressed when the person entered their pen (Hemsworth et al., 1987). The number of people present may also have been associated with the aversiveness of the situation. Only one experimenter was present during retraining whereas during the choice test nine people were present. The experimenter who conducted the retraining was the person who provided the feed reward in Grandin et al. This person never participated in chasing, grabbing or electro immobilizing the sheep. Sheep in the choice test were given a grain reward. They entered the race easily with little resistance prior to electro-immobilization so it is doubtful that driving them through the race was aversive.

CONCLUSIONS

Sheep can definitely be trained to voluntarily accept repeated restraint in a relatively comfortable restraint device. The tilt table had no sharp edges or bars which dug into the sheep. Sheep which have had a previous aversive experience can be retrained to voluntarily reenter the same race and be restrained in a tilt table. Training sheep to voluntarily accept restraint would be practical for sheep used in research and valuable breeding animals. Labor requirements would be reduced because one person can easily restrain the animals. Stress on animals may be reduced because sheep which voluntarily accept restraint seldom struggle.

REFERENCES

Grandin, T., Curtis, S.E., Widowski, T.M. and Thurmon, J.C., 1986. Electro-immobilization versus mechanical restraint in an avoid-avoid choice test for ewes. *J. Anim. Sci.*, 62: 14691480.

Hemsworth, P.H., Barnett, J.L. and Hansen, C., 1987. The influence of inconsistent handling by humans on the behaviour, growth and corticosteroids of young pigs. *Appl. Anim. Behav. Sci.*, 17:245-252.

Hutson, G.D., 1985. The influence of barley food rewards on sheep movement through a handling system. *Appl. Anim. Ethol.*, 14: 263-273.

Rushen, J., 1986. Aversion of sheep to electro-immobilization and physical restraint. *Appl. Anim. Behav. Sci.*, 15: 315-324.

SIX

Effect of Rearing Environment and Environmental Enrichment on the Behavior and Neural Development in Young Pigs

T. Grandin Doctoral Dissertation (1989) Effect of Rearing Environment and Environmental Enrichment on the Behavior and Neural Development in Young Pigs, University of Illinois, Urbana, Ill, USA.

I included the summary of my dissertation because it contained one of the first studies that showed that environmental enrichment affected both the behavior and nervous system development of young pigs. Environmental enrichment reduced excitability and the tendency to startle. Soft objects that pigs could chew were preferred.
It was never published in a journal. When I started at Colorado State in 1990, I was too busy with both research and developing the center track restrainer for cattle. Another reason is that my advisor would do up to twenty copies of grammar revisions on all his student's papers, and at the time, I was too busy to do multiple revisions.

INTRODUCTION

There has been increasing societal concern about the welfare of farm animals residing in certain modern systems which provide less varied sensory input than natural surroundings (Harrison, 1964; Singer, 1976; Hason and Singer, 1980; Fox, 1984). Environmental complexity affects both central nervous system anatomical development and behavior

(Bennett et al., 1964; Diamond et al., 1964; Greenough and Chang, 1985). Barren environments which restrict sensory input in terms of quantity, quality, or variety tend to increase both behavioral and central nervous system excitability. One question is: Do farm animals residing in these systems have symptoms of sensory restriction? Another is: Will simple, inexpensive methods of environmental enrichment such as objects or extra contact with people have beneficial effects on the animals' wellbeing? One aim of the research reported in this thesis was to answer these and other questions as they pertain to the pig.

A second aim was to determine if environmental enrichment would have beneficial effects on productivity and handling behavior. Animals which move easily during handling at the meat packing plant will be less likely to get bruised or have stress induced meat quality problems.

A third aim was to quantify the pig's interactions with environmental enrichment objects and determine if environmental enrichment would reduce fighting in newly mixed pigs.

SUMMARY OF RESULTS AND CONCLUSIONS

Dendritic growth in somatosensory region of brain cortex in pigs residing in a simple or a complex environment

Rearing environment had a significant effect in dendritic growth and soma size in the somatosensory cortex of young pigs. It had no effect on these parameters in the visual cortex. Pigs residing in pairs in small barren indoor pens (simple environment, SE) had greater dendritic growth and larger somas than 12 pigs living together in a large outdoor pen with straw, play objects, and positive daily contact with a person (complex environment, CE).

The pigs in the SE engaged in greater amounts of belly nosing compared to the CE pigs. The CE pigs had greater overall rooting activity toward a variety of objects, but 95% of that activity was directed toward objects. Even though the SE pigs had less overall rooting activity, only some 40% of their rooting was directed at objects, and approximately 60% of the time was spent rooting on the other pig. Casual observations indicated that the SE pigs were more excitable and aggressive toward the experimenter than the CE pigs.

ENVIRONMENTAL ENRICHMENT REDUCES EXCITABILITY IN PIGS

Experiment I

Environmental enrichment reduced excitability in young weanling pigs which resided in small barren pens. Control pigs were rated more excitable than pigs which had continuous access to hanging cloth strips or 5 mm of daily petting from a person. Pigs which had two environmental enrichment treatments were rated less excitable than pigs which had only one.

Experiment II

Smaller amounts of environmental enrichment also reduced excitability in older finishing pigs. Controls were rated more excitable than animals which received continuous access to hanging rubber hoses, 5 min. of weekly petting from a person, or a weekly drive in the aisle. A combination of environmental enrichment treatments was more effective than a single treatment. Environmental enrichment treatments had no effect on weight gain.

ENVIRONMENTAL EFFECTS ON EASE OF HANDLING

In Trial 1, environmental enrichment treatments reduced the force required to move finishing pigs through a single-file chute. The treatments were continuous access to hanging rubber hose objects (objects), gentle petting for 10 min weekly (mingle), weekly drives in the aisle (drive) or a combination of these treatments. In Trial 2, control pigs required the least amount of force to drive them through the chute.

Trial 2 pigs were calmer and tamer at the start of the trial compared to Trial 1 pigs. There may be an optimum level of handling and contact which will produce a calm market animal that is easy to drive, but not so tame that driving becomes difficult. Trial 2 controls were calm and tame due to a change in farrowing house manager between trials and increased pen washing in Trial 2. Pen washing is a form of handling, and in this case, it served to tame the controls. Trial 2 control pigs were further tamed by the new farrowing house manager. Pigs in the mingle treatment in Trial 2 became so tame that they refused to move through the chute. Control pigs in Trial 1 were excessively excitable, which made driving more difficult. In conclusion, for animals that will be marketed for slaughter, it is desirable to have a calm animal that is easy to drive, but not so tame that driving is difficult.

OBJECT INTERACTION IN WEANLING PIGS

Experiment I

A short-term 15min. preference test indicated that weanling pigs prefer suspended cloth strips. Chain was the least preferred object. In three trials, duration of touching and duration of biting indicated that pigs preferred cloth objects. Duration of touching and biting of hanging rubber hoses was reduced because the pigs had difficulty grabbing them with their mouths.

Experiment II

A short-term 15min. preference test similar to Experiment I indicated that pigs had individual preferences for different cloth textures. In general, there were no group preferences but there were distinct individual preferences.

Experiment III

Chewing and pulling suspended cloth objects decreased over a period of 3 weeks from an average of approximately 1 hour to 20 to 30 minutes. Even though play decreased, the pigs still actively played with the cloth strips at the end of the trial.

Experiment IV

Object interactions had a tendency to increase when the pigs were most active. Play and fighting both occurred more often when the pigs were standing or eating.

Experiment V

A 7-day preference test indicated that object preferences changed over the time. For the first three days of the experiment, cloth objects were preferred. At the end of the day period, however, hanging rubber hoses were preferred. Hanging chains were the least preferred objects all along. The change from cloth to hose preference may be due to the development of motor coordination. Observations showed that small pigs often had difficulty grabbing the hose until they had had some experience with it.

Experiment VI

The presence of hanging cloth strips during regrouping reduced the number of 5min. periods which contained fighting. The presence or absence of hanging cloth objects for seven days prior to regrouping had no effect on fighting. Providing hanging cloth strips for seven days prior to regrouping reduced chewing and pulling of the strips during the 24-hour regrouping period.

GENERAL CONCLUSIONS

Rearing environment has strong effects on both the central nervous system and pig behavior. Small amounts of environmental enrichment reduced excitability and fighting. The provision of hanging cloth strips during regrouping of young pigs reduced fighting. Suspended objects and small amounts of contact with a person in the pen also reduced excitability.

Environmental enrichment uniformly reduced excitability in all the experiments, but it affected behavior during handling in a complex manner. There may be an optimum level of contact with people for animals which will be marketed for slaughter. Both excessively excitable and overly tame animals are difficult to drive and handle. In one trial, regrouping and objects reduced the force required to drive pigs through a single file chute.

But, in a second trial, the controls were easiest to drive. In Trial 2, the pigs in the mingle group were overly tame and thus difficult to drive. The recommended amount of environmental enrichment for finishing pigs will vary depending on genetics and the animals' previous experiences. Management procedures which make pigs more amenable to handling at the slaughter plant would help prevent costly bruises. They also would help improve meat quality because excitement in the stunning chute increases PSE and other meat quality problems (Barton-Gade, 1985; Grandin, 1986).

There has been increasing societal concern about the welfare of farm animals residing in certain modern systems which provide less varied sensory input than natural surroundings (Harrison, 1964; Singer, 1976; Mason and Singer, 1980; Fox, 1984). Some pigs kept in barren pens on modern farms may be exhibiting symptoms of sensory restriction. The animals are excitable, as may be judged, for example, by their jumping up suddenly when a door slams.

Excitability and increased irritability is a symptom of sensory restriction. Pairs of puppies residing in barren kennels become unusually aroused and excited when exposed to something new (Melzack, 1954). Environmental restriction may have long-term detrimental effects on the central nervous system. The young dogs still had abnormal electroencephalographic patterns 6 months after being removed from the barren kennel (Heizack and Burns, 1965). Animals living in an enriched environment are usually less excitable than animals in barren surroundings (Walsh and Cummins, 1975).

The experiments on dendritic branching and soma size in young pigs indicate that rearing environment has significant effects on nervous system development. The results were contrary to the original hypothesis that animals in the enriched environment (complex environment, CE) would have greater dendritic branching. Volkmar and Greenough (1972) found that rats residing in an enriched environment with objects and other rats had more dendritic branching in the visual cortex compared to pairs or isolates in plain cages.

Pigs reared in the simple environment (SE) had greater dendritic branching in the somatosensory cortex. There were no significant differences in dendritic growth in the visual cortex.

The behavior of the SE pigs differed significantly from the CE pigs. The SE pigs engaged in greater amounts of belly nosing. Increased belly nosing may have stimulated the somatosensory cortex. As the environment is made increasingly barren, nibbling, and massaging of other pigs will increase (Stolba, 1981). The SE pigs also were more excitable and aggressive toward the experimenter.

Rearing environment affects dendritic development in complex ways. More dendritic development Is not necessarily beneficial. Monkeys reared in the colony had less dendritic branching in the Motor 1 cortex compared to isolates in a cage with ladders and swings (Stell and Riesen, 1987). Rats exposed to continuous lighting had greater spine density in the visual cortex, but their retinas were damaged (Parnavelas et al. 1973; Bennett et al., 1972; O'Steen, 1970).

Pigs in the SE condition appeared to be actively seeking stimulation. Walsh and Cummins (1975) state that an organism becomes increasingly sensitized to stimulation in an attempt to restore balance. Lack of normal sensory input causes the areas of the brain that receive sensory input to become more excitable. Trimming the whiskers of baby rats causes the receptive field in the brain to become more excitable and enlarged (Simons and Land, 1987). The receptive fields were still enlarged three months after the whiskers had regrown.

Increased belly nosing may have been an attempt to obtain more stimulation and reduce arousal. The SE condition was more barren than any environment which would exist in nature. Perhaps the SE environment deprived the animals beyond their ability to cope. The increased dendritic branching in the somatosensory cortex of the SE pigs may be highly abnormal.

Experimental results indicate that simple and inexpensive environmental enrichment procedures such as suspended cloth strips or hoses and small amounts of increased positive contact with people will reduce both excitability and fighting in young pigs. Reducing excitability is advantageous for animal welfare reasons. Excessive excitability may be a sign of detrimental effects of sensory restriction in the nervous system.

REFERENCES

Barton-Gade, P. 1985. Developments in the pre-slaughter handling of slaughter animals, Proc. European Meeting of Meat Res. Workers. Paper 1:1, pp.1-6.

Bennett, E.L., M.C. Diamond, D., Kerch, and M.R. Rosenzweig, 1964. Chemical and anatomical plasticity of the brain, *Science* 146:610-619.

Diamond, M.C., D. Krech, and M.R. Rosenzweig, 1964. The effect of an enriched environment on the histology of the rat cerebral cortex, *J. Comp. Neural,* 123:111-120.

Diamond, M.C., F. Law, H. Rhodes, B. Lindner, M.R. Rosenzweig, D. Krech, and E.L. Bennett. 1966. Increases in cortical depth and glia numbers in rats subjected to an enriched environment, *J. Comp. Neural.* 128:117-126.

Fox, M.W. 1984. *Farm Animals, Husbandry Behavior and Veterinary Practice*, University Park Press, Baltimore, MD.

Grandin, T. 1986. Good pig handling improves pork quality, Proc. European Meeting of Meat Res. Workers, Paper 2:12, pp. 105-108.

Greenough, W.T. and F.F. Chang. 1985. Synaptic structural correlates of information storage in mammalian nervous systems, In: *Synaptic Plasticity*, pp. 335-363. Guilford Press, New York, NY.

Harrison, R. 1964. *Animal Machines*, Stuart, London, England.

Mason, J. and P. Singer. 1980. *Animal Factories*, Crown, New York, NY.

Melzack, R. 1969. The role of early experience in emotional arousal, *Ann. N.Y. Acad. Sci.* 159:721-730.

Melzack, R. 1954. The genesis of emotional behavior: An experimental study of the dog, *J. Comp. Physical. Psych*, 47:166-168.

Melzack, R. and S.K. Burns. 1965. Neurophysiological effects of early sensory restriction, *Exp. Neural*, 13:163-175.

O'Steen, W.K. 1970, Retinal and optic nerve serotonin and retinal degeneration as influenced by photoperiod, *Exp. Neural*, 27:194-205.

Parnaveias, J. G., A. Gloous, and P. Kaups. 1973. Continuous illumination from birth affects spine density of neurons in the visual cortex of the rat, *Exp. euro*. 40:742-747.

Simons, D., and P. Land. 1987. Early experience of tactile stimulation influences organization of somatic sensory cortex, *Nature*, 326:694-697.

Singer, P. 1976. *Animal Liberation*, Random House, New York, NY.

Stell, M. and A. Risen. 1987. Effects of early environments on monkey cortex neuroanatomical changes following somatomotor experience: Effects on layer III pyramidal cells in monkey cortex, *Behave. Neurosci.* 101:341-346.

Stolva, A. and D.G.M. Wood-Gush, 1980. Arousal and exploration in growing pigs in different environments, *Apple. Animal. Ethol.* 6:382-383.

Volkmar, F.R. and W.T. Greenough, 1972. Rearing complexity affects branching of dendrites in the visual cortex of rat. *Science*, 176:1445-1447.

Walsh, R.N. and R.A. Cummins, 1975. Mechanisms mediating production of environmentally induced brain changes, *Bull.* 82:986-1000.

SEVEN

Behavioral Agitation during Handling of Cattle Is Persistent over Time

T. Grandin (1993) Behavioral agitation during handling of cattle
is persistent over time, *Applied Animal Behavior Science*, Vol. 36, No, 1, pp. 1-9.

I get asked all the time why I used the word *agitation*, when the behavior was caused by fear. The reviewers of this paper made it very clear that *fear* was an emotional word that should never be used in a scientific paper. Today the word *fear* is a recognized scientific term. At the time, I was forced to change "fear" to "behavioral agitation" to be more objective.

ABSTRACT

Cattle which become extremely behaviorally agitated during restraint and handling are dangerous to handlers and are more likely to become stressed. Fifty-three Gelbvich×-Simmental×Charolais cross bulls and 102 steers were restrained for blood testing every 30 days in a squeeze chute (crush). At the same time, they were also weighed in a single animal scale. Out of four consecutive restraint sessions. five (9%) of the bulls became extremely behaviorally agitated every time they were restrained. During three consecutive restraint sessions, six (6%) of the steers were always behaviorally agitated. Of the bulls, 13 (25%) were very calm and stood still in the squeeze chute. Of the steers. 40 (40%) were always very calm. The implications of the study are that behaviorally agitated behavior is very persistent over a series of handling and restraint sessions, and cattle which repeatedly become agitated during restraint should be culled. There was also a relationship between balking and temperament rating. Agitated bulls balked less during entry into the squeeze chute or scale. Cattle also balked less while entering the scale. This indicates that restraint in the squeeze chute was more aversive than the scale.

INTRODUCTION

Cattle with a bad temperament are more difficult to handle and create a safety hazard for handlers. Both previous experiences and genetic factors affect the behavior of livestock during handling. Cattle and sheep can remember an aversive experience for many months (Hutson, 1985; Pascoe, 1986). Sheep which had been inverted in a sheep handling machine were more difficult to work through the corrals the following year (Hutson, 1985). Cattle which had previous handling experiences in a livestock market settled down more quickly at the slaughter plant stockyard facility compared with cattle which had come directly from the farm. Possibly, the slaughter plant stockyard appeared less novel and frightening to the cattle which had been in a livestock market (Cockram, 1990).

Genetics also have an effect on behavior and stress levels during handling. Tulloh (1961) reported that Angus cattle are more excitable than Herefords. Brahman cross cattle were more restless and moved around more in a squeeze chute compared with Shorthorns (Fordyce et al., 1988). In Brahman heifers, excitable animals had higher serum cortisol levels (Stahringer et al., 1989). Grandin (1991a) reports that there are increasing problems with excitable pigs which are difficult to handle at the slaughter plant. Genetics may be a major contributor to this problem. Observations on farms by Grandin (1991a) indicated that different genetic lines of sows housed in the same room varied greatly in startle response.

The purpose of this study was to determine if temperament problems would remain persistent over a series of handling and restraint sessions. This information would be useful in making culling decisions.

ANIMALS, MATERIALS, AND METHODS

Behavioral observations were made of 53 bulls and 102 steers while they were restrained for blood testing. The animals were Gelbvieh, Charolais and Simmental crosses which had been raised on a commercial ranch under extensive pasture conditions. Both the steers and the bulls were the same age. The animals weighed 260 kg at the start, and they were housed in three outdoor feedlot pens with a fence line feed bunk. Every 30 days they were restrained in a Powder River manual squeeze chute (crush) for blood testing and weighed on a single animal scale. The head of each animal was restrained by a stanchion (head bail) clamped around its neck and the body was held between two squeeze panels. A rope halter was used to position the head for blood testing from the jugular vein. Blood testing was conducted as part of another experiment. The single animal scale was located in the single file race prior to the squeeze chute. The cattle had to pass through it to enter the squeeze chute. During all handling and restraint sessions, care was taken to handle animals gently. An electric prod was used on only one stubborn animal.

TEMPERAMENT RATINGS

A rater stood by the squeeze chute and rated each animal on a five-point scale. The rating was made after the head was clamped by the stanchion. The ratings were: (1) calm, no movement; (2) slightly restless; (3) squirming, occasionally shaking the squeeze chute; (4) continuous, very vigorous movement and shaking of the squeeze chute; (5) rearing, twisting of the body and struggling violently.

Ratings of 4 or 5 were classified as behaviorally agitated. On the Fordyce et al. (1988) movement (MOY) scale, a rating of 4 or 5 would be equivalent to Fordyce's 6 or 7. Fordyce et al. (1988) describes a (6) as very disturbed and continuous, very vigorous movement and a (7) as struggling violently and attempting to jump out. A five-point rating scale was used for my observations because it is very difficult for an observer to accurately differentiate between seven different ratings.

BALKING RATINGS

The same rater also rated balking behavior at the entrance to both the squeeze chute and the scale. Cattle were classified into balkers and non-balkers: (1) Non-Balker: entered voluntarily when the gate opened, or a light tap on the rump was required to induce the animal to enter the scale or the squeeze chute. To be classified as a non-balker, the animal had to enter both the scale and the squeeze chute without balking. (2) Balker: a hard slap on the rump or tail twisting was required to induce an animal to enter the scale or squeeze.

RESTRAINT SESSIONS

There were five handling and restraint sessions spaced 30 days apart. Temperament data for bulls was collected and tabulated for Sessions 1, 2, 3 and 4. Temperament data for steers was tabulated for Sessions 2, 3 and 4. Temper ament data was not tabulated for Session 5 because a hydraulic squeeze was used. The tight pressure of the hydraulic squeeze masked the animals' movements. Balking data was collected for all five sessions. During Session 5, only 33 bulls and 68 steers were rated because some of the animals had to be slaughtered as part of another experiment.

RESULTS

In both the bulls and the steers, there was a group of animals which remained calm during all of the restraint sessions and another group which became behaviorally agitated during all the sessions. There was also a large group which had mixed temperament ratings. Animals in this group were calm during some restraint sessions and behaviorally agitated during others. Table 1 shows the percentage of bulls in each temperament rating group.

There were five bulls (9%) that were behaviorally agitated for all four temperament-rated restraint sessions. They had ratings of 4 or 5 for four out of four restraint sessions (Table 1). Two different bulls reared over the top of the squeeze chute and one of these animals escaped. These incidents occurred during restraint Sessions 1 and 2. Four bulls (8%) had a rating of 3 or greater for three out of four temperament-rated sessions. None of these animals had a rating of 1. One of these animals was caught around the

TABLE 1: Behavior of 53 Bulls during Four Restraint Sessions Spaced 30 Days Apart

	N (%)	Temperament rating means and standard errors for each restraint session				Comments
		1	2	3	4	
(Behaviorally agitated) Rating of 4 or 5 for all restraint sessions	5 (9)	4.4±0.27	4.4±0.27	4.4±().27	4.4±0.27	Two different animals reared over the top of the squeeze and one escaped
(Behaviorally agitated) Rating of 3 or greater for three out of four restraint sessions	4 (8)	3 ±0.40	3 ±0.66	3.5±0.8	4.5 ± 0.28	One animal was caught around the midsection during Session I and charged handlers during Session 3. No ratings of I
(Mixed ratings) Rating of 3 or greater for two out of four restraint sessions	6 (11)	3 ±0.43	1.3 ± 0.19	2.3±0.91	2.5±0.98	One animal charged handlers during Session I
(Mixed ratings) Rating of 3 or greater for one out of four restraint sessions	11 (21)	1.3± 0.20	1.7 ± 0.20	2.1=0.37	2.5±0.30	No ratings of 5
(Calm or restless rating) Rating of 2 or less for all restraint sessions	14 (26)	1.4±0.14	1.5±0.14	1.5:t:0.15	1.3±0.11	
(Calm rating) Rating of 1 for all restraint sessions	13 (25)	1±0	1±0	1±0	1±0	

Temperament rating system on a 1 to 5 scale. 1=calm; 5=extremely excited.

TABLE 2: Behavior of 102 Steers during Three Restraint Sessions Spaced 30 Days Apart

	N (%)	Temperament rating means and standard errors for each restraint session			Comments
		1	2	3	
(Behaviorally agitated) Rating of 4 or 5 for all restraint sessions	3 (3)	4 ±0	4 ±0	4.3± 0.32	
(Behaviorally agitated) Rating of 3 or greater for three out of three restraint sessions	3 (3)	3 ±0	3.3±0.32	4.3±0.32	
(Mixed ratings) Rating of 3 or greater for two out of three restraint sessions	10 (10)	2.5±0.41	2.8±0.28	3.5±0.44	Includes 3 different animals with a 5 rating in Session 4. Also includes 4 animals with a 1 rating
(Mixed ratings) Rating of 3 or greater for one out of three restraint sessions	22 (22)	1.8±0.14	2.4±0.23	2.8±0.28	Two animals with rating of 5 for Session 3. Both animals had animals with a rating of 4 or 5 ahead of them. Both had I and 2 ratings for Sessions I and 2. No ratings of 5
(Calm or restless rating) Rating of 2 or less for all restraint sessions	24 (24)	1.3±0.10	1.6±0.10	1.3±0.10	
(Calm rating) Rating of 1 for all restraint sessions	40 (40)	1 ±0	1 ±0	1 ±0	

Temperament rating system on a 1 to 5 scale. 1=calm; 5=extremely excited.

TABLE 3: Relationship between Balking and Temperament in Bulls		
	Balker	**Non-balker**
(Behaviorally agitated) Rating of ≥ 4 on three out of four restraint sessions	3 (33%)	6 (67%)
(Calm and calm restless group) Rating of I or 2 on all restraint sessions	20 (65%)	7 (35%)
X^2 = < 0.05.		

TABLE 4: Relationship between Balking and Temperament in Steers		
	Balker	**Non-balker**
(Behaviorally agitated) Rating of ≥ 3 for all restraint sessions	4 (66%)	2 (33%)
(Calm and calm restless group) Rating of I or 2 for all restraint sessions	41 (64%)	23 (36%)
X^2 = 1.6 NS.		

midsection by the head stanchion during the first session. This same animal charged the handlers during Session 3. Seventeen bulls (32%) had mixed ratings which varied from 1 to 5 (Table 1). Only one bull in the mixed ratings group received a rating of 5 for one session. All other bulls in the mixed rating groups had ratings which varied from 1 to 4. Fourteen bulls (26%) received a rating of 1 or 2 for all rated restraint sessions and 13 (25%) were always calm and received a rating of 1 for all sessions.

The steers had similar results. A higher percentage of the steers received a rating of 1. During all temperament rated restraint sessions, 40% of the steers received a rating of 1 whereas only 25% of the bulls had a rating of 1. Steers were significantly calmer than

TABLE 5: Comparison of Balking at the Squeeze with Balking at the Scale

	Scale	Squeeze
	Number (%) balking	**Number (%) balking**
Bulls		
Session 1	7 (13%)	7 (13%)
Session 2	9 (17%)	14 (26%)
Session 3	7 (13%)	15 (28%)
Session 4	4 (7%)	8 (15%)
Session 5	0 (0%)	14 (42%)
Average Percentage	10%	25%
Steers		
Session 1	24 (24%)	21 (21%)
Session 2	21 (21%)	22 (22%)
Session 3	18 (18%)	40 (40%)
Session 4	21 (21%)a	25 (25%)a
Session 5	0 (0%)	15 (22%)
Average Percentage	17%	26%

Only 33 bulls and 68 steers were rated during Session 5. $X^2 = 6.5 > 0.05$. Scale less aversive than squeeze.

bulls: X2 = 6.72>0.01. Three steers (3%) were behaviorally agitated with ratings of 4 or 5 for the three temperament rated sessions (Table 2). Three additional steers (3%) had ratings of 3 or greater for all rated restraint sessions. This makes a total of 6% behaviorally agitated steers. A total of 32 steers (32%) fell into the two mixed rating groups. Both of the mixed groups contained animals with a rating of 5. Twenty-four steers (24%) received a rating of 1 or 2 for all the rated restraint sessions and 40 (40%) were always calm with a rating of 1.

Tables 3 and 4 show the relationship between balking and temperament. An animal was classified as a balker if he balked during any of the restraint sessions. To be classified as a non-balker, he had to enter both the scale and squeeze chute without balking. In the bulls, behaviorally agitated animals (rating of 3 or greater for all rated sessions) balked significantly less than the animals with a calm score (Rating 1) or a calm restless score (Rating 1 or 2): $X^2 = 4.8 > 0.05$. In the steers, there were no significant differences.

In both the bulls and the steers, the squeeze chute was more aversive than the scale: $X^2 = 6.5 > 0.05$ (Table 5). During the first restraint session, balking percentages for bulls were the same at both the scale and the squeeze chute. For steers, they were slightly different. Gradually, the scale became less aversive. During Session 5, none of the animals balked at the scale entrance. In the steers balking at the squeeze chute did not steadily escalate (Table 5). Twenty-one percent balked during the first session and 22% balked during the last session. In bulls balking at the squeeze chute was 13% for Session I and 42% for Session 5. However, balking was only 15% during Session 4.

DISCUSSION

The implication of these results is that in certain individuals, the extreme tendency to become behaviorally agitated is stable over time. These individuals should be culled. In a breeding program, it would be advisable to select animals from the group which always had a calm rating of 1. The bulls and steers fell into three basic behavior groups: (a) always behaviorally agitated with ratings of 4 or 5 for all rated restraint sessions; (b) mixed ratings varying from 1 to 5; (c) always calm with ratings of 1. These results indicate that culling decisions should not be based on a single evaluation of temperament, because there is a high percentage of cattle with highly variable ratings. Three sessions, to evaluate temperament, would provide a more accurate assessment.

It is probably necessary to handle each animal individually to evaluate temperament. When the cattle were housed as a group in the feedlot pens, there were no observable signs of agitated behavior. The cattle which always had ratings of 4 or 5 during restraint could not be distinguished from the other cattle when they were undisturbed in the feedlot pens. Temperament differences were also not evident when the cattle were observed during handling in a 400-head-per-hour high speed slaughter plant. They moved at a steady walking speed through a double rail restrainer system (Grandin, 1988, 1991b) in a continuous single file line. Both the steers and the bulls walked quietly into the restrainer. The inability to detect temperament differences at the slaughter plant may possibly be explained by the stressful effects of being separated from other animals while they were restrained in the squeeze chute. The cattle in the slaughter plant maintained both visual and physical contact with each other during handling and stunning.

Both genetics and previous experience can affect the behavior of cattle during handling. Previous studies by Tulloh (1961) and Fordyce et al. (1988) indicate differences in temperament between different cattle breeds. Anecdotal reports from ranchers reveal increasing complaints about temperament problems in cattle breeds such as Charolais, Limousin and Salers. Producers also report that there are excitable genetic lines within breeds.

An animal's early experiences can affect its responses towards people later in life. Hemsworth et al. (1986) found that early handling experiences affected behavior later in the life of a pig. Cattle which originated from a feedlot with gentle handling had fewer bruises (Grandin, 1981). Casual observations by the author indicate that cattle from feedlots which had a reputation for rough handling were wilder and more difficult to handle at the slaughter plant compared with cattle from feedlots with gentle handling.

The cattle observed in this study were wild and raised under extensive pasture conditions. Untamed, extensively raised beef cattle will have higher cortisol levels during restraint compared with tame dairy cows (Lay et al., 1992). The implication of this study is that the tendency to become behaviorally agitated is stable over time in certain individual animals. More research will be needed to accurately determine the relative contribution of genetics and previous experience.

The balking data indicated that cattle are capable of discriminating between the less aversive scale and the more aversive squeeze chute. In bulls, there was a significant tendency for the animals with ratings of 4 and 5 to balk less.

REFERENCES

Cockram. M.S. 1990. Some factors influencing the behavior of cattle in a slaughterhouse lair age. *Anim. Prod.* 50: 475-481.

Fordyce. G. Dodt, R.M. and Wythcs. J.R. 1 LJ88. Cattle temperaments in northern Queensland. I. Factors affecting temperament. *Aust. J. Exper. Agric.* 28: 683-687.

Grandin. T. 1981. Bruising on southwestern feedlot cattle. *J. Anim. Sci.* (Abstract) 53 (Suppl. I): 2 I 3.

Grandin. T. I 988. Double rail restrainer for livestock handling. *Int. J. Agric. Eng.* 41: 32 7 -338.

Grandin, T. 1991a. Handling problems caused by excitable pigs. 37th International Congress of Meat Science and Technology, September 1991. Federal Center for Meat Research. Kulmbach. Germany. Paper 2.8. pp. 249-252.

Grandin. T. 1991 b. Double rail restrainer system for beef cattle. Am. Soc. Agric. Eng. Technical Paper 915004. St. Joseph, Ml.

Hemsworth. P.H. Barnett. J.L, Hansen, C. and Gonyou. H.W. 1 LJ86. The influence of early contact with humans on subsequent behavioral response of pigs to humans. *Appl. Anim. Behav. Sci.* 15: 55-63.

Hutson. G.D. 1985. The influence of barley food rewards on sheep involvement through a handling system. *Appl. Anim. Behav. Sci.* 14: 263-273.

Lay. D.C. Friend. T.H. Bowers. C.L. Grissom. K.K. and Jenkins. O.C. 1992. A comparative and physiological and behavioral study of freeze and hot iron branding using dairy cows. *J. Anim. Sci.* 70: 1121-1125.

Pascoe. P.J. 1986. Humaneness of electrical immobilization unit for cattle. *Am. I. Vet. Res.* 10: 2252-2256.

Stahringer. R.C. Randall, R.D. and Neuendorff. D.A. 1989. Effect of naloxone on scrum luteinizing hormone and cortisol concentrations in seasonally anestrous Brahman heifers. *J. Anim. Sci.* (Abstract) 67 (Suppl. I): 359.

Tulloh. N.M. 1961. Behavior of cattle in yards: II. A study in temperament. *Anim. Beha.* LJ: 25-30.

EIGHT

Teaching Principles of Behavior and Equipment Design for Handling Livestock[1]

T. Grandin (1993) Teaching Principles of Behavior and Equipment Design for Handling *Livestock, Journal of Animal Science*, Vol. 71, no. 4, pp. 1065-1070.

When I arrived at Colorado State University in 1990, I started teaching a short course on Livestock Handling, which covered behavior during handling and the design of handling facilities.

ABSTRACT

A course is described in which students are taught principles of livestock behavior and how an understanding of behavior can facilitate handling. Some of the principles that are covered in the course are livestock senses, flight zone, herd behavior during handing, and methods to reduce stress during handling. To teach problem solving and original thinking, the students design three different types of handling facilities. Design of restraint equipment and humane slaughter procedures are also covered. Both existing systems and ideas for future systems are discussed. Students are provided with information from both scientific studies and practical experience. Key Words: *Behavior, Teaching, Handling, Restraint*

1. Presented at a symposium titled "The Impact of the Animal Rights Welfare Movement on Animal Science Teaching Programs" at the ASAS 83rd Annual Mtg., Laramie, WY.

INTRODUCTION

To improve animal welfare and reduce stress, there is a need for students to learn about livestock behavior during handling. My course is titled "Livestock Behavior and Handling," and it is offered in the Department of Animal Sciences at Colorado State University. Most of the students who take the course are either animal science or equine science majors. Students taking the course learn the principles of livestock behavior and how they can be used to facilitate handling during truck loading, feedlot processing, veterinary procedures, sorting, and slaughter. Information from both scientific studies and practical experience is presented. A major objective of this course is to stimulate students to think in an original manner. The emphasis is on learning how to use facts rather than reciting them back on an exam. The major areas covered in the course are livestock senses, flight zone principles, facility layout and design, restraint, stress and handling, humane slaughter, and welfare aspects of handling and various procedures. To stimulate problem-solving abilities and original thought, students design three different types of livestock handling facilities and design restraint equipment for a fictitious animal.

CONTENT OF THE COURSE

Livestock senses

It is important for students to learn some basic information on how cattle, pigs, and sheep perceive the world. All species of livestock have wide-angle panoramic vision (Prince, 1977). This fact explains why practical experience has shown that the use of solid sides on loading ramps, chutes, and crowd pens facilitates handling and reduces agitation (Rider et al., 1973; Grandin, 1980, 1982). Research has shown that sheep have depth perception when they are standing still (Lehman and Patterson, 1964). Hutson (1985a) suggested that there may be a blind area at ground level and sheep may not be able to use motion parallax or retinal disparity to perceive depth. Contrary to popular belief, livestock have color vision (Hebel and Sambraus 1976; Munkenbeck, 1982; Klopfer and Butler, 1984; Gilbert and Arave, 1986).

Observations in the field have shown that livestock will often balk at puddles, shadows, and moving objects (Lynch and Alexander, 1973; Grandin, 1980). Differences in illumination will also affect livestock movement (Lynch and Alexander, 1973; Van Putten and Elshof, 1978; Hutson, 1981; Grandin, 1987, 1989a). Students are shown numerous slides of handling facilities, many taken from a cow's perspective. The slides show both well-designed and poorly designed facilities, and a full explanation is given for each slide. The slides also show things that make cattle balk, such as shadows, puddles, and swinging chains in chutes and alleys. Exposure to many pictures helps students identify and correct problems in the field.

Cattle and sheep are more sensitive to high frequency noise than humans (Ames and Arehart, 1972). Excessively loud noise is stressful, but animals can adapt to reasonable noise levels (Ames, 1974). Sheep slaughtered in a noisy commercial slaughter plant had elevated cortisol levels compared to sheep slaughtered in a quiet research abattoir (Pearson et al., 1977). During lecture, students are informed about ways to reduce noise in livestock handling facilities. More detailed information based on practical experiences has been published previously (Grandin, 1987, 1989a).

Flight zone and behavior during handling

It is very important for students to understand the concept of flight zone and point of balance. Knowledge of these principles will enable students to handle livestock safely, humanely, and efficiently. Students need to learn to stay on the edge of the animal's flight zone. The size of the flight zone is affected by both the size of the enclosure an animal is held in and whether the animal has had previous contact with people (Hutson, 1982; Hargreaves and Hutson, 1990a).

Inexperienced handlers often make the mistake of standing in front of the point of balance at the shoulder and prodding an animal on the head to make it go forward. A complete practical explanation of these principles was described by Grandin (1980, 1987, 1989a) and Kilgour and Dalton (1984). Students are taught flight zone principles with lectures, videotapes, and practice with live animals. Point of balance is taught by confining several cattle or sheep in a single-file chute. Progressing from head to tail, when the student walks past the point of balance of the animal, it will move forward.

All livestock are herd animals, and they will become stressed or agitated when they are separated from their herd-mates. Students must be warned that a steer that is calm in its home pen with a group of animals may charge and run over them when it is separated from its herd-mates. Many serious cattle handling accidents are caused by a lone animal.

Many students do not realize that behavior during handing is affected by previous experiences. Animals remember painful or frightening experiences for many months (Hutson, 1985b; Pascoe, 1986). The author has observed that cattle that came from feedlots with rough handling were wilder and had more bruises than cattle from feedlots with gentle handling. There is also some evidence that sheep can remember specific people who participated in painful surgery (Fell and Shutt, 1989).

Observations by the author indicate that certain individual cattle within a herd may become very agitated in the squeeze chute. Cattle with an excitable temperament should be culled from a teaching program because they are likely to injure students.

For both humane and safety reasons, a college campus must have adequate handling facilities for teaching students. A bare minimum for cattle would be a squeeze chute, single-file chute, crowd pen, and one or two holding pens. For safety, it is essential that the equipment is kept in good repair. Latching mechanisms on the squeeze chute must be replaced when they become worn to prevent accidents. For maximum educational value, a modern facility with a curved chute with solid sides and a round crowd pen is recommended. It is important to expose students to well-designed facilities. Many college campuses are sadly lacking.

Facility layout and design

In my class, I have students design and lay out three different handling facilities for a packing plant, a feed-yard, and a ranch. For added realism, the assignments are based on real projects from my consulting practice. This assignment is given in the middle of the course after the students have been exposed to many different types of facilities in both lectures with slides and by visiting and critiqueing the cattle handling facilities at Colorado State University. During the slide lectures, both poor and good layouts are shown with full explanation. The students are also provided with a book of drawings of layouts of 30 different cattle handling facilities.

Some of the design principles that are taught are the use of solid sides on chutes and crowd pens to prevent animals from seeing out with their wide-angle vision and layout of curved chutes and round crowd pens. A circular crowd pen and a curved chute reduced the time spent moving cattle by up to 50% (Vowles and Hollier, 1982).

Planning livestock flow through a facility is good training in problem solving. Students have to plan sufficient pen space to gather the cattle and then have enough different pens and space for sorting cattle. They are taught to draw a flow chart of the sequence of handling, weighing, and sorting events that will be conducted in the corral and then make sure the system they design will be able to accommodate those needs. Further information on cattle facility layout that can be used in class has been published previously (Grandin 1980, 1982,1984a, 1989a, 1990a; Midwest Plan Service, 1987; Meat and Livestock Commission [Cattle Handling; Meat and Livestock Commission, Milton Keynes, U.K.1).

It is also important for students to learn behavioral differences between species that should be taken into account when a facility is laid out. Practical experience has shown that funnel shaped crowd pens with one side straight and the other side at a 30° angle work well for directing cattle and sheep into a single-file chute or loading ramp. However, pigs will jam the funnel, and a crowd pen for them should have an abrupt entrance to the chute.

Restraint equipment and use

Slides are used extensively in the discussion of restraint equipment. Both existing systems and ideas for new systems are covered. The advantages and disadvantages of different squeeze chute designs are reviewed. For example, a curved bar stanchion head gate that is suitable for a chute with squeeze sides is not suitable for use on the end of an alley that does not have squeeze sides. Squeeze sides are required with a curved bar stanchion headgate to prevent the animal from lying down and choking. A straight bar stanchion that does not put pressure on the carotid arteries is safer for use on the end of an alley. It is essential to emphasize to students that if an animal begins to lose consciousness due to pressure on the carotid arteries, the headgate must be released immediately to prevent death (White, 1961; Fowler, 1978). It must also be stressed that an animal should never be left unattended in a squeeze chute. If a hydraulic chute is used, the pressure relief valve must be properly adjusted to prevent serious injuries to the cattle. Experience in the field indicates that a properly adjusted hydraulic chute is safer than a manual chute for both cattle and students. Practical recommendations for squeeze chutes and headgates were provided by Grandin (1975, 1983a). The chapter by Ewbank (1968) contains useful information about animal behavior.

For some procedures, a squeeze chute is not required. Ranchers have found from practical experience that unruly cows will stand still for pregnancy checking and artificial insemination when they are confined in a dark box chute. The cow is held in a narrow enclosure with solid sides, solid front, and solid top (Parsons et al., 1969). Experiments with poultry and cattle indicate that sight restriction in a dark enclosure reduces stress (Douglas et al., 1984; Hale et al., 1987). The dark box is a good example of the use of behavioral principles to reduce an animal's tendency to resist restraint. There is a constant emphasis throughout the course on using an animal's natural behavior patterns to make handling and restraint more humane and efficient. For example, pigs and sheep can be trained to enter a relatively comfortable restraining device voluntarily (Panepinto et al., 1983; Grandin, 1989). Pigs will relax and go to sleep in a padded chute that presses on their sides (Grandin et al., 1989). Voluntary acceptance of restraint can be facilitated with food rewards. This is especially useful with research animals. Animals can readily learn to discriminate between an extremely aversive and a less aversive method of restraint in a avoid-avoid choice test (Grandin et al., 1986).

I have learned from practical experience that cattle are extremely averse to nose tongs. Observations indicated that cattle repeatedly restrained for blood testing with a halter were more cooperative during successive blood tests. The use of nose tongs made head restraint more difficult in the future. Students are repeatedly reminded that restraint should be done as gently as possible and the use of aversive restraint methods such as nose tongs should be avoided.

One of the assignments or exam questions I give to my students is to design a humane restraint device for either a fictitious animal with strange behavioral traits or an unconventional animal, such as a jellyfish or tomato worm. Some students have devised very inventive schemes, such as suspending the tomato worm in gelatin.

The purpose of this exercise is to stimulate problem solving and original thinking. It also helps students gain better understanding of the behavior of cattle, sheep, or pigs. Before the assignment, new ideas for improving or replacing squeeze chutes are discussed. One approach is to redesign or modify existing systems that are used in packing plants. Some examples are the conveyorized double rail restrainer (Giger et al., 1977; Grandin, 1988), V restrainer (Schmidt, 1972), or the upright holding chute that is used for kosher slaughter (Marshall et al., 1963). The kosher chute designed by Marshall et al. (1963) has closed in solid sides and the author has observed that cattle are often more calm in this chute than in a conventional squeeze chute. A possible explanation is that the solid sides removed the chute operator from inside the animal's flight zone and a solid panel in front of the headgate blocked a view of a pathway of escape. I also encourage the students to invent totally new concepts that are unlike anything used by the industry.

Reducing stress

During lectures I explain that reducing stress during handling provides benefits of both improved productivity and welfare. A major concept that I want my students to understand is that the degree of stress imposed on an animal during handling or restraint can vary greatly, depending on factors such as the animal's previous experiences, genetics, tameness, painfulness of the procedure, and skill of the handlers.

Animals can become habituated to nonpainful handling procedures such as weighing of cattle or taking pigs for walks in the aisle during finishing (Peischel et al., 1980; Grandin, 1989d). When an animal first experiences a handling procedure, it may be very stressed because of the novelty and then have little or no stress after it has become accustomed to the procedure. However, animals do not habituate to severely aversive procedures (Hargreaves and Hutson, 1990b; Coppinger et al., 1991).

Livestock that are accustomed to close contact with people are calmer and less stressed by handling than livestock that seldom see people. Ried and Mills (1962) found that sheep that were raised in a barn in close contact with people had a less intense physiological response to handling than did sheep raised on pasture. Students need to learn that isolation is a strong stressor to both cattle and sheep (Kilgour and DeLangen, 1970; Rushen, 1986).

A major point made in my lectures is that gentle treatment of animals is good for both animal welfare and productivity. Reproductive function is especially sensitive to stress. Sows that are fearful of humans and react by withdrawing farrow fewer pigs than sows that are not fearful (Hemsworth et al., 1981). In cattle and sheep handling, stresses were detrimental to reproductive performance (Doney et al., 1976; Hixon et al., 1981). Mistreatment of growing animals is also detrimental. Occasional mistreatment of growing pigs reduced weight gains (Hemsworth et al., 1987). Handling and transport stresses can also impair rumen function and immune function (Galyean et al., 1981; Kelly et al., 1981; Mertsching and Kelly, 1983; Blecha et al., 1984). The section on stress is concluded with information on the importance of the relationship between humans and animals on productivity and stress. Good reviews on this subject have been published by Seabrook (1984) and Hemsworth and Barnett (1987).

Slaughter

Handling at the slaughter plant and humane slaughter procedures are also covered. All procedures involving live animals in a federally inspected slaughter plant are covered by the Humane Methods of Slaughter Act and its regulations (USDA, 1979). Slaughtering is one of the few areas in animal agriculture that has strict regulations to protect welfare.

The regulations require that animals be rendered insensible to pain before hoisting, bleeding, and cutting. The proper use of captive bolt or electrical stunning equipment is essential to ensure instant, painless insensibility. Recommended practices can be found in the *Recommended Animal Handling Guidelines for Meat Packers* published by the American Meat Institute (Grandin, 1991a) and in the instruction manuals for commercial stunning equipment. Further detailed information on electrical stunning can be found in the publications by Hoenderken (1978), Kirton et al. (1980-81), Blackmore and Newhook (1981), Lambooy (1982), Hoenderken (1983), Gregory and Wotton (1984), Grandin (1985-86), and Gregory (1988). An electric stunner must pass sufficient amperage through the brain to induce a *grand-mal* seizure (Croft, 1952). Insufficient amperage will cause suffering. The course also covers design of restraining devices for holding animals during slaughter (Schmidt, 1972; Giger et al., 1977; Grandin, 1980, 1989c, 1990a, 1991b, 1991).

Animal welfare

The animal welfare issue is discussed and students are informed that this issue will become increasingly important. Increasing public interest in animal welfare will increase the need for teaching students good animal handling methods. Some of the major points made in the lectures are that the livestock industry must improve some of its practices, and the public must be educated about agriculture. A high percentage of animal welfare problems that occur during handling are due to poor management.

The handling of non-ambulatory livestock and how to prevent crippling injuries is a major problem area. Emphasis is put on prevention of downed animals through the use of nonslip flooring, prompt marketing, or euthanatizing of sick or debilitated animals. Welfare concerns about branding, castration, dehorning, and other husbandry procedures are also included in lectures.

The controversial topic of kosher slaughter is also discussed. Religious slaughter is exempt from the Humane Slaughter Act. Based on the author's own experiences in many kosher slaughter plants, the major problem with kosher slaughter is the cruel methods of restraint used in some plants. Because religious slaughter is exempt from the regulation, some plants refuse to spend money on humane restraint devices. Further information on religious slaughter and humane restraint equipment can be found in the publications by Marshall et al. (1963), Giger et al. (1977), Dunn (1990), and Grandin (1990b).

At the conclusion of the course, the importance of good management is stressed. Well-designed equipment provides the tools that make humane handling possible, but the equipment must have good management to go with it. I have observed many cases of animal abuse in good facilities that resulted from the managers' failure to supervise the employees. I tell my students that they are the managers for tomorrow and that they will be in a position to enforce high standards of animal welfare.

Current events involving the actions of people who label themselves "animal rights" proponents are discussed in class, along with case histories of how the industry reacted. Both industry successes and mistakes are covered. Students are encouraged to write letters to the editors of newspapers that carry negative stories and to carry on active discussion in class.

IMPLICATIONS

Educating undergraduate students in the behavioral principles of animal handling will help produce leaders in the livestock industry who recognize both the ethical and the productivity benefits of good animal handling and restraint practices.

REFERENCES

Ames, D. R. 1974. Sound stress and meat animals. In: *Proc. Int. Livest. Environ. Symp. Am. Soc. Agric. Eng.* SP0174. p 324.

Ames, D. R., and L. A. Arehart. 1972. Physiological response of lambs to auditory stimuli. *J. Anim. Sci.* 34:994.

Blackmore, D. K., and J. C. Newhook. 1981. Insensibility during slaughter of pigs in comparison to other domestic stock. *N. Z. Vet. J.* 29:219.

Blecha, F., S. L. Boyles, and J. G. Riley. 1984. Shipping suppresses lymphocyte blastogenic responses in Angus and Brahman x Angus feeder calves. *J. Anim. Sci.* 59:576.

Coppinger, T. R., J. E. Minton, P. G. Reddy, and F. Blecha. 1991. Repeated restraint and isolation stress in lambs increases pituitary adrenal secretions and reduces cell mediated immunity. *J. Anim. Sci.* 69:2808.

Croft, P. G. 1952. Problems with electrical stunning. *Vet. Rec.* 64: 255.

Doney, J.M., R. G. Smith, and R. G. Gunn. 1976. Effects of post-mating environmental stress on administration of ACTH on early embryonic loss in sheep. *J. Agric. Sci.* (Camb.) 87:133.

Douglas, A. G., M. D. Darre, and D. M. Kinsman. 1984. Sight restriction as a means of reducing stress during slaughter. In: *Proc. 30th Eur. Mtg. of Meat Res. Workers*. pp 1011. September 914, Bristol, U.K.

Dunn, C. S. 1990. Stress reactions of cattle undergoing ritual slaughter using two methods of restraint. Vet. Rec. 126:522. Ewbank, R. 1968. The behavior of animals in restraint. In: M. W. Fox (Ed.) *Abnormal Behavior in Animals*. p 159. W. B. Saunders, Philadelphia, PA.

Fell, L. R., and D. A. Shutt. 1989. Behavioral and hormonal response to acute surgical stress in sheep. *Appl. Anim. Behav. Sci.* 22: 283.

Fowler, M. E. 1978. *Restraint and Handling of Wild and Domestic Animals.* Iowa State University Press, Ames.

Galyean, M. L., R. W. Lee, and M. E. Hubbert. 1981. Influence of fasting and transit on ruminal and blood metabolites in beef steers. *J. Anim. Sci.* 53:7.

Giger, W., R. P. Prince, R. G. Westervelt, and D. M. Kinsman. 1977. Equipment of low stress small animal slaughter. *Trans. ASAE* 20:571.

Gilbert, B. J., and C. W. Arave. 1986. Ability of cattle to distinguish among different wavelengths of light. *J. Dairy Sci.* 69:825.

Grandin, T. 1975. Survey of behavioral and physical events which occur in hydraulic restraining chutes for cattle. M.S. Thesis. Arizona State University, Tempe.

Grandin, T. 1980. Observations of cattle behavior applied to the design of cattle handling facilities. *Appl. Anim. Ethol.* 6:19.

Grandin, T. 1982. Pig behavior studies applied to slaughter plant design. *Appl. Anim. Ethol.* 9:141.

Grandin, T. 1983a. Design of ranch corrals and squeeze chutes for cattle. Great Plains Beef Cattle Handbook, Bull. GPE5251, Regional Cooperative Extension Project GPE9.

Grandin, T. 1983b. Welfare requirements of handling facilities. In: S. H. Baxter, M. R. Baxter, and J.A.C. McCormack (Ed.) *Farm Animal Housing and Welfare.* p 137. Martinus Nijhoff, Boston, MA.

Grandin, T. 1984a. Race system for cattle slaughter plants with 1.5 m radius curves. *Appl. Anim. Behav. Sci.* 13:295.

Grandin, T. 1984b. Reduce stress of handling to improve productivity of livestock. *Vet. Med.* 79:827.

Grandin, T. 198586. Cardiac arrest stunning of livestock and poultry. In: M. W. Fox and L. D. Mickley (Ed.) *Advances in Animal Welfare Science.* p l. Martinus Nijhoff, Boston, MA.

Grandin, 1987. Animal handling. In: E. 0. Price (Ed.) *Vet. Clin. North Am.* 3:323.

Grandin, T. 1988. Double rail restrainer for livestock handling. *Int. J. Agric. Eng.* 41:327.

Grandin, T. 1989a. Behavioral principles of livestock handling. *Prof. Anim. Sci.* 5(2):1.

Grandin, T. 1989b. Voluntary acceptance of restraint by sheep. *Appl. Anim. Behav. Sci.* 23:257.

Grandin, T. 1989c. Effect of rearing environment and environmental enrichment on behavior and neural development in young pigs. Ph.D. Dissertation. University of Illinois, Champaign.

Grandin, T. 1990a. Design of loading facilities and holding pens. *Appl. Anim. Behav. Sci.* 28:187.

Grandin, T. 1990b. Humanitarian aspects of Shehitah in the United States. *Judaism* 39:536.

Grandin, T. 1991a. *Recommended Animal Handling Guidelines for Meat Packers.* American Meat Institute, Washington, DC.

Grandin, T. 1991b. Double rail restrainer for handling beef cattle. ASAE paper 915004. Am. Soc. Agric. Eng., St. Joseph, MI.

Grandin, T. 1991c. Principles of abattoir design to improve animal welfare. In: J. Matthews (Ed.) *Progress in Agricultural Physics and Engineering.* CAB International, Wallingford, U.K.

Grandin, T., S. E. Curtis, T. M. Widowski, and J. C. Thurmon. 1985. Electro-immobilization versus mechanical restraint in an avoid-avoid choice test for ewes. *J. Anim. Sci.* 62:1469.

Grandin, T., N. Dodman, and L. Shuster. 1989. Effect of naltrexone on relaxation induced by flank pressure in pigs. *Pharmacol. Biochem. Behav.* 23:839.

Gregory, N. G. 1988. Humane Slaughter. 34th Int. Cong. of Meat Science and Technology, Workshop on Stunning. CSIRO Meat Laboratory, Brisbane, Australia.

Gregory, N. G., and S. B. Wotton. 1984. Sheep slaughtering procedures III. Head to back electrical stunning. *Br. Vet. J.* 140:281.

Hale, R. L., T. H. Friend, and A. S. Macaulay. 1987. Effect of method of restraint of cattle on heartrate, and cortisol and thyroid hormones. *J. Anim. Sci.* 65(Suppl. 1):217 (Abstr.).

Hargreaves, A. L., and G. D. Hutson. 1990a. The effect of gentling on heart rate, flight distance, and aversion of sheep to handling procedure. *Appl. Anim. Behav. Sci.* 26:243.

Hargreaves, A L., and G. D. Hutson. 1990b. Some effects of repeated handling on stress responses in sheep. *Appl. Anim. Behav. Sci.* 26:253.

Hebel, R., and H. H. Sambraus. 1976. Are domestic animals color blind? *Berl. Muench. Tierarerztl. Wochenschr.* 89(16):321.

Hemsworth, P. H., and J. L. Barnett. 1987. Human animal interactions. In: E. 0. Price (Ed.) *Vet. Clin. North Am.* 3:339.

Hemsworth, P. H., J. L. Barnett, and C. Hansen. 1987. The influence of inconsistent handling by humans on behavior, growth and corticosteroids of young pigs. *Appl. Behav. Sci.* 17: 245.

Hemsworth, P. H., A. Brand, and P. G. Willems. 1981. The behavioral response of sows to the presence of human beings and its relation to productivity. *Livest. Prod. Sci.* 8:67.

Hixon, D. L., D. J. Kesler, T. R. Troxel, D. L. Vincent, and B. S. Wiseman. 1981. Reproductive hormone secretions and first service conception rate subsequent to ovulation control with Syncro-Mate B. *Theriogenology* 16:219.

Hoenderken, R. 1978. Electrical stunning of slaughter pigs. Thesis. State University of Utrecht, The Netherlands.

Hoenderken, R. 1983. Electrical and carbon dioxide stunning of pigs for slaughter. In: G. Eikelenboom (Ed.) *Stunning of Animals for Slaughter.* p 59. Martinus Nijhoff, Boston, MA.

Hutson, G. D. 1981. Sheep movement on slotted floors. *Aust. J. Exp. Agric. Hush.* 21:474.

Hutson, G. D. 1982. Flight distance in Merino sheep. *Anim. Prod.* 35:231.

Hutson, G. D. 1985a. Sheep and cattle handling facilities. In: B. L. Moore and P. J. Chenoweth (Ed.) *Grazing Animal Welfare.* pp 124136. Aust. Vet. Assoc. Queensland, Australia.

Hutson, G. D. 1985b. The influence of barley food rewards on sheep movement through a handling system. *Appl. Anim. Behav. Sci.* 14:263.

Kelley, K. W., C. Osborn, J. Evermann, S. Parish, and D. Hinrichs. 1981. Whole blood leukocytes vs separated mononuclear cell blasto-genesis in calves, time dependent changes after shipping. *Can. J. Comp. Med.* 45:249.

Kilgour, R., and C. Dalton. 1984. *Livestock Behaviour, a Practical Guide.* Westview Press, Boulder, CO.

Kilgour, R., and H. DeLangen. 1970. Stress in sheep resulting from management practices. Proc. *N. Z. Soc. Anim. Prod.* 30:65.

Kirton, A. H., L. F. Frazerhurst, E. G. Woods, and B. B. Chrystall. 198081. Effect of electrical stunning method and cardiac arrest on bleeding efficiency and residual blood and blood splash in lambs. *Meat Sci.* 5:347.

Klopfer, F. D., and R. L. Butler. 1964. Color vision in swine. *Am. Zool.* 4:294.

Lambooy, E. 1982. Electrical stunning of sheep. *Meat Sci.* 6:123.

Lehman, W. B., and G. H. Patterson. 1964. Depth perception in sheep: Effects of interrupting the mother-neonate bond. *Science* (Washington, DC) 145:835.

Lynch, J. J., and G. Alexander. 1983. *The Pastoral Industries of Australia.* p 371. University Press, Sydney, Australia.

Marshall, M., E. E. Milbury, and E. W. Shultz. 1963. Apparatus for holding cattle for humane slaughtering. U. S. Patent No. 3,029,871. Washington, DC.

Mertsching, H. J., and K. W. Kelley. 1983. Restraint reduces size of thymus gland and PHA swelling in pigs. *J. Anim. Sci.* 57(Suppl. 1):175 (Abstr.).

Midwest Pan Service. 1987. *Beef Housing and Equipment Handbook.* Midwest Plan Service. Iowa State University, Ames.

Munkenbeck, N. W. 1982. Color vision in sheep. *J. Anim. Sci.* 55(Suppl. 1):129 (Abstr.).

Panepinto, L. M., R. W. Phillips, S. Norden, P. C. Pryor, and R. Cox. 1983. A comfortable minimum stress method of restraint for Yucatan miniature swine. *Lab. Anim. Sci.* 33(1):95.

Parsons, R. A., and W. N. Helphinstine. 1969. Rambo A.I. breeding chute for beef cattle. *One-Sheet-Answers*, University of California Agricultural Extension Service, Davis.

Pascoe, P. J. 1986. Humaneness of an electro-immobilization unit for cattle. *Am. J. Vet. Res.* 10:22-52.

Pearson, A. M., R. Kilgour, H. de Langen, and E. Payne. 1977. Hormonal responses of lambs to trucking, handling and electrical stunning. *Proc. N. Z. Soc. Anim. Prod.* 37:243.

Peischel, A., R. R. Schalles, and C. E. Owenby. 1980. Effect of stress on calves grazing Kansas Hills range. *J. Anim. Sci.* 51(Suppl. 1):245 (Abstr.).

Prince, J. H. 1977. The eye and vision. In: M. J. Swenson (Ed.) *Dukes Physiology of Domestic Animals.* p 696. Cornell University Press, New York.

Reid, R. L., and S. C. Mills. 1962. Studies of the carbohydrate metabolism of sheep, XVI. The adrenal response to physiological stress. *Aust. J. Agric. Res.* 13:282.

Rider, A., A. F. Butchbaker, and S. Harp. 1974. Beef working, sorting and loading facilities. Technical Paper No. 744523. *Am. Soc. Agric. Eng.*, St. Joseph, Ml.

Rushen, J. 1986. Aversion of sheep for handling treatments paired choice studies. *Appl. Anim. Behav. Sci.* 16:363.

Schmidt, C. 0. 1972. Cattle handling apparatus. U. S. Patent No. 3,657,767. Washington, DC.

Seabrook, M. F. 1984. The psychological interaction between the stockman and his animals and its influence on the performance of pigs and cows. *Vet. Rec.* 115:84.

USDA. 1979. Part 313 through 313.50. Regulations on Humane Slaughter of Livestock in Accordance with the Humane Methods of Slaughter Act of 1978.

van Putten, G., and W. J. Elshof. 1978. Observations on the effect of transport and slaughter on the wellbeing and lean quality of pigs. *Anim. Reg. Stud.* 1:247.

Vowles, W. J., and T. J. Hollier. 1982. The influence of yard design on the movement of animals. *Proc. Aust. Soc. Anim. Prod.* 14: 597.

White, J. B. 1961. Letter to the Editor. *Vet. Rec.* 73:935.

NINE

Behavioral Principles of Handling Beef Cattle and the Design of Corrals, Lairages, Races and Loading Ramps

T. Grandin (2019) Behavioral Principles of Handling Beef Cattle and the Design of Corrals, Lairages, Races and Loading Ramps. In: T. Grandin (Editor) *Livestock Handling and Transport*, 5th Edition, CABI Publishing Wallingford, Oxfordshire, UK, pp. 80-109.

I included this chapter out of chronological order because I was invited to edit the first edition of this book in 1993. It now has five editions. It is an updated version, written in 2019, of two chapters I wrote for the first edition.

SUMMARY

Beef cattle and other grazing animals have behaviour patterns during handling that are influenced by vision. A prey species animal has wide-angle vision that will detect rapid motion and possible danger. The eight basic behavioural patterns during handling in grazing animals are: (i) flight zone; (ii) turn and look at people who are outside their flight zone; (iii) point of balance that controls direction of movement; (iv) natural following the leader behaviour; (v) return to where they came from; (vi) soft-bunching; (vii) milling when a predator attacks; and (viii) isolation alone can be highly stressful.

Good stock people understand the principle of pressure and release. When cattle move in the desired direction, the handler reduces pressure on the flight zone. Tame cattle can be led instead of being driven. This chapter also contains drawings of popular layouts for races, yards, lairages and corrals for handling cattle. There is also a list of common design mistakes that can cause balking and refusal to move through a handling system. Cattle will move more easily through a race if visual distractions are removed. Common distractions are: shadows, sunbeams, reflections from shiny vehicles, and seeing people up ahead. Covering the outer perimeter fence of a facility will help block distractions.

INTRODUCTION

Dylan Biggs, one of the founders of modern low stress livestock handling, said, "Every step of every animal should be voluntary." (Biggs, 2013) Bud Williams and Dylan Biggs started teaching low stress handling methods in the USA in the late 1970s and 1980s. The first edition of this volume recognized Williams's innovative work (Grandin, 1993). More and more ranchers and feedlot managers have adopted calmer, improved handling methods. Today many stockmanship classes are available. It is likely that these methods may be rediscoveries of the ways of the stockmen of bygone years. In the late 1800s, cowboys handled and trailed cattle quietly on the great cattle drives from Texas to Montana. In A Cowboy's Diary, Andy Adams wrote: "Boys, the secret of trailing cattle is to never let your herd know that they are under restraint. Let everything that is done be done voluntarily by the cattle." (Adams, 1903) Unfortunately, the quiet methods of the early 1900s were forgotten and some of the more modern cowboys were rough (Wyman, 1946; Hough, 1958; Burri, 1968). There is an excellent review of the history of herding in Smith (1998).

Progressive producers of cattle know that reducing stress will improve both productivity and safety. Effective low-stress methods of handling cattle will take time to learn, and people have to have a positive attitude (Smith, 2018).

Animals can remember both good and bad previous experiences. If they are treated well, they will be easier to handle in the future. Cattle poked with electric prods will be harder to move through a handling facility in the future (Goonewardene et al., 1999). A sudden novel event can frighten cattle. In the 1800s and 1900s, stampedes were often caused by a horse with a saddle under its belly, or a flapping raincoat (Harger, 1928; Linford, 1977). Stampedes were more likely to occur at night (Ward, 1958).

PERCEPTION OF GRAZING LIVESTOCK

Vision

To assist in the avoidance of predation, cattle and other grazing ungulates have wide-angle (360°) panoramic vision (Prince, 1977), and vision has dominance over hearing (Uetake and Kudo, 1994). Bovines can discriminate colours (Thines and Soffie, 1977; Darbrowska et al., 1981; Gilbert and Arave, 1986; Arave, 1996). Cattle, sheep, and goats are all dichromats (only two of the three primary colours can be discerned), with cones that are more sensitive to yellowish green (wavelength 552–555 nm) and blue purple light (444–455 nm) (Jacobs et al., 1998). The horse is most sensitive to light at wavelengths of 539 nm and 428 nm (Carroll et al., 2001). Dichromatic vision may provide better vision at night and aid in detecting motion (Miller and Murphy, 1995). The visual acuity of bulls may be worse than that of younger cattle or sheep (Rehkamper and Gorlach, 1998).

Grazing livestock can see depth, but they may have to stop and put their heads down to perceive it. This may explain why they balk at shadows on the ground. Observations by Smith (1998) indicate that cattle do not perceive objects that are overhead unless they move. Research with horses and sheep indicates that they have a horizontal band of sensitive retina, instead of a central fovea as in the human (Saslow, 1999; Shinozaki et al., 2010). This enables them to scan their surroundings while grazing. Grazing cattle have a visual system that is very sensitive to motion and contrasts of light and dark. They are able to scan the horizon constantly while grazing and they may have difficulty in quickly focusing on nearby objects, owing to weak eye muscles (Coulter and Schmidt, 1993). This may explain why grazing animals spook at nearby objects that suddenly move.

Wild ungulates, domestic cattle, and horses respect a solid fence and will seldom ram or try to run through a solid barrier. Sheets of opaque plastic can be used to corral wild ungulates (Fowler, 1995), and portable corrals constructed from canvas have been used to capture wild horses (Wyman, 1946; Amaral, 1977). Excited cattle will often run into a cable or chain-link fence because they cannot see it. A 30 cm wide, solid, belly rail installed at eye height, or ribbons attached to the fence, will enable the animal to see the fence and prevent fence ramming (Ward, 1958).

REMOVE VISUAL DISTRACTIONS FROM HANDLING FACILITIES

Visual distractions that cause animals to back up or refuse to move must either be removed from a handling facility or blocked by solid walls. Vehicles parked alongside a fence often cause problems. Some of the most common distractions are dangling chains, reflections on shiny metal or a wet floor, drain gratings, shadows, moving people, vehicles and flapping objects (Lynch and Alexander, 1973; Grandin, 1980a, 1987, 1989, 1990a,b, 2015; Grandin and Deesing, 2008). To locate visual distractions, people need to get into the race and look at it from the point of view of the bovine eye. Cattle will often stop and balk at shadows and high contrast. Highway departments found they could stop cattle from crossing the paved road by painting stripes on it (*Western Livestock Journal*, 1973). The animals may also stop if the flooring surface changes. Some examples are moving from dirt to concrete or from concrete flooring to a metal floor. If the visual distraction cannot be removed, the lead animal must be allowed to stop and look at it. If livestock stop at a drain grate, or a change in flooring, the lead animal will often stop and lower its head to look at it. The handler should wait for the lead animal to raise its head back up before attempting to move the group forward. If they are rushed towards, they may either balk and refuse to move or turn back.

LIVESTOCK ATTRACTED TO LIGHT

Research conducted by Joe Stookey at the University of Saskatchewan showed that when cattle were given a choice, they preferred to move into a chute where they could see light. Fig. 1 shows a Y shaped chute that was used to test this principle. Adding a lamp to illuminate a dark race entrance would significantly improve both cattle and pig movement (Van Putten and Elshof, 1978; Grandin, 1982, 2001). Ranchers have found that illuminating

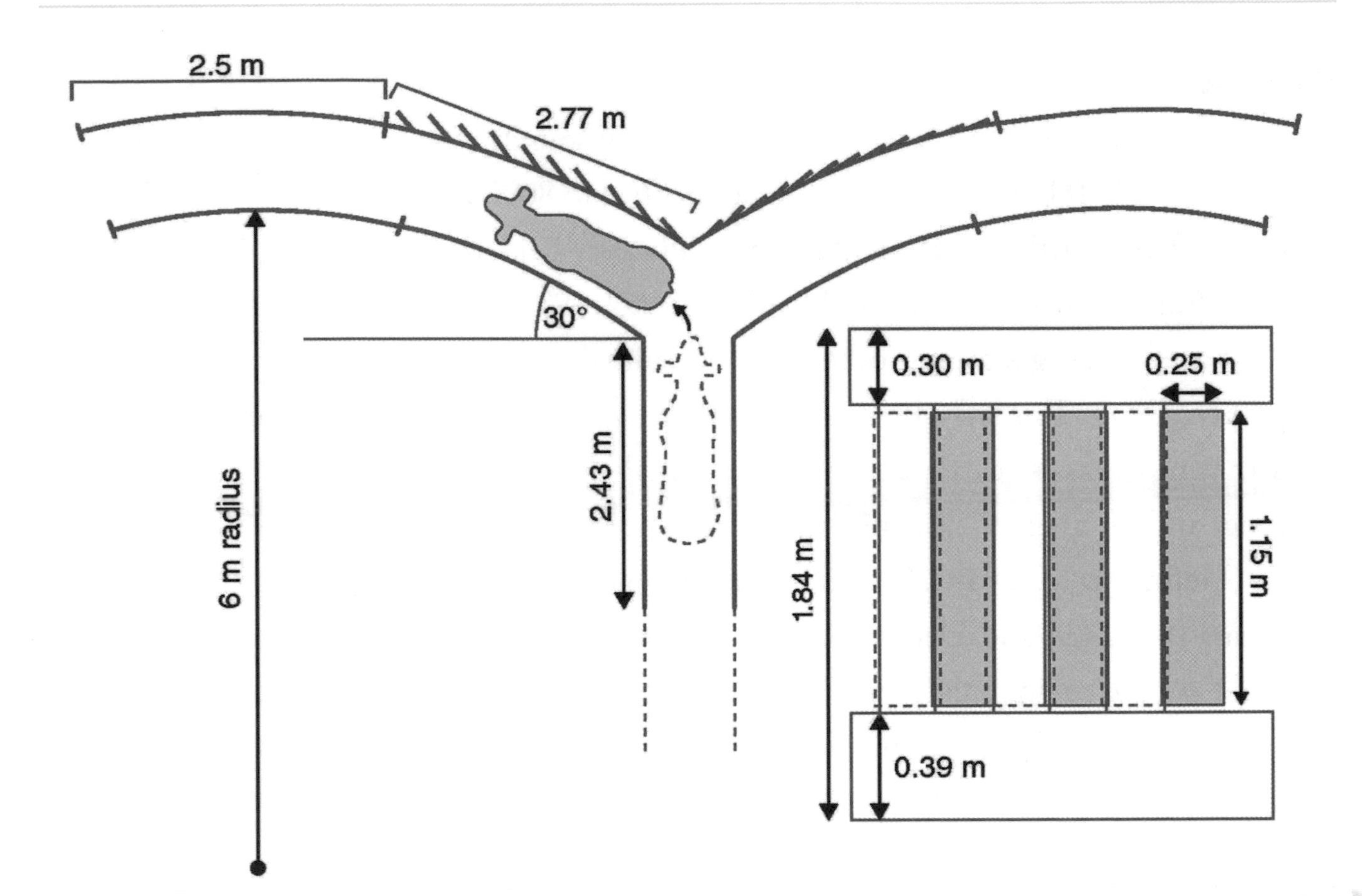

Fig. 1.
Diagram of a Y-maze used to test preference for either a lighted or a darker single-file race. Opening or closing the louvres either illuminated or darkened either the righthand or the lefthand side. Twenty-four out of 27 cattle exited via the arms with the open louvres, which enabled the cattle to see light ($p<0.001$). (From Stookey and Watts, 2014)

the inside of a truck or trailer will attract animals into it when they are loaded at night. It is often difficult to get animals that are outside in bright sunlight to enter a dark building. They will enter more easily if they can see daylight through it. Another alternative is equipping it with translucent white plastic skylights to let in lots of diffuse shadow-free natural daylight. Even though cattle and other livestock are attracted towards a more brightly illuminated area (Grandin 1980a,b), they will not move towards a blinding rising or setting sun. To solve this problem, either change the time of day the cattle are handled or avoid facing races and chutes towards the sun.

HEARING

Grazing ungulates are sensitive to high frequency sounds. The human ear is more sensitive at 1000– 3000 Hz, but cattle are most sensitive to 8000 Hz (Ames, 1974; Heffner and Heffner, 1983). Cattle, horses, and alpacas can easily hear sounds of up to 10,000 Hz frequency (Algers, 1984). Heffner and Heffner (1992, 1983) and Heffner et al. (2014) found that cattle, alpacas, goats, and other grazing animals have a poor ability to localize high frequency sound. This may be related to their wide-angle vision (Heffner et al., 2014). Ears that move assist cattle and other mammals to determine the location of a sound (Heffner and Heffner, 2018). This is due to a trait in prey species animals with wide-angle vision (Heffner et al., 2014). Cattle and horses will 'watch' people and other animals with their ears (Grandin and Deesing, 2008); they will point each ear independently at two different people or animals. Noise is stressful to grazing animals (Pajor et al., 2003). The sounds of people yelling or whistling is more stressful to cattle than the sounds of gates banging (Waynert et al., 1999).

NEVER YELL AT CATTLE

Cattle are able to differentiate between the threatening sound of a person yelling at them and machinery noise that is not directed at them. When people yell at livestock, it has emotional intent. The animals know that yelling is directed at them. Shouting close to the ear of a cow may be as aversive as an electric prod (Pajor et al., 2003). Yelling at cattle raised cortisol, but normal talking had no effect (Hemsworth et al., 2011). Cattle may be able to habituate to steady machinery sounds that are not directed at them. Commercial manual squeeze chutes are noisy and can reach levels of 99–115 dB (Lyvers, 2013), but heart rate habituated when loud recorded sounds of the squeeze chute were played over several trials. Movement of the cattle in the squeeze chute increased at a noise level of 100 dB (Lyvers, 2013). Loud sounds over 85 dB increased heart rate more than sounds at 65 dB (Johns et al., 2017). Subjecting cattle to loud recorded noise during handling raised cortisol levels but had no effect on weight gain in relatively tame Angus cattle at a research station (Bauer et al., 2012). In both of these situations, the cattle may have perceived that the noise was not directed at them. The cattle at the research station may also have been less stressed by the treatment because they had become accustomed to many students handling them. Lanier and Grandin (2002) found that cattle that became agitated in an auction ring were more likely to flinch or jump in response to sudden, intermittent movements or sounds. Intermittent movements or sounds appear to be more frightening than steady stimuli.

Talling et al. (1998) found that pigs were more reactive to intermittent sounds than to steady sounds. High-pitched sounds increased a pig's heart rate more than low-pitched sounds (Talling et al., 1996). Sudden movements have the greatest activating effect on the amygdala (LeDoux, 1996), the part of the brain that controls fearfulness (LeDoux, 1996; Rogan and LeDoux, 1996).

BASIC CATTLE AND LIVESTOCK BEHAVIOURS THAT ARE GOVERNED BY VISION

There are eight natural instinctual behaviours that cattle and other grazing animals have during handling. In completely tame cattle that are trained to lead, these behaviours may either disappear or become less obvious. The eight behaviours are:

1. Flight Zone – The animals move away when the flight zone is entered (Fig. 2).
2. Turn and Face People – 'Cattle want to see you' according to Ron Gill, Texas A&M University (effectivestockmanship.com – n.d.). There is a zone just outside the flight zone where cattle and other grazing livestock are aware of the handler's presence, and they will turn and face the stockperson. There are many different names for this zone such as recognition zone (Smith, 1998), pressure zone, zone of awareness or zone of influence. Livestock handling specialist Curt Pate explains that groups of cattle can be easily turned or sorted by using a combination of 'driving pressure' when their flight zone is entered and 'drawing pressure' where the handler is just outside the flight zone (Detering, 2010). The cattle will want to watch and walk towards the handler. A person who is skilled at sorting groups of cattle will carefully alternate between driving pressure and drawing pressure to sort individual cattle from others in the group (curtpatestockmanship.com – n.d.).

3. Point of Balance – The animal will change direction of movement depending on how the handler is positioned in relation to the point of balance (Fig. 2). Livestock will usually move forward when the stockperson is behind the point of balance at the shoulder.
4. Natural Following Behaviour – Cattle and other livestock will follow the leader. Skillful handlers will use following behaviour by timing bunches of cattle when they are working in races (chutes). They wait until the race is empty so that they can utilize natural following behaviour to refill a partially empty race. Natural following behaviour is also used to move large groups of cattle in a low stress manner.
5. Return to Where They Came from – Well-designed handling facilities and skillful sorting of cattle takes advantage of the natural behaviour of cattle, pigs or sheep to go back to where they came from. Handling systems, such as round tub crowd pens or Bud boxes take advantage of this natural behaviour (Figs. 11, 12, and 14). To facilitate flow through a handling facility, do not store livestock in the crowd pen that leads up to the single file race. Keep them moving to prevent them from turning around and attempting to go back to where they came from. When sorting cattle, move them into another pen and then sort them back into the pen they just left. This will take advantage of the natural behaviour of going back to where they came from.
6. Soft Bunching – Cattle and other grazing animals that live in areas where there are many predators will graze together in soft bunches. Laporte et al. found that when wolves were present, cattle grouped more closely together during grazing.
7. Milling – Cattle or other livestock will form a tight circle with the dominant animals pushing into the centre. This is a behaviour that occurs when predators attack. Good stock people never want to see this behaviour. If milling occurs, the handlers should wait 20–30 minutes for the cattle to calm down before attempting to handle or move them.
8. Cattle and Sheep Do Not Like Being Alone – Many people get injured by a lone animal that wants to get back to its herd mates. Put some other animals in with

it. Do not get into a small pen with an agitated lone bovine, bison, deer or other large grazing ungulate.

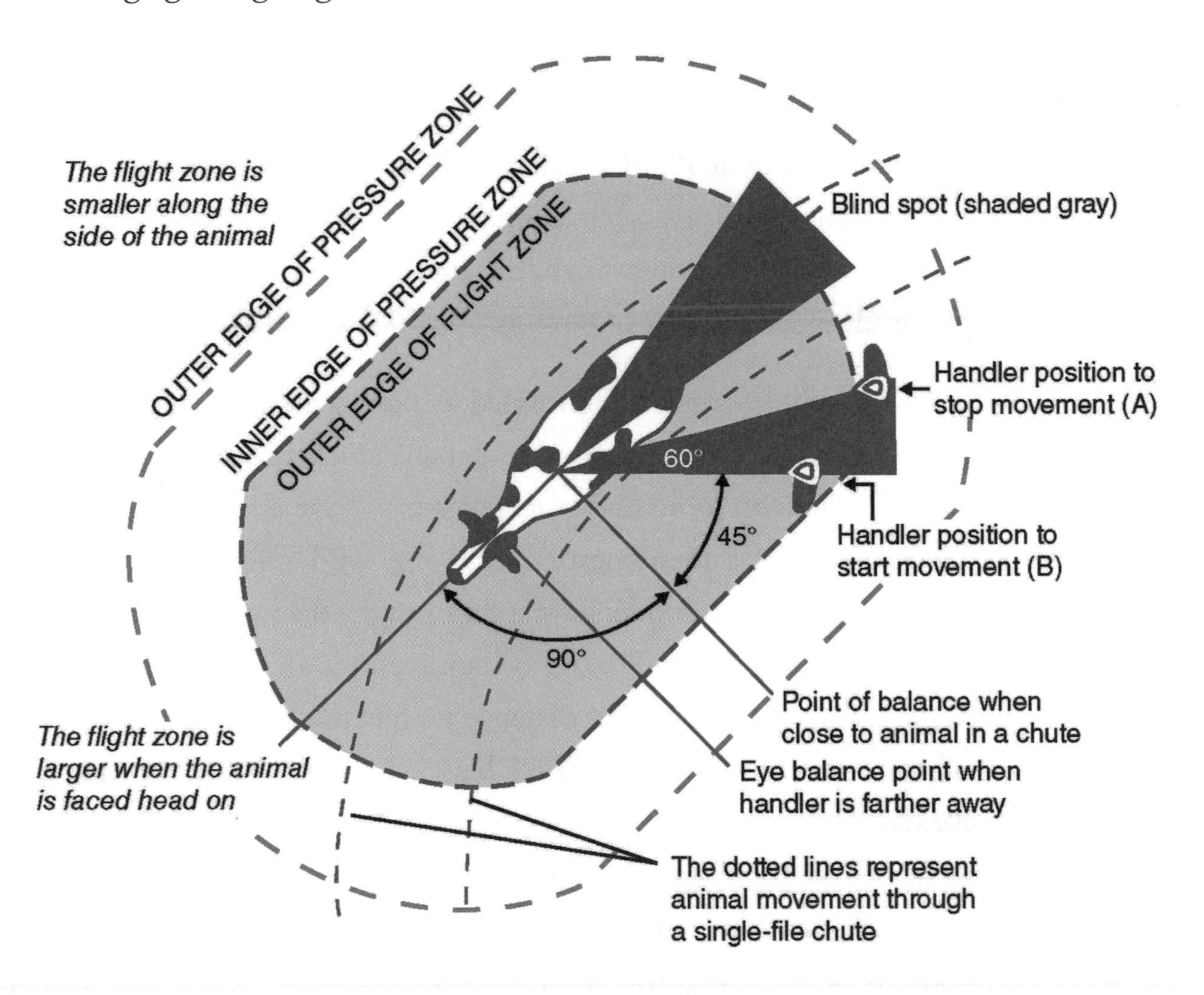

Fig. 2.
There are two zones around cattle. They are the flight zone and the pressure zone ('cattle turn and face you' zone). When the handler moves outside the pressure zone, the cattle will ignore them. When the flight zone is entered, the animal will move away. This diagram shows the correct position for moving a single bovine through a race (chute). To move a single animal forward, the handler must be behind the point of balance and stay out of the blind spot directly behind the animal. When the handler is close to the animal, the point of balance is at the shoulder. When the handler is further away, the point of balance may move forward to just behind the eye. When the handler is on the outer edge of the pressure zone, the animal becomes aware of the handler's presence and turns around and looks. When the outermost edge of the flight zone is penetrated, the animal moves away. (www.grandin.com/principles/flight. zone.html). (Diagram: Temple Grandin)

FLIGHT ZONE

The concept of flight distance was originally applied to wild ungulates. Hedigar (1968) states:

> By intensive treatment, i.e. by means of intimate and skilled handling of the wild animals, their flight distance can be made to disappear altogether, so that eventually such animals allow themselves to be touched. This artificial removal of flight distance between animals and man is the result of the process of taming.

The size of a bovine's flight zone is determined by both genetics and previous handling experiences. Previous experiences with calm, patient stock-people will reduce the size of the flight zone. Cattle with more flighty genetics may have a larger flight zone.

This same principle applies to domestic cattle and wild ungulates. Extensively raised wild cows on an Arizona ranch may have a 30 m flight distance, whereas feedlot cattle may have a flight distance of 1.5–7.61 m (Grandin, 1980a). Cattle with frequent contact with people will have a smaller flight distance than cattle that seldom see people, and cattle subjected to gentle handling will usually have a smaller flight zone than cattle subjected to abusive handling.

Fearfulness will enlarge the flight zone. Totally tame dairy cattle may have no flight zone and people can touch them. The edge of the flight zone can be determined by slowly walking up to a group of cattle. When the flight zone is entered, the cattle will turn and start to move away and, if the person stands still, the cattle will turn and face the handler but keep their distance. When a person reenters the edge of the flight zone, the animals will turn around and move away. When the flight zone of a group of bulls was invaded by a moving mechanical trolley, the bulls moved away and maintained a constant distance between themselves and the moving trolley (Kilgour, 1971). The flight distance was determined by the size of a piece of cardboard attached to the trolley.

Cattle remain further away from a larger object (Smith, 1998). When a person approaches full face, the flight zone will be larger than when approaching with a small, sideways profile. The author has also observed that large objects are more threatening to flighty antelope. Tame, hand reared prong horn antelope panicked and hit the fence of their enclosure when a large, novel object such as a wheelbarrow was brought into their pen. They had to be carefully habituated to each new large object. Small novel objects, such as coffee cups, that had been brought into their pens had no effect.

If the handler works on the edge of the flight zone, cattle can be moved more efficiently (Grandin, 1980a, 1987). The animals will move away when the flight zone is penetrated and stop when the handler retreats. Smith (1998) and Kosako et al. (2008) report that the edge of the flight zone is not distinct and that approaching an animal quickly will enlarge the zone.

If an animal rears up in a race, handlers should back up to remove themselves from the animal's flight zone. Handlers should not attempt to push a rearing animal down, because deep penetration of the flight zone causes increased fear and further attempts to escape. If cattle attempt to turn back in an alley, the handler should back up and remove him/herself from deep inside the flight zone. The angle of approach and the size of the animal's enclosure will also affect the size of the flight zone.

Experiments with sheep have indicated that animals confined in a narrow alley had a smaller flight zone than animals confined in a wider alley (Hutson, 1982). Cattle will have a larger flight zone when they are approached head-on. A basic principle is that the flight zone is smallest along the sides of the animal and greatest in front of and behind it (Cote, 2003). Extremely tame cattle are often hard to drive because they no longer have a flight zone. These animals should be led. More information and flight zones can be found in Smith (1998) and Grandin and Deesing (2008).

POINT OF BALANCE

To make an animal move forward in a single file race, the handler should stand behind the point of balance at the shoulder, and to make the animal back up, the handler should stand in front of the point of balance. When cattle are calm, the point of balance will move forward and be closer to the eye. It will never be in front of the eye. When cattle are being handled in groups, the point of balance may keep changing. Curt Pate, a low-stress handling specialist, recommends carefully working cattle to move the point of balance forward. The caption of Fig. 2 provides additional information about the point of balance. A common mistake made by many handlers when attempting to move an animal forward in a single-file race is to stand in front of its eyes and poke its rear with a stick. This gives the animal confusing directional signals. To turn an animal located in an open area, start on the edge of the flight zone at the point of balance and approach the animal's rear at an angle (Cote, 2003).

Another principle is that grazing animals, either singly or in groups, will move forward when a handler quickly passes the point of balance at the shoulder in the opposite direction to the desired movement (Grandin, 1998) (Fig. 3). The principle is to move inside the flight zone in the opposite direction to the desired movement and outside the

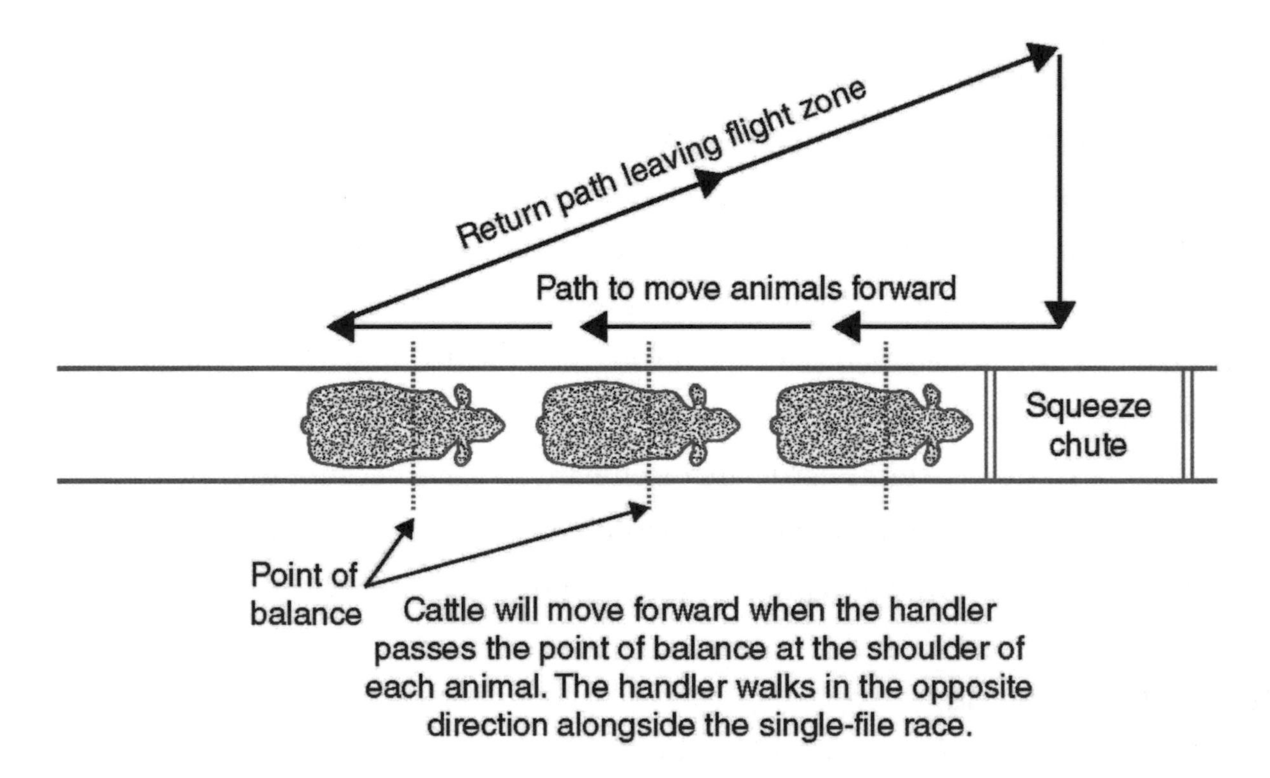

Fig. 3.

Movement pattern to move cattle forward in a chute by walking inside the flight zone in the opposite direction to the desired movement. The handler should quickly walk by the point of balance at the shoulder. (Grandin, 2000)

Fig. 4.
Curved single-file chute with a partially open inner radius. The building's wall creates a completely solid outer fence to block distractions. The curved race takes advantage of the natural tendency of cattle to go back to where they came from. The space close to the open-sided race must remain free of people except when the flight zone is entered to move a steer forward. (Photo courtesy of Temple Grandin)

flight zone in the same direction as the desired movement. Using these movements to induce cattle to enter a squeeze chute makes it possible to greatly reduce or eliminate the use of an electric prod (see Fig. 3). Wait until the tailgate on the squeeze chute is open so the animal has a place to go before doing the movement pattern.

When an animal is approached head-on, it will turn right if the handler moves left, and vice versa (Kilgour and Dalton, 1984; Holmes, 1991). Calm cattle in crowd pens and other confined areas can be easily turned by moving plastic strips on a stick next to the animal's head. For example, when a cow's vision is blocked on the left side by the plastic strips, she will turn right. Handlers should avoid deep penetrations of the flight zone, because this may cause cattle to panic. Fig. 4 shows a curved single-file race where the solid wall of the building blocks outside distractions. A handler walking on the ground can easily perform the movement pattern shown in Fig. 3 through the partially open side.

MOVE IN TRIANGLES FOR MOVING GROUPS

Ron Gill at Texas A&M University, Guy Glossom at Mesquite, Texas, and many other people who teach good stockmanship explain that when groups are moved, herders should walk in triangles alongside a group of cattle (Fig. 5). The handler moves inside

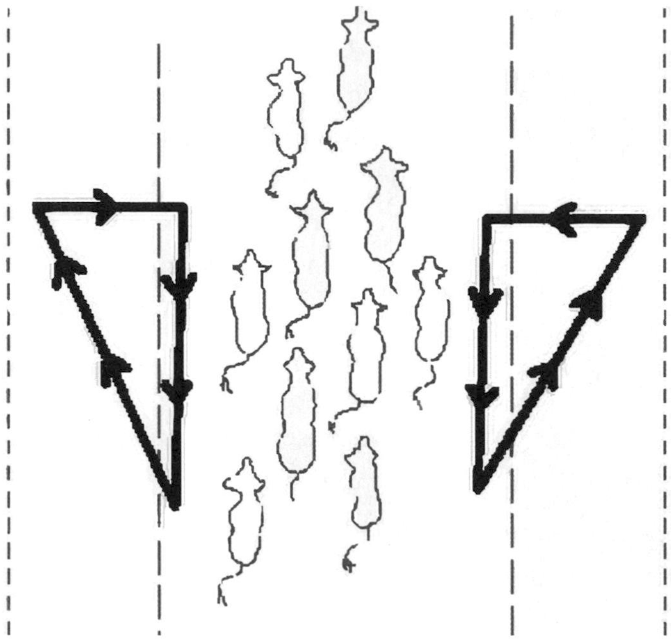

Fig. 5.
The diagram shows the movement pattern when two people are moving a group of cattle. To keep the group moving, the triangle pattern is repeated multiple times. The dotted line with long dashes represents the outer edge of the collective flight zone. The dotted line with small dashes represents the outer edges of the animal's zone of awareness (pressure zone). When the pressure zone is entered, the animals become aware of the handler's presence. The handler is inside the outer edge of the collective flight zone when he walks in the opposite direction to the desired movement to speed the herd up and move them forward. The handler is outside the collective flight zone but still inside the pressure zone where he walks in the same direction to the desired movement. This double triangle pattern diagram is adapted from the work of Guy Glossom, Mesquite, Texas. He warns that it is essential to keep the angles on the triangle sharp. Never allow the triangles to turn into circles. (Diagram: Temple Grandin – www.grandin.com//principles/flight.zone.html)

the flight zone in the opposite direction to the desired movement to speed up forward movement and outside the flight zone in the same direction to slow the group of cattle down. Bud Williams, a specialist in low stress cattle-handling, developed these principles. Grandin (1993) worked with Bud Williams and made some of the first diagrams to show his principles. Dylan Biggs in Canada recommends a small version of the same movement pattern when cattle are being moved through a gate (Stookey and Watts, 2014) (Fig. 6). Curt Pate, a cattle stockmanship specialist, says: "Concentrate on directing the animal's nose." When sorting, induce the animal to move when it is paying attention to the stockperson with its nose pointed in the right direction.

LET LIVESTOCK FLOW AROUND THE STOCKPERSON

When a person positions themselves at the correct location at a gate, the cattle or sheep will flow around them (Fig. 7). The cattle will move around the handler just slightly outside the flight zone but still inside the pressure zone. Cattle have a natural behaviour to flow around on the edge of a bubble formed on the edge of the flight zone (Grandin, 2004). It is like a 'force field' in a science fiction movie that is formed around the stockperson. The collective flight zone of a large flock of sheep is shown in a photo in Grandin (1987).

GATHERING CATTLE AND OTHER LIVESTOCK ON PASTURE

Wild and semiwild cattle can be easily gathered on pasture by initiating a natural loose bunching pat tern. Figure 8 shows a windscreen wiper pattern in which the handler walks on the edge of the group's collective flight zone. The handler moves at a slow walk and must be careful not to circle around the animals. Bud Williams was the originator of this principle and early published versions are in Smith (1998) and Grandin (2000). Many livestock specialists recommend walking in straight lines. Circling around livestock mimics the behaviour of a predator and must be avoided. The stockperson must also resist the urge to chase stragglers. When the bunching instinct is triggered, the herd will come together and line up. The stragglers will join the other cattle. Care must be taken to be

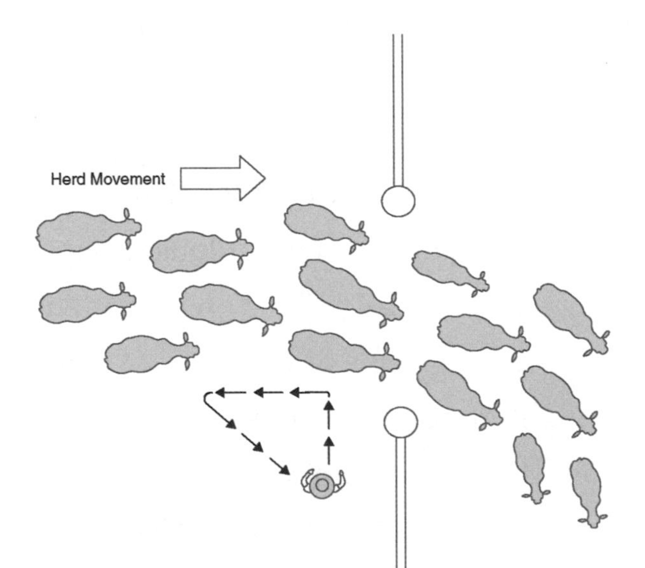

Fig. 6.
Movement pattern to quietly move cattle out of a gate. (Adapted from Dylan Biggs – Stookey and Watts, 2014)

Fig. 7.
When a handler stands at the correct position at a gate, the cattle will flow along the stockperson on the edge of the flight zone. (Photo courtesy of Temple Grandin)

quiet and keep the animals moving at a walk. The principle is to induce bunching before any attempt is made to move the herd. The animals will move towards the pivot point of the windscreen wiper. If too much pressure is applied to the collective flight zone before bunching, the herd will scatter. More information can be found in Grandin (2000), Smith (1998), Grandin and Deesing (2008), Grandin (2017), Ruechel (2006), and Beaver and Hoagland (2016), and *Stockmanship Journal* (2012–2016). This method will not work on completely tame animals with little or no flight zone. Leading is often the best low-stress method to move really tame cattle.

Cattle bunching is less intense than sheep bunching. Sheep have extreme bunching behaviour. A flock of sheep will often immediately bunch when they see a dog. Inducing soft bunching activates a mild anxiety, but induction of high stress milling and circling should be avoided. The least stressful handling procedure would be entirely voluntary.

Smith (1998) states that there is no black and white dividing line between herding, leading and training. It is likely that cows gathered with the windscreen wiper method (Fig. 8) may have slight anxiety at first, but then become completely trained and have diminishing anxiety. Figure 9 shows a movement pattern developed by Bud Williams to bring a split herd back together (Grandin, 1993).

TRAINING CATTLE TO TRUST PEOPLE DURING HERDING ON THE RANGE

Bud Williams further developed his herding methods so that they are no longer dependent on triggering hardwired, instinctive behaviour patterns. Cote (2003) has written a book that explains these methods. The principle is to train groups of cattle on a pasture to calmly respond to pressure on their collective flight zone. It is a process of training them to move quietly rather than simple habituation. This may be similar to settling cattle before a cutting horse contest. You want them to be "feeling a little bit scared" (Rice, n.d.). When cattle and calves learn to trust the handler, they will move straight and away from the handler. They will stop attempting to circle around to look at the person (Cote, 2003). Cattle handling specialist Ron Gill at Texas A&M University explains how a handler can teach cattle to stop when pressure on the flight zone is removed. The handler will need to

Fig. 8.
Stockperson movement pattern for gathering cattle that are not completely tame. The handler walks back and forth on the edge of the flight zone but well inside the pressure zone (zone of recognition). Imagine that the leaders are on the pivot point of a windscreen wiper. Do not circle around the cattle. Do not enter the flight zone until the herd moves together. If the flight zone is entered too quickly, the herd may scatter. The width of the zigzag narrows when the herd starts to move forward. (From Grandin, 2000)

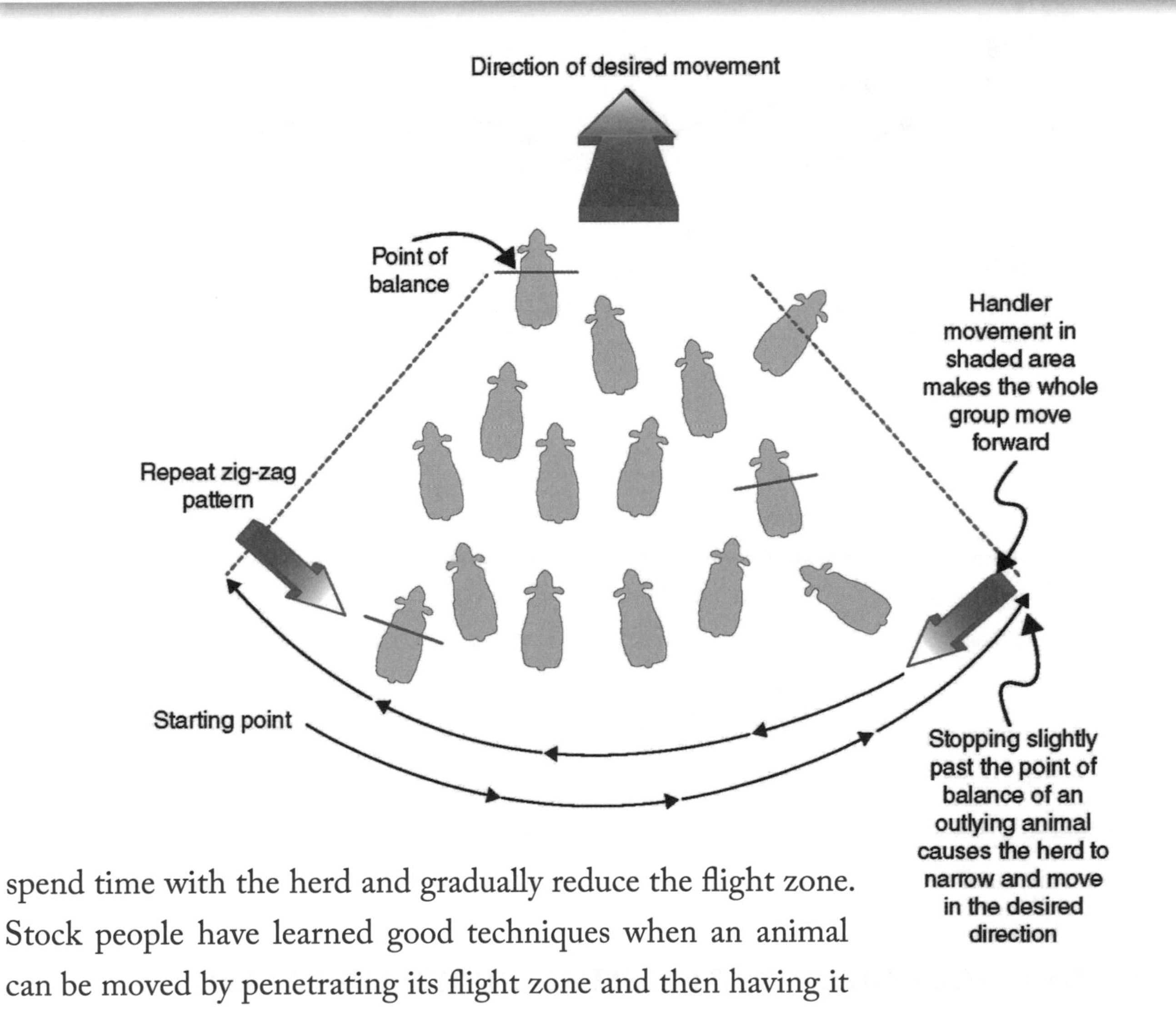

spend time with the herd and gradually reduce the flight zone. Stock people have learned good techniques when an animal can be moved by penetrating its flight zone and then having it stop when the handler retreats from the flight zone.

Instinct is now overridden by trust and learning, but the handler must never work in the blind spot because cattle will turn if they cannot see the handler. All handler movements are done at a walk. The three main principles are:

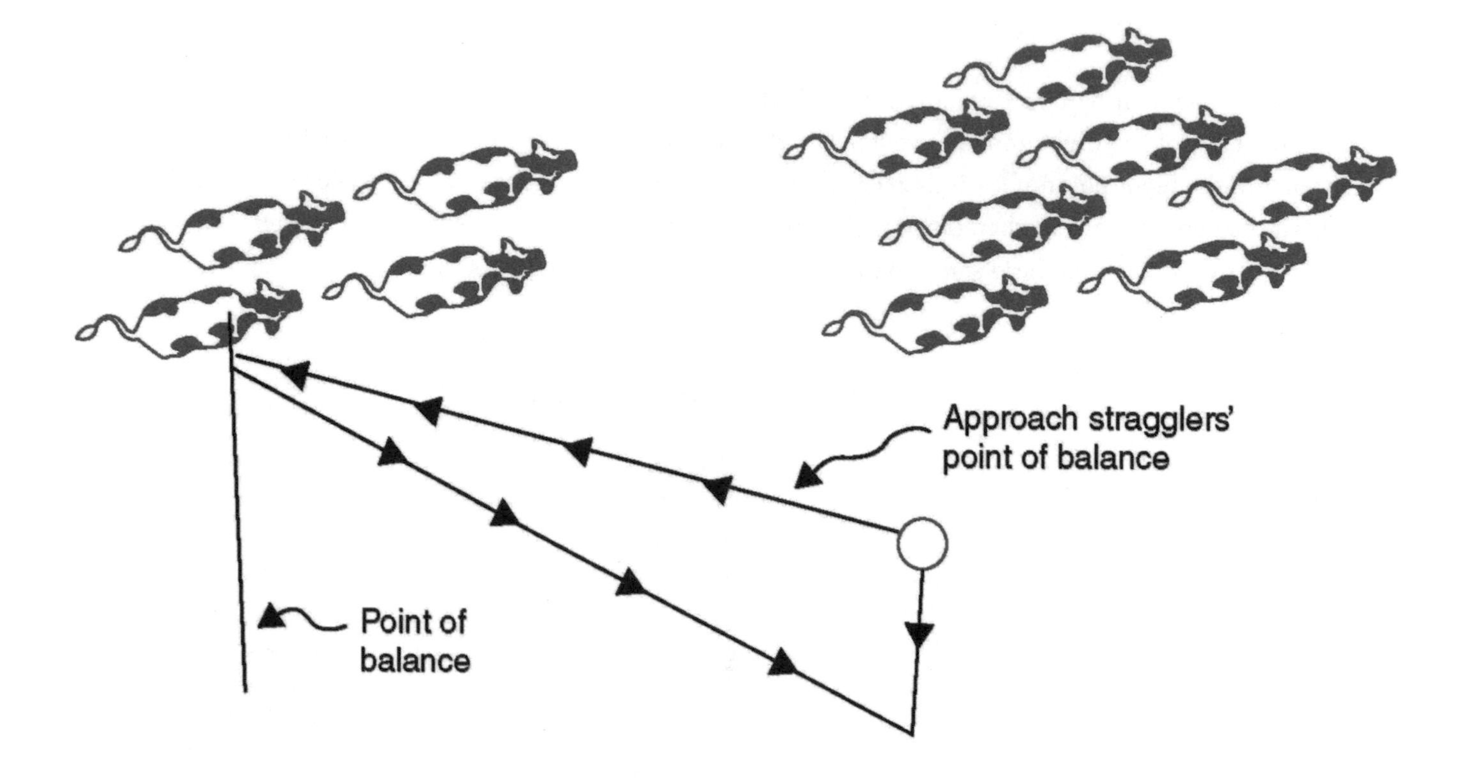

Fig. 9.
Handler movement pattern for bringing a split herd back together. The handler moves inside the flight zone in the opposite direction to the desired movement and outside the flight zone in the same direction. This diagram was developed after the author worked with Bud Williams. (From Grandin, 1993)

- never apply pressure to the flight zone when an animal is moving where you want it to go;
- release pressure when cattle move where you want them to go;
- reapply pressure only when the cattle stop moving where you want them to go.

REWARD CATTLE BY RELIEVING PRESSURE ON THE FLIGHT ZONE

Stock people must avoid the temptation to put extra pressure on the flight zone when cattle or other livestock are close to their destination. At this time, the handler should back up and relieve the pressure. Dylan Biggs warns that when cattle are getting close to a gate, this is not the time to apply more pressure (Biggs, 2013). He has successfully loaded bulls into an empty stock trailer that was in the middle of the pasture. This required patience. If the bull faced the trailer, he retreated from the flight zone. If the bull looked away from the trailer, he slightly penetrated the flight zone. Gradually he rewarded the bull for any behaviour directed towards the trailer. All of the stockperson's movements were at a slow walk. Cattle and other livestock need time to make decisions. Give them time to process the incoming information. Curt Pate stated, "You want them to think instead of just reacting." (Curt Pate, personal communication, 2018)

LEADERS AND LEADING CATTLE

The natural following of cattle can be used to facilitate cattle movements. On the old cattle drives in the USA, the value of calm leaders was recognized. The same leaders would lead a herd of 1,000 cattle every day (Harger, 1928). A good leader was usually a sociable cow, and she was not the most dominant animal. Smith (1998) contains excellent information on the effect of social behaviour on handling.

In frontier days, excitable, nervous animals that became leaders were usually destroyed and calm leaders were kept (Harger, 1928). If the cattle herd refused to cross a bridge or brook, a calf would be roped and dragged across to encourage the other cattle to follow (Ward, 1958). In Australia, a herd of tame 'coacher' cattle is used to assist in gathering wild feral cattle (Roche, 1988), and similar methods have also been used with wild horses (Amaral, 1977).

Fordyce (1987) also recommends mixing a few quiet old steers in with *Bos indicus* calves to facilitate training in handing procedures. Dumont et al. (2005) found that to determine which animal is the true leader, one should observe spontaneous long-distance movements to a new feeding site. The leader cannot be determined by watching cattle slowly graze through a field.

Cowboys and stock people working in large feed lots in Australia are using leading to move cattle in feedlots. Kev Sullivan from Queensland, Australia, encourages handlers to lead cattle every time they have the opportunity (Smith, 2018). A skilled handler uses 'drawing pressure', which has been described by Curt Pate. According to Tom Noffsinger, a skilled handler has a positive attitude and develops a 'presence' that attracts the cattle (Smith, 2018).

Bos indicus cattle have a much stronger following instinct than Bos taurus cattle. Observations by the author indicate that tame, purebred Brahmans are difficult to drive but will often follow a person or a trained lead animal. Progressive stock people in Brazil are leading Nelore cattle with a person on foot. No driving aids are used. The person finds a leader who is looking in the right direction who will slowly follow them. When they get near the destination gate, the handler relieves pressure on the flight zone. To stop the person from arm waving, hands are held behind the back. In Australia, they have been trained to follow lead dogs.

Tina Williams, the daughter of Bud Williams, warns that cattle that have never learned how to be driven can become highly stressed when they are taken to a new location where driving is the standard handling method.

CALLING CATTLE

Cattle reared under extensive conditions can easily be trained to come when called. The animals learn to associate a vehicle horn with feed (Hasker and Hirst, 1987). Jason C. Holt of the USDA NRCS (US Department of Agriculture Natural Resources Conservation Service) and David Kitner at Duval County Ranch in Texas (personal communication, 2012) trained cattle to come when called from huge 6000-acre pastures. The cattle were trailed to a series of successively larger paddocks. In the northern USA, when snow is on the ground, cattle will come running when they see a feed truck. However, the animals can become a nuisance if they always chase a truck for feed. They should be trained to associate the vehicle horn with feed, and then the truck can be driven in a pasture without the animals running after it. Young calves are less likely to become stressed and separated from their mothers if the mothers have learned not to constantly chase vehicles.

Many ranchers have adopted intensive grazing systems in which cattle are switched to a new pasture every few days (Savory, 1978; Smith et al., 1986). The cows quickly learn to make the switch. Calves are sometimes stressed when the cows rap idly run into the new pasture and leave them behind. To prevent calf stress, handlers should either stand near the gate of the new pasture and make the cows walk by them at a walk or lead them slowly into the pasture. To prevent cattle from mobbing around a gate, the gate should remain shut until the cows have settled down. They have to learn that they must walk calmly when pastures are switched. The goal is to have them walk past a person at a gate in an orderly single file. Ranchers who carry out intensive rotational grazing in which cattle are moved every few days usually lead cattle instead of driving them. Cattle can also be easily trained to come into corrals by feeding them in the corrals.

INTENSIVE GRAZING WITHOUT FENCES

There is increasing interest in practicing intensive grazing methods without the expense of fences on vast extensive ranges in the western USA. Herding methods are being used to both keep cattle groups in a specific area and to move them to new grazing sites. When cows and calves are moved, it is important to move them slowly so that the calves do not become separated from the cows. On long moves, the animals should have time to graze. Relaxed, calm cattle will stay in a new location. Frightened, stressed cattle will return to the old location. Bailey (2004), a grazing specialist in Montana, suggests a combination of herding and other methods to encourage cattle to stay in a new location. When cows with calves are moved long distances, the move should be timed so that they arrive at their destination in the late afternoon when it is time for the calves to bed down. This will encourage the cows to stay in the new location. Tasty supplements placed in the pasture also encourage cows to stay. Research on extensive rangelands in Montana indicated that low-stress herding methods reduced the percentage of the herd grazing in riparian areas (Bailey et al., 2008). Keeping cattle out of these areas will prevent damage to fragile lands along streams.

One of the big problems is that some cattle are ‘bunch quitters’ and do not want to stay with the herd (Nation, 1998). These are most likely to be high headed, nervous cows. Selling the bunch quitting cow is often the best option. Herding works best with uniform groups of cattle that have all been raised on the same ranch. Bunch quitters are more likely to be a problem in cows obtained from several different ranches. The principle of herding without fences is to relieve pressure on the collective flight zone when cows stay where you want them and to apply pressure when they go to where you do not want them to be. To encourage cattle to stay in a new location, wait for them to relax and graze in random orientations (Tina Williams, personal communication, 2014). This is the same principle that is used when settling cattle for sorting in an open field. Settled cattle will graze in a bunch while facing in different directions (Whitehurst, 2017).

Learning also has a huge effect on grazing. Livestock prefer the feeds that they ate with their mother when they were little (Provenza, 2003). Bringing adult cows to an area that has unfamiliar feeds can cause big problems. When replacement animals are purchased, they should come from areas that have similar pasture (Nation, 2003). Calves can be taught to eat a new feed in the future by feeding this to both the cow and the calf. Cattle that are being brought in from a different part of the country can be taught as calves to eat feeds that they will encounter when they grow up.

Herders have to spend many hours with their herds and have lots of patience. More information on pasture herding can be found in Biggs and Biggs (1996, 1997), Herman (1998), Nation (1998), Smith (1998), Williams (1998), and Lanier and Smith (2006).

SELF-MUSTERING SYSTEMS

In extremely extensive operations, it may be difficult to gather and muster all the cattle. For years, ranchers have successfully used trap gates which allow cattle to enter a corral to eat or drink but they cannot leave. These systems are described in Stookey and Watts (2014), Anderson and Smith (1980), Webber (1987), Ward (1958), Connelly et al. (2000), and Adcock et al. (1986).

HERDING BY PASTORAL PEOPLE

The herding methods that have been described in this chapter are a relearning of old pastoral herding methods that have been used for thousands of years. In all of these methods, a great deal of time is spent with the animals. Norwegian reindeer herders are in close contact with their animals and the animals associate the smells and noises of camp with serenity (Paine, 1994). The Fulani African tribesmen have no horses, ropes, halters, or corrals (Lott and Hart, 1977, 1982). Their cattle are completely tame and have no flight zone. Instead of chasing the cattle, the herdsman becomes a member of the herd, and the cattle follow him (Lott and Hart, 1979).

HANDLING DOMINANT ANIMALS

The nomadic Fulani tribesmen use the animals' natural following, dominance, submission and grooming to control the overall herd. If a bull makes a broadside threat, the herdsman yells and raises a stick. The herdsman charges at the bull and hits it with a stick if it attempts a charge. Similar methods have also been used successfully in other species. Raising a stick over the handler's head has been used to exert dominance over bull elk (Bud Williams, personal communication; Smith, 1998). The stick is never used to hit the bull elk.

The author has used similar methods to control aggressive pigs. Pigs exert dominance by biting and pushing against another pig's neck (Houpt and Wolski, 1982). The aggressive pig was stopped by shoving a board against the neck to simulate the bite of another pig. Using the animal's natural method of communication was more effective than slapping it on the rear. Exerting dominance is not beating an animal into submission. The handler used the animal's own patterns to become the 'boss' animal. The aversion that cattle have for manure can be used to keep them away from crops, by smearing the borders of a field with faeces (Lott and Hart, 1982). Manure is also smeared onto the cow's udder to limit milk intake by the calf.

STROKING AND LEADING CATTLE

The Fulani stroke their cattle in the same areas where a mother cow licks her calf (Lott and Hart, 1979), so adult cattle will approach and stretch out their necks to be stroked under the chin (Lott and Hart, 1982).

Similar methods are used at the J.D. Hudgins Ranch in Hungerford, Texas, and at the J. Carter Thomas Ranch in Cuero, Texas. Purebred Brahmans are led to the corrals and will eat out of the rancher's hand. Cows and bulls in the pasture will come up to Mr. Thomas for stroking and brushing (Julian, 1978).

Small herds of Zebu cattle raised in the Philippines have no flight zone and are easily led by small children. The author's observations indicate that taming crossbreeds of Brahman and *Bos taurus* is more difficult. This may be partially due to a lower level of inquisitiveness, and a desire for stroking and following. The cattle herding methods of the Fulani are also practised by other African tribes such as the Dinka (Deng, 1972; Schwabe and Gordon, 1988), and the Nuer (Evans-Pritchard, 1940). The less nomadic tribes do use corrals and tethers, but the cattle are still completely tame with no flight zone. Surplus bulls are castrated and kept as steers by all tribes.

The cattle-handling practices of African tribes date back to before the great dynasties of Egypt (Schwabe, 1985; Schwabe and Gordon, 1988). It is also noteworthy that the religion of the Nuer and Dinka tribes centres around cattle (Seligman and Seligman, 1932; Evans-Pritchard, 1940). One factor that makes African tribal handling methods successful is that relatively small herds are handled and each tribe has many herdsmen. Therefore, each herdsman has time to develop an intimate relationship with each animal.

HERDING AND OBSERVING CATTLE WITH DRONES

Some ranchers have used a drone to move cattle. A review of several online videos indicated that moving cattle with drones ranged from excellent, calm working on the edge of the flight zone to terrible running and chasing of cattle. These videos can be easily found by using the keywords 'Videos', 'cattle drone' and 'herding'. Tillman-Brown (2018) explains that some drones may malfunction near cell phone towers. Some of the problems may be due to other noncell (mobile) phone equipment that is mounted on the same tower.

BULL BEHAVIOUR

Dairy bulls have had a bad reputation for attacking humans, possibly due to the differences in the way beef bulls and dairy bulls are raised. Bulls are responsible for about half of the fatal accidents with cattle (Drudi, 2000). Dairy bull calves are often removed from the cow shortly after birth and raised in individual pens, whereas beef bull calves are reared by the cow.

Price and Wallach (1990) found that 75% of Hereford bulls reared in individual pens from one to three days of age threatened or attacked the handlers, while only 11% threatened handlers when they were hand-reared in groups. These authors also report that they have handled over 1,000 dam reared bulls and have experienced only one attack. Bull calves that are hand-reared in individual pens may fail to develop normal social relations with other animals, and they possibly view humans as a sexual rival (Reinken, 1988).

Both dairy and beef breed bulls will be safer if bull calves are raised on a cow and kept in groups with other cattle. This provides socialization with their own species, and they will be less likely to direct attacks towards people. Similar aggression problems have also been reported in hand-reared male llamas (Tillman, 1981). Fortunately, hand rearing does not cause aggression problems in females or castrated animals. In contrast, it will make these animals easier to handle. More information on bulls can be found in Smith (1998). Ron Gill (2012, personal communication) warns to never turn your back on a bull. Also, when bulls are being driven, do not deeply invade their flight zone, especially in small pens or pastures. Bulls are more likely to become aggressive towards handlers when they are overcrowded than are cows or steers.

VISION AND DESIGN OF CATTLE RESTRAINT DEVICES

One of the reasons why cattle become agitated in a squeeze chute is due to the operator being deep inside their flight zone. They can see him/her through the open bar sides.

Cattle will remain calmer in a restraining device that has solid sides and a solid barrier around the headgate to block the animal's vision (Grandin, 1992; Muller et al., 2007). Cattle also struggled less in a restraint device if their vision was blocked until they had been completely restrained (Grandin, 1992), and they are less likely to attempt to lunge through the head opening if there is a solid barrier 1.2 m in front of the head opening. This prevents them from seeing a pathway of escape.

Restraint device designs that have been successfully used in slaughter plants could be adapted for handling on the ranch and feedlot (Marshall et al., 1963; Grandin, 1992). Most cattle stood quietly when the Marshall et al. (1963) restraint device was slowly tightened against their bodies (Grandin, 1992). Solid sides prevent the animals from seeing the operator or other people inside their flight zone. Observations also indicate that cattle unaccustomed to head restraint will remain calmer if body restraint is used in conjunction with head restraint. Head restraint without body restraint can cause stress (Ewbank et al., 1992).

Breeders of American bison prevent injuries and agitation by covering the open-barred sides of squeeze chutes and installing a solid gate (crash barrier) about 1.0–1.5 m in front of the headgate. Covering the sides of a squeeze chute so that the animal does not see the operator standing beside it will keep animals with a large flight zone calmer. When bison are handled, the top must also be covered to prevent rearing. Squeeze chutes are now commercially available with rubber louvres on the sides to block the animal's vision. The louvres are mounted at a 45° angle and the drop bars on the width of the chute can still be opened. A cow's eye view of a squeeze chute equipped with louvres may be found in Grandin (1998). Covering the open barred sides of a squeeze chute with cardboard will also result in calmer cattle because they will not see the squeeze chute operator who is deep in their flight zone.

Cattle will remain calmer if they 'feel restraint'. Sufficient pressure must be applied to hold the animal snugly, but excessive pressure will cause struggling due to pain. There is an optimal amount of pressure. If an animal struggles due to excessive pressure, the pressure should be slowly and smoothly reduced. Many people mistakenly believe that the only way to stop animal movement is by greatly increasing the pressure. Sudden, jerky movements of the apparatus will cause agitation, whereas smooth, steady movements help to keep the animal calm (Grandin, 1992). Handlers must be careful to avoid rewarding struggling. Cattle should not be released from a squeeze chute until they have stopped struggling (Stookey and Watts, 2014). Fumbling a restraining procedure will also cause excitement (Ewbank, 1968). It is important to restrain the animal properly on the first attempt.

DARK-BOX RESTRAINT

For artificial insemination (AI) and pregnancy testing, mechanical holding devices can be eliminated by using a dark box race (Parsons and Helphinstine, 1969; Swan, 1975). This consists of a narrow stall with solid sides, a solid front and a solid top. Very wild cattle will stand still in the darkened enclosure. A cloth can be hung over the cow's rump to darken the chamber completely. Comparisons between a dark box and a regular squeeze chute with open-bar sides indicated that cows in the dark box were less stressed (Hale et al., 1987). Further experiments indicated that cortisol levels were lower in the dark box, but heart rate data were highly variable due to the novelty of the box (Lay et al., 1992).

Blindfolding of both poultry and cattle reduces heart and respiration rates (Douglas et al. 1984; Jones et al., 1998; Stookey and Watts, 2014; Don Kinsman, personal communication). Mitchell et al. (2004) reported that blindfolding Hereford × Angus × Charolais heifers with several layers of opaque dark towel resulted in less struggling in the squeeze chute compared with control nonblind folded animals. Observations by Jennifer Lanier in the author's laboratory showed that blindfolds on American bison had to be opaque to provide the greatest calming effect. Installation of a solid top on a squeeze chute also kept bison calmer.

If large numbers of cattle are inseminated, two or three dark boxes can be constructed in herringbone configuration (McFarlane, 1976; Canada Plan Service, 1984) (Fig. 10). The outer walls are solid, with open-barred partitions between the cows. Side-by-side bodily contact helps to keep cattle calmer (Ewbank, 1968). To prevent cattle from

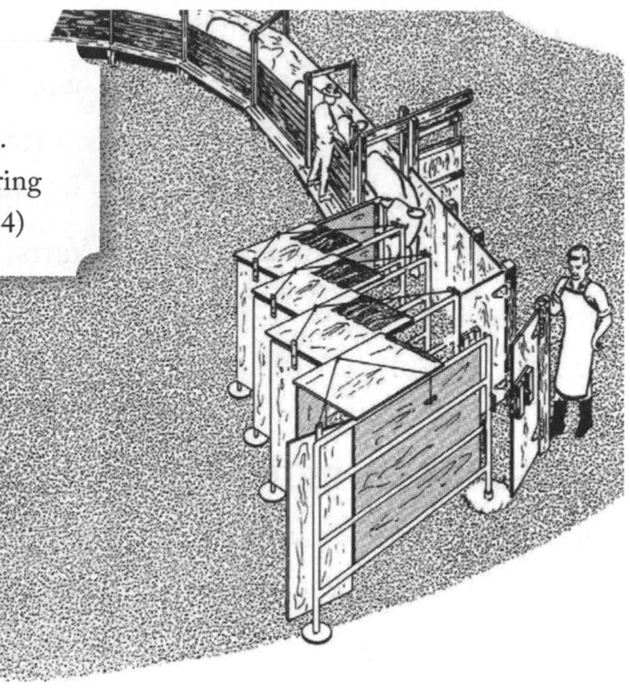

Fig. 10.
Dark-box restraint races in a herringbone configuration. The solid sides restrict vision and keep cattle calmer during artificial insemination. (From Canada Plan Service, 1984)

being frightened by a novel dark box, the animals should be handled in the box before insemination. The effectiveness of dark-box restraint is probably due to a combination of factors, such as blocking the view of an escape route and preventing the animal from seeing people that are outside its flight zone. Darkness, however, has a strong calming effect. Wild ungulates remain much calmer in a totally dark box. Small light leaks sometimes cause animals to become agitated. A well-designed dark box for domestic cattle has small slits in the front to admit light. Cattle will enter easily because they are attracted to the light. For wild ungulates, it may be desirable to block the slits after the animal is in the box.

USE OF SOLID OR OPEN SIDES ON RACES

Many stockmanship specialists advocate the use of open sides on single-file races so that cattle can see the position of the stockperson and move away in a controlled manner. For years the author has recommended the use of solid sides on all the races. Who is correct? Both, according to the situation. Before temperament selection started in the USA, extensively raised cattle would remain calmer in chutes and races with solid sides. Today, in many parts of the USA, *Bos taurus* cattle are much calmer due to 20 years of genetic selection for a calmer temperament. A skilled stockperson can easily use a race with open sides on pastures that are free of visual distractions. In slaughter plants and other places with lots of activity around the single-file race, adding a solid side reduced electric prod use and improved cattle movement (Ercolano, 2018).

Depending on the size of the animal's flight zone, handlers must maintain a people-free zone around a race with open sides. This zone is only entered when an animal has to be moved. If people stand in the 'force field' alongside the race, the cattle will become agitated. If an animal rears up in the race, the handler must back up and retreat from the flight zone.

COVER THE OUTER PERIMETER TO BLOCK VISION

Many new systems on ranches and feedlots have been built with a partially open fence along the inner side where the people work, and a completely solid high fence on the outer perimeter to block the view of vehicles and other distractions outside the facility. The use of the open fence eliminates the need for catwalks and prevents the problem of a person suddenly appearing when the cattle do not expect to see them. Figure 11 is a simple design with an open inner fence and solid outer perimeter fence (Grandin, 2017).

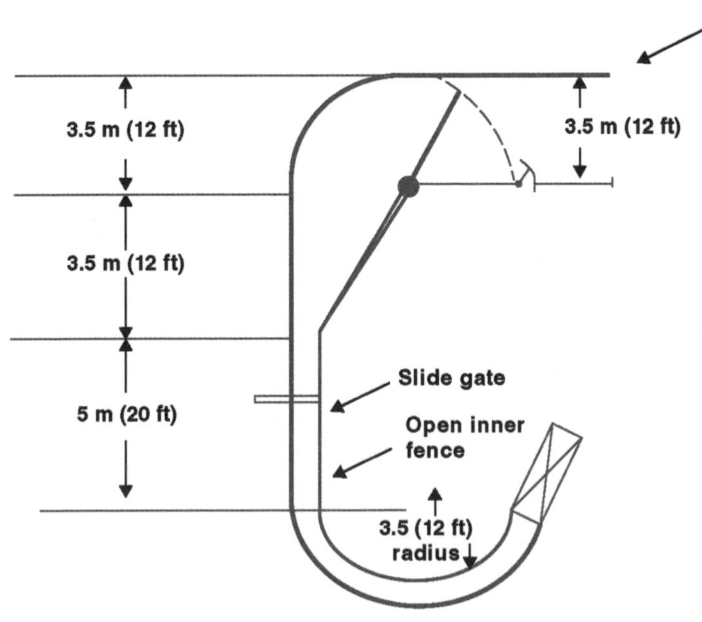

Fig. 11.
Single-quarter round crowd pen and curved single-file race that can be easily constructed from portable panels. The handler works at the pivot point of the crowd pen and the cattle circle around. The design takes advantage of the natural tendency of cattle to circle around the handler. All catwalks are eliminated. The outer perimeter fences are solid and the inner fence is open on the top. (Courtesy of Temple Grandin)

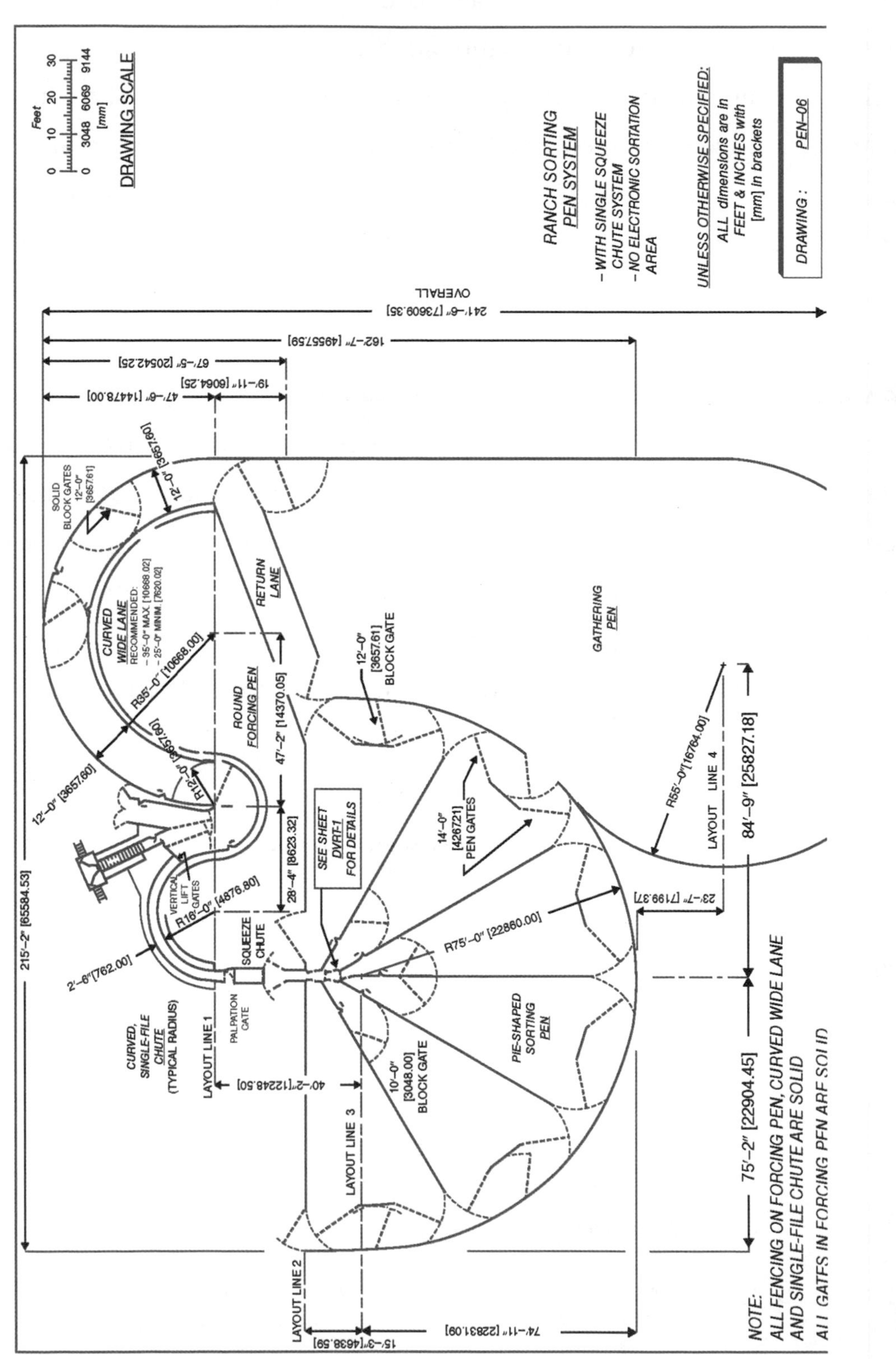

Fig. 12.
Large corral design suitable for drafting cattle with individual IDs into the pieshaped pen. Managers of large ranches often need numerous sorting pens that cattle can be directed into after they have been individually weighed and classified. (Grandin, 2000)

DESIGN OF CORRALS, YARDS AND LAIRAGES

Modern curved races and round crowd pens evolved independently in Australia (Daly, 1970), New Zealand (Kilgour, 1971; Diack, 1974), and the USA (Oklahoma State University, 1973; Rider et al., 1974; Thompson, 1987). In the mid 1960s and early 1970s, many large feedlots were built with curved single-file races, round crowd pens and diagonal pens (Paine et al., 1976). Grandin combined (1980a, 1983a,b) the best features of many different designs and developed layout principles to make the systems work more efficiently. Vowles and Hollier (1982) and Vowles et al. (1984) reported that cattle moved more efficiently through curved single-file races.

Figure 12 shows a large system for a ranch with numerous pens for sorting cattle after they have been weighed and evaluated. Numerous sorting pens are strongly recommended when cattle have to be marketed to programmes that have strict specifications. Variations of this design are also shown in a computerized cattle sorting system (Pratt 2007, 2009). Fig. 13 clearly illustrates why blocking a bovine's vision of outside distractions is important. Cattle moving through this single-file race could see trucks being loaded. In a feedlot that had two identical working areas, this explained why this one worked poorly and the other one worked well.

Fig. 13.
Single-file race with a partially open side. This system would have worked better if the outer perimeter was completely solid. The cattle moving through the race are able to see trucks being loaded. This distraction can cause cattle to balk. (Photo courtesy of Temple Grandin)

The more efficient facility with the solid outer wall is shown in Fig. 14. Large beef abattoirs should have stockyard/lairage designs that have one-way traffic flow through the yards. Angled pens eliminate sharp corners (Grandin, 1997, 2008). The pens should be on a 60–80° degree angle (Fig. 14). Do not use a 45°angle. A high-capacity truck loading ramp is shown in Fig. 15. This design will work well when many trucks have to be loaded. Fig. 16 shows an old Australian yard design that is still popular.

STRATEGIC LOCATION OF SOLID BARRIER TO BLOCK VIEW OF A PERSON UP AHEAD

Cattle and other livestock will often balk and refuse to approach people they can see up ahead.

Solid barriers for stock people to hide behind will often facilitate cattle movement. They are especially important in abattoirs where cattle have to go up ramps.

PORTABLE CATTLE CORRALS ON RENTED LAND

Since the fourth edition, more and more people need portable cattle-handling corrals that can be used on rented or leased pasture. They need to have facilities that can be easily moved to a new location. In many countries, cattle have been selected for a calmer temperament. This has made it easier for less skilled people to use simpler facilities constructed from portable panels. Figure 11 in this chapter is a design that can be easily constructed from portable components. There are two types of portable yards and corrals. The first type is designed to be easily moved from pasture to pasture on a trailer. The second type is a semi-permanent installation that would only be moved when the rancher stops leasing the land. These systems are described in Grandin (2017).

AVOID LAYOUT MISTAKES WHEN DESIGNING CATTLE-HANDLING FACILITIES

Layout mistakes can reduce the efficiency of cattle movement through a handling facility. Some of the most common mistakes are:

- **Dead-ending the race** – bending the junction between the single-file race (chute) entrance and the crowd pen too sharply. When an animal stands at the single-file race (chute) entrance, it must be able to see two body lengths up the race; they

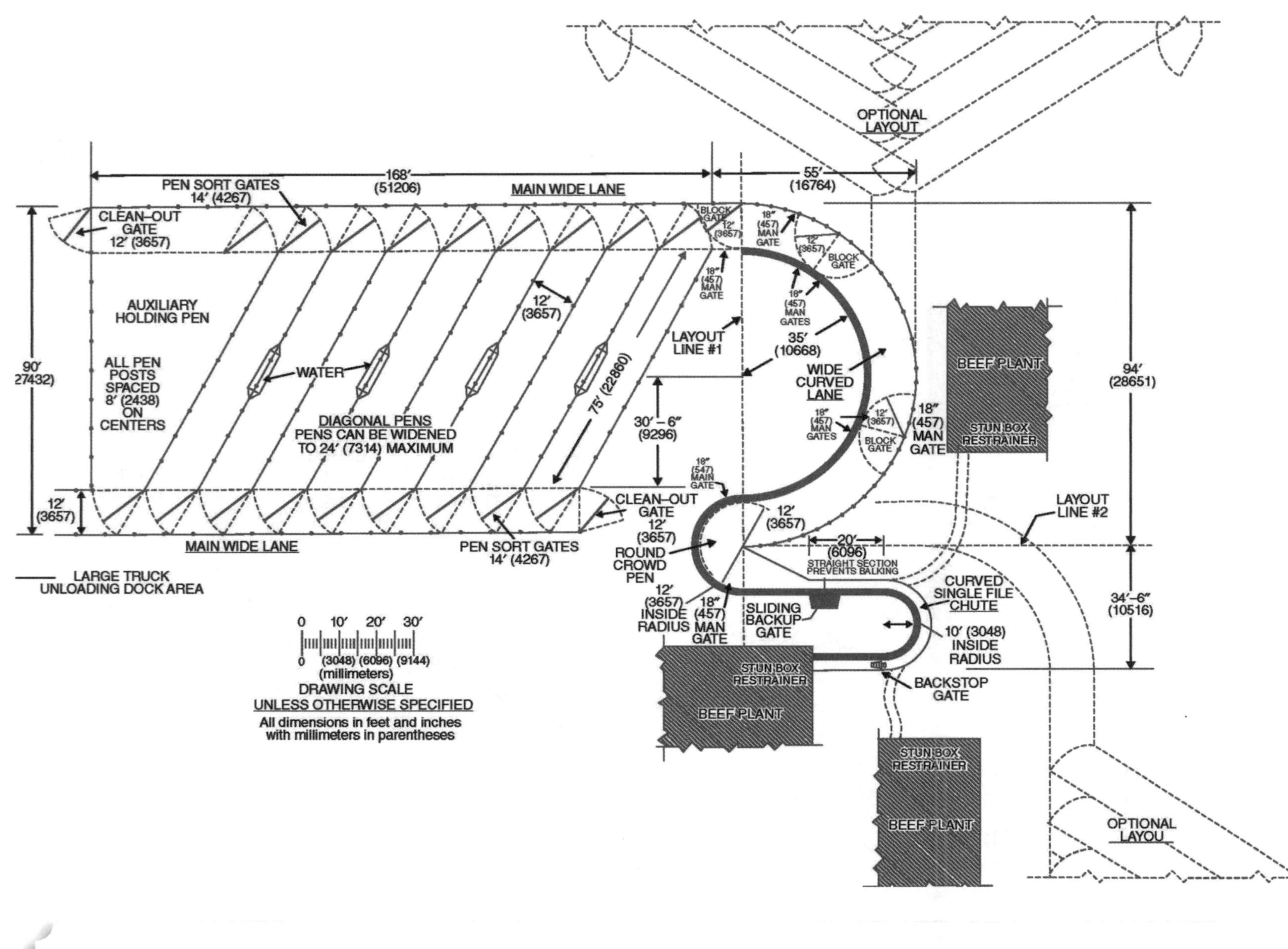

Fig. 14.
Large lairage stockyard layout for a beef abattoir. The angled pens in the stockyard facilitate oneway cattle traffic flow through the yard (Drawing by Mark Deesing, courtesy of Grandin Livestock Handling Systems, Inc.)

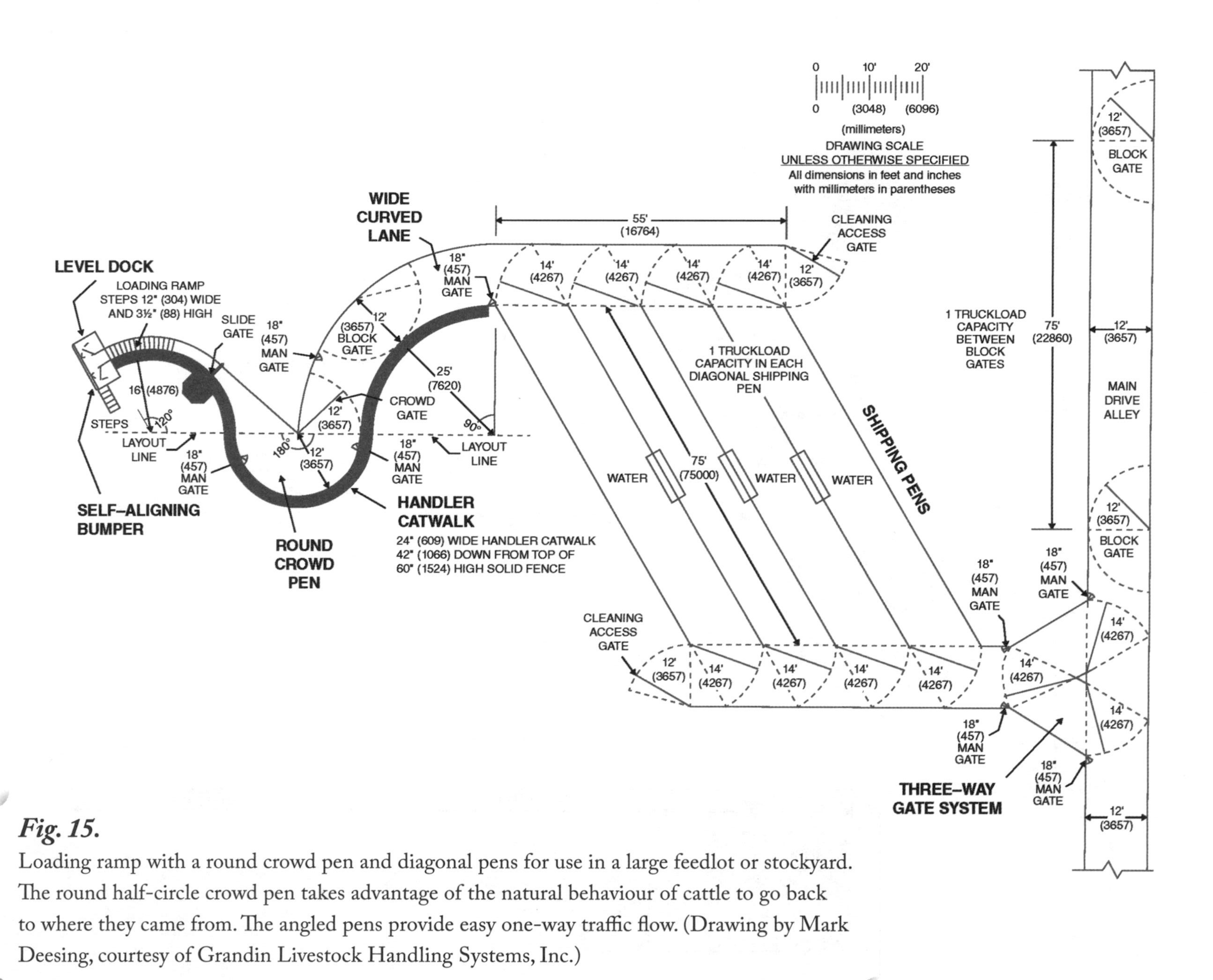

Fig. 15.
Loading ramp with a round crowd pen and diagonal pens for use in a large feedlot or stockyard. The round half-circle crowd pen takes advantage of the natural behaviour of cattle to go back to where they came from. The angled pens provide easy one-way traffic flow. (Drawing by Mark Deesing, courtesy of Grandin Livestock Handling Systems, Inc.)

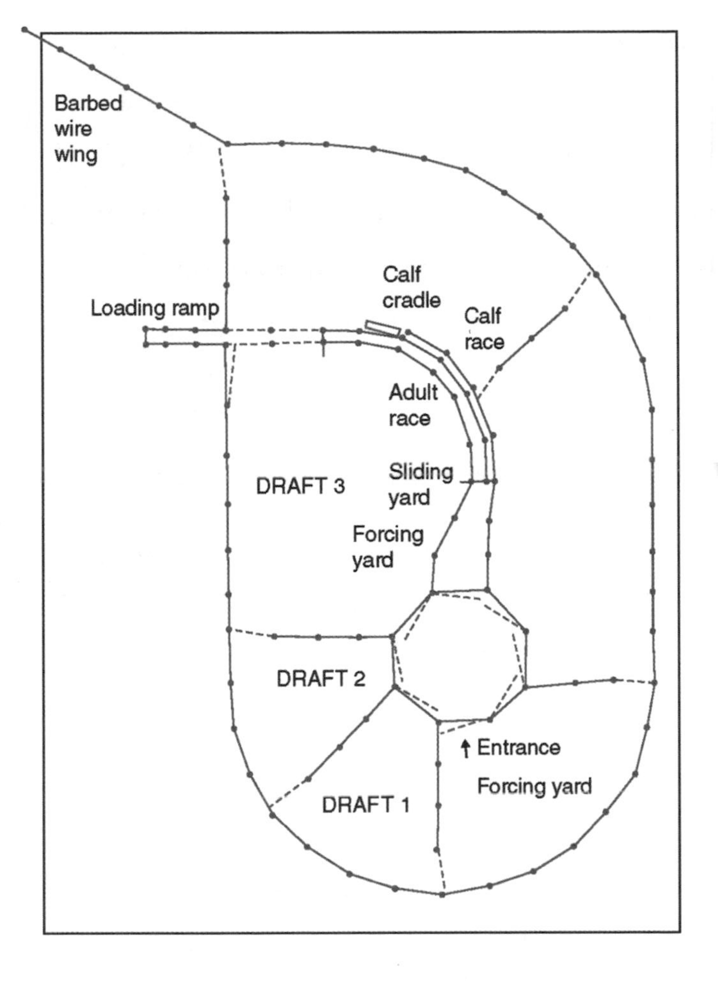

Fig. 16.
Australian Weean cattle yard. This is an old design that is still very popular. This diagram originally appeared in the 1993 version of this book. Cattle are sorted (drafted) in the round pound yard that is located in the middle of the yard. This layout can be easily constructed from portable steel panels. In 2018, several different Australian equipment companies had the Weean yard on their websites.

have to see a place to go before going around a bend.

- **Round crowd pen (tub) is either too big or too small** – on a circular crowd pen the recommended radius for the crowd (forcing) gate is 3 m (10 ft) minimum and 3.5 m (12 ft) maximum (Grandin, 1997).
- **Bud box is too narrow** – the width should be 3.5 m (12 ft) minimum and preferably 4.2 m (14 ft.). The cattle have to be able to go around the person. A Bud box is always a rectangle. It should be 6 m (20 ft) to 7.6 (25 ft) long.
- **Crowd pen angle leading to a single-file race is wrong** – when a funnel-type crowd pen is built, one side should be straight, and the other side should be on a 30° angle. The stockperson should work on the angled side.
- **Never put a crowd pen on a ramp** – if a system requires a ramp, it should be in the single-file race portion of the system. Never hold or store groups of cattle on wide ramps because they may pile up on the rear gate.
- **Flooring is too slippery** – cattle become agitated on a slick floor. Animals standing in a race or scale doing numerous little slips will become agitated. Provide good nonslip flooring (OIE 2017; Grandin and Deesing, 2008; Grandin, 2017).

Ramp surface

- On concrete ramps, stairsteps are recommended. The steps should have a 10 cm (3.5 in) rise and 30 cm (12 in) to 45 cm (18 in) tread length (Grandin, 1987). Cattle and other large livestock can easily walk on stairsteps even when they became worn. On steel or wood ramps, space the cleats (battens) so that an animal's hoof will fit comfortably between them. The space should be 20 cm (8 in) for cattle (Mayes, 1978; Grandin, 2008).

CATTLE-RESTRAINING CRUSHES

Before the invention of squeeze chutes (crushes), range cattle were caught and restrained with a lariat (Ward, 1958), which is a rope that is thrown to catch and restrain cattle. The invention of mechanized restraint devices both improved animal welfare and reduced labour requirements. These devices also required less skill to operate than a lariat. Good lariat use requires tremendous skill to maintain an acceptable level of animal welfare. The lariat is a good example of a method that is low-cost but highly skill dependent. One of the first mechanical devices for restraining cattle was patented by Reck and Reck (1903). It had squeeze sides that pressed against the animal and a stanchion to hold its head. In the 1920s, Thompson (1931) developed a head catching gate designed for wild, horned cattle, which was installed on the end of a single-file race. Squeeze chutes are more expensive than roping animals, but they greatly improve employee safety. Good equipment for restraining cattle can also help to reduce injuries. An Australian study showed that 57% of the injuries to veterinarians occur either during pregnancy checking or the examination of cattle (Lucas et al., 2012).

A squeeze chute restrains an animal with two devices—a stanchion (head bail) is closed around the animal's neck and side panels that press against its body to control movement. This device is available with either manual or hydraulic controls. The best squeeze chutes have two side panels that close in evenly on both sides of the animal's body. A double-sided squeeze enables the animal to stand in a balanced position. An animal may struggle, and resist being restrained if pressure is applied to only one side of the body.

Struggling occurs because the animal is thrown off balance. Nonslip flooring is essential in squeeze chutes because animals often panic if they start to slip. Today, most adult range cattle are restrained in a squeeze chute. Lariats are still used on some large ranches for the restraint of calves for branding and vaccination. One of the first mechanical devices for holding calves was invented by Thompson and Gill (1949) and there are some good designs by Kratky (1976) and Priefert (1990).

HEADGATE DESIGN

There are six different types of headgates (head bails or stanchions) for restraining cattle. These can be used either alone at the end of a single-file race or in conjunction with a complete squeeze chute. The six types are: self-catch (Pearson, 1965), positive (Thompson, 1931; Heldenbrand, 1955), scissors stanchion, pivoting horizontal, sliding doors, and rotating headgates (see Cummings, 1993, 1999; Mollhagen, 1994, 2010; Daniels and Schmitz, 2003; and Daniels et al., 2012).

The advantage of pivoting, rotating headgates or horizontal sliding doors is that they open up to the full width of the squeeze and the animals can exit more easily. These designs may also help to reduce shoulder injuries. A picture of the rotating headgate may be found in Grandin (1998). The Cummings (1993, 1999) headgate requires very little force to restrain the animal compared with scissors stanchion headgates. Many of the references given here are to US patents, and the diagrams can be found by searching Google Patents or the US Patent and Trademark Office (USPTO) database.

TYPES OF HEADGATES

All types of headgates are available with either straight vertical neck bars or curved neck bars. Stanchions with straight neck bars are recommended for general-purpose uses because they are less likely to choke an animal. Pressure on the carotid arteries exerted by a neck stanchion will quickly kill cattle (White, 1961; Fowler, 1995). Stanchions with straight vertical neck bars are the safest because cattle can lie down and there is no pressure on the carotid arteries.

Curved-bar stanchions provide a good compromise between control of head movement and safety for the animal. Positive headgates, which clamp tightly around the neck, provide better head control but have an increased risk of choking the cattle (Grandin, 1975, 1980b). Both positive type headgates and a curved bar stanchion must be used with squeeze sides or some other apparatus to prevent the animal from lying down and choking in the headgate.

TYPES OF SQUEEZE SIDES

There are two commonly used designs to prevent animals from lying down in squeeze chutes. They are: (i) pivoting of the two squeeze sides on the floor so that they form a V shape to support the animal; or (ii) brisket bars, which are used with squeeze sides that remain perpendicular to the floor when the squeeze is tightened (Daniels and Schmitz, 2003; Daniels et al., 2012). Perpendicular squeeze sides open up wide at the floor so that it is easier for large, fat cattle to enter and exit. The disadvantage is that cattle can easily lie down unless a brisket bar (sternum support bar) is installed on the floor. Brisket bars must never be used when the squeeze sides are hinged at the floor to form a V because the space between the sides is too narrow. Sides that form a V-shape are mechanically simpler to construct than sides that remain perpendicular. They work well unless cattle are very large and fat. Nonslip flooring is essential in all types of restraint devices. Cattle become agitated when they slip on the floor.

HYDRAULIC SQUEEZE CHUTES

Hydraulically activated squeeze chutes have become increasingly popular. To reduce noise, the pump and motor should be removed and located away from the chute. Some squeeze chutes have plastic inserts to reduce noise and prevent metal to metal noise when the gates open and close. A properly adjusted hydraulic chute is safer for both people and cattle. Operator safety is improved because long, protruding lever arms are eliminated. However, the pressure relief valve must be properly set to prevent severe injury from excessive pressure. Injuries may include broken ribs and internal ruptures. Additional information on the proper adjustments can be found in Grandin (1980b, 1983b, 1990b). If an animal vocalizes (moos or bellows) immediately after it is squeezed in a hydraulic powered restrainer, the pressure setting must be reduced (Grandin, 2001; Bourquet et al., 2011). The valve must be set so that the squeeze sides automatically stop squeezing before excessive pressure is applied. This will prevent a careless operator from applying excessive pressure. Other indicators of excessive pressure are laboured breathing and straining.

If the chute has an additional hydraulic device attached to the headgate (Mollhagen, 2010) to hold the animal's head still, it must have its own separate pressure relief valve. Head-holding devices must be set at a pressure that is much lower than the pressure required to operate the squeeze.

Carelessness and rough handling are the major causes of injuries to cattle in squeeze chutes (Grandin, 1980a). Bruises directly attributable to the squeeze chute occur in 2–4% of cattle. In one study, bruises occurred in five out of seven feed lots, and 1.6–7.8% of the cattle had increased bruising compared with animals that had not been handled in the squeeze chute (Brown et al., 1981). Observations have also indicated that cattle can be injured when the headgate is suddenly closed around the neck of a running animal. Cattle should be handled quietly so that they walk into a squeeze chute and walk out. Hitting the headgate too hard can cause heamatomas and bruises. The shoulder meat may still be damaged when cattle are slaughtered.

DIPPING VATS

In many places, dipping vats have been replaced with pour-ons, sprays and other types of pharmaceuticals. There are still some places where dip vats are used. Grandin (1978, 1980a,c) developed an improved dip-vat entrance. Drawings can be found on www.grandin.com. Other useful publications are Hewes (1975); Texas Agricultural Extension Service (1979); Fairbanks et al. (1980); Sweeten (1980); Sweeten et al. (1982); Kearnan et al. (1982); and Midwest Plan Service (1987). The dip-vat system designed by the author can be seen in the HBO movie *Temple Grandin*. An exact working replica was built from the original drawings.

CONCLUSIONS

Low-stress handling of cattle and other livestock requires both skilled stock people and good equipment. Handlers need to understand the eight natural cattle behaviours that can be used for effective handling. Corrals and races should be designed to use natural behaviours such as cattle wanting to go back to where they came from. Restraint device operation requires extra care to prevent injuries and stress.

REFERENCES

Adams, A. (1903) *Log of a Cowboy: A Narrative of the Old Trail Days.* Houghton Mifflin, New York. Available at: https://archive.org/details/thelogofacowboy12797gut (accessed 15 April 2019).

Adcock, D., Kingstone, T. and Lewis, M. (1986) Trapping cattle with hay. *Queensland Agricultural Journal* 112, 243–246.

Algers, B. (1984) A note on responses of farm animals to ultrasound. *Applied Animal Science* 12, 387–391.

Amaral, A. (1977) Mustang: *Life and Legend of Nevada's Wild Horses.* University of Nevada Press, Reno, Nevada.

Ames, D.R. (1974) Sound stress and meat animals. *Proceedings of the International Livestock Environment Symposium*, SP0174. American Society of Agricultural Engineers, St. Joseph, Michigan.

Anderson, D.M. and Smith, J.N.C. (1980) A single bayonet gate for trapping range cattle. *Journal of Range Management* 33, 317.

Arave, C.W. (1996) Assessing sensory capacity of animals using operant technology. *Journal of Animal Sciences* 74, 1996–2009.

Bailey, D.W. (2004) Management strategies for optimal grazing distribution and use of arid rangelands. *Journal of Animal Science* 82 (esuppl.) (E147–E153).

Bailey, D.W., Van Wagoner, H.C., Weinmeister, R. and Jensen, D. (2008) Evaluation of low stress herding and supplement placement for managing grazing in riparian and upland areas. *Rangeland, Ecology and Management* 61, 26–37.

Bauer, J.M. et al. (2012) *Impact of Different Handling Styles (Good vs. Aversive) and Growth Performance, Behavior, and Cortisol Concentrations in Cattle.* Arkansas Animal Science Departmental Report.

Beaver, B.V. and Hoagland, D.L. (2016) *Efficient Livestock Handling: The Practical Application of Animal Welfare and Behavioral Science.* Academic Press, Elsevier, San Diego, California.

Biggs, D. (2013) www.dylanbiggs.com (accessed March 25, 2018).

Biggs, D. and Biggs, C. (1996) Using the magnetic nature of herd movement. *Stockman Grass Farmer*, Sept., 15–17. Biggs, D. and Biggs, C. (1997) Gatekeepers: low stress tactics for moving cattle. Stockman Grass farmer, September, 36–37.

Bourquet, C., Deiss, V., Tannigi, C.C. and Terlouw, E.M.C. (2011) Behavioral and physiological reactions of cattle in a commercial abattoir: relationships with organizational aspects of the abattoir and animal characteristics. *Meat Science* 88, 158–168.

Brown, H., Elliston, N.G., McAskill, J.W. and Tonkinson, L.V. (1981) The effect of restraint of fat cattle prior to slaughter on the incidence and severity of injuries, resulting in carcass bruises. *Western Section, American Society of Animal Science* 33, 363–365.

Burri, R. (1968) *The Gaucho.* Crown Publishing, New York.

Canada Plan Service (1984) Herringbone, A.I. Breeding Chute Plan, 1819. Agriculture Canada, Ottawa.

Carroll, J., Murphy, C.J., Neitz, M. and Van Hoere, J.N. (2001) Photopigment basis for dichromatic vision in the horse. *Journal of Vision* 1, 80–87.

Connelly, P., Horrocks, D., Pahl, L. and Warman, K. (2000) Cost-effective and multipurpose self-mustering enclosures for stock. Department of Primary Industries, Brisbane, Australia.

Cote, S. (2003) Stockmanship: a powerful tool for grazing management. USDA National Resources Conservation Service, Boise, Idaho.

Coulter, D.B. and Schmidt, G.M. (1993) Special senses. 1: Vision. In: Swenson, M.J. and Reece, W.O. (eds) *Duke's Physiology of Domestic Animals.* Comstock Publishing, Ithaca, New York.

Cummings, W.D. (1993) Cattle headgate, US Patent 5263.438. US Patent Office, Washington, DC.

Cummings, W.D. (1999) Cattle headgate, US Patent 5908.009. US Patent Office, Washington, DC.

Curtpatestockmanship.com (n.d.) Contains videos of cattle handling and horsemanship (accessed 2 November, 2018).

Daly, J.J. (1970) Circular cattle yards. *Queensland Agricultural Journal* 96, 290–295.

Daniels, D.D. and Schmitz, D. (2003) Squeeze chute apparatus. US Patent 6609.480 B2. US Patent Office, Washington, DC.

Daniels, D.D., Carpenter, P.O., Dutrow, W.L. and Raymond, J. (2012) Squeeze chute apparatus. US Patent 8240.276 B1 (2012). US Patent Office, Washington, DC.

Darbrowska, B., Harmata, W. and Lenkiewiez, Z. (1981) Color perception in cows. *Process* 6, 1–10.

Deng, F.M. (1972) *The Dinka of Sudan*. Holt, Rinehart, and Winston, New York.

Detering, H. (2010) Low stress handling pays. *The Cattleman* (Texas and Southwestern Cattle Raisers Association, Fort Worth, Texas), February, 57–70.

Diack, A. (1974) Design of round cattle yards. *New Zealand Farmer*, 26 September, 25–27.

Douglas, A.G., Darre, M.D. and Kinsman, D.M. (1984) Sight restriction as a means of reducing stress during slaughter. In: *Proceedings of the 30th European Meeting of Meat Researcher Workers*. Bristol, UK, pp. 10–11.

Drudi, D. (2000) Are animals occupational hazards? Compensation and working conditions. U.S. Department of Labor, Washington, D.C., pp. 15–22.

Dumont, B., Boissey, A. and Archard, C. (2005) Consistency of order in spontaneous group movements allows the measurement of leadership in a group of grazing heifers. *Applied Animal Science* 95, 55–66.

Effectivestockmanship.com (n.d.) Contains videos on cattle handling (accessed 2 November 2018).

Ercolano, H. (2018) Presentation at Animal Care and Handling Conference, North American Meat Institute, 18–19 October, Kansas City, Kansas.

Evans-Pritchard, E.E. (1940) *The Nuer*. Clarendon Press, Oxford, UK.

Ewbank, R. (1968) The behavior of animals in restraint. In: Fox, M.W. (ed.) *Abnormal in Animals*. Saunders, Philadelphia, Pennsylvania, pp. 159–178.

Ewbank, R., Parker, M.J. and Mason, C.W. (1992) Reactions of cattle to head restraint at stunning: a practical dilemma. *Animal Welfare* 1, 55–63.

Fairbanks, W.C., Grandin, T., Addis, D.G. and Loomis, E.C. (1980) *Dip Vat Design and Management*. Leaflet 21190. Division of Agricultural Sciences, University of California, Riverside, California.

Fordyce, C. (1987) Weaner training. *Queensland Agricultural Journal* 113, 323–324.

Fowler, M.E. (1995) *Restraint and Handling of Wild and Domestic Animals*, 2nd edition. Iowa State University Press, Ames, Iowa.

Gilbert, B.J. and Arave, C.W. (1986) Ability of cattle to distinguish among different wavelengths of light. *Journal of Dairy Science* 69, 825–832.

Goonewardene, L.A., Price, M.A., Okine, E. and Berg, R.T. (1999) Behavioural responses to handling and restraint on dehorned and polled cattle. *Applied Animal Science* 643, 159–167.

Grandin, T. (1975) Survey of behavioural and physical events which occur in restraining chutes for cattle. Master's thesis, Arizona State University, Tempe, Arizona.

Grandin, T. (1978) Dipping is easy with the right vat design. *Beef Magazine*, September.

Grandin, T. (1980a) Observations of cattle applied to the design of cattle handling facilities. *Applied Animal Ethology* 6, 19–31.

Grandin, T. (1980b) Good cattle restraining equipment is essential. *Veterinary Medicine and Small Animal Clinician* 75, 1291–1296.

Grandin, T. (1980c) Safe design and management of cat tle dipping vats. American Society of Agricultural Engineers, Paper No. 805518, St. Joseph, Michigan.

Grandin, T. (1982) Pig students applied to slaughter plant design. *Applied Animal Ethology* 9, 141–151.

Grandin, T. (1983a) Design of ranch corrals and squeeze chutes for cattle. In: *Great Plains Beef Cattle Handbook*. Cooperative Extension Service, Oklahoma State University, Stillwater, Oklahoma.

Grandin, T. (1983b) Handling and processing feedlot cattle. In: Thompson G.B. and O'Mary, C.C. (eds) *The Feedlot*. Lea and Febiger, Philadelphia, Pennsylvania, pp. 213–236.

Grandin, T. (1987) Animal handling. In: Price, E.O. (ed.) *The Veterinary Clinics of North America, Food Animal Practice* 3(2), 323–338.

Grandin, T. (1989) Behavioural principles of livestock handling. *Professional Animal Scientist*, 5(2), 1–11.

Grandin, T. (1990a) Design of loading facilities and holding pens. *Applied Animal Science* 28, 187–201.

Grandin, T. (1990b) Chuting to win. *Beef* 27(4), 54–57.

Grandin, T. (1992) Observations of cattle restraint device for stunning and slaughtering. *Animal Welfare* 1, 85–91.

Grandin, T. (1993) Behavioral principles of cattle handling under extensive conditions. In: Grandin, T. (ed.) *Livestock Handling and Transport*. CAB International, Wallingford, UK.

Grandin, T. (1997) The design and construction of facilities for handling cattle. *Livestock Production Science* 49, 113–119.

Grandin, T. (1998) Handling methods and facilities to reduce stress on cattle. *Food Animal Practice* 14, 325–341.

Grandin, T. (2000) Behavioural principles of handling cattle and other grazing animals under extensive conditions. In: Grandin, T. (ed.) *Livestock Handling and Transport*, 2nd edition. CAB International, Wallingford, UK, pp. 63–86.

Grandin, T. (2001) Cattle vocalizations are associated with handling and equipment problems at beef slaughter plants. *Applied Animal Science* 71, 191–201.

Grandin, T. (2004) Principles for the design of handling facilities and transport systems. In: Benson, G.J. and Rollin, B.E. (eds) *The Wellbeing of Farm Animals*. Blackwell Publishing, Ames, Iowa.

Grandin, T. (2008) Engineering and design of holding yards, loading ramps and handling facilities for land and sea transport of livestock. *Veterinaria Italiana* 44, 235–246.

Grandin, T. (2015) *Improving Animal Welfare: A Practical Approach*. CABI International, Wallingford, UK.

Grandin, T. (2017) *Temple Grandin's Guide to Working with Farm Animals*. Storey Publishing, North Adams, Massachusetts.

Grandin, T. and Deesing, M. (2008) Humane Livestock Handling. Storey Publishing, North Adams, Massachusetts. Hale, R.H., Friend, T.H. and Macaulay, A.S. (1987) Effect of method of restraint of cattle on heart rate, cortisol, and thyroid hormones. *Journal of Animal Science* (Suppl. 1) abstract.

Harger, C.M. (1928) *Frontier Days*. Macrae Smith, Philadelphia, Pennsylvania.

Hasker, P.J.S. and Hirst, D. (1987) Can cattle be mustered by audio conditioning? *Queensland Agriculture Journal* 113, 325–326.

Hedigar, H. (1968) *The Psychology and Behavior of Animals in Zoos and Circuses*. Dover Publications, New York.

Heffner, H.E. and Heffner, R.S. (2018) The evolution of mammalian hearing. *AIP Conference Proceedings*. DOI: 10.1063/1.5038516

Heffner, R.S. and Heffner, H.E. (1983) Hearing in large mammals: horse (*Equus caballus*) and cattle (*Bos taurus*). *Behavioural Neuroscience* 97(2), 209–309.

Heffner, R.S. and Heffner, H.E. (1992) Hearing in large mammals: sound localization acuity in cattle (*Bos taurus*) and goats (*Capra hircus*) *Journal of Comparative Psychology* 106, 107–113.

Heffner, R.S., Koay, G. and Heffner, H.E. (2014) Hearing in alpacas (Vicugna pacos): audiogram localization acuity and use binaural locus cues. *Journal Acoustical Society of America* 13.

Heldenbrand, L.E. (1955) Headgate for cattle chute. US Patent 2.714,872. US Patent and Trademark Office, Washington, DC.

Hemsworth, P.H., Rice, M., Karlen, M.G., Calleja, L., Barnett, J.L., Nash, J. and Coleman, G.J. (2011) Human–animal interactions at abattoirs: relationships between handling and animal stress in sheep and cattle. *Applied Animal Science* 135, 24–33.

Herman, J. (1998) Patience and silence are the major virtues in successful stock herders. *Stockman Grass Farmer* (October) 55(10).

Hewes, F.W. (1975) Stock dipping apparatus. US Patent No. 3,916,839. US Patent Office, Washington, DC. Holmes, R.J. (1991)

Cattle. In: Anderson, R.S. and Edney, A.T.B. (eds) *Practical Animal Handling*. Pergamon Press, Oxford, UK, pp. 15–38.

Hough, E. (1958) The roundup. In: Neider, C. (ed.) *The Great West*. Coward McCann, New York.

Houpt, K. and Wolski, T.R. (1982) *Domestic Animal Behavior for Veterinarians and Animal Scientists*. Iowa State University Press, Ames, Iowa.

Hutson, G.D. (1982) Flight distance in Merino sheep. *Animal Production* 35, 231–235.

Jacobs, G.H., Deegan, J.F. and Neitz, J. (1998) Photopigment basis for dichromatic colour vision in cows, goats, and sheep. *Visual Neuroscience* 15, 581–584.

Johns, J., Masneuf, S., Patt, A. and Hillmann, E. (2017) Regular exposure to cowbells affects the behavioral reactivity to a noise stimulus in dairy cows. *Frontiers in Veterinary Science*. DOI: 10.3389/fvets.201700153

Jones, R.B., Satterlee, D.G. and Cadd, G.G. (1998) Struggling responses of broiler chickens shackled in groups on a moving line: effects of light intensity hoods and curtains. *Applied Animal Science* 38, 341–352.

Julian, S. (1978) Gentle on the range. *Western Livestock Journal*, Denver, Colorado, 24–30 November.

Kearnan, A.F., McEdwin, T. and Reid, T.J. (1982) Dip vat construction and management. *Queensland Journal of Agriculture* 108, 25–47.

Kilgour, R. (1971) Animal handling in works: pertinent studies. In: *Proceedings of the 13th Meat Industry Conference*, Hamilton, New Zealand, 14–15 July, pp. 9–12.

Kilgour, R. and Dalton, C. (1984) Livestock Behavior: A Practical Guide. Granada Publishing, St Albans, UK. Kosako, T., Fukasawa, M., Kohari, D., Oikawa, K. and Tsukada, H. (2008) The effect of approach direction and pace on flight distance of beef cows. *Animal Science Journal* 79, 722–726.

Kratky, F. (1976) Animal restraint. US Patent 3,960,113. US Patent Office, Washington, DC.

Lanier, J.L. and Grandin, T. (2002) The relationship between *Bos taurus* feedlot cattle temperament and cannon bone measurements. *Proceedings*, Western Section, American Society of Animal Science, Fort Collins, Colorado, June 19–21.

Lanier, J.L. and Smith, B. (2006) Slow is fast: low-stress livestock handling. *The Stockman Grass Farmer*, February, 30–31.

Laporte, I., Muhly, T.B., Pitt, J.A., Alexander, M. and Musiani, M. (2010) Effects of wolves on elk and cattle behavior: implications for livestock production and wolf conservation. *PLOS ONE* 5(8): E11954. DOI: 10.1371/journal.pone.0011954

Lay, D.C., Friend, T.H., Grisson, K.K., Hale, R.L. and Bowers, C.C. (1992) Novel breeding box has variable effect on heart rate and cortisol response of cattle. *Journal of Animal Science* 70, 1–10.

LeDoux (1996) *The Emotional Brain*. Simon and Schuster, New York.

Lemmon, W.B. and Patterson, G.H. (1964) Depth perception in sheep: effects of interrupting the mother–neonate bond. *Science* 145, 835–836.

Linford, L. (1977) Stampede. *New Mexico Stockman* 42(11), 48–49.

Lott, D.F. and Hart, B. (1977) Aggressive domination of cattle by Fulani herdsmen and its relation to aggression in Fulani culture and personality. *Ethos* 5, 172–186.

Lott, D.F. and Hart, B. (1979) Applied ethology in a nomadic cattle culture. *Applied Animal Ethology* 5, 309–319.

Lott, D.F. and Hart, B. (1982) The Fulani and their cattle applied behavioural technology in a nomadic cattle culture and its psychological consequences. *National Geographic Research Reports* 14, 425–430.

Lucas, M., Day, L. and Fritsch, L. (2012) Serious injuries to Australian veterinarians working with cattle. *Australian Veterinary Journal* 91, 57–60.

Lynch, J.J and Alexander, G. (1973) *The Pastoral Industries of Australia*. University Press, Sydney, Australia, pp. 371–400.

Lyvers, C.M. (2013) Evaluation of handling equipment sound pressure levels as stressors in beef cattle. Theses and dissertations – biosystems and agricultural engineering, University of Kentucky, paper 13. https://uknowledge.uky.edu/

Marshall, M., Milberg, E.E. and Shultz, E.W. (1963) Apparatus for holding cattle in position for humane slaughtering. US Patent 3,092,871. US Patent Office, Washington, DC.

Mayes, H.F. (1978) Design criteria for livestock loading chutes. Paper No. 786014, American Society of Agricultural Engineers, St. Joseph, Michigan.

McFarlane, I. (1976) Rationale in the design of housing and handling facilities. In: Ensminger, M.E. (ed.) *Beef Cattle Science Handbook* 13. Agri-services Foundation, Cloves, California.

Midwest Plan Service (1987) *Beef Housing and Equipment Handbook*, 4th edition. Midwest Plan Service, Iowa State University, Ames, Iowa.

Miller, P.E., and Murphy, C.J. (1995) Vision in dogs. *Journal of the American Veterinary Medical Association* 12, 1623–1634.

Mitchell, K.D., Stookey, J.M., Laturnas, D.K., Watts, J.M., Haley, D.B. and Huyde, T. (2004) The effects of blind folding on and heart rate in beef cattle during restraint. *Applied Animal Science* 85, 233–240.

Mollhagen, J.D. (1994) Portable squeeze chute apparatus and method. US Patent 5,331,923. US Patent Office, Washington, DC.

Mollhagen, J.D. (2010) Apparatus for securing livestock. US Patent 7,770,542 B2. US Patent Office, Washington, DC.

Muller, R., Schwartzkoph, K.S., Shah, M.A. and von Keyserlingk, M.A.G. (2007) Effect of neck injection and handler visibility on behavioural reactivity of beef steers. *Journal of Animal Science* 86, 1215–1222.

Nation, A. (1998) Western grazers practice management intensive grazing without fences. *Stockman Grass Farmer*, September, 1–12.

Nation, A. (1999) Herding is not as easy as it looks. *Stockman Grass Farmer*, September, 1–13.

Nation, A. (2003) Allan's observations. *Stockman Grass Farmer*, November, 8–11.

Oklahoma State University (1973) Expansible Corral with Pie Shaped Pens. Plan No. Ex. Ok. 724–75. Cooperative Extension Service, Stillwater, Oklahoma.

OIE (2017) Transport of animals by land. Chapter 7.3 in *Terrestrial Animal Health Code*. World Organization of Animal Health, Paris.

Paine, M., Teter, N. and Guyer, P. (1976) Feedlot layout. In: *Great Plains Beef Cattle Handbook*. Cooperative Extension Service, Oklahoma State University, Stillwater, Oklahoma, pp. 5201.1–5201.6.

Paine, R. (1994) *Herds of the Tundra: A Portrait of Saami Reindeer Pastoralism*. Smithsonian Institute Press, Washington, DC.

Pajor, E.A., Rushen, J. and de Passillé, A.M.B. (2003) Dairy cattle's choice of handling treatments in a Y-maze. *Applied Animal Science* 80, 93–107.

Parsons, R.A. and Helphinstine, W.M. (1969) Rambo, A.I. breeding chute for beef cattle. One Sheet Answers, University of California Agricultural Extension Service, Davis, California.

Pearson, L.B. (1965) Automatic livestock headgate. US Patent No. 3, 221,707. US Patent Office, Washington, DC.

Pratt, W.C. (2007) Cattle management method and system. US Patent 7,464,667. US Patent Office, Washington, DC.

Pratt, W.C. (2009) Cattle management system. US Patent 8261694B2. US Patent Office, Washington, DC.

Price, E.O. and Wallach, S.J.R. (1990) Physical isolation of hand-reared Hereford bulls increases their aggressiveness towards humans. *Applied Animal Science* 27, 263–267.

Priefert, W.D. (1990) Livestock handling device. US Patent 4,930,450. US Patent and Trademark Office, Washington, DC.

Prince, J.H. (1977) The eye and vision. In: Swenson, M.J. (ed.) *Duke's Physiology of Domestic Animals*. Cornell University Press, New York, pp. 696–712.

Provenza, F.D. (2003) *Foraging Behavior Managing to Survive a World of Change*. USDA, Utah State University, Logan, Utah.

Reck, E. and Reck, J.P. (1903) Cattle stanchion. US Patent 733,874. US Patent Office, Washington, DC.

Rehkamper, G. and Gorlach, A. (1998) Visual identification of small sizes by adult dairy bulls. *Journal of Dairy Science* 81, 1574–1580.

Reinken, G. (1988) General and economic aspects of deer farming. In: Reid, H.W. (ed.) *The Management and Health of Farmed Deer*. Springer, London, pp. 53–59.

Rice, B. (n.d.) Settling cattle. In: Turner, C. (host) National Cutting Horse Association, Fort Worth, Texas.

Rider, A., Butchbaker, A.F. and Harp, S. (1974) Beef working, sorting and loading facilities. Paper No. 744523. American Society of Agricultural Engineers, St. Joseph, Michigan.

Roche, B.W. (1988) Coacher mustering. *Queensland Agricultural Journal* 114, 215–216.

Rogan, M.T. and LeDoux, J.E. (1996) Emotional systems, cells, synaptic plasticity. *Cell* 85, 469–475.

Ruechel, J. (2006) *Grassfed Cattle*. Storey Publishing, North Adams, Massachusetts.

Saslow, C.A. (1999) Factors affecting stimulus visibility in horses. *Applied Animal Science* 61, 273–284.

Savory, A. (1978) Ranch and range management using short duration grazing. In: *Beef Cattle Science Handbook*. Agri-services Foundation, Clovis, California, pp. 376–379.

Schwabe, C.W. (1985) A unique surgical operation on the horns of African bulls in ancient and modern times. *Agricultural History* 58, 138–156.

Schwabe, C.W. and Gordon, A.H. (1988) The Egyptian W3ssceptor and its modern analogue: uses in animal husbandry, agriculture and surveying. *Agricultural History* 62, 61–89.

Seligman, C.G. and Seligman, B.Z. (1932) *Pagan Tribes of the Nilotic Sudan*. Routledge and Kegan Paul, London.

Shinozaki, A., Hosaka, Y., Imayawa, T. and Uehara, M. (2010) Topography of ganglion cells and photoreceptors in the sheep retina. *Journal of Comparative Neurology* 518, 2305–2315.

Smith, B. (1998) *Moving 'Em: A Guide to Low Stress Animal Handling*. Graziers Hui, Kamuela, Hawaii.

Smith, T. (2018) Follow the leader. *Angus Beef Bulletin*, November, 60–68.

Smith, B., Leung, P and Love, G. (1986) *Intensive Grazing Management Forage: Animals, Men, Profits, Graziers*. Hui Kamuela, Hawaii.

Stookey, M., and Watts, J.M. (2014) Low stress restraint, handling, and sorting of cattle. In: Grandin, T. (ed.) *Livestock Handling and Transport*, 4th edition. CAB International, Wallingford, UK, pp. 65–76.

Stockmanship Journal 2012–2016, volumes 1–5. (Accessed 2 November 2018)

Swan, R. (1975) About AI facilities. *New Mexico Stockman* 11, February, 24–25.

Sweeten, J.M. (1980) Static screening of feedlot dipping vat solution transactions. American Society of Agricultural Engineers 23(2), 403–408.

Sweeten, J.M., Winslow, R.B. and Cochran, J.C. (1982) Solids removal from cattle dip pesticide solution and sedimentation tanks. Paper No. 824083. American Society of Agricultural Engineers, St. Joseph, Michigan.

Talling, J.C., Waran, N.K. and Wathes, C.M. (1996) Behavioural and physiological responses of pigs to sound. *Applied Animal Behaviour Science* 48, 187–202.

Talling, J.C., Waran, N.K., Wathers, C.M. and Lines, J.A. (1998) Sound avoidance by domestic pigs depends upon characteristics of the signal. *Applied Animal Behaviour Science* 58(34), 255–266.

Texas Agricultural Extension Service (1979) Portable cattle dipping vat, drawing No. 600. Texas A&M University, College Station, Texas.

Thines, G. and Soffie, M. (1977) Preliminary experiments on color vision in cattle. *British Veterinary Journal* 133, 97–98.

Thompson, A.C. (1931) Cattle holding and dehorning gate. US Patent 1,799,073. US Patent Office, Washington, DC.

Thompson, A.C. and Gill, C. (1949) Restraining chute for animals. US Patent 477,888. US Patent Office, Washington, DC.

Thompson, R.J. (1987) Radical new yard proven popular. *Queensland Agricultural Journal* 113, 347–348.

Tillman, A. (1981) *Speechless Brothers: The History and Care of Llamas*. Early Winters Press, Seattle, Washington.

Tillman Brown, D. (2018) Using drones as the herd dog of the future. *The Stockman Grass Farmer*, June 12–13.

Uetake, K. and Kudo, Y. (1994) Visual dominance over hearing in feed acquisition procedure of cattle. *Applied Animal Behaviour Science* 42, 1–9.

Van Putten, C. and Elshof, W.J. (1978) Observation of the effects of transport on the wellbeing and lean quality of slaughter pigs. *Animal Regulation Studies* 1, 247–271.

Vowles, W.J. and Hollier, T.J. (1982) The influence of yard design on the movement of animals. *Proceedings of the Australian Society of Animal Production* 14, 597.

Vowles, W.J., Eldridge, C.A. and Hollier, T.J. (1984) The behavior and movement of cattle through single file handling races. *Proceedings of the Australian Society of Animal Production* 15, 767.

Ward, F. (1958) *Cowboy at Work*. Hastings Press, New York.

Waynert, D.E., Stookey, J.M., Schwartzkopf-Gerwein, J.M., Watts, C.S. and Waltz, C.S. (1999) Response of beef cattle to noise during handling. *Applied Animal Behaviour Science* 62, 27–42.

Webber, R.J. (1987) Molasses as a lure for spear trap mustering. *Queensland Agricultural Journal* 113, 336–337.

Western Livestock Journal (1973) Put on foils escape minded cattle. *Western Livestock Journal*, Denver, Colorado, July, 65–66.

White, J.B. (1961) Letter to the editor. *Veterinary Record* 73, 935.

Whitehurst, B. (2017) Settle, sort: the key to open field sorting. *Progressive Cattlemen*, 24 May. Available at: www.progressivecattle.com (accessed 27 March 2018).

Williams, E. (1998) How to place livestock and have them stay where you want them. *Stockman Grass Farmer*, September, 6–7.

Wyman, W.D. (1946) *The Wild Horse of the West*. Caxton Printers, Caldwell, Idaho.

Videos of Handling and Livestock

Due to licensing restrictions on YouTube, a number of these videos are not currently available outside the US. To access them from other countries, please search for the video title followed by the word 'video' using your preferred search engine.

National BQA, *Cattle Handling Tips: Facilities* – it shows three different types of handling facilities. They are: (i) small farm facility; (ii) round crowd tub; and (iii) Bud Box. Temple Grandin and Curt Pate demonstrate the round crowd tub. https://www.youtube.com/watch? v=qLW1vh39m9k

Temple Grandin, *Cattle Handling in Chutes and Races* – shows how to move cattle forward in a single-file race (chute) by walking in the opposite direction of desired movement. To motivate the animal to move forward, the handler needs to walk briskly past it, to prevent it from backing up. https://www.youtube.com/watch? v=RV10Cyfx18

Temple Grandin, *Yard Demo at Beef Works* – shows quiet movement of cattle through a curved single-file race that has an open inner fence and a solid outer fence. It demonstrates how to work the flight zone. https://www.youtube.com/watch?V= cFa8k2rW2Ro

Farm Quip – Argentina has many new videos of cattle being handled in yards with both curved races and round crowd tubs, and portable systems with a straight single file race. Three notable videos were uploaded in 2017. https://www.youtube.com/watch?v=e8BE3MF78c

Instalacionen San Jorge Santa Fe Completo – shows aerial view of a large-curved corral. https://www.youtube. com/watch?v=cFA8kznW2Ro

Corral Desmontable Farmquip – shows a portable corral with a simple straight race. https://www.youtube.com/ watch?v=qRlgAktr6to

Farmquip Uruguay, *Installation ganadera* – shows a smaller curved system. https://www.youtube.com/ watch?v=f8_LFJF24L4

Temple Grandin, *Proper Operation of Cattle Squeeze Chutes* – shows methods to improve the operation of squeeze chute (crushes). https://www.youtube.com/ watch?v=5cUk9RH61Q

TEN

The Reluctance of Cattle to Change a Learned Choice May Confound Preference Tests

T. Grandin, Odde, K.G., Schutz, D.N., and Behrns, L.M. (1994) The Reluctance of Cattle to Change a Learned Choice May Confound Preference Tests, *Applied Animal Behavior Science*, Vol. 39, pp. 21-28.

This study was my first research collaboration with other faculty at Colorado State University. Ken Odde, a professor, and David Schutz, the manager of our experiment station, collaborated with me on this study. It was performed in a new handling facility I designed. The facility could be easily changed to a choice testing facility by removing one gate.

ABSTRACT

Choice testing utilizing a Y-maze has been successfully used to test animal preferences. In this experiment, 12 female Angus X Hereford X Simmental X Charolais heifers were given a choice of walking through a squeeze chute (crush) or being restrained in a squeeze chute. The objective of the study was to determine if previously learned choices in a Y-maze would confound future choices. A start box led to two races in a Y configuration. There was a hydraulic squeeze chute at the end of each race. Animals that chose the right sidc wcrc allowed to walk through the squeeze chute and animals that chose the left side were restrained in the squeeze chute for 30 s. During eight choice trials, the heifers had a definite preference for the 'walk' side. There were 64 walk choices and 32 'restraint' choices. For six additional trials, the restraint and walk sides were switched. Walk choices dropped to 16 and restraint choices rose to 56. The resistance to switching effect was significant (P<0.01). Significantly more heifers vacillated (looked back and forth) at the decision point after the sides were switched (P< 0.01). The switch had been perceived by the animals. There is a tendency for cattle to resist changing a choice once they are accustomed to a treatment being associated with a specific side.

INTRODUCTION

Choice tests and preference tests are important for answering many animal welfare questions. and T-maze choice tests have been used to study preferences in farm animals (Hughes, 1976; Hitchcock and Hutson, 1979; Hutson, 1981; Grandin et al., 1986; Pollard et al., 1993). They are especially useful for determining the relative aversiveness of different husbandry or handling procedures. Since preference testing may be used to make legislative decisions concerning animal welfare, it is essential that preference test results are not confounded by variables such as previous learning. Previous experiences can affect choices. For example, grazing preferences in sheep are affected by previous experience (Arnold and Mailer, 1977). Another example is that rearing environment can affect flooring preferences in caged hens (Hughes, 1976). The purpose of this experiment was to determine if previous experiences in a Y-maze testing facility affects future choices.

ANIMALS, MATERIALS AND METHODS

Twelve 365kg heifers that were crossbreds of Angus X Hereford X Charolais X Simmental were used. The heifers were housed in outdoor feedlot pens adjacent to the choice testing facility. The test facility was constructed from 1.52m high solid steel fences (Fig. 1). It consisted of a crowding pen, single file race, start box and two races in a Y configuration which led to two identical hydraulic squeeze chutes (crushes; Bowman Livestock Equipment, Council Grove, KS, USA; Fig. 2). The squeeze chutes had hydraulically activated squeeze sides and head stanchions. The start box was used to admit each animal one at a time into the Y decision point. Solid sliding gates on each end of the start box prevented animals that were waiting in line from observing the choices made by an animal leaving the start box. All animals were allowed to voluntarily leave the start box and they were never touched by a handler until after they had made a decision and had moved into one of the races. This prevented the decision-making process from being confounded by a handler driving an animal. Great care was taken during the entire experiment to prevent the activities of people from affecting the animals' choices. No electric prods were used and heifers that voluntarily entered one of the squeeze chutes were not touched in accord with good industry practice. If an animal balked at the squeeze chute entrance, it was tapped on the rump. If tapping failed to move an animal, its tail was twisted to induce it to enter the squeeze chute. To control for the effect of animals seeing people, a person was stationed beside each squeeze chute with his hands on the controls. These people stood completely still until each animal entered the squeeze chute.

The testing procedure was similar to the procedure in Grandin et al. (1986) It consisted of a series of training and choice trials. After all 12 heifers had passed through the facility, they were immediately returned to the crowd pen for the next trial. Both squeeze

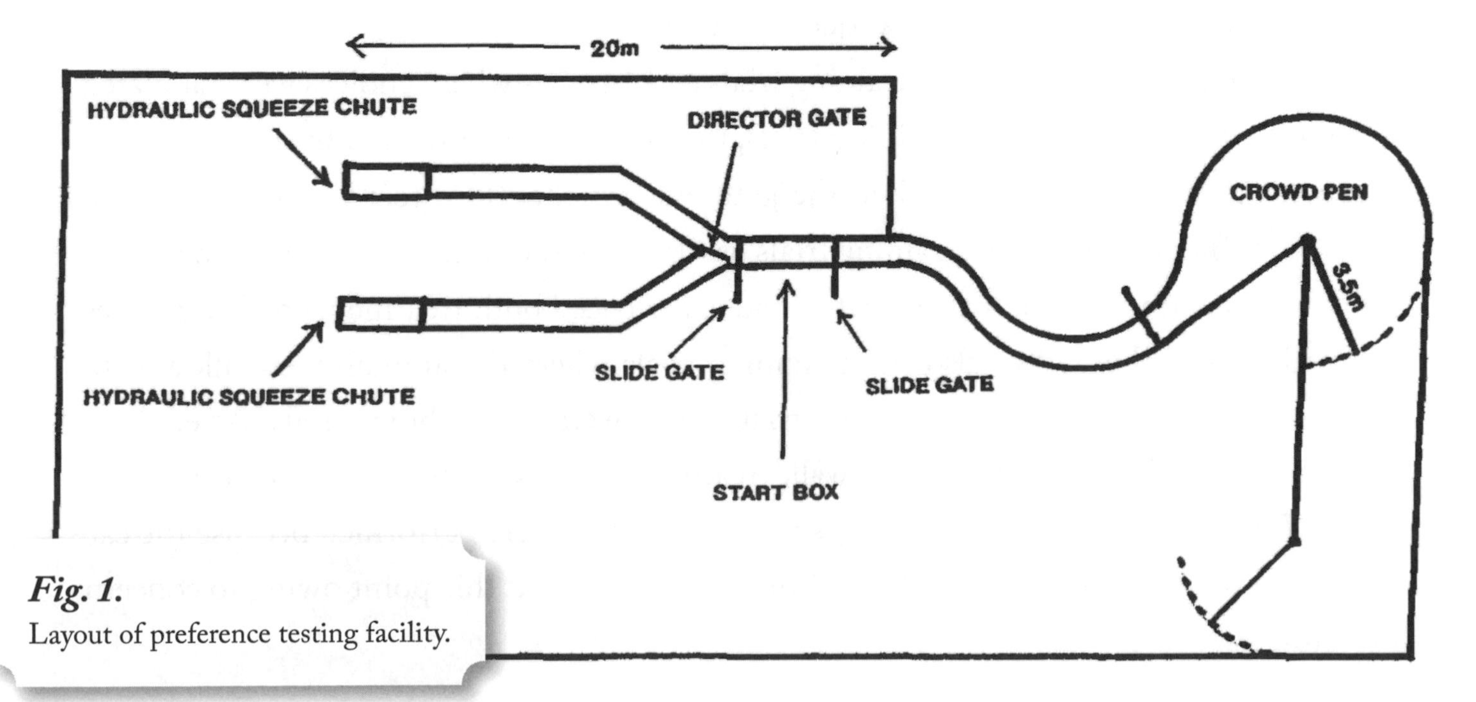

Fig. 1.
Layout of preference testing facility.

Fig. 2.
Hydraulic squeeze chutes used to restrain cattle in the preference test.

chute operators were present for all training and choice trials, and the person that drove balky animals into the squeeze chute stood in the exact center between the two races. This person stood absolutely still and did not move from this position unless a heifer balked and stopped near the entrance of a squeeze chute.

There were a total of seven training trials and 14 trials where choices were tabulated. For the first two and last two training trials, the animals were directed by a gate to walk through one side of the Y and then the gate was switched to direct them through the other side. The purpose of the training trials was to teach the animals that the Y-maze had two choices and to ensure that the cattle had experienced both treatments before choices were tabulated. There were also three training trials where the animals were allowed to choose a race. After the seven training trials, there were eight choice trials. After these eight choice trials, the 'restraint' and 'walk' treatments were switched to the opposite sides of the Y maze for six more trials. Only six switched trials were performed because the cattle started to appear stressed. The experiment was stopped at this point owing to concern for animal welfare.

Below is a listing of the sequence and procedures for training and choice trials. Training trial 1. A director gate was installed at the decision-point in the Y maze (Fig. 1). Each heifer was individually released from the start box and directed by the director gate to move through the righthand walk side of the facility. No restraint was applied and each heifer was allowed to walk through the squeeze chute.

Training trial 2. The director gate was switched and each heifer was forced to move through the lefthand restraint side of the facility. She was allowed to walk through the squeeze chute and no restraint was applied.

Training choice trials 3, 4, 5. The director gate was removed, and each heifer was allowed to make a choice, and walk through the race and squeeze chute of her choice. No restraint was applied, and the animals were allowed to walk through the two squeeze chutes.

Training treatment directed trial 6. The director gate was reinstalled, and every other heifer was forced down the restraint side of the Y and restraint was applied gently and carefully. The other six heifers were directed through the squeeze chute on the opposite side of the Y and were allowed to walk through the chute.

Training treatment trial 7. The director gate was kept in place and heifers which had experienced restraint were directed down the walk side and heifers that had experienced the walk treatment were directed down the restraint side. Restraint was applied to heifers on the restraint side.

Choice trials 1 through 8. The director gate was removed, and each heifer was allowed to make eight choices. Each choice was tabulated. There was a 1.5h break after Choice Pass 4.

Switched choice trials 9 through 14. The restraint and walk sides were switched. Heifers which had successfully avoided restraint by choosing the walk side now experienced restraint in the walk side. Each choice was tabulated.

Choice trials 2 weeks later. Each animal was admitted one at a time to the Y decision-point and its choice was recorded. The restraint side was the same side as in switched trials nine through 14. Only one choice trial was possible at this time because the cattle were entering a physiological experiment and there were concerns that more than one trial would confound this experiment.

RESULTS

The cattle expressed a preference for the walk side of the Y race before the restraint treatment side was switched (Table 1). During the first eight choice trials, there were 32 squeeze choices and 64 walk choices. After the restraint side was switched, there were 56 restraint choices and 16 walk choices. The resistance to switching effect was significant. A paired T-test with arcsine transformation of the data was used to determine whether heifers chose the restraint side more after the switch ($T=3.76<0.01$). Two weeks later, the cattle were returned to the facility and ten out of 12 animals chose the restraint side. Seven heifers were extremely resistant to switching. They neither chose the walk side after the restraint side was switched, nor 2 weeks later. All seven chose the restraint side (Table 1). Only two heifers completely avoided the restraint side for all six switched choice trials. One of these heifers (1184) was accidentally caught around the shoulders by the head stanchion during the fifth-choice trial prior to switching. She also avoided the restraint side 2 weeks later. Animal number 1200 completely avoided the restraint side for all 14 choice trials, but she chose restraint 2 weeks later.

The number of times each animal looked back and forth before it made a choice (vacillations) was significantly higher after restraint was switched to the other side. Prior to switching sides, there was a total of 15 vacillations and after switching sides, vacillations rose to 72. Heifer 1002 had a very high number of vacillations (46) and may possibly be an outlier that would confound the results. When her data were removed from the analysis, vacillations were still higher after switching (29 vs 9). A paired T-test indicated that vacillations were significantly greater after switching ($T = 3.40<0.01$). Six animals never vacillated during the first eight choice trials and after the sides were switched, all animals except one vacillated ($x^2=5.03<0.05$). Eight out of 12 animals never balked and entered the squeeze chute voluntarily. Heifer 1042 balked during choice trials 1 through 8, but she entered voluntarily for the switched choice trials 9 through 14. On three trials, tail

TABLE 1: Choices before and after Squeeze Side Was Switched

Animal Number	Pre-choice training passes (walk through)		Chocies passes 1–8		Vacillations	Chocies passes with choices switched		Vacillations	Choice 2 weeks later[1]	Resistant to switching[2]
	Squeeze	Walk	Squeeze	Walk		Squeeze	Walk			
1002	0	3	3	5	6	5	1	46	S	
1038	0	3	0	8	1	6	0	4	S	R
1042	2	1	7	1	0	4	2	2	W	
1045	2	1	4	4	0	6	0	2	S	R
1150	2	1	2	6	3	6	0	2	S	R
1161	1	2	6	2	0	6	0	4	S	R
1184	3	0	6	2	0	0	6	4	W	
1189	1	2	2	6	1	6	0	1	S	R
1190	3	0	1	7	2	6	0	3	S	R
1192	0	3	0	8	0	6	0	0	S	R
1197	0	3	1	7	0	5	1	3	S	
1200	0	3	0	8	2	0	6	4	S	
Totals	14	22	32	64	15	56	16	75		

1. Choice 2 weeks later: S, chose race leading to squeeze; W, chose race leading to walk. Squeeze was tabulated for choices leading to the squeeze after the sides were switched.
2. Cattle were labelled resistant to switching if they chose the squeeze side for all six choices after the choice sides were switched and they chose the squeeze side 2 weeks later.

twisting was required to induce her to enter the squeeze chute. Animals 1150 and 1197 balked during two trials. Animal 1197 had her tail twisted on choice trial number 13 as she entered the walk side. On the 14th trial, she chose the restraint side after entering the walk side and then backing out of it.

DISCUSSION

The results of this experiment indicate that the tendency of cattle to resist changing a choice once they are accustomed to a treatment being associated with a specific side, could severely confound choice tests. For accurate choice testing results, a new group of naive cattle should be used when the sides of a choice test are switched. Even though most animals persisted in entering the restraint side after the sides were changed, the increase in the number of vacillations indicates that the animals had perceived the switch. Vacillation (vicarious trial and error) occurs when an animal is unsure or learning to discriminate (Muenzinger, 1938; Goss and Wischer, 1956). Even though the animals perceived that conditions had changed, most chose the previously learned safe route.

Stewart et al. (1992) found that cattle could quickly learn a maze. Performance deteriorated when they had to learn a new maze (C.W. Arave, personal communication, 1992). This observation is a further illustration of the bovine's resistance to changing a learned behavior. C.W. Arave (personal communication, 1992) also observed that some heifers persisted in following the pattern of a maze after the partitions were removed. Bailey et al. (1989) reported a similar resistance to change. Steers quickly learned to choose the arm in a five-arm parallel maze which contained the most grain. When the location of the largest grain reward was changed it took somewhat longer for steers to learn the new location.

The confounding effects of resistance to switching in choice tests is likely to be greatest when the choices are only mildly aversive. All cattle in the experiment were handled gently and care was taken to avoid banging them on the head with the head stanchion. The restraint and handling procedure was only mildly aversive because most animals moved voluntarily through the system at a slow walk. A previous study by Grandin et al. (1986) in a similar Y-maze choice test indicated that sheep quickly learned to avoid highly aversive electro-immobilization.

They were given a choice between immobilization or a tilt squeeze table. During the pre-choice nontreatment training trials, the sheep preferred walking through the righthand immobilizer race. Possibly, the sheep initially preferred the immobilizer race because it was wider and easier to walk through than the tilt table. When the treatments were applied, most sheep immediately switched sides to avoid the immobilizer.

A severely aversive treatment may possibly overcome the reluctance of the cattle to switch sides. Heifer 1184 had a more severe aversive experience in the squeeze chute than all of the other cattle because she was accidentally caught around the shoulders. When the sides were switched, she avoided the restraint side. When she returned 2 weeks later, she still avoided the restraint side. The question of laterality in ruminants also needs to be addressed. For example, cattle prefer to lie on the left side (Uhrbrock, 1969). In both Grandin et al. (1986) and this experiment, there was an initial tendency during the training for the animals to prefer the righthand side. During the three training choice trials, five out of 12 heifers always chose the righthand side and two heifers always chose the left side. The initial choice of sides tended to persist during the choice trials. However, all heifers except two made different choices during the 14 choice trials.

There was also a tendency for calmness or behavioral agitation to affect choices. Animals 1038 and 1192 were the most resistant to switching sides in the entire experiment. These animals switched sides during neither the pre-choice training trials nor all the choice trials. Both of these animals displayed escape behavior and agitation. Animal 1038 attempted to jump out of the start box several times and 1192 was difficult to chase out of the crowd pen into the start box. She also jammed her head under the start gate. There appeared to be a tendency for all the animals to become slightly more agitated as the experiment progressed. Possibly, calmer animals were able to make more accurate choices to avoid aversive treatment. After the 1.5h lunch break, animals 1161 and 1045 appeared calmer compared with the fourth-choice trial before lunch. They avoided the squeeze chute for the next two to three trials. Calmer animals also had a tendency to vacillate more at the decision-point. The one animal that vacillated 46 times was a calm heifer with a small flight zone. Spacing each choice trial several hours apart instead of 5 min apart may help alleviate this problem. An increased interval between the choice trials could allow the animals to calm down before the next trial. In conclusion, it appears that previously learned choices may affect future choices in Y-maze s for cattle. Another area that needs to be researched is the effects of a mildly aversive treatment versus a severely aversive treatment on the tendency of a bovine to resist changing a learned choice. The effects of arousal level and excitement on a bovine's choice behavior also need to be investigated.

ACKNOWLEDGMENTS

The authors wish to thank the following people who assisted with this experiment: M.J. Daniel, A. Early, G.R. Oss, T. Harrington and J.P. Sonderman.

REFERENCES

Arnold, G.W. and Maller, R.A., 1977. Effects of nutritional experience in early adult life on the performance and dietary habits of sheep. *Appl. Anim. Ethol.*, 3: 526.

Bailey, D.W., Rittenhouse, L.R., Hart, R.H., Swift, D.M. and Richards, R.W., 1989. Association of relative food availabilities and locations by cattle. *J. Range Manage.*, 42: 480-482.

Goss, A.E. and Wischner, G.J., 1956. Vicarious trial and error and related behaviour. *Psychol. Bull.*, 53: 335.

Grandin, T., Curtis, S.E., Widowski, T.M. and Thurman, J.C., 1986. Electro-immobilization versus mechanical restraint in an avoid-avoid animal choice test for ewe. *J. Anim. Sci.*, 62: 1469-1480.

Hitchcock, D.K. and Hutson, G.D., 1979. The movement of sheep on inclines. *Aust. J. Exp. Agric. Anim. Husb.*, 19: 176-182.

Hughes, B.O., 1976. Preference decisions of domestic hens for wire or litter floors. *Appl. Anim. Ethol.*, 2: 155-165.

Hutson, G.D., 1981. Sheep movement on slatted floors. *Aust. J. Exper. Agri. Anim. Husb.*, 21: 474-479.

Muenzinger, K.F., 1938. Vicarious trial and error at the point of choice. A general survey of its relation to learning efficiency. *J. Gen. Psychol.*, 53: 75.

Pollard, J.C., Littlejohn, R.P. and Suttie, M., 1993. Responses of red deer to restraint versus no restraint in a Y maze preference test. Australian Society for the Study of Animal Behaviour, Department of Psychology, University of Canterbury, Christchurch, New Zealand, (Abstract).

Stewart, P.J., Arave, C.W. and Dias, B., 1992. The effect of observation on Holstein heifer's learning ability in a closed field maze. *J. Anim. Sci.*, 70(Suppl. 1): 155 (Abstract).

Uhrbrock, R.S., 1969. Bovine laterality. *J. Genetic Psychol.*, 115: 77-79.

ELEVEN

Euthanasia and Slaughter of Livestock

T. Grandin (1994) Euthanasia and Slaughter of Livestock,
Journal of the American Veterinary Medical Association, Vol. 204, pp. 1354-1394.

This paper is a review of scientific literature and observations I made during startups of restraint equipment I had designed for Kosher Slaughter. Kosher is religious slaughter that is conducted without first rendering the cattle unconscious. The cattle did not appear to react to the supersharp kosher knife. This observation was made under PERFECT conditions, which required a custom designed, upright restraint box that minimized pressure applied to the animal. I was the operator of the box, and one of the best shochets in the industry performed the cut. When slaughter without stunning for religious slaughter is conducted under poorly managed conditions, welfare is likely to be seriously compromised. Good design of the hydraulic or pneumatic controls of the restraint box is very important. The operator must be able to control both the position of the head holder and the speed of movement. Jerky motion and excessive pressure applied to the animal must be avoided. A poorly designed or operated restraint box will cause a high percentage of cattle to vocalize (bellow). This is an indicator of severe stress. I did extensive work in the 1980s and early 1990s on improving restraint methods for religious slaughter. The U.S. Humane Slaughter Act requires that all livestock be rendered insensible to pain before slaughter. For Religious Slaughter, Kosher (Jewish), and Halal (Muslim), there is an exemption in the regulations to protect religious freedom. Many cattle were restrained using highly stressful methods, such as suspension by one leg. In the '80s and '90s, I worked with five different plants to eliminate this cruel method of restraint.

Euthanasia is defined as a humane death that occurs without pain and distress.[1] Physical methods of euthanasia are used in slaughter plants and in clinical practice. These methods are unpleasant to watch, but humane when used properly. In some cases, the most humane euthanasia method for an injured cow or horse is by gunshot, because fear and anxiety are minimized.

During the past 20 years, the author has worked in more than 100 slaughter plants as a consultant for improving handling techniques and equipment used for slaughter in an effort to improve the humaneness of slaughter. When slaughter is performed properly, it is euthanasia; however, when it is performed improperly, animal suffering may result. There is a tremendous need for veterinarians to increase their knowledge of physical methods of euthanasia and humane slaughter. Observations by the author at slaughter plants indicate that some veterinarians lack such practical and scientific knowledge. It is the responsibility of veterinarians to enforce the Humane Methods of Slaughter Act of 1978.[2] The purpose of this report is to provide veterinarians with practical and scientific information on the use of captive bolt guns, electrical stunning, CO_2 anesthesia, and ritual slaughter.

CAPTIVE BOLT AND GUNSHOT

Gunshot and penetrating captive bolt are acceptable methods of euthanasia, according to AVMA guidelines.[1] The captive-bolt gun or firearm must be aimed at the correct location on the animal's forehead (Fig. 1). It should not be aimed between the eyes. The hollow behind the poll also should be avoided, because it is less effective than the forehead position.[3] In sheep, the shot must be aimed at the top of the head because the front of the skull is thick. A captive-bolt gun must be placed firmly against the skull and a firearm must be held 5 to 20 cm away from the skull. Penetrating captive-bolt guns that are actuated by a blank cartridge can be obtained.[a] Practical experience has shown that a 22-caliber firearm is sufficient for cattle and horses. A larger caliber should be used on large bulls, boars, or buffalo.

A captive-bolt gun kills the animal by concussive force and penetration of the bolt.[4,5] A nonpenetrating, mushroom head captive bolt only stuns the animal, thus, it cannot be used as the sole method of euthanasia.[1] A nonpenetrating captive bolt must be followed by an adjunctive method, such as exsanguination.[1] The gun must be carefully cleaned and maintained to achieve maximal hitting power. Observations by the author in slaughter plants indicate that poor gun maintenance is a major cause of poor stunning, in which more than one shot is required. In large, highspeed plants, maintenance personnel must be dedicated to servicing captive-bolt guns.[6] Many large slaughter plants use pneumatic powered captive-bolt guns. To obtain maximal hitting power, the gun must be supplied with sufficient air pressure and air volume per the manufacturer's recommendations.

When a standing animal is shot, it should instantly drop to the floor. In cattle, the neck contracts in a spasm for 5 to 10 seconds. Hogs have violent convulsions. Observations by the author indicate that this can occur even when the brain and part of the spinal cord have been destroyed. These are normal reactions. Rhythmic breathing must be absent, and the animal must not moan, bellow, or squeal.[7] All eye reflexes should be absent.[7] Gasping or gagging reflexes are permissible because they are signs of a dying brain.[7] Within 10 seconds, the neck and head should be completely relaxed. In a clinical situation and in a slaughter plant, the animal's limbs may make uncoordinated movements for several minutes.

Fig. 1.
Correct position for euthanasia of livestock by use of a captive-bolt gun or firearm

ELECTRIC STUNNING AND ELECTROCUTION

Electric stunning is not recommended for use in research facilities or on farms. Stunning and euthanasia by electrocution is acceptable only conditionally, because special skills and equipment are required.[1] Restraint equipment is required to hold the animal so that the electrodes can be placed in the correct position. Use of an electrical cord plugged into 115 V house current to kill piglets or other livestock is not acceptable. Piglets less than three weeks old should be euthanatized by appropriate administration of pharmaceutical agents or by application of blunt trauma to the forehead.[8] Captive bolt or gunshot can be used on older pigs.

Properly applied electrical stunning in a slaughter plant induces instantaneous unconsciousness and is approved under the Humane Methods of Slaughter Act of 1978. Sufficient amperage must be applied through the brain to induce a grand-mal epileptic seizure.[9,10] The animal's brain must be in the current path between the two electrodes.

Hogs in small locker plants or meat science laboratories often are stunned with head only reversible electric stunning. To prevent retum to sensibility, the animal must be exsanguinated within 30 seconds,[10] and some researchers recommend 10 to 15 seconds.[11] The author has observed that delays between reversible head-only stunning and exsanguination are common in small locker plants because the hoist is too slow. lnadequate equipment causes severe welfare problems.

In large pork and sheep slaughter plants, animals are held in a conveyor restrainer, and stunning that induces cardiac arrest (cardiac arrest stunning) is used. The interval between stunning to exsanguination is less critical. One electrode is placed on the forehead or in the hollow behind the ears, and the second electrode is placed on the back, side of the body, or forelimb. In pigs and sheep, this method simultaneously induces instantaneous insensibility and cardiac arrest.[10,12] The head electrode must not be placed on the neck, because failure to induce an epileptic seizure causes suffering.[12] Properly and poorly stunned animals look similar, because cardiac arrest masks the clinical signs of the grand-mal seizure.[12] The animal may be conscious but appears to be properly stunned.

When cardiac arrest stunning is used, electrode positions and electrical settings must be verified by measures of electrical or neurotransmitter activity in the brain.[9,13] Only scientifically verified electrode positions and electrical settings should be used. Visual assessment must not be used to verify new settings for cardiac arrest stunning. New settings must be verified by scientific research to ensure that instantaneous insensibility occurs.

For large (108 kg) market weight hogs, a minimum of 1.25 A at 300 V for one second should be used.[10] For slightly smaller pigs, the voltage can be dropped to 250 V.[7] For sheep, a minimum of 1 A at 375 V for three seconds should be used.[14] In New Zealand, electrical stunning is being successfully used in cattle.[15-17] Electrical stunning of cattle requires the use of a restraint device to hold the head.[17] Unlike pigs or sheep, cattle must have a stunning current (2.5 A) passed through the brain before the head to body cardiac arrest current is applied. In one study,[15] a single, 400 V 1.5 A current passed from the neck to the brisket failed to induce epileptic form changes in the electroencephalographic recordings. Equipment manufacturers have found that a minimum of 400 to 450 V is required to achieve insensibility. Practical experience indicates that greater amperages and voltages are required to achieve insensibility than to achieve cardiac arrest.

Amperage (current) is the most important factor in inducing unconsciousness.[9] Modern stunning circuits use a constant amperage power source, and voltage (electrical potential) varies with animal resistance. Some slaughter plants attempt to reduce meat quality defects by lowering the amperage or using high frequencies. This must not be permitted. Petechial hemorrhages can be minimized by the use of a constant amperage power supply.[18] Electrical frequencies of over 200 Hz must not be used unless they are scientifically verified.[9] In one report, a frequency of 500 Hz failed to induce unconsciousness. Most electrical stunning devices operate at 50 to 60 Hz.

ASSESSMENT OF STUNNING EFFICACY IN SLAUGHTER PLANTS

Head-only reversible electric stunning of pigs and sheep causes an initial spasm (tonic phase), which lasts approximately 10 seconds. After the tonic phase, kicking begins (clonic phase).[9] Animals stunned with head-only reversible stunning kick more vigorously than animals in cardiac arrest. Animals are unconscious during the tonic and clonic phases. After a stunned animal is hung upside-down prior to exsanguination, the field methods for verifying insensibility are similar for captive bolt arid electric stunning. Eye reflexes and blinking must be absent. In electrically stunned animals, eye reflexes should be checked 20 to 30 seconds after stunning. Prior to this time, eye reflexes are masked by the epileptic seizure. Veterinarians also need to check amperages, voltages, and electrode positioning to ensure that the stunner is being operated correctly.

In animals that have been shot with a captive bolt or stunned electrically, the limbs may move. Random limb movement should be ignored, but a limb that responds vigorously in response to a stimulus is a possible sign of return to sensibility. After the animal is hung on the overhead rail, the head must hang straight down, and the neck should be limp. The tongue should hang out and the ears should droop down. Gasping and gagging reflexes are permissible,[7] but rhythmic breathing and vocalization must be absent. The animals must not have an arched back righting reflex. Fully conscious animals suspended upside-down arch their backs in an attempt to lift their heads.

CARBON DIOXIDE STUNNING

Carbon dioxide stunning is an approved method for inducing insensibility under the Humane Methods of Slaughter Act of 1978.[2] The AVMA states that CO_2 is an acceptable method of euthanasia, but other methods are preferable because large animals, such as swine, appear to be more distressed than small laboratory animals. In both pigs and human beings, there is great variability in reactions to CO_2.[19-23] Swedish Yorkshire pigs react well to CO_2, and the motoric excitation phase begins after the electroencephalograph, indicating second stage anesthesia.[24] Hoenderken found that the excitation phase started before the pig was unconscious.[10] Visual assessment by other investigators has shown that alothane-positive pigs have a greater amount of excitation than halothane negative pigs.[21] Administration of halothane is used as a test to detect pigs that have porcine stress syndrome. Such pigs react to halothane by becoming rigid. There is concern that some of the pigs sensitive to halothane may be conscious during an initial excitation phase.[21] Observations by the author revealed that some pigs quietly lost consciousness when exposed to CO_2, whereas other pigs violently struggled when they first sniffed the gas. The encephalographic measurements that have been performed on the Swedish Yorkshire breed should be performed on pigs of various breeds that are sensitive and non-sensitive to halothane. At this time, available research data suggest that CO_2, is a good euthanasia method for certain genetic types of pigs but may possibly cause discomfort in others.

To reduce excitation during anesthetic induction, pigs should be rapidly exposed to 80 to 90% CO_2.[21] Veterinarians at slaughter plants should monitor CO_2 concentrations because plant management may be tempted to lower concentrations to save money. Observations in the field have indicated that pigs that walk quietly into the CO_2 chamber have a milder excitation phase than agitated, excited pigs. A new CO_2 stunning system in Denmark appeared to greatly reduce excitement and squealing during handling because groups of five pigs were moved into the chamber at one time. The author observed that these pigs had little reaction when they first contacted the gas, and the motorific excitation phase appeared to occur after they became unconscious. One possible explanation for this observation is that Denmark has a low prevalence of pigs sensitive to halothane.

PRESLAUGHTER STRESS

Properly performed slaughter induces cortisol concentrations equal to or less than that induced by on farm handling and restraint when a captive bolt is used.[25-29] When preslaughter handling is performed properly, cattle should move through the chute at a slow walk and calmly enter the stunning area without balking. To reduce stress, cattle should be stunned immediately after they enter the stunning box or restrainer. Cattle should not have signs of visible agitation, such as bellowing or rearing. In a well-designed handling system, the author has been able to move 8 of 10 cattle into the stunning pen or restrainer without use of an electric prod.

Findings in studies[28-30] have indicated that cortisol concentrations can double or triple when cattle slip on slick floors, are restrained in poorly designed equipment, or are over prodded. When this occurs, cortisol concentrations may greatly exceed on-farm handling concentrations. Epinephrine and norepinephrine are of limited value for evaluating preslaughter stress, because electric and captive-bolt stunning trigger massive release of these substances.[31,32] When stunning is performed correctly, the animal does not feel any discomfort because it is unconscious when the hormones are released.

In large (1,000 head/h) pork slaughter plants, it is likely that hogs experience more stress than from on-farm handling because they squeal and jam together as they move through the single-file chute. To improve conditions for the hogs' welfare and for pork quality, two restrainer systems may be required in high-speed plants. Providing confinement hogs with rubber hoses to chew on and additional contact with people during finishing results in calmer hogs that are easier to handle. Indiscriminate genetic selection for leanness and rapid growth tends to produce nervous, excitable hogs.[33]

BEHAVIORAL PRINCIPLES

People who handle animals must be trained to use behavioral principles. They need to understand the animal's flight zone (Fig. 2) and point of balance.[34,35] To make an animal move forward, the handler must stand behind the point of balance. Handlers also should work on the edge of the animal's flight zone because deep penetration of the flight zone may cause panic or vigorous escape reactions. Tame animals have a smaller flight zone than wild animals.

In new and old facilities, distractions that cause animals to balk must be eliminated. The author has observed that distractions, such as shadows, puddles, light reflections on the floor, and visible people ahead, can ruin the performance of a well-designed system. Moving an overhead lamp to eliminate a water reflection or installing a shield to prevent approaching animals from seeing a person ahead facilitates animal movement. Veterinarians need to look up chutes and see what the animals see. Lamps can be used to attract animals into dark chutes and restrainers. The light must not glare in the eyes of approaching

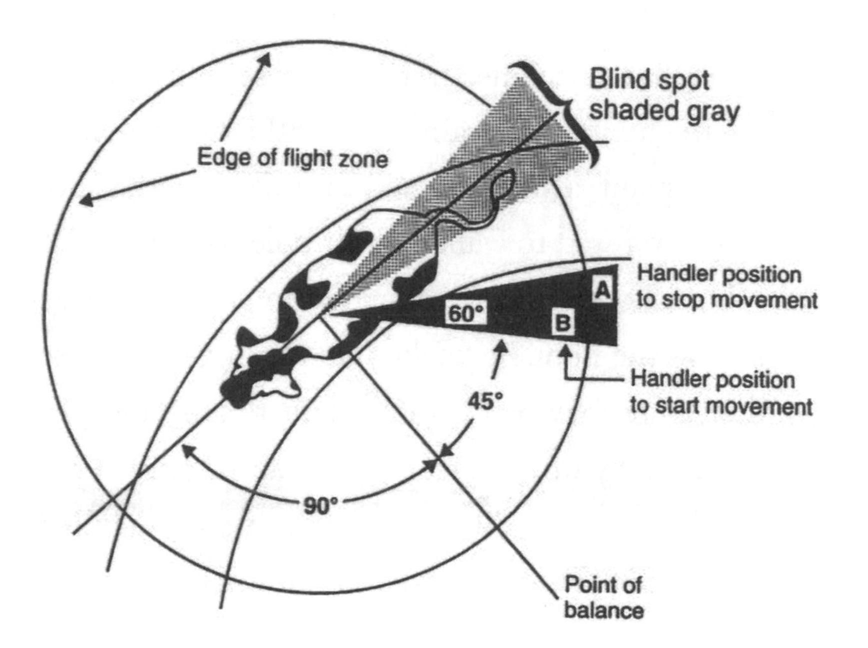

Fig. 2.
Diagram of an animal's flight zone (personal space). To move the animal forward, the handler must stand behind the point of balance at the shoulder and should work on the edge of the flight zone. Moving to position B causes the animal to move forward, whereas moving to position A usually causes the animal to stop.

animals or cause reflections off the floor or chute walls. Equipment should be designed to minimize noise. High pitched motor sounds, air hissing, and metal clanging and banging are more likely to cause excitement than the low rumble of a conveyor. Ventilation systems must not blow slaughter or rendering smells into the faces of approaching animals. The novelty of a smell or shadow causes animals to balk. Novelty is highly stressful when livestock are being handled in an unfamiliar environment. The author has observed that in a familiar environment, such as a feedlot pen, animals initially fear a novel stimulus, such as a loader used for pen cleaning. After they learn that it will not hurt them, it becomes environmental enrichment and the animals may approach and lick a parked machine. A spot of blood on the floor of the chute sometimes impedes animal movement. This appears to be attributable to visual contrast. The author has observed that throwing a piece of paper in a chute has the same effect.

EFFECT OF BLOOD

Observations by the author during new restraint equipment start-ups in many plants indicate that blood from relatively calm cattle does not appear to frighten the next animal that enters a restrainer. The animal usually voluntarily enters a restrainer that is covered with blood. Some cattle may lick the blood. Blood or saliva from a highly stressed animal, however, appears to upset other cattle. If an animal becomes frenzied for several minutes, the cattle next in line often balk and refuse to enter the restrainer. After the equipment is washed, however, the cattle will enter. in one plant, a steer refused to walk over the spot where he had flipped over backward, and then refused to walk over dribbles of saliva that were smeared on the floor where he had flipped over. He voluntarily reentered the chute three times, but when he reached his saliva on the floor, he backed up through the chute for over 15 m. There is some evidence that there may be a "smell-of-fear" substance. In one study, blood from stressed rats was avoided by other rats, but human or guinea pig blood had no effect.[37] According to animal behaviorist, Eibl-Eibesfeldt, if a rat is killed instantly by a trap, the trap can be used again, but if the trap fails to kill instantly, it will be avoided by the other rats.[38]

Possibly, the substance that the cattle are smelling is cortisol or some other substance that is secreted in conjunction with cortisol. Cortisol is present in the blood and saliva of cattle.[39] Cortisol is a time-dependent measure, up to 20 minutes is required to reach peak values.[40] The time course of cortisol secretion fits the author's observations. If an animal is stressed for only a few seconds by an electric prod, the next animal usually remains calm and walks into the restrainer. The most serious balking and refusals to enter occur after an animal has become seriously stressed by becoming jammed in a piece of equipment. The other cattle often balk and refuse to enter for several hours.

DESIGN AND OPERATION OF RESTRAINT DEVICES

Observations by the author in more than 100 slaughter plants indicate that the attitude of management is the single most important factor that determines how animals are treated.[41] Plants with good animal welfare practices have a manager who acts as their conscience. He or she is involved enough to care but not so involved that he or she becomes numb and desensitized to animal suffering. The author's observations also indicate that abuses, such as excessive prodding, dragging downed crippled animals, or running animals over the top of a downed animal, often occur when management is lax. The author has observed that in a few poorly managed plants, up to 10% of the cattle must be shot more than once with a captive bolt to render them insensible. It is the responsibility of the manager to enforce high standards of animal welfare. Good managers take the time to incrementally improve livestock handling. Perfecting handling techniques can take several months of sustained effort. There have been great improvements in equipment to handle and euthanatize livestock in slaughter plants. Unfortunately, in the United States, advances in equipment have not been paralleled by similar advances in management. In many plants, management attitude toward animal treatment has improved, but in some plants animal handling has become rougher. This is attributable to an overemphasis on speed or management personnel who do not care.

Large slaughter plants use a variety of restraint devices for holding animals during stunning and slaughter.[35,42,43] Proper operation is essential for good animal welfare. The best equipment causes stress and suffering if it is operated roughly, and animals are poked repeatedly with electric prods.

It is beyond the scope of this article to discuss equipment design in detail, but some basic principles should be mentioned. Solid sides should be installed on chutes, crowd pens, and restraint devices.[34] Solid sides keep animals calmer because they block outside distractions and prevent animals from seeing people deep inside their flight zone. Solid sides also make an animal feel more secure because there is a solid barrier between it and a threatening person. A basic principle is that an animal remains calmer in a restraint device if its vision is blocked until it has the feeling of restraint.[43,44] On conveyor restrainers, the solid hold-down over the entrance must be long enough to block the animal's vision until its rear feet are off the entrance ramp and it is completely settled down on the conveyor.[42]

A second basic principle is that sudden jerky motions of equipment or people excite animals, and slow steady movements have a calming effect. A third principle is the concept of optimal pressure. A restraint device should apply enough pressure to provide a feeling of restraint, but excessive pressure, which would cause pain, must be avoided. If an animal struggles because of excessive pressure, the pressure should be slowly released. When head restraint equipment is used, the animal should be stunned or ritually slaughtered immediately after the head is restrained. On restraint devices with moving parts that press against the animal, pressure limiting valves must be installed to prevent discomfort. A pressure limiting valve automatically prevents a careless operator from applying excessive pressure.

RITUAL SLAUGHTER

Ritual slaughter is slaughter performed according to the dietary codes of Jews or Muslims. Cattle, sheep, or goats are exsanguinated by a throat cut without first being rendered unconscious by preslaughter stunning. Ritual slaughter is exempt from the Humane Methods of Slaughter Act of 1978 to protect religious freedom.[2]

Because ritual slaughter is exempt, some plants use cruel methods of restraint, such as suspending a conscious animal by a chain wrapped around one hind limb. In other plants, the animal is held in a restrainer that holds it in an upright position.[43,44,45] Whether or not ritual slaughter conforms to the requirements of euthanasia is a controversial question. When ritual slaughter is being evaluated, the variable of restraint method must be separated from the act of throat cutting without prior stunning. Distressful restraint methods mask the animal's reactions to the cut.

The author designed and operated four state-of-the-art restraint devices that hold cattle and calves in a comfortable upright position during kosher (Jewish; Fig. 3) slaughter.[35,42,43,46] To determine whether cattle feel the throat cut, at one plant the author deliberately applied the head restrainer so lightly that the animals could pull their heads out. None of the 10 cattle moved or attempted to pull their heads out. Observations of hundreds of cattle and calves during kosher slaughter indicated that there was a slight quiver when the knife first contacted the throat. Invasion of the cattle's flight zone by

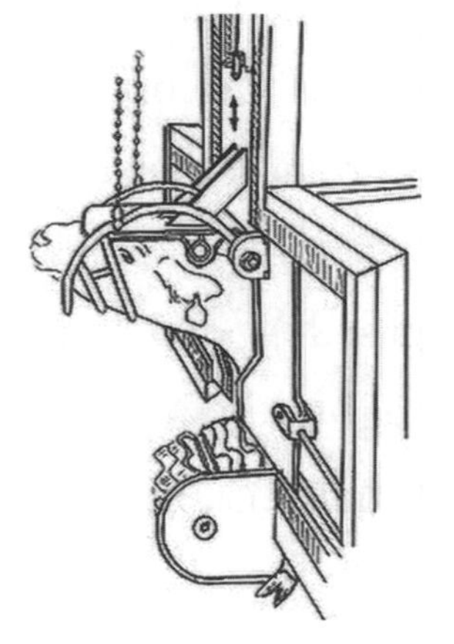

Fig. 3.
Head holding device for ritual slaughter mounted on a double rail (center track) restrainer. The chin lift and forehead brachet should be equipped with pressure-limiting valves. The 15 cm wide forehead brachet with a 7.5 cm round pipe behind the animal's poll holds the animal's head securely with little pressure. The pipe prevents the animal from pulling its head out. The pipe and forehead bracket are covered with rubber belting.

touching its head caused a bigger reaction. In another informal experiment, mature bulls and Holstein cows were gently restrained in a head holder with no body restraint. All of them stood still during the cut and did not appear to feel it. Disturbing the edges of the incision or bumping it against the equipment, however, is likely to cause pain. Observations by the author also indicated that the head must be restrained in such a manner that the incision does not close back over the knife. Cattle and sheep struggle violently if the edges of the incision touch during the cut.

The design of the knife and the cutting technique appeared to be critical in preventing the animal from reacting to the cut. In kosher slaughter, a straight, razor-sharp knife that is twice the width of the throat is required, and the cut must be made in a single continuous motion. For halal (Muslim) slaughter, there are no knife design requirements. Halal slaughter performed with short knives and multiple hacking cuts resulted in a vigorous reaction from cattle. Fortunately, many Muslim religious authorities accept pre-slaughter stunning. Muslims should be encouraged to stun the cattle or use long, straight, razor-sharp knives that are similar to the ones used for kosher slaughter.

Investigators agree that kosher slaughter does not induce instantaneous unconsciousness.[47,48,b] In some cattle, consciousness is prolonged for over 60 seconds. Observations by the author indicated that near immediate collapse can be induced in over 95% of cattle if the ritual slaughterer makes a rapid, deep cut close to the jawbone.[45] Further observations indicated that calm cows and bulls lose sensibility and collapse more quickly than cattle with visible signs of agitation. The author has observed that cattle that fight restraint are more likely to have prolonged sensibility. Gentle operation of restraint devices facilitates rapid loss of sensibility.

Cattle do not appear distressed even when the onset of unconsciousness is delayed. Pain and distress cannot be determined by measurements such as an electroencephalogram. Behavioral observations, however, are valid measures for assessing pain.[49] The author has observed that cattle appear unaware that their throat is cut. Investigators in New Zealand have made similar observations.[50] Immediately after the cut, the head holder should be loosened slightly to allow the animal to relax. The author also has observed that after the head restraint is released, the animal collapses almost immediately or stands and looks around like a normal, alert animal. Within 5 to 60 seconds, cattle go into a hypoxic spasm and sensibility appears to be lost. The spasms are similar to those that occur when cattle become unconscious in a headgate that is used for restraint in feedlots. Practical experience has shown that pressure on the carotid arteries and surrounding areas of the neck from a V-shaped headgate stanchion can kill cattle within 30 seconds.

Even though exsanguination is not an approved method of euthanasia by the AVMA,[1] the author has observed that kosher slaughter performed with the long, straight, razor-sharp knife does not appear to be painful. This is an area that needs further research. One can conclude that it is probably less distressful than poorly performed captive-bolt or electrical stunning methods, which release large amounts of epinephrine.[31,32]

Welfare can be greatly improved by use of a device that restrains the animal in a comfortable upright position. For cattle and calves, a conveyor restrainer or an upright restraint pen can be used.[42,43,46] In small plants, sheep or goats can be held by a person. If an upright pen is used, vertical travel of the lift under the animal's belly should be restricted to 71 cm to prevent the animal from being lifted off the floor. A pressure limiting valve must be installed on the head holder and rear pusher gate.[43,45] Many existing upright restraint boxes apply excessive pressure. To prevent excessive bending of the neck, the head holder should position the animal's forehead parallel to the floor. Equipping the head holder with a 15 cm wide, rubber covered forehead bracket will make the head holder more comfortable (Fig 3). The animal should stand in the box with its back level. An arched back is a sign of excessive pusher gate pressure. In some plants, animals are removed from the restrainer before they become unconscious. Discomfort to the animal can be minimized by allowing it to lapse into unconsciousness before it is removed from the restrainer.

During the past five years, many large kosher slaughter plants for cattle have replaced shackling and hoisting with upright restraint. Large numbers of veal calves and sheep, however, are still shackled and hoisted. Progressive plant owners have installed upright restraint equipment, but unfortunately there are some plant owners who still refuse to install humane restraint equipment because they are not legally required to do so. Animal handling guidelines published by the American Meat institute recommend the use of upright restraint.[6]

CONCLUSIONS

The technology exists that allows slaughter and euthanasia to be one. Although some slaughter plants maintain high animal welfare standards, there are others in which management allows abuses to occur. After adequate equipment has been installed, the single most important determinant of good animal welfare is the attitude of management. Good equipment provides the tools that make humane slaughter and handling possible, but it is useless unless it has good management to go with it.

a. Koch Supplies, Kansas City, MO

b. Nangeroni LL, Kennett PD. Department of Physiology, Cornell University, Ithaca, NY: Unpublished data, 1963.

REFERENCES

1. Andrews EJ, Bennett TB, Clark JD, et al. Report of the AVMA panel on euthanasia. J Am Vet Med Assoc 1993, 202:229-249.
2. Federal meat inspection publication 95445, Part 313. Humane slaughter of livestock 3.3.2. In: Humane methods of Slaughter Act of 1978.
3. Daly CC. Proceedings symposium on humane slaughter of animals. Herts, UK: Universities Federation for Animal Welfare, 1987:15.
4. Blackmore DK. Energy requirements for penetration of heads of domestic stock and developments of a multiple projectile. Vet Rec 1985, 116:3640.
5. Daly CC, Whittington PF. Investigation into the principle determinants of effective captive bolt stunning of sheep. Res Vet Sci 1989, 46:406-408.
6. Grandin T. Recommended animal handling guidelines for meat packers. Washington, DC: American Meat Institute, 1991.
7. Gregory NG. Humane slaughter, in Proceedings. 34th Int Cong Meat Sci Technol, Workshop on Stunning Livestock, 1988.
8. Blackburn PW. The casualty pig. Malmesbury, Wilts. UK: Pig Veterinary Society Grove Center, 1993.
9. Croft PS. Problems of electrical stunning. Vet Rec 1952, 64:255-258.
10. Hoenderken R. Electrical and carbon dioxide stunning of pigs for slaughter. In: Eikelenboom G, ed. Stunning of Animals for Slaughter. Boston: Martinus Nijhoff Publishers, 1982, 59-63.
11. Blackmore DK, Newhook JC. Insensibility during slaughter of pigs in comparison to other domestic stock. N Z Vet J 1981, 29:219-222.
12. Gilbert KV, Cook CJ, Devine CE, et al. Electrical stunning in cattle and sheep: Electrode placement and effectiveness. In Proceedings, 37th Int Congr Meat Sci Technol 1991:245-248.
13. Devine CE, Cook CJ, Maasland SA, et al. The humane slaughter of animals. In Proceedings, 39th Int Congr Meat Sci Technol 1993:223-228.
14. Gregory NG, Wotton SB. Sheep slaughtering procedures. I11. Head-to-back electrical stunning. Br Vet J 1984, 140:570-575.
15. Cook CJ, Devine CE, Gilbert KV, et al. Electroencephalograms and electrocardiograms in young bulls following upper cervical vertebrae to brisket stunning. N Z Vet J 1991, 39:121-125.
16. Cook CJ. Stunning Science-A guide to better electrical stunning. Meat Focus 1993, 2(3):128-131.

17. Gregory NG. Slaughter technology, electrical stunning of large cattle. Meat Focus 1993;2(1):32-36.

18. Grandin T. Cardiac arrest stunning of livestock and poultry. In: Fox MW, Mickley L.D, eds. Advances in Animal Welfare Science. Boston: Martinus Nijhoff Publishers, 1985/1986, 130.

19. Grandin T. Possible genetic effect in pig`s reaction to CO_2 stunning. In Proceedings, 34th Int Congr Meat Sci Technol 1988, 23-24.

20. Dodman NH. Observations on the use of the Wemberg dip-lift carbon dioxide apparatus for preslaughter anesthesia of pigs. Br Vet J 1977, 133:71-80.

21. Troeger K, Waltersdorf W. Gas anesthesia of slaughter pigs. Fleischwirtsch Int 1991, 4:4349.

22. Griez E, Zandbergen J, Pols H, et al. Response to 35% CO_2 as a marker of panic and severe anxiety. Am J Psychiatry 1990, 147:796797.

23. Clark DH. Carbon dioxide therapy for neuroses. J Ment Sci 1954, 100:722726.

24. Forslid A. Transient neocortical, hippocampal and amygdaloid EEG silence by one minute inhalation of high concentrations of CO_2 in swine. Acta Physiol Scand 1987, 130:110.

25. Grandin T. Farm animal welfare during handling, transport, and slaughter. J Am Vet Med Assoc 1994, 204:372377.

26. Mitchell G, Hattingh J, Ganhao M. Stress in cattle assessed after handling, transport and slaughter. Vet Rec 1988, 123:201205.

27. Zavy MT, Juniewicz PE, Phillips WA, et al. Effect of initial restraint, weaning, and transport stress on baseline and ACTH-stimulated cortisol responses in beef calves of different genotypes. Am J Vet Res 1992, 53:551-557.

28. Ewbank R, Parker MJ, Mason CW. Reactions of cattle to head restraint at stunning; a practical dilemma. Anim Welf 1992, 1:55-63.

29. Tume RK, Shaw FD. Beta-endorphin and cortisol concentrations in plasma of blood samples collected during emanguination of cattle. Meat Sci 1992, 31:211217.

30. Cockram MS, Corley KTT. Effect of pre-slaughter handling on the behavior and blood composition of beef cattle. Br Vet J 1991, 147:444-454.

31. Althen TGK, Ono GK, Topel DG. Effect of stress susceptibility or stunning method on catecholamine levels in swine. J Anim Sci 1977, 44:985-989.

32. Pearson AM, Kilgour R, de Langen H. Hormonal responses of lambs to trucking, handling and electric stunning. In Proceedings, N Z Soc Anim Prod 1977, 37,243-248.

33. Grandin T. Environmental and genetic factors which contribute to handling problems at slaughter plants. In: Eldridge C, Boon C, eds. Livestock Environment IV. St joseph, Mich: American Society of Agriculture Engineers, 1993,64-68.

34. Grandin T. Animal handling. Vet Clin North Am Food Anim Pract 1987, 3:323-338.

35. Grandin T. Welfare of livestock in slaughter plants. In: Grandin T, ed. Livestock Handling and Transport. Wallingford, Oxon, UK: CAB International, 1993, 289-311.

36. Hombuckle PA, Beall T. Escape reactions to the blood of selected mammals by rats. Behav Biol 1974, 12:573-576.

37. Stevens DA, Gerzog-Thomas DA. Fright reactions in rats to conspecific tissue. Physiol Behav 1977, 18:47-51.

38. Stevens DA, Saplikoski NJ. Ratsë reactions to conspecific muscle and blood evidence for alarm substances. Behav Biol 1973, 8:75-82.

39. Fell LR, Shutt DA. Adrenocortical response of calves to transport stress as measured by salivary cortisol. Can J Anim Sci 1986, 66:637-641.

40. Lay DC, Friend TH, Randel RD, et al. Behavioral and physiological effects of freeze and hot iron branding on crossbred cattle. J Anim Sci 1991, 70:330336.

41. Grandin T. Behavior of slaughter plant and auction employees towards animals. Anthrozoo 1988, 1:205-213.

42. Grandin T. Double rail restrainer for handling beef cattle. Paper No. 915004. St Joseph, Mich: American Society of Agricultural Engineers, 1991.

43. Grandin T. Observations of cattle restraint devices for stunning and slaughtering. Anim Welf 1992, 1:85-91.

44. Grandin T. The effect of previous experience on livestock behavior during handling. Agri-Pract 1993, 14:15-20.

45. Regenstein JM, Grandin T. Religious slaughter and animal welfare, in Proceedings. 45th Annu Reciprocal Meat Conf 1992;155-160.

46. Grandin T. Double rail restrainer conveyor for livestock handling. J Agric Eng Res 19B8, 41:327-338.

47. Daly CC, Kallweit E, Ellendorf F. Conventional captive bolt stunning followed by exsanguination compared to shechitah slaughter. Vet Rec 1988; 122:325-329.

48. Blackmore DK. Differences between sheep and cattle during slaughter. Vet Sci 1984; 37:223-226.

49. Fraser AF, Broom DM. Farm Animal Welfare. London: Bailliere Tindall, 1990.

50. Bager F, Braggins TJ, Devine CF, et al. Onset of insensibility in calves: effects of electropletic seizure and exsanguination on the spontaneous electrocortical activity and indices of cerebral metabolism. Res Vet Sci 19S4, 37:223-226.

TWELVE

Factors That Impede Animal Movement at Slaughter Plants

T. Grandin (1996) Factors that impede animal movement at slaughter plants, Journal of the American Veterinary Medical Association, Vol. 209, pp. 757-759.

I was invited by the Canadian Food and Inspection Association to conduct an animal welfare assessment in twenty-nine Canadian beef, pork, and lamb slaughter plants. This was my first major project where I assessed animal welfare. On this trip, I noted problems that would compromise animal welfare, such as slipping and falling on slick floors and distractions that impeded animal movement.

ABSTRACT

Summary: Factors that impede animal movement in slaughter plants and that are likely to cause excitement, stress, or bruises are major mistakes in the design of chutes and stockyard pens; lack of training or poor supervision of employees; distractions that impede animal movement, such as sparkling reflections on a wet floor, air hissing, high pitched noise, or air drafts blowing down the chute toward approaching animals; poor maintenance of facilities, such as worn out or slick floors that cause animals to fall; and animals from genetic lines that have an excitable temperament.

Veterinarians need to be aware of these factors because such factors can cause animals to balk and become excited, which may result in excessive prodding. When a handling system is being evaluated, one must be careful to discriminate between a major design mistake and small distractions that can be easily corrected, but that can ruin the performance of the best systems.

A survey of 29 Canadian slaughter plants revealed that:

- 21% (6 plants) had slick floors that would cause animals to slip and fall
- 27% (8 plants) had high-pitched motor noise or hissing air that caused animals to balk
- Air drafts blowing down the chutes, which will often impede animal movement, were a problem in 10 % (3) of the plants.

Simple modifications of lighting and elimination of air drafts and hissing will often greatly improve animal movement.

SUMMARY

Factors that impede animal movement in slaughter plants and that are likely to cause excitement, stress, or bruises are major mistakes in the design of chutes and stockyard pens; lack of training or poor supervision of employees; distractions that impede animal movement, such as sparkling reflections on a wet floor, air hissing, high-pitched noise, or air drafts blowing down the chute toward approaching animals; poor maintenance of facilities, such as worn out or slick floors that cause animals to fall; and animals from genetic lines that have an excitable temperament. Veterinarians need to be aware of these factors because such factors can cause animals to balk and become excited, which may result in excessive prodding. When a handling system is being evaluated, one must be careful to discriminate between a major design mistake and small distractions that can be easily corrected, but that can ruin the performance of the best systems. A survey of 29 Canadian slaughter plants revealed that 21% (6 plants) had slick floors that would cause animals to slip and fall, and 27% (8 plants) had high-pitched motor noise or hissing air that caused animals to balk. Air drafts blowing down the chutes, which will often impede animal movement, were a problem in 10% (3) of the plants. Simple modifications of lighting and elimination of air drafts and hissing will often greatly improve animal movement (*J Am Vet Med Assoc* 1996; 209:757–759).

Stress during preslaughter handling at the abattoir is detrimental to meat quality and animal welfare. Swine slaughtered in more stressful handling systems had a progressive increase in the potential for pale, soft, exudative meat (PSE) or dark, firm, and dry meat.[1] The sound level of squealing and blood lactate concentration and creatine kinase activity also were correlated in these swine. Excitement and overexertion prior to slaughter will cause PSE in swine.[2] Danish researcher Patricia Barton-Gade found that careful, considerate handling prior to slaughter reduced PSE in swine that were heterozygotes or free of the halothane gene.[3]

Practical experience and observations by the author in more than 200 beef and pork slaughter plants in the United States, Canada, Mexico, Europe, Australia, and New Zealand indicate that careful, quiet handling in the stunning chute will reduce PSE; dark, firm, and dry meat; and bruises. Cattle handled roughly had almost twice as many bruises, compared with those that had been handled gently.[a] In US and Canadian pork slaughter plants, the author has observed that moving swine quietly and reducing use of electric prods reduced PSE.

When a chute system design is being evaluated, one must determine whether handling problems such as balking and refusing to move are caused by a basic design mistake or by an easily correctable problem, such as untrained employees or small distractions. The author has observed that easily correctable faults in lighting or ventilation, or excessive high-pitched noise, can ruin the performance of a well-designed handling system. When animals balk, handlers are more likely to repeatedly poke them with electric prods. Electric prodding is stressful. A pig's heart rate will increase with each successive application of an electric prod.[4] Elimination of electric prods reduced petechial hemorrhages in pigs. Slick floors that cause animals to slip and fall also make humane handling difficult. A British study revealed that slick floors increased stress in cattle.[5]

The author has observed increasing problems with excitable swine and cattle.[6,7] These animals are more difficult to drive at the slaughter plant than are animals from calmer genetic lines. Observations by the author in the United States, Canada, Ireland, and the Netherlands indicate that there may be a relationship between indiscriminate selection for leanness and an excitable temperament. During these plant visits, the author worked with the employees driving thousands of animals. Some of the most difficult swine to drive are hybrids from halothane-positive hybrid boars. Although practical experience and research has revealed that swine that are homozygous or heterozygous for the halothane gene will be more likely to have PSE,[3] some breeders use boars with the halothane gene because the offspring will have larger loineye area and thinner backfat.[8] Unfortunately, the price for increased pork quantity is poorer quality. Interestingly, the author has never observed an excitability problem in Danish swine viewed in 7 Danish abattoirs. As a population, lean swine in Denmark have been bred to be almost free of the halothane gene. Environmental factors during finishing can also affect excitability. Providing environmental enrichment to swine raised on slatted concrete floors resulted in calmer swine and facilitated their movement through a chute.[9-11] Environmental enrichment consisted of providing toys to chew and manipulate or people walking in the finishing pens every day during the entire finishing period. Practical experience on several farms indicates that swine will be easier to drive if the person walking through the pens moves in a different direction each day. Varying the routine during pen walking appears to train the swine to tolerate novelty.

Poor design can make quiet, calm animal handling difficult.[9,12,13] The most serious design mistake is constructing a single-file chute to the stunner that looks like a dead end. An animal approaching the entrance of the chute must be able to see that there is a place to go. When an animal stands in the forcing pen, it must be able to see at least 3 body lengths up the single-file chute. For cattle, curved chutes will work more efficiently than straight chutes, but they must be laid out correctly, so that the entrance to the chute does not appear to be a dead end.[9] Facilities for swine will work more efficiently if they are completely level. Ramps to the stunner will work well for cattle and sheep. The ramp must be located in the single-file chute. The forcing pen must never be built on a ramp or slope, or the animals will pile up against the rear crowd gate. Animal movement will be facilitated by installing solid sides on chutes, forcing pens, and restraint devices to prevent the animals from seeing outside the facility. Solid sides also prevent the animals from seeing people deep in their flight zones.[12,14]

The author has observed 5 important variables that impede animal movement in the slaughter plant. They are major mistakes in the design of chutes, restrainers, stockyards, or other equipment; rough handling, caused by a lack of employee training or poor supervision of employees; distractions that cause balking, such as easily correctable faults in lighting or ventilation, or air hissing or high-pitched noise; poor maintenance of facilities or worn, smooth floors that cause animals to fall; and animals with an excitable temperament. Most of these faults can be corrected at minimal expense.

To determine the prevalence of easily correctable faults in animal handling facilities, the author visited 29 Canadian plants for slaughter of cattle, swine, lambs, horses, or sheep. The sample included the 11 largest plants in Canada and a representative sample of small and medium-sized plants. Data were collected on 33 slaughter lines in 29 plants. The number of plants with slick floors that would cause animals to fall was tabulated. The prevalence of distractions that caused obvious balking and impeded animal movement, including lighting at the stunning box or restrainer that was too dim or too bright, ventilation systems blowing air through the chute toward approaching animals, balking caused by seeing movement of people or movement of reflections on water or metal equipment, and high-pitched motor noise or hissing air exhausts, were determined in 33 slaughter lines.

Slick floors and distractions that impeded the movement of animals were evident in a high percentage of the plants (Tables 1 and 2). Slick floors that caused animals to slip were located in the cattle stunning box or in high—traffic areas where the rough finish had worn off the concrete. Approximately a quarter of the abattoirs surveyed had an easily correctable fault that caused balking. Ventilation air blowing down the chute into the faces of approaching animals caused swine and cattle to back up, which resulted in excessive use of electric prods. In 1 large beef plant, cattle balked at the entrance of the conveyor restrainer because the sun shone in their eyes and air blew out the entrance. An electric prod had to be used on almost every animal. In the afternoon at that plant, the wind changed direction and negated the effects of the positive-pressure ventilation system, and the setting sun no longer shone into the eyes of approaching cattle. The combination of a change in direction of the sun and the air flow greatly facilitated cattle movement, and electric prodding was greatly reduced. The author has corrected distractions that caused balking and has improved animal movement in many plants in the United States and Canada. In 1 surveyed beef plant and 2 un-surveyed pork plants with new, state-of-the-art chute systems, there was severe balking at the restrainer conveyor entrance because of air blowing toward the animals. Reversing air flow at the restrainer entrance greatly improved animal movement. In another beef plant, multiple electric prods were used on each animal before, and a single prod on < 10% of the cattle after, the ventilation system was modified.

TABLE 1: Distractions That Impeded Movement of Livestock in 33 Slaughter Lines *

Type of Distraction	Number of Affected Slaughter Lines (%)
Lighting problems (too dim or too bright)	5 (15)
Ventilation blowing toward approaching animals	3 (9)
Seeing movement of people or moving, sparkling reflections	8 (24)
High-pitched motor noise or hissing air exhausts	8 (24)

* Plants that slaughtered more than 1 species were tabulated as separate lines

In 2 other beef plants, cattle movement was facilitated by increasing lighting at the restrainer entrance. When the system was new, it had adequate illumination from overhead sodium lamps. As the lamps dimmed with age, balking increased. The addition of more light improved animal movement. Cattle and swine tend to move from a darker

TABLE 2: Conditions of Floors in 29 Slaughter Plants	
Floor Condition	**Number of Affected Slaughter Lines (%)**
Excellent, nonslip	8 (27)
Acceptable	15 (52)
Slick, not acceptable	6 (21)

place to a more brightly illuminated place.[9,13] In 1 beef plant and 2 pork plants, moving an overhead light sideways so that it was off the center line of the single-file chute eliminated sparkling reflections that caused balking.

Animals may balk if they see movement: moving people or moving, sparkling reflections. At 2 beef plants and 1 pork plant, animal movement was facilitated by installing shields that prevented the animals from seeing people ahead. Animals may also balk at jiggling objects. In 2 plants, swine and cattle balked at a jiggling gate and a moving chain that hung down in the chute. Removal of the wiggling chain and bracing a jiggling gate reduced balking. High-pitched noise causes agitation and increases balking. Animals are more sensitive to high-pitched noise than are human beings.[15,16] Sheep slaughtered in a noisy commercial abattoir had higher blood cortisol concentrations than sheep slaughtered in a quiet research abattoir.[17] Observations indicate that high-pitched noise is more disturbing to animals than is the low-pitched rumble of conveyor chains. In 1 beef plant, replacement of undersized plumbing in a hydraulic system facilitated animal movement because a high-pitched whine was eliminated. In another plant, cattle movement was improved, and visible signs of agitation were almost eliminated when silencers were installed on a hissing air valve located on a stunning box gate.

Improvements in lighting and elimination of air hissing also will improve the operation of head-restraint devices. In 1 beef plant, placement of a light over a head-restraint yoke attracted the animal so that it voluntarily placed its head in the restraint yoke. In England, head-restraint devices are required by legislation to hold a bovid's head for captive bolt stunning. In some circumstances, this restraint can increase stress.[18] Blood cortisol concentrations were higher in cattle with use of a head-restraint device, compared with those with use of a conventional single animal stunning box.[18] A mean time of 32 seconds was needed to induce the animal to enter the head-restraint device. Low cortisol blood concentrations were found in cattle with use of a well-designed head restraint, in which the animal entered without balking and was stunned immediately after the head was caught.

The author has observed that detecting the distractions that cause balking in calm animals is easy. A calm pig or cow will stop and look directly at a jiggling chain or a sparkling reflection. Small distractions appear to be a greater impediment to animal movement when animals with an excitable temperament are driven.

Plant design and staff expertise affected stress levels in swine.[20] The author has observed that untrained, poorly supervised employees will often handle animals roughly. High stress levels will be observed in well-designed facilities if supervision is lax. The most common mistake made by handlers is putting too many animals in the forcing pen. Forcing pens will work best when filled only half full. Another factor is line speed. A design that works well at a slow line speed may work poorly at a high line speed. In British abattoirs, swine were more stressed in abattoirs with single-file races, compared with those in plants where small groups were electrically stunned on the floor1. Electrical stunning on the floor is most practical for abattoirs that slaughter < 240 pigs/h. The author has observed that floor stunning becomes rough and sloppy at higher speeds. The author also has observed that stunning swine with CO_2 in a group avoided the stress of lining them up in single file. In such a system in Denmark, each group of 5 pigs quietly entered the stunning compartments, without squealing. Single-file chutes and restrainers work well for cattle because they have the behavioral trait of walking in single file. During cattle drives in "the Old West," the animals were led, rather than driven, and they often moved in single file.[21] On range land in the western United States, single-file cow paths are worn into the pastures.

To make an accurate determination of the cause of preslaughter stress, veterinarians must differentiate between stress caused by equipment design and stress caused by other factors. A number of plants may have easily correctable problems that would impede animal movement, such as slick floors; sudden, loud noises; or ventilation and lighting mistakes.

a. Grandin T. Bruises on southwestern feedlot cattle (abstr). *J Anim Sci* 1981;53(Suppl 1):213.

b. Calkins CR, Davis GW, Cole AB, et al. Incidence of blood-splashed hams from hogs subjected to certain antemortem handling methods (abstr). *J Anim Sci* 1980,50:15

REFERENCES

1. Warris PD, Brown SN, Adams SJM, et al. Relationship between subjective and objective assessments of stress at slaughter and meat quality in pigs. Meat Sci 1974; 38:329-340.
2. Sayre RN, Brisky EJ, Hockstra WG. Effect of excitement, fasting and sucrose feeding on porcine phosphorylase and postmortem glycolysis. J Food Sci 1963; 28:472-477.
3. Barton-Gade P. Influence of halothane genotype on meat quality in pigs subjected to various preslaughter treatments, in Proceedings. 30th Intern Congr Meat Sci Technol 1984;89.
4. Van Putten G, Elshof WJ. Observations on the effect of transport on the wellbeing and lean quality of pigs. Anim Reg Stud 1978,1:247271.
5. Cockrum MS, Corley KTT. Effect of preslaughter handling on the behavior and blood composition of beef cattle. Br Vet J 1991; 147:444-454.
6. Grandin T. Environmental and genetic factors which contribute to handling problems at pork slaughter plants. In: Collins E, Boon C, eds. Livestock Environment IV. St Joseph, Mo: American Society of Agriculture Engineers, 1993,6468.
7. Grandin T. Solving livestock handling problems. Vet Med 1994; 89:989-998.
8. Aalhus JL, Jones SDM, Robertson AKW, et al. Growth characteristics and carcass composition of pigs with known genotypes for stress susceptibility over a weight range of 70 to 120 kg. Anim Prod 1991; 52:347-353.
9. Grandin T. Handling and welfare in slaughter plants. In: Grandin T, ed. Livestock Handling and Transport. Wallingford, Oxon, UK: CAB International, 19933289311.
10. Grandin T. Environment and genetic effect on hog handling. Paper No. 894514. St Joseph, Mo. American Society of Agriculture Engineers, 1989.
11. Pedersen BK, Curtis SE, Kelley KW, et al. Well—being in growing finishing pigs: environmental enrichment and pen allowance. In: Collins E, ed. Livestock Environment IV. St Joseph, Mo. American Society of Agriculture Engineers, 1993,143.
12. Grandin T. Slaughter. Meat Focus 1994; 3:115-123.
13. Grandin T. Observations of cattle behavior applied to the design of cattle handling facilities. Appl Anim Behav Sci 1980; 6:19-31.
14. Grandin T. Euthanasia and slaughter of livestock. J Am Vet Med Assoc 1994; 204:1354-1360.
15. Algers B. A note on responses of farm animals to ultrasound. Appl Anim Behav Sci 1984; 12:387-391.
16. Ames DR. Sound stress in meat animals, in Proceedings. Intern Livest Envir Symp Am Soc Agric Eng 1974;324.
17. Pearson AJ, Kilgour R, deLangen H, et al. Hormonal responses of lambs to trucking, handling and electric stunning. N Z Soc Anim Prod 1977; 37:243-249.
18. Ewbank R, Parker NM, Mason CW. Reactions of cattle head restraint at stunning: a practical dilemma. Anim Welfare 1992; 1:55-63.
19. Tume RK, Shaw FD. Beta endorphin and cortisol concentrations in plasma blood samples collected during exsanguination of cattle. Meat Sci 1992; 31:211-217.
20. Weeding CM, Hunter EJ, Guise HJ, et al. Effects of abattoir and slaughter handling systems on stress indicators in pig blood. Vet Rec 1993; 133:10-13.
21. Adams RF. Old-Time Cowhand. Lincoln, Neb: University of Nebraska Press, 1948.

THIRTEEN

Habituating Antelope and Bison to Veterinary Procedures

T. Grandin (2000) Habituating Antelope and Bison to Veterinary Procedures, *Journal of Applied Animal Welfare Science*, Vol. 3, No. 3, pp. 253-261.

This was the first study that showed that it was possible to train flighty animals, such as nyala antelope, to cooperate with blood sampling. Nancy Irlbeck, Denver Zoological Gardens Nutritionist and a faculty member at Colorado State University asked me, "How can I get blood samples from antelope without stressing them?" I told her that we would have to train them to cooperate. In a later study, we trained Bongo antelope and were able to obtain very low cortisol (stress hormone) levels. Many researchers who were accustomed to forceful methods restraint did not want to believe that very low cortisol levels were obtained.

COMMENTARY

Undomesticated, wild, nonhuman animals with an excitable temperament frequently sustain injuries when handled for veterinary procedures such as blood sampling or injections. Both bison and antelope may become agitated when they are restrained, often resulting in broken horns. Also, antelope may panic when restraint is forced on them suddenly. This research group has successfully trained both Bongo and Nyala antelope to enter a box voluntarily for food rewards. While quiet, non-sedated antelope were standing in the box, blood samples were taken from their rear legs. Three Bongo antelope who had been conditioned to stand in the box had low cortisol levels that averaged 4.4 ± 3.4 ng/ml in a male and 8.5 ± 4.7 ng/ml in two females. The un-sedated animals were held in the box for 10–20 min to allow cortisol levels to rise. These values were much lower than cortisol levels reported in the literature for restrained un-sedated wild ruminants and domestic cattle who are not habituated to restraint. Because antelope are flighty animals, each new step in the conditioning process had to be introduced in very small steps to prevent panic. In another experiment, bison calves were trained to walk through the handling chutes for feed rewards. In both species, avoiding panic is important, especially in the early stages of training.

There is a need for methods to prevent the serious injuries that often occur when wild, undomesticated, ruminant, grazing nonhuman animals are handled for veterinary procedures. I have observed that antelope and bison often are injured and break horns during routine handling. Even when these animals are handled carefully, they are likely to ram into fences and struggle violently when restrained. Bison confined in a small pen in a handling facility sometimes gore each other.

My colleagues and I have developed practical methods for habituating and conditioning antelope and bison to being handled. Animals with a flighty, excitable temperament are more likely than those of calmer temperament to become fearful when a novel experience is suddenly forced on them (Grandin, 1997; Grandin & Deesing, 1998). The difference between a domestic ruminant such as cattle and a wild ruminant such as antelope or bison is the intensity of their reaction when they are handled. Price (1998) stated that part of the process of domestication is culling of the most ex- citable, difficult-to-handle individuals. This process of domestication of deer al- ready has started in New Zealand. Mathews (2000) observed that deer have become calmer and calmer since the beginning of deer farming 30 years ago.

Animals with an excitable temperament will not habituate to a forced, nonpainful handling procedure. Even in domestic cattle, new experiences and novelty are stressful

(Dantzer & Mormede, 1983; Moberg & Wood, 1982; Stephens & Toner, 1975). E. K. T. Lanier et al. (1995) found that some pigs habituated to being placed repeatedly in a swimming pool, and others remained fearful with elevated adrenaline levels. Both scientific studies and practical experience have shown that animals with a flighty, excitable temperament do not habituate readily to forced handling and restraint.

After seeing numerous broken horns and several serious injuries, I concluded that the only way to prevent stress and injury to bison and antelope was to condition them to cooperate voluntarily during handling, restraint, and veterinary procedures.

Many different species have been conditioned successfully to cooperate with veterinary procedures. Primates, dolphins, elephants, pigs, and sheep will either extend a limb for injections or enter a restraining device voluntarily (Calle & Bornmann, 1988; Grandin, 1989; Panepinto, 1983; Wienker, 1986). One of the problems with bison and antelope, compared with the species just listed, is that they are more excitable and flighty. We had to develop conditioning methods that would not trigger a massive destructive panic reaction.

CONDITIONING METHODS

Successful methods for conditioning Nyala and Bongo antelope to enter a wood crate voluntarily for blood sampling already have been reported (Grandin et al., 1995; Phillips, Grandin, Graffam, Irlbeck, & Cambre, 1998). Table 1 outlines the conditioning procedures. During our work, we observed some important basic principles. To prevent a massive fear response early in the conditioning process, the antelope had to be conditioned slowly and carefully to all the sounds associated with the crate (see Figure 1).

On the first day, the vertical slide door on the box was moved about 2 cm. This caused the antelope to turn their heads toward the sound. The instant the antelope oriented toward the sound; the door movement was stopped. (It took 10–17 days to habituate the antelope to the door movement.) Each day the door was moved more. Care was taken not to push the animals past the orienting behavior, because when these animals panic, they react explosively and ram into walls.

TABLE 1: The Advantages of Training Antelope to Cooperate with Veterinary Procedures		
Habituation and Conditioning Steps	Time Required (Days)	
	Nyala	Bongo
Acclimate and habituate the animals to the presence of a shipping or transfer crate in the stall. At this stage of training, the animals can walk by the crate to go outside to the exhibit.	7	7
Habituate to handling crate and movement of the vertical slide doors. Barn door closed against the crate. Animals must walk through the crate to go to outdoor exhibit.	10	17
Train to enter crate and eat treats with rear door open and front door closed.	21	21
Initial capture in crate and release immediately.	7	7
Habituate to stand captured in the crate for up to 10 min while continually being fed treats.	10	24
Habituate to open bleed hole door and being touched on the rear leg.	14	14
Condition to simulated blood sampling (increasingly hard pinches). Halfway through this period, treats were withheld and food was given only as a reward after the animal had stood still.	14	14
Successful blood sampling of all animals.	14	14
Total training days	97	up to 118

As the animal's excitability increases, new experiences, sights, and sounds must be introduced in smaller and smaller steps to avoid panic. The first phase of conditioning an excitable animal is habituation to the procedures to avoid panic. The second phase of the conditioning procedure uses standard operant conditioning methods. With all species, great care must be taken to ensure that an animal's first experience with something new is either positive or neutral. Both practical experience and research show that if an animal's first experience with something new is frightening or painful, the animal will actively avoid the place where the aversive event occurred. Miller (1960) found that if a rat was shocked the first time it entered a new arm in a maze, it would never enter that area again. First experiences with new places make a big impression on animals.

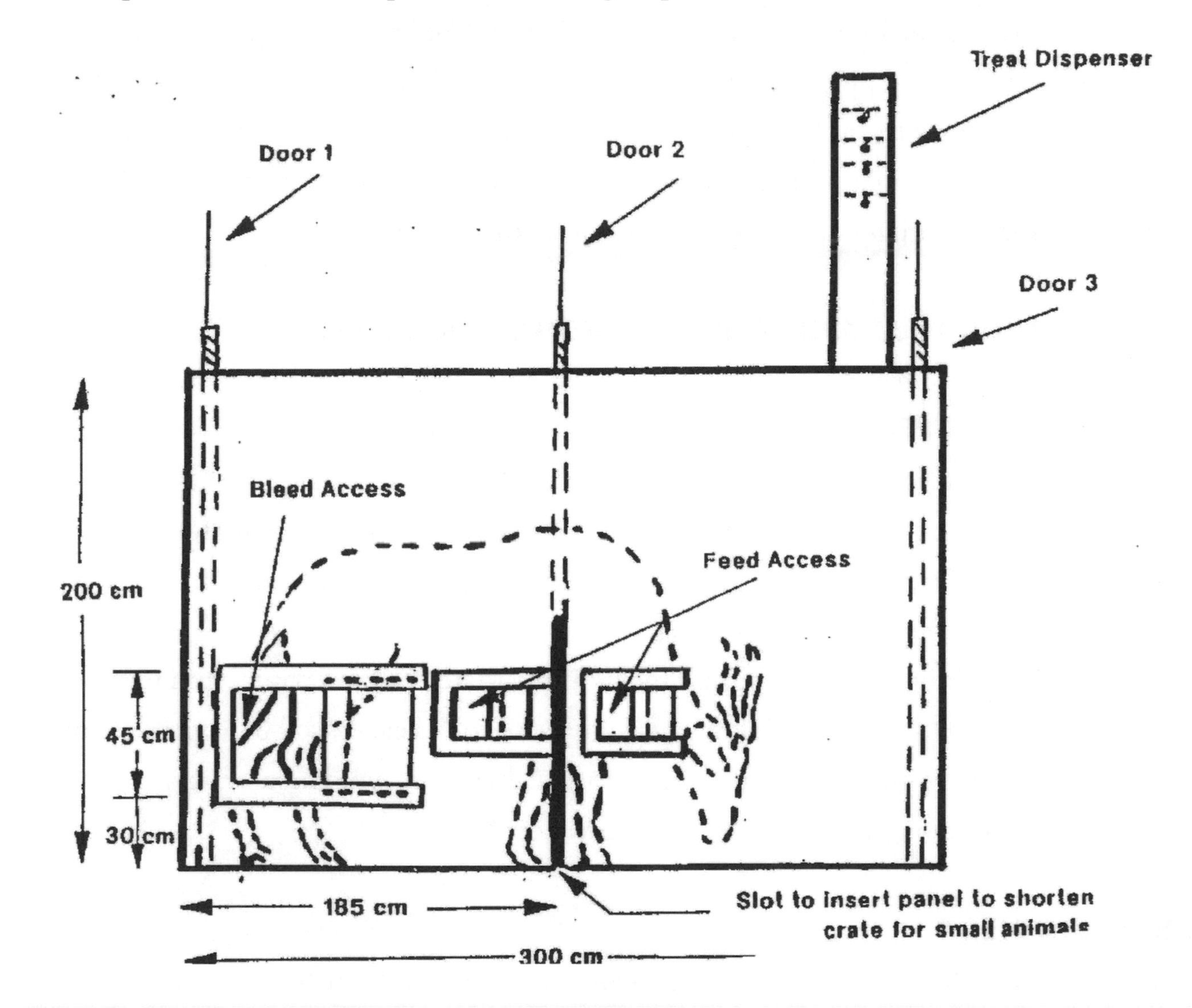

Fig. 1.
Handling crate for Bongo antelope. The 20 × 20-cm square dispenser was made of plywood and stood 120 cm high. It had four flip-down shelves that operated like those of a candy vending machine. Pulling cables that were attached to pins supporting each shelf dispensed spin- ach. Foam tape on the leading edge of each shelf reduced noise when a shelf was released. The crate was created to hold a Bongo antelope who has been conditioned for blood sampling from the rear leg.

CHOOSING FOOD TREATS

The food used to reward the animals must be a highly palatable treat that is not part of their regular diet. We tried many foods to find the animal's favorite treats. For Nyala, we used yams and carrots. Bongo liked spinach, and bison liked pretzels, peanut butter, and a molasses-based cattle supplement. Early in the conditioning procedure, the treat was used as a "bribe" to induce the ante- lope to enter the crate. Later in the process, the treat was withheld until the animal performed the desired behavior such as standing still. Timing is critical for the animal to associate the treat with the desired behavior.

CONDITIONING TO BLOOD SAMPLING FROM THE REAR LEG

The antelope first were conditioned to having their rear legs and other parts of their hind-quarters touched. At first, they were given treats continually to keep them distracted from being touched. After a few successful blood samplings, they learned that they could kick the needle away. At this point, the treats were withheld until the animals stood still even for an instant. The treat was then given the instant they did the desired behavior. Figure 2 shows blood sampling from the rear leg.

People working with the animals must differentiate between panic- and fear-based kicking and learned avoidance kicking away of the needle. Fear-based behaviors are most likely to occur early in the conditioning process. Learned avoidance occurs when the animal is relaxed enough not to have a panic response.

TRAINING BISON

My student Jennifer Lanier trained bison calves to move voluntarily through a handling facility that consisted of holding pens, a single file chute, and a squeeze chute. It was similar to a cattle-handling facility. She had 18 days to train bison before shipment to the National Western Livestock Show in Denver.

Bison are more flighty and fearful than cattle but less flighty than antelope (J. Lanier, Grandin, Chaffin & Chaffin, 1999). The entire training period for bison was shorter than the training period for antelope. The first step was to introduce the animals to operant

Fig. 2.
Blood sampling from the rear leg of a cooperative, non-sedated Nyala antelope. Each new step in the habituation process had to be introduced in small increments to avoid a massive panic reaction. The use of a needle with a plastic tube attached makes it easier to collect the blood into a syringe.

conditioning procedures. They learned that a whistle signaled a treat reward. The reason for using the whistle was to improve safety for the handler. Unlike the antelope, who were inside a box, the bison were loose in a pen. It is safer for the handler if the animals associate treats with the whistle. This helps pre- vent the bison from mobbing the handler. The bison then were taught that standing still would be rewarded with treats—after they had walked through the handling facility. Over the period of 18 days, bison were walked through the facility and trained to stand in the squeeze chute. During the 18-day period, new sights and sounds gradually were introduced. The animals were habituated to the sounds of the squeeze chute clanking and banging. They were then walked through the chute and then given treats. Approximately halfway through the 18-day training period, they were stopped briefly in the squeeze chute by shutting the gates. Unfortunately, there was not enough time to train them voluntarily to accept having their necks clamped in the head gate.

RESULTS

Antelope

Blood sampling from the rear leg and injections could be done easily in cooperative Nyala and Bongo. Three Bongo antelope who had been conditioned to stand in the box had low cortisol levels that averaged 4.4 ± 3.4 ng/ml in a male and 8.5 ± 4.7 ng/ml in two females. The un-sedated animals were held in the box for 10–20 min to allow cortisol levels to rise. Glucose and creatine phosphokinase levels also were obtained and compared to levels in the animals' zoo hospital records. The mean glucose value for crate-conditioned animals was 61.25 ± 19.45 mg/dl versus 166.5 ± 54.59 mg/dl in individuals immobilized with a dart or pole injection. The mean creatine phosphokinase values were 70.75 ± 17.9 IU in crate-conditioned and 288.75 ± 194 IU in immobilized animals. Other veterinary procedures that have been con- ducted include injections in the muscle and tuberculin testing.

Bison

Approximately half of 13 bison calves received training, and the other half served as controls. There was no difference in weight gain between trained and untrained animals. After training, all bison calves were brought to the National Western Stock Show in Denver. Observers who did not know the animals observed both trained and untrained calves. When the animals were handled at Denver, the trained calves were definitely calmer. Untrained calves attempted to jump over fences and were harder to drive into the squeeze chute. However, there were no observable differences between trained and untrained bison when their necks were clamped in the head gate.

DISCUSSION

Conditioning and habituating flighty animals to handling procedures will reduce stress. The conditioned Bongo antelope had lower cortisol values when com- pared with untrained domestic cattle restrained in a squeeze chute. Typical cortisol values for untrained domestic cattle for squeeze chute restraint range from 20 to 63 ng/ml (Crookshank, Elisalde,

White, Clanton, & Smalley, 1979; Grandin, 1997; Lay et al., 1992; Mitchell, Hattingh, & Ganhao, 1988; Zavy, Juniewiez, Phillips, & von Tungelin, 1992). Morton, Anderson, Foggin, Kock, and Tiran (1995) obtained plasma cortisol levels for both physical restraint and chemical immobilization in six antelope species in Zimbabwe. They ranged from a high of 120.4 ng/ml in un-sedated physically restrained waterbuck to 20.04 ng/ml in sedated Eland.

Both bison and antelope can make very specific sound and sight discriminations. Novel stimuli can trigger a massive panic response. Nyala antelope who had become fully habituated to zoo visitors along the fence of their exhibit panicked and crashed into the fence when they saw people on the roof of their barn. They had learned that people along the fence were not a threat, but people on the roof were frightening.

After the antelope in this study were fully conditioned, a handling session once a month was sufficient to maintain conditioning. We also observed that new people were able to handle the animals, but the veterinarian who previously had shot them with the dart was not able to handle them. They became agitated if they heard either his voice or footsteps.

Bison were able to differentiate between the sound of opening and closing gates on the squeeze chute from the sound of a person shaking the chute like a struggling animal. Bison who had become fully habituated to the noisy ratchet latches on the gates became visibly agitated when a person shook the chute like a struggling animal. It took several days to habituate them to the new sound. Bongo antelope were very much aware of any strange noise. They immediately turned their heads and oriented toward the sound of a video camera. Megan Phillips, their trainer, re- ported that she was able to calm them when they heard her familiar voice.

With both bison and antelope, having a solid barrier between the animal and the person was important. Although the animal still could see the person, the presence of the barrier helped keep the animal calm. Observations indicate that ruminant animals attempt to place a barrier between themselves and a perceived threat. Both cattle and bison will circle when threatened, and the dominant, strongest animals will push themselves to the middle of the mob. Domestic cattle remain calmer in chutes with solid sides (Grandin, 1980, 2000).

Because of their flighty nature, bison and antelope will remain calm only when they are subjected to the procedures for which they have been conditioned specifically. The bison were trained to walk through the chutes and stand in the squeeze chute with both the head gates tailgate closed. Due to time constraints, there was not enough time to gradually introduce them to having their necks restrained in the head gate. When they went to the Denver Stock Show, they remained calm while walking through the chute but became highly agitated when their necks were clamped in the head gate. People working with flighty animals must be aware that their learning is very specific. Although their learning generalized to a new but similar chute in Denver, they became highly agitated when suddenly subjected to a totally novel procedure.

REFERENCES

Calle, P. P., & Bornmann, J. C. (1988). Giraffe restraint habituation and desensitization at Cheyenne Mountain Zoo. *Zoo Biology*, 7, 243–252.

Crookshank, H. R., Elisalde, M. H., White, R. G., Clanton, D. C., & Smalley, H. E. (1979). Effect of transportation and handling of calves upon blood serum composition. *Journal of Animal Science*, 48, 430–435.

Dantzer, R., & Mormede, P. (1983). Stress in farm animals: A need for re-evaluation. *Journal of Animal Science*, 57, 6–18.

Grandin, T. (1980). Observations of cattle behavior applied to the design of cattle handling facilities. *Applied Animal Ethology*, 6, 19–31.

Grandin, T. (1989). Voluntary acceptance of restraint by sheep. *Applied Animal Behavior Science*, 23, 257–261.

Grandin, T. (1997). Assessment of stress during handling and transport. *Journal of Animal Science*, 75, 249–257.

Grandin, T. (2000). *Livestock Handling and Transport* (2nd ed.). Wallingford Oxon, England: CAB International.

Grandin, T., & Deesing, M. J. (1998). Behavioral genetics and animal science. In T. Grandin (Ed.), *Genetics and the Behavior of Domestic Animals* (pp. 1–30). San Diego, CA: Academic.

Grandin, T., Rooney, M. B., Phillips, M., Cambre, R. C., Irlbeck, N. A., & Graffam, W. (1995). Conditioning of Nyala (*Tragelaphus angasi*) to blood sampling in a crate with positive reinforcement. *Zoo Biology*, 14, 261–273.

Lanier, E. K. T., Friend, T. H., Bushong, D. M., Knabe, D. A., Champney, T. H., & Lay, D. G. (1995). Swim habituation on a model for stress and distress in the pig Abstract. *Journal of Animal Science*, 73(Suppl. 1), 126.

Lanier, J., Grandin, T., Chaffin, A., & Chaffin, T. (1999, October–December). Training American Bi- son (Bison bison) calves. *Bison World*, pp. 94–99.

Lay, D. C., Friend, T. H., Randel, R. D., Bowers, C. L., Grisson, K. K., & Jenkins, O. C. (1992). Behavioral and physiological effects of freeze and hot iron branding on crossbred cattle. *Journal of Animal Science*, 70, 330.

Mathews, L. (2000). Deer handling and transport. In T. Grandin (Ed.), *Livestock Handling and Transport* (2nd ed., pp. 331–362). Wallingford Oxon, England: CAB International.

Miller, N. E. (1960). Learning resistance to pain and fear effects of overlearning exposure and rewarded exposure in context. *Journal of Experimental Psychology*, 60, 137–145.

Mitchell, G., Hattingh, J., & Ganhao, M. (1988). Stress in cattle assessed after handling, transport and slaughter. *Veterinary Record*, 122, 201–205.

Moberg, G. P., & Wood, V. A. (1982). Effect of differential rearing on the behavioral and adrenocortical response of lambs to a novel environment. *Applied Animal Ethology*, 8, 269.

Morton, D. J., Anderson, E., Foggin, C. M., Kock, M. D., & Tiran, E. P. (1995). Plasma cortisol as an indicator of stress due to capture and translocation in wildlife species. *Veterinary Record*, 136, 60–63.

Panepinto, L. M. (1983). A comfortable minimum stress method of restraint for Yucatan miniature swine. *Laboratory Animal Science*, 33, 95–97.

Phillips, M., Grandin, T., Graffam, W., Irlbeck, N. A., & Cambre, R. C. (1998). Crate conditioning of Bongo (Tragelephus eurycerus) for veterinary and husbandry procedures at the Denver Zoological Gardens. *Zoo Biology*, 17, 25–32.

Price, E. O. (1998). Behavioral genetics and the process of animal domestication. In T. Grandin (Ed.), *Genetics and the Behavior of Domestic Animals* (pp. 31–66). San Diego, CA: Academic.

Stephens, D. B., & Toner, J. N. (1975). Husbandry influences on some physiological parameters of emotional responses in calves. *Applied Animal Ethology*, 1, 233–243.

Wienker, W. R. (1986). Giraffe squeeze cage procedure. *Zoo Biology*, 5, 371–377.

Zavy, M. T., Juniewiez, P. E., Phillips, W. A., & von Tungelin, D. L. (1992). Effects of initial restraint, weaning and transport stress on baseline and ACTM stimulated cortisol responses in beef calves of different genotypes. *American Journal of Veterinary Research*, 53, 552–557.

FOURTEEN

Assessment of Stress During Handling and Transport

T. Grandin (1997) Assessment of Stress During Handling and Transport, *Journal of Animal Science*, Vol. 75, pp. 249-257.

This was one of the first papers where neuroscience literature about fear was introduced to the animal science research community. These were new concepts for both the veterinary and animal science world. In this paper, I reviewed old neuroscience studies that showed that animals experienced the emotion of fear. I also discussed the effects of novelty, which could be either fear-provoking or attractive to animals. New things are frightening if they suddenly appear and attractive when animals are allowed to voluntarily approach them. This paper has over 1,000 citations. It also showed the importance of communicating research from neuroscience researchers to animal scientists and veterinarians. Cross-disciplinary research is very important for scientific progress.

ABSTRACT

Fear is a very strong stressor, and the highly variable results of handling and transportation studies are likely to be due to different levels of psychological stress. Psychological stress is fear stress. Some examples are restraint, contact with people, or exposure to novelty. In many different animals, stimulation of the amygdala with an implanted electrode triggers a complex pattern of behavior and autonomic responses that resemble fear in humans. Both previous experience and genetic factors affecting temperament will interact in complex ways to determine how fearful an animal may become when it is handled or transported. Cattle trained and habituated to a squeeze chute may have baseline cortisol levels and be behaviorally calm, whereas extensively reared animals may have elevated cortisol levels in the same squeeze chute. The squeeze chute is perceived as neutral and nonthreatening to one animal; to another animal, the novelty of it may trigger intense fear. Novelty is a strong stressor when an animal is suddenly confronted with it. To accurately assess an animal's reaction, a combination of behavioral and physiological measurements will provide the best overall measurement of animal discomfort.

INTRODUCTION

Studies to determine the amount of stress on farm animals during routine handling and transport often have highly variable results and are difficult to interpret from an animal welfare standpoint. This paper will cover some of the factors that influence how an animal may react during handling. Much of the variability between handling studies is likely to be due to different levels of psychological stress. Animals can be stressed by either psychological stress:

- restraint
- handling
- or novelty

or physical stresses:

- hunger
- thirst
- fatigue
- injury
- or thermal extremes

Procedures such as restraint in a squeeze chute do not usually cause significant pain, but fear may be a major psychological stressor in extensively raised cattle. Many apparently conflicting results of different studies may be explained if the varying amounts of psychological stress and physical stress within each study are considered. Fear responses in a particular situation are difficult to predict because they depend on how the animal perceives the handling or transport experience. The animal's reactions will be governed by a complex interaction of genetic factors and previous experiences. For example, animals with previous experiences with rough handling will remember it and may become more stressed when handled in the future than animals that have had previous experiences with gentle handling. Previous handling experiences may interact with genetic factors. Rough handling may be more detrimental and stressful to animals with an excitable temperament compared to animals with a more placid temperament. For example, Brahman cross cattle had higher cortisol levels when restrained in a squeeze chute than English crosses (Zavy et al., 1992). An animal's social rank within the group can also affect stress levels. McGlone et al. (1993) found that subordinate submissive pigs were more stressed by 4 hours of transport than dominant pigs. This paper will only address short-term stressors such as handling and transport. The measurement of chronic stress imposed by the environment or different housing systems is much more complex.

IMPORTANCE OF FEAR AND EFFECTS OF NOVELTY

Fear is a universal emotion in the animal kingdom and motivates animals to avoid predators. All vertebrates can be fear-conditioned (LeDoux, 1994). The amygdala in the brain is probably the central fear system that is involved in both fear behavior and the acquisition of conditioned fear (Davis, 1992). Davis (1992) cited over 20 animal studies from many different laboratories that showed that electrical stimulation of the amygdala with an implanted electrode triggers a complex pattern of behaviors and changes in autonomic responses that resembles fear in hum ans. In humans, electrical stimulation of the amygdala elicits feelings of fear (Gloor et al., 1981). Studies have also shown that electrical stimulation of the amygdala will increase plasma corticosterone in cats (Setekleiv et al.,

1961; Matheson et al., 1971) and in rats (Redgate and Fahringer, 1973). Lesioning of the amygdala will block both unconditioned and conditioned fear responses (Davis, 1992). Large lesions in the amygdala will reduce emotionality in wild rats as measured by flight distance (Kemble et al., 1984). Kemble et al. also noted that lesioning of the amygdala had a taming effect on wild rats. LeDoux (1994) explains that fear conditioning takes place in a subcortical pathway and that extinguishing a conditioned fear response is difficult because it requires the animal to suppress the fear memory via an active learning process. A single, very aversive event can produce a strong conditioned fear response, but extinguishing this fear response is much more difficult.

Observations by the author on cattle ranches have shown that to prevent cattle and sheep from becoming averse and fearful of a new squeeze chute or corral system, painful or highly aversive procedures should be avoided the first time the animals enter the facility. The same principle also applies to rats. Rats that receive a strong electrical shock the first time they enter a novel alley will refuse to enter it again (Miller, 1960). However, if the rat is subjected to a series of shocks of gradually increasing intensity, it will continue to enter the alley to get a food reward. Therefore, Hutson (1993) recommends that stress in sheep during routine handling could be reduced if the animals were conditioned gradually to handling procedures. Less severe procedures should be done first (Stephens and Toner, 1975; Dantzer and Mormede, 1983).

Novelty is a very strong stressor (Stephens and Toner, 1975; Moberg and Wood, 1982; Dantzer and Mormede, 1983). This is especially true when an animal is suddenly confronted with it. In the wild, novelty and strange sights or sounds are often a sign of danger (Grandin, 1993a). Cattle will balk at shadows or differences in flooring during movement through handling facilities (Grandin, 1980). Pigs that have been trained to tolerate laboratory procedures will respond to deviations in their daily routine with a rise in blood pressure (Miller and Twohill, 1983). Reid and Mills (1962) have suggested that livestock can be trained to accept changes in management routines that would cause a significant increase in physiological measurements in animals that had not been trained. Gradual exposure of animals to novel experiences enables them to become accustomed to non-painful stimuli that had previously evoked a flight reaction. Grandin et al. (1995b) reported that training nyala antelope to cooperate during blood sampling had to be done very slowly to avoid triggering a massive flight reaction. The animals are very vigilant and will react to any unfamiliar sights and sounds.

There are some situations in which novelty is attractive to animals. Cattle and pigs often approach and manipulate a piece of paper dropped on the ground. The author has observed that the same piece of paper will cause animals to balk and jump away if they are being forced to walk toward it. Therefore, the paper may be perceived as threatening in one situation and nonthreatening in another. The author has observed that cattle in the Philippines seldom react to cars, trucks, and other distractions when they graze on the highway median strip. Cars and trucks are no longer novel because they have seen them since birth. In the nyala antelope, animals born after the adults had been trained to blood sampling procedures learned to cooperate more quickly (Grandin et al., 1995b).

Cattle can become accustomed to repeated non-aversive procedures such as weighing or drawing blood through an indwelling catheter (Peischel et al., 1980; Alam and Dobson, 1986). Sheep, pigs, and giraffes have been trained to voluntarily enter a restraint device (Panepinto, 1983; Wienker, 1986; Grandin, 1989).

However, animals do not habituate to procedures that are very aversive (Hargreaves and Hutson, 1990a). A procedure can be highly aversive without being painful. Full inversion to an upside-down position is extremely aversive to sheep. The time required to drive sheep down a race into a restraint device that inverted them increased the following year (Hutson, 1985). Cortisol levels did not decrease with experience when cattle were subjected to repeated truck trips during which they fell down (Fell and Shutt, 1986). Hargreaves and Hutson (1990a) found that repeated trials of a sham shearing procedure failed to reduce the stress response. Sheep also did not habituate to 6 hours of restraint with their legs tied (Coppinger et al., 1991). Apple et al. (1995) found that in sheep, 6 hours of restraint stress caused dark cutting meat and very high (>110 ng/mL) levels of cortisol. Epidural blockage with lidocaine, which prevents the animals from contracting their muscles and straining against the restraint, failed to inhibit glycogen metabolism. This experiment indicates that psychological stress was probably a significant factor.

Cattle are very sensitive to the relative aversiveness of different parts of handling procedures. When they were handled every 30 days in a squeeze chute and a single animal scale, balking at the scale decreased with successive experience and balking at the squeeze chute increased slightly (Grandin, 1992). The animals learned that the scale never caused discomfort. Cattle that had been mishandled in a squeeze chute and struck hard on the head by the headgate were more likely to resist entry into the chute in the future (Grandin et al., 1994) compared with cattle that had never been hit with the headgate.

EFFECTS OF ADAPTATION TO HANDLING ON STRESS

Tame animals that are accustomed to frequent handling and close contact with people are usually less stressed by restraint and handling than animals that seldom see people. Binstead (1977), Fordyce et al. (1985), and Fordyce (1987) report that training weanling heifer calves produced calmer adult animals that were easier to handle. Training these extensively raised calves involved walking quietly among them, teaching them to follow a lead horseman and quiet walking through chutes. How an animal is handled early in life will have an effect on its physiological response to stressors later in life. Calves on a university experiment station that had become accustomed to petting by visitors had lower cortisol levels after restraint than calves that had less frequent contact with people (Boandle et al., 1989). Lay et al. (1992a) found that restraint in a squeeze chute was almost as stressful as hot iron branding for extensively reared beef cattle. In hand-reared dairy cows, branding was much more stressful than restraint (Lay et al., 1992b).

Taming may reduce the physiological reactivity of the nervous system. Hastings et al. (1992) found that hand-reared deer had lower cortisol levels after restraint compared with free-ranging deer. Even though the physiological response to restraint was lower in the tame animals, hand-reared deer struggled just as violently as free-range deer. Associations that animals make seem to be highly specific. Mateo et al. (1991) found that tame sheep approached a person more quickly, but behavioral measurements of struggling indicated that taming did not generalize to other procedures. Similar findings by Hargreaves and Hutson (1990a,b) showed that gentling and reduction of the sheep's flight zone failed to reduce aversion to shearing. Tame animals can sometimes have an extreme flight reaction when suddenly confronted with novelty that is perceived as a threat. Reports from ranchers and horse trainers indicate that horses and cattle that are calm and easy to handle at their home farm sometimes become extremely agitated when confronted with the novelty of a livestock show or auction. The animal's behavioral reaction seems to be less likely to generalize to other procedures than its physiological reaction. Moberg and Wood (1982) found that experiences during rearing greatly affected behavior in an open field test but had little effect on adrenocortical response of lambs. Exposing piglets to novel noises for 20 min. increases both heart rate and motor activity. Heart rate habituated to a recording of abattoir sounds more quickly than motor activity (Spensley et al., 1995).

The effects of previous experience on an animal's fear response may provide onc explanation for the often variable results in handling and transport studies. For example, extensively raised animals may have more psychological or fear stress during loading and unloading for transport compared to more intensively reared animals. British researchers have found that loading and unloading of sheep and calves was the most stressful part of the journey (Trunkfield and Broom, 1990; Knowles, 1995). Kenney and Tarrant (1987) reported that for Irish cattle, the actual journey was more stressful than loading and unloading. The physical stresses of the trip, such as jiggling, were more stressful than the psychological stresses of loading or unloading. A possible explanation for this discrepancy between these two studies may be the amount of contact the animals had with people. There may be a big difference in the degree of fear stress between U.S. cattle reared on range land where they seldom see people and European pasture-reared cattle. Differences in the degree of psychological stress may explain why too many rest stops during long-distance transport is detrimental to the health of weaner calves raised under U.S. conditions. Cattle feeders have learned from practical experience that 200 to 300kg calves shipped from the southeast to Texas will have fewer health problems if they are transported nonstop for the entire 32-hour trip. For these extensively reared calves, rest stops may possibly turn into stress stops. Research is needed to conclusively determine what factors cause the rest stops to be stressful. Legislating too many rest stops may be detrimental to welfare. One possibility is fear stress during loading and unloading at rest stops and the second possibility is that the calves become infected with diseases at the rest stop. Many of the calves shipped on these trips are not properly vaccinated. There may be an interaction between rest stops and disease. Frequent rest stops may be more beneficial to fully vaccinated calves.

GENETICS

Genetic factors such as temperament interact in complex ways with an animal's previous handling experiences and learning to determine how it will react during a particular handling procedure. Wild species are usually more reactive to novel stimuli than domesticated animals. Price (1984) maintains that the domestic phenotype have reduced responses to changes in the environment. Domesticated animals are more stress-resistant because they have been selected for a calm attitude toward people (Parsons, 1988). When deer or antelope are tamed, the flighty temperament is masked until they are confronted with a novel stimulus that is perceived as threatening. A tame deer or antelope can have an explosive reaction to a novel event. A wild species has a more intense flight response because this enables it to flee from predators.

Temperament in cattle is a heritable trait that may affect the animal's reaction to handling (Le Neindre et al., 1995). There are differences in temperament both between and within cattle breeds. Within the Brahman breed, temperament is heritable (Hearns haw et al., 1979; Fordyce et al., 1988). Temperament differences between breeds have also been reported by Stricklin et al. (1980) and Tulloh (1961). Genetics also affects an animal's response to stress. Brahman cross cattle had higher cortisol levels while restrained in a squeeze chute compared to English crosses (Zavy et al., 1992). Recent research by Grandin et al. (1995a) and replicated by H. Randle (1995, personal communication, University of Plymouth, U.K.) indicated that the spiral hair whorl on a bovine's forehead is an indicator of temperament. Cattle with spiral hair whorls above the eyes became more agitated while restrained than animals with hair whorls below the eyes.

Temperament may be under genetic control in many different animals. Research with rats has shown that they can be selected for either high or low emotionality (Fujita et al., 1994) or for reduced fear induced aggressiveness toward humans (Popova et al., 1993). Phenotypic characteristics are also related to temperament. Interestingly, it seems that different genetic factors control fear-induced aggression and intermale aggression. Selection for reduced fear induced aggression had no effect on aggressive behavior toward other male rats.

Temperament is a trait that seems to be stable over time. In European Continental-cross cattle, certain individuals became extremely agitated every time they were handled in a squeeze chute and others were always calm (Grandin, 1992). The agitated animals failed to adapt to being held in the squeeze chute during four handling sessions spaced 30 days apart. Cattle with a very excitable temperament may have greater difficulty adapting to repeated non painful handling procedures than cattle with a calmer temperament. The two types of animals may have differing physiological and behavioral reactions to the same procedure. Animals with a calm temperament may adapt more easily and become less stressed with repeated handling treatments and animals with a very excitable temperament may become increasingly stressed with each repeated handling treatment. Lanier et al. (1995) found that some pigs habituated to a swimming task and maintained near baseline levels of epinephrine and norepinephrine and other animals failed to habituate and never adapted.

At five slaughter plants in the United States, Holland, and Ireland, the author has observed increasing problems with very excitable pigs and cattle from certain genetic lines that become highly agitated. It is almost impossible to drive them quietly through a high-speed slaughter line. These animals seem to have a much stronger startle reaction to novelty, are more likely to balk at small distractions such as shadows or reflections in the race and are more likely to bunch together. Observations at slaughter plants and reports from ranchers also indicate that excitable cattle are more likely to injure themselves when they are confronted with the novel, unfamiliar surroundings of an auction market or slaughter plant. The appearance of greater numbers of more excitable pigs and cattle may possibly be related to the increasing emphasis of the livestock industry on lean beef and pork. In both cattle and pigs, the author has observed that excessive excitability occurs most often in animals bred for leanness that have a slender body shape and fine bones. Cattle and pigs bred for large, bulging lean muscles usually have a calmer temperament. This is an area that needs to be researched. Practical experience indicates that the excitable animal problem needs to be corrected because excessive excitability creates serious animal welfare problems during handling at auction markets and slaughter plants.

Cattle and pig producers need to select animals with a calm temperament, but care must be taken not to over-select for any one particular trait. A good example of over selection for a single trait is the halothane gene in pigs. Pigs with this gene have increased meat production, but the price for this increased production is poor meat quality (Pommier and Houde, 1993). Over-selection for calm temperament may possibly have detrimental effects on economically important traits, such as maternal ability. Researchers in Russia found that selecting foxes for calmness over 80 years produced animals that lost their seasonal breeding pattern and had strange piebald black and white colored coats (Belyaev, 1979; Belynev and Borodin, 1982). The foxes turned into animals that acted and looked like Border collies.

FEAR PHEROMONES

Another factor that could confound handling stress studies is fear pheromones. Vieville-Thomas and Signoret (1992) found that urine from a stressed gilt caused other gilts to avoid a feed dispenser and urine from an unstressed animal had no effect. Both the results of this experiment and observations by the author indicate that it takes 10 to 15 min. for the fear pheromone to be secreted. Observations by the author indicate that cattle will voluntarily walk into a restraining chute that is covered with blood, but if an animal becomes extremely agitated for several minutes, the other animals refused to enter (Grandin, 1993b). In a laboratory setting pigs witnessing slaughter had no increases in either beta endorphins or cortisol. These were calm animals fitted with jugular catheters (Anil et al., 1995). Eibl-Eibesfeldt (1970) observed that if a rat is instantly killed by a trap, the trap will remain effective and can be used again. Rats will avoid a trap that failed to instantly kill. Research with rats indicates that blood may contain a fear pheromone (Stevens and Gerzog-Thomas, 1977). Stevens and Saplikoski (1973) found that blood and muscle tissue from stressed rats was avoided in a choice test, whereas brain tissue and water had no effect. Blood from guinea pigs and people also had little effect (Hornbuckle and Beall, 1974).

SHORT-TERM STRESS MEASUREMENTS

This discussion will be limited to measuring short-term stress induced by handling procedures such as being held in a squeeze chute. Assessment of stress and discomfort should contain both behavioral and physiological measures. Behavioral indicators of discomfort are attempting to escape, vocalization, kicking, or struggling. Other behavioral measures of how an animal perceives a handling procedure are choice tests and aversion tests. Common physiological measures of stress are cortisol, beta endorphin, and heart rate. Cortisol is a useful indicator of short-term stresses from handling or husbandry procedures such as castration. Researchers must remember that cortisol is a time-dependent measure that takes 10 to 20 min to reach peak values (Lay et al., 1992a).

A review of many studies indicates that cortisol levels in cattle fall into three categories:

1. baseline
2. levels that occur during restraint in a headgate, and
3. extreme stress (Table 1)

Cortisol levels are highly variable and absolute comparisons should not be made between studies, but the figures on Tables 1 and 2 would make it possible to determine whether a handling or slaughter procedure was either very low stress or very high stress. One could tentatively conclude that a mean value of >70 ng/mL in either steers or cows

TABLE 1: Mean Cortisol Values in Cattle during Handling

Cortisol level, ng/mL	Breed	Gender	Study
Baseline			
.5 to 2	Friesian	Bulls	Tennessen et al., 1984
2	Friesian	Cows	Alam and Dobson, 1986
3	Angus cross	Bull calves	Henricks et al., 1984
6	Angus cross	Heifer calves	Henricks et al., 1984
9	Friesland and Nuguni	Cows	Mitchell et al., 1988
Restraint in Headgate			
13	Holstein cows	Hand-reared	Lay et al., 1992a
24 (weaned 2 wk before test)	Unknown British or European	Weanlings, mixed genders	Crookshank et al., 1979
27	Brahman cross	Steers	Ray et al., 1972
28	Angus X Hereford	Steers	Zavy et al., 1992
30	Simmental X Hereford X Brahman	83% Steers	Lay et al., 1992b
36	Angus X Brahman	Steers	Zavy et al., 1992
46 (weaned day of test)	Unknown British or European	Mixed genders	Crookshank et al., 1979
63	Brahman X Hereford X Afrikander	Steers and heifers	Mitchell et al., 1988
Extreme Value			
93	Unknown British or European	Mixed	Dunn, 1990

TABLE 2: Mean Cortisol Values during Slaughter		
Cortisol level, ng/mL	**Handling Methods**	**Study**
	Baseline Quiet Research Abattoir	
15	Held in head restraint, shot immediately with captive bolt	Tume and Shaw, 1992
	Commercial Slaughter Plant	
24	Handled quietly in conventional stunning box	Ewbank et al., 1992*
32	Unknown	Mitchell et al., 1988
44	Conventional stunning box	Tume and Shaw, 1992
45	Conventional stunning box	Dunn, 1990*
51	Poorly designed head restraint only 14% of cattle voluntarily entered it	Ewbank et al., 1992*
63 (median)	Electric prod all cattle, 38% animals slipped, conventional stun box	Cockram and Corley, 1991*
	Extreme Stress	
93	Inverted on back for 103 s	Dunn, 1990*

* Conducted in either England or Ireland with *Bos taurus* cattle.

would possibly be an indicator of either rough handling or poor equipment, and low values close to the baseline values would indicate that a procedure was either low stress or was very quick. Quick procedures would be completed before cortisol levels could rise. Restraint in a headgate for blood sampling and slaughter produced similar values (Tables 1 and 2). Mature bulls have much lower cortisol levels than steers, cows, or heifers (Tennessee et al., 1984). In one study, there was an extreme mean of 93 ng/mL for inverting cattle on their backs for 103 seconds (Dunn, 1990). This very high figure is not due to differences in assay methods because this same researcher obtained more reasonable values of 45 ng/mL for upright restraint. Properly performed cattle slaughter seems to be no more stressful than farm restraint (Tables 1 and 2).

Less clear-cut ranges have been obtained in sheep. Pearson et al. (1977) found that slaughter in a quiet research abattoir produced lower cortisol levels than slaughter in a noisy commercial plant. The values were 40 vs 61 ng/mL. Values for shearing and other on-farm handling procedures were 73 ng/mL (Hargreaves and Hutson, 1990c,d) and 72 ng/mL (Kilgour and de Langen, 1970). Prolonged restraint and isolation for 2 hours increased cortisol levels up to 100 ng/mL (Apple et al., 1993).

Creatine phosphokinase (CPK) and lactate seem to be useful measures for assessing handling stresses in pigs (Warris et al., 1994). Warris et al. (1994) found that the sound level of squealing pigs in a commercial abattoir was highly correlated with CPK measurements. White et al. (1995) also reported that vocalizations in pigs were indicative of stress and were correlated with other measures of acute stress, such as heart rate. Cattle that become behaviorally agitated have higher cortisol levels (Stahringer et al., 1989). Heart rate in cattle during restraint in a squeeze chute was highly correlated with cortisol levels (Lay et al., 1992a,b). Stermer et al. (1981) found that rough handling in poorly designed facilities resulted in greater heart rates than quiet handling in well-designed facilities.

Isolation is also a factor in handling stress. During restraint for routine husbandry procedures, animals are often separated from their conspecifics. Stookey et al. (1994) found that cattle became less behaviorally agitated during weighing on a single animal scale if they could see another animal in the chute less than 1 meter away in front of the scale. Agitation was measured electronically by measuring movement and jiggling via the scale load cell system. Numerous studies have shown that isolation from conspecifics will raise cortisol and other physiological measures (Kilgour and deLangen, 1970; Whittlestone et al., 1970; Arave et al., 1974).

AVERSION TESTS

Aversion to a handling procedure can be measured by either choice testing or measuring aversion. One measure of aversion is the time required to induce an animal to reenter a chute where it was previously handled (Rushen, 1986a,b 1995). In a choice test, the animals are allowed to choose between two different chutes that lead to different procedures (Grandin et al., 1986; Rushen and Congdon, 1986a,b). Another useful measure is the degree of force required to induce an animal to move through a race. In some cases, measuring the degree of force provides a more accurate assessment of aversion than time. Examples of force are the number of pats on the rump or number of electrical prods. Experience and genetic factors can confound aversion tests. Rushen (1996) warns that to accurately measure aversion in a race, the animal must experience the aversive procedure more than once. Observations by the author indicate that excitable cattle sometimes run through a single file chute quickly in an attempt to escape. Research (in progress by Bridgette Voisinet and the author) reveals that bulls trained to move through a race to a squeeze chute exhibit no aversion in the race after a single noxious treatment. After one aversive treatment, they continued to voluntarily walk through the race into the squeeze chute but balking and turning back in the crowd pen at the entrance to the race greatly increased. At this point, the animals may perceive that they may be able to avoid reentering the race. In aversion studies, balking and other behaviors indicative of aversion must be measured in both the single file race and in the pens and alleys that lead up to the entrance of the single file race. This is especially important if the aversive procedure is performed only once. After the animal is forced to enter the chute that leads to the squeeze, it may perceive that it may be able to escape by running quickly through it toward the squeeze chute. Under certain conditions, choice tests may be unreliable for measuring choices between mildly aversive procedures. Research conducted by Grandin et al. (1994) showed that cattle are reluctant to change a previously learned choice if the two choices in a choice test are only mildly aversive. Other research showed that sheep immediately switched sides to avoid highly aversive electro-immobilization (Grandin et al., 1986).

IMPLICATIONS

Both researchers and people making decisions about animal welfare must understand that fear during nonpainful routine handling and transport can vary greatly. Fear is a very strong stressor. Cattle that have been trained and habituated to a handling procedure may be completely calm and have baseline cortisol and heart rate measurements during handling and restraint. Extensively reared cattle with an excitable disposition may have very high cortisol levels and show extreme behavioral agitation during the same procedure. For one animal, a squeeze chute may be perceived as neutral and nonthreatening, but to another it may trigger an extreme fear response. The animal's response will be determined by a complex interaction of genetics and previous experience. Studies to assess animal welfare during handling and transport should contain both behavioral and physiological measurements.

REFERENCES

Alam, M.G.S., and H. Dobson (1986) Effect of various veterinary procedures on plasma concentrations of cortisol, luteinizing hormone and prostaglandin E2 metabolite in the cow. *Vet. Rec.* 118:7

Anil, M. H., J. L. McKinstry, M. Field, M. Bracke, and R. G. Rodway (1995) Assessment of distress experienced by witnessing slaughter in pigs. *Proc. Br. Soc. Anim. Sci.*, paper 190

Apple, J. K., M. E. Dikeman, J. E. Minton, R. M. McMurphy, M. R. Fedde, D. E. Leith, and J. A. Unrah (1995) Effects of restraint and isolation stress and epidural blockade on endocrine and blood metabolite status, muscle glycogen depletion, and incidence of dark cutting longissimus muscle in sheep. *J. Anim. Sci.* 73:2295

Apple, J. K., J. E. Minton, K. M. Parsons, and J. A. Unrah (1993) Influence of repeated restraint and isolation stress and electrolyte administration on pituitary adrenal secretions, electrolytes and other blood constituents of sheep. *J. Anim. Sci.* 71:71

Arave, C. W., J. L. Albright, and C. L. Sinclair (1974) Behaviour, milk yield and leucocytes of dairy cows in reduced space and isolation. *J. Dairy Sci.* 59:1497

Belyeev, D. K. (1979) Destabilizing selection as a factor in domestication. *J. Hered.* 70:301

Belyaev, D. K, and P. M. Borodin (1982) The influence of stress on variation and its role in evolution. *Biol. Zentbl.* 100:705

Binstead, M. (1977) Handling cattle. *Queensland Agric. J.* 103:293

Boandle, K. E., J. E. Wohlt, and R. V. Carsia (1989) Effect of handling, administration of a local anesthetic and electrical dehorning on plasma cortisol in Holstein calves. *J. Dairy Sci.* 72:2193

Cockram, M. S., and K.T.T. Corley (1991) Effect of preslaughter handling on the behavior and blood composition of beef cattle. *Br. Vet. J.* 147:444

Coppinger, T. R., J. E. Minton, P. G. Reddy, and F. Blecha (1991) Repeated restraint and isolation stress in lambs increases pituitary-adrenal secretions and reduces cell-mediated immunity. *J. Anim. Sci.* 69:2808

Crookshank, H. R., M. H. Elissalde, R. G. White, D. C. Clanton, and H. E. Smalley (1979) Effect of transportation and handling of calves upon blood serum composition. *J. Anim. Sci.* 48:430

Dantzer, R., and P. Mormede (1983) Stress in farm animals: A need for reevaluation. *J. Anim. Sci.* 57:6

Davis, M. (1992) The role of the amygdala in fear and anxiety. *Annul Rev. Neurosci.* 15:353

Dunn, C. S. (1990) Stress reactions of cattle undergoing ritual slaughter using two methods of restraint. *Vet. Rec.* 126:522

Eibl-Eibesfeldt, I. (1970) *Ethology: The Biology of Behavior*. p 236. Holt Rhinehart and Winston, New York

Ewbank, R., M. J. Parker, and C. W. Mason (1992) Reactions of cattle to head restraint at stunning: A practical dilemma. *Anim. Welfare* 1:55

Fell, L. R., and D. A. Shutt (1986) Adrenal response of calves to transport stress as measured by salivary cortisol. *Canad. J. Anim. Sci.* 66:637

Fordyce, G. (1987) Weaner training. *Queensland Agric. J.* 113:323

Fordyce, G., R. M. Dodt, and J. R. Wythes (1988) Cattle temperaments in extensive herds in northern Queensland. *Aust. J. Exp. Agric.* 28:683

Fordyce, G., M. E. Goddard, R. Tyler, C. Williams, and M. A. Toleman (1985) Temperament and bruising in Bos indicus cattle. *Aust. J. Exp. Agric.* 25:283

Fujita, O., Y. Annen, and A. Kitaoka (1994) Tsukuba Highland low emotional strains of rats (*Rattus norvegicus*): An overview. *Behav. Gen.* 24:389

Gloor, P., A. Olivier, and L. F. Quesney (1981) The role of the amygdala in the expression of psychic phenomena in temporal lobe seizures. In: Y. Ben Avi (Ed.) *The Amygdaloid Complex*. Elsevier, New York

Grandin, T. (1980) Observations of cattle behavior applied to the design of cattle handling facilities. *Appl. Anim. Ethol.* 6:19

Grandin, T. (1989) Voluntary acceptance of restraint by sheep. *Appl. Anim. Behav. Sci.* 23:257

Grandin, T. (1992) Behavioral agitation during handling is persistent over time. *Appl. Anim. Behav. Sci.* 36:1

Grandin, T. (1993a) Handling facilities and restraint of range cattle. In: T. Grandin (Ed.) *Livestock Handling and Transport*. p 43. CAB International, Wallingford, Oxon, U.K.

Grandin, T. (1993b) Handling and welfare of livestock in slaughter plants. In: T. Grandin (Ed.) *Livestock Handling and Transport*. p 289. Wallingford, Oxon, U.K.

Grandin, T., S. E. Curtis, T. M. Widowski, and J. C. Thurmon (1986) Electro-immobilization versus mechanical restraint in an avoid-avoid choice test for ewes. *J. Anim. Sci.* 62:1469

Grandin, T., M. J. Deesing, J. J. Struthers, and A. M. Swinker (1995a) Cattle with hair whorl patterns above the eyes are more behaviorally agitated during restraint. *Appl. Anim. Behav. Sci.* 46:117

Grandin, T., M. B. Rooney, M. Phillips, R. C. Cambre, N. A. Irlbeck, and W. Graffam (1995b) Conditioning of nyala (*Tragelaphus angasi*) to blood sampling in a crate with positive reinforcement. *Zoo Biol.* 14:261

Grandin, T., K G. Odde, D. N. Schutz, and L. M. Beherns (1994) The reluctance of cattle to change a learned choice may confound preference tests. *Appl. Anim. Behav. Sci.* 39:21

Hargreaves, A. L., and G. D. Hutson (1990a) Some effects of repeated handling on stress responses in sheep. *Appl. Anim. Behav. Sci.* 26:253

Hargreaves, A. L., and G. D. Hutson (1990b) The effect of gentling on heart rate, flight distance, and aversion of sheep to a handling procedure. *Appl. Anim. Behav. Sci.* 26:243

Hargreaves, A. L., and G. D. Hutson (1990c) The stress response in sheep during routine handling procedures. *Appl. Anim. Behav. Sci.* 26:83

Hargreaves, A. L., and G. D. Hutson (1990d) Changes in heart rate, plasma cortisol and haematocrit of sheep during a shearing procedure. *Appl. Anim. Behav. Sci.* 26:91

Hastings, B. E., D. E. Abbott, and L. M. George (1992) Stress factors influencing plasma cortisol levels and adrenal weights in Chinese water deer (*Hydropotes inermis*). *Res. in Vet. Sci.* 53: 375

Hearnshaw, H., R. Barlow, and G. Want (1979) Development of a 'temperament' or handling difficulty score for cattle. *Proc. Aust. Assoc. Anim. Breed. and Genet.* 1:164

Henricks, D. M., J. W. Cooper, J. C. Spitzer, and L. W. Grimes (1984) Gender differences in plasma cortisol and growth in the bovine. *J. Anim. Sci.* 59:376

Hornbuckle, P. A., and T. Beall (1974) Escape reactions to the blood of selected mammals by rats. *Behav. Biol.* 12:573

Hutson, G. D. (1985) The influence of barley food rewards on sheep movement through a handling system. *Appl. Anim. Behav. Sci.* 14:263

Hutson, G. D. (1993) Behavioral principles of sheep handling. In: T. Grandin (Ed.) *Livestock Handling and Transport*. CAB International, Wallingford, Oxon, U.K

Kemble, E. D., D. C. Blanchard, R. J. Blanchard, and R. Takushi (1984) Taming in wild rates following medial amygdaloid lesions. *Physiol. Behav.* 32:131

Kenny, F. J., and P. V. Tarrant (1987) The physiological and behavioral responses of crossbred steers to short haul transport by road. *Livest. Prod. Sci.* 17:63

Kilgour, R., and H. de Langen (1970) Stress in sheep resulting from management practices. *Proc. N. Z. Soc. of Anim. Prod.* 30:65

Knowles, T. G. (1995) The effects of transport in slaughter weight lambs. Br. Soc. Anim. Sci., Winter Meeting (Summary), Paper 43

Lanier, E. K, T. H. Friend, D. M. Bushong, D. A. Knabe, T. H. Champney, and D. G. Lay, Jr. (1995) Swim habituation as a model for eustress and distress in the pig. *J. Anim. Sci.* 73(Suppl. 1):126 (Abstract)

Lay, D. C., Jr., T. H. Friend, C. L. Bowers, K. K Grissom, and O. C. Jenkins (1992a) A comparative physiological and behavioral study of freeze and hot-iron branding using dairy cows. *J. Anim. Sci.* 70:1121

Lay, D. C., Jr., T. H. Friend, R. D. Randel, C. L. Bowers, K K Grissom, and O. C. Jenkins (1992b) Behavioral and physiological effects of freeze and hot-iron branding on crossbred cattle. *J. Anim. Sci.* 70:330

LeDoux, J. E. (1994) Emotion, memory and the brain. *Sci. Am.* 271: 50

Le Neindre, P., G. Trillet, J. Sapa, F. Menissier, J. N. Bonnet, and J. M. Chupin (1995) Individual differences in docility of Limousin cattle. *J. Anim. Sci.* 73:2249

Mateo, J. M., D. Q. Estep, and J. S. McCann (1991) Effects of differential handling on the behavior of domestic ewes. *Appl. Anim. Behav. Sci.* 32:45

Matheson, B. K., B. J. Branch, and A. N. Taylor (1971) Effects of amygdaloid stimulation on pituitary adrenal activity in conscious cats. *Brain Res.* 32:151

McGlone, J. J., J. L. Salak, E. A. Lumpkin, R. L. Nicholson, M. Gibson, and R. L. Norman (1993) Shipping stress and social status effects on pig performance, plasma cortisol, natural killer cell activity and leukocyte numbers. *J. Anim. Sci.* 71:888

Miller, N. E. (1960) Learning resistance to pain and fear effects of overlearning, exposure, and rewarded exposure in context. *J. Exp. Psychol.* 60:137

Miller, K N., and S. Twohill (1983) A method for measuring systolic blood pressure in conscious swine (Sus scrofa). *Lab Anim.* 12(6):51

Mitchell, G., J. Hattingh, and M. Ganhao (1988) Stress in cattle assessed after handling, transport and slaughter. *Vet. Rec.* 123: 201

Moberg, G. P., and V. A. Wood (1982) Effect of differential rearing on the behavioral and adrenocortical response of lambs to a novel environment. *Appl. Anim. Ethol.* 8:269

Panepinto, L. M. (1983) A comfortable minimum stress method of restraint for Yucatan miniature swine. *Lab. Anim. Sci.* 33:95

Parsons, P. A. (1988) Behavioral stress and variability. *Behav. Gen.* 18:293

Pearson, A. J., R. Klgour, H. de Langen, and E. Payne (1977) Hormonal responses of lambs to trucking, handling and electric stunning. *N.Z. Soc. Anim. Prod.* 37:243

Peischel, A., R. R. Schalles, and C. E. Owensby (1980) Effect of stress on calves grazing Kansas Flint Hills range. *J. Anim. Sci.* 51(Suppl. 1):245 (Abstract)

Pommier, S. A., and A. Houde (1993) Effect of genotype formalignant hypothermia as determined by a restriction endonuclease assay on the quality characteristics of commercial pork loins. *J. Anim. Sci.* 71:420

Popova, N. J., E. M. Nikulina, and A. V. Kulikov (1993) Genetic analysis of different kinds of aggressive behavior. *Behav. Gen.* 23:491

Price, E. O. (1984) Behavioral aspects of domestication. *Q. Rev. Biol.* 59:1

Ray, D. E., W. J. Hansen, B. Theurer, and G. H. Stott (1972) Physical stress and corticoid levels in steers. *Proc. West. Sect. Am. Soc. Anim. Sci.* 23:255

Redgate, E. S., and E. E. Fahringer (1973) A comparison of pituitary adrenal activity elicited by electrical stimulation of preoptic amygdaloid and hypothalamic sites in the rat brain. *Neuroendocrinology* 12:334

Reid, R. I., and S. C. Mills (1962) Studies of carbohydrate metabolism in sheep. XVI. The adrenal response to physiological stress *Aust. J. Agric. Res.* 13:282

Rushen, J. (1986a) Aversion of sheep to electro-immobilization and physical restraint. *Appl. Anim. Behav. Sci.* 15:315

Rushen, J. (1986b) Aversion of sheep for handling treatments: Paired-choice studies. *Applied Animal Behavioural Science* 16:363

Rushen, J. (1996) Using aversion learning techniques to assess the mental state, suffering, and welfare of farm animals. *Journal of Animal Science* 74:1990

Rushen, J., and P. Congdon (1986a) Relative aversion of sheep to simulated shearing with and without electro-immobilization. *Australian Journal of Experimental Agriculture* 26:535

Rushen, J., and P. Congdon (1986b) Sheep may be more averse to electro-immobilization than to shearing. *Australian Veterinary Journal* 63:373

Setckleiv, J., O. E. Skaug, and B. R. Kaada (1961) Increase in plasma 17hydroxycorticosteroids by cerebral cortical and amygdaloid stimulation in the cat. *Journal of Endocrinology* 22:119

Spensley, J. C., C. M. Wathes, N. K Waran, and J. A. Lines. (1995) Behavioral and physiological responses of piglets to naturally occurring sounds. *Applied Animal Behavioural Science* 44:277 (Abstract)

Stahringer, R. C., R. D. Randel, and D. A. Neuenforff (1989) Effect of nalexone on serum luteinizing hormone and cortisol concentration in seasonally anestrous Brahman heifers. *Journal of Animal Science* 67(Suppl.1):359 (Abstract)

Stephens, D. B., and J. N. Toner (1975) Husbandry influences on some physiological parameters of emotional responses in calves. *Applied Animal Ethology* 1:233

Stermer, R., T. H. Camp, and D. G. Stevens (1981) Feeder cattle stress during transportation Paper No. 816001. American Society of Agricultural Engineers, St. Joseph, MO

Stevens, D. A., and N. J. Saplikoski (1973) Rats reaction to conspecific muscle and blood evidence for alarm substances. *Behavioural Biology* 8:75

Stevens, D. A., and D. A. Gerzog-Thomas (1977) Fright reactions in rats to conspecific tissue. *Physiology of Behaviour* 18:47

Stooky, J. M., T. Nickel, J. Hanson, and S. Vandenbosch (1994) A movement measuring device for objectively measuring temperament in beef cattle and for use in determining factors that influence handling. *Journal of Animal Science* 72(Suppl.1):207 (Abstract)

Stricklin, W. R., C. E. Heisler, and L. L. Wilson (1980) Heritability of temperament in beef cattle. *Journal of Animal Science* (Suppl.1) 51:109 (Abstract)

Tennessen, T., M. A. Price, and R. T. Berg (1984) Comparative responses of bulls and steers to transportation. *Canadian Journal of Animal Science* 64:333

Trunkfield, H. R., and D. M. Broom (1990) Welfare of calves during handling and transport. *Applied Animal Behavioural Science* 28:135

Tulloh, N. M. (1961) Behavior of cattle in yards. II. A study of temperament. *Animal Behaviour* 9:25

Tume, R. K., and F. D. Shaw (1992) Beta-endorphin and cortisol concentrations in plasma of blood samples collected during exsanguination of cattle. *Meat Science* 31:211

Vieville-Thomas, C., and J. P. Signoret (1992) Pheromonal transmission of an aversive experience in domestic pigs. *Journal of Chemical Endocrinology* 18:1551

Warriss, P. D., S. N. Brown, and M. Adams (1994) Relationships between subjective and objective assessments of stress at slaughter and meat quality in pigs. *Meat Science* 38:329

White, R.G., J.A. DeShazer, C.J. Tressler, G.M. Borcher, S. Davy, A. Waninge, A.M. Parkhurst, M.J. Milanuk, and E.T. Clemens (1995) Vocalization and physiological response of pigs during castration with and without a local anesthetic. *J. Anim. Sci.* 73:381

Whittlestone, W.G., R. Kilgour, H. de Langen, and G. Duirs (1970) Behavioral stress and cell count of bovine milk. *J. Milk Food Technol.* 33:217

Wienker, W.R. (1986) Giraffe squeeze cage procedures. *Zoo Biol.* 5:371

Zavy, M.T., P.E. Juniewicz, W.A. Phillips, and D.L. Von Tungeln (1992) Effects of initial restraint, weaning, and transport stress on baseline and ACTH stimulated cortisol responses in beef calves of different genotypes. *Am. J. Vet. Res.* 53:551

FIFTEEN

Feedlot Cattle with Calm Temperaments Have Higher Average Daily Gains Than Cattle with Excitable Temperaments

Voisinet, B.D., Grandin, T., Tatum, J.D., O'Connor, S.F. and Struthers, J.J. (1997) Feedlot Cattle with Calm Temperaments Have Higher Average Daily Gains Than Cattle with Excitable Temperaments, *Journal of Animal Science*, Vol 75, pp. 892-896.

Bridget Voisinet was my first graduate student. At the time this study was conducted, the idea that cattle temperament would influence weight gain was radical and new. Fearful cattle that struggled when restrained for veterinary work gained less weight. We were proven right and now many studies show that genetic influences on temperament have a significant effect on the productivity of many farm animals. At the time we worked on this paper, I did not know that the major professor for a student was supposed to be the last author.

ABSTRACT:

This study was conducted to assess the effect of temperament on the average daily gains of feedlot cattle. Cattle (292 steers and 144 heifers) were transported to Colorado feedlot facilities. Breeds studied included Braford (n = 177), Simmental x Red Angus (n = 92), Red Brangus (n = 70), Simbrah (n = 65), Angus (n = 18), and Tarentaise x Angus (n = 14). Cattle were temperament rated on a numerical scale (chute score) during routine weighing and processing. Data were separated into two groups based on breed, Brahman cross

(2 25% Brahman) and non-Brahman breeding. Animals that had Brahman breeding had a higher mean temperament rating (3.45 .09) or were more excitable than animals that had no Brahman influence (1.80 .10); ($P < .001$). These data also show that heifers have a higher mean temperament rating than steers ($P < .05$). Temperament scores evaluated for each breed group also showed that increased temperament score resulted in decreased average daily gains ($P < .05$). These data show that cattle that were quieter and calmer during handling had greater average daily gains than cattle that became agitated during routine handling.

INTRODUCTION

"No one likes wild cattle, so why raise them?" This quote, from *The Lasater Philosophy of Cattle Raising* (Lasater, 1972), seems obvious due to animal and handler safety concerns. Some beef producers do, in fact, consider temperament to be an important trait when selecting cattle for purchase (Elder et al., 1980). Often, however, the economic implications of livestock temperament have been unrecognized. Reports of very excitable cattle that become highly agitated and excited when restrained or handled are increasing (Grandin, 1994). This trend could possibly be counterproductive for the beef industry.

Few experiments have attempted to identify links between temperament and various measures of productivity. One study reported that cows with calm temperaments had a 25 to 30% increase in milk production (Drugociu et al., 1977). Observations tend to show that more excitable cattle with higher temperament scores have lower live weights and (v) weight gains (Tulloh, 1961; Fordyce and Goddard, 1984), though few data have been presented. The present study was conducted to identify the relationship between temperament and productivity as measured by daily weight gain.

MATERIALS AND METHODS

Cattle. Four hundred thirty-six cattle (7 to 11 months old), 292 steers and 144 heifers, were transported to feedlot facilities near Fort Collins, Colorado, for finishing. Breeds studied included:

- Braford (3/8 Brahman x 5/8 Hereford or 1/2 Brahman x 1/2 Hereford)
- Simmental x Red Angus
- Red Brangus (3/8 Brahman x 5/8 Red Angus or 1/4 Brahman x 3/4 Red Angus)
- Simbrah (3/8 Brahman x 5/8 Simmental) Angus
- Tarentaise x Angus.

Braford, Red Brangus, and Simbrah cattle will be referred to as *Bos indicus*-cross; Simmental x Red Angus, Angus, and Tarentaise x Angus cattle will be referred to as *Bos taurus*.

All cattle were received at the feedlot from October through December 1994 and acclimated to feedlot conditions for 2 to 3 weeks before the start of the trial. The *B. indicus*-cross cattle were obtained from Florida, Simmental x Red Angus were obtained from Nebraska, and Angus and Tarentaise x Angus cattle were obtained from Wyoming. All cattle, regardless of origin, were produced on extensive operations with minimal human interaction. While in the feedlot, cattle were housed in groups of approximately 20 to 50 cattle, with group allotments determined by ranch and thus breed, gender, and weight. Cattle were fed to acquire a constant subcutaneous fat thickness of 9 to 13 mm (target = 11 mm) over the 12th rib, as determined by visual indices and ultrasound measurements.

All cattle received a diet consisting primarily of whole corn and corn silage. For the complete diet, see O'Connor et al. Growth implants were administered at the start of the finishing period and after approximately 120 days on feed. Implant protocols were as follows: steers were given an initial implant of Synovex-S (Syntex Animal Health, St. Louis, MO, 1994) and a second implant of Revalor-S (Hoechst Roussel AgriVet, Somerville, NJ); heifers received Finaplix-H (Hoechst Roussel AgriVet) for the initial and the second implants. Each heifer received .4 mg/d of melengestrol acetate (MGA) for the entire feeding period.

Experimental procedure

Approximately every 28 days, weight gain assessment and ultrasound determination of subcutaneous fat thickness data were recorded for all cattle. During processing, two independent observers assessed the temperament of each animal. A single temperament rating was recorded for each animal by each observer. The number of cattle prohibited temperament observations for all cattle from being completed on a single day. Observer 1 scored cattle after they had four to eight previous experiences with the handling facility at the feed yards. Observer 2 scored cattle during the animals first encounter with the handling facilities. Observers temperament scored the same cattle using slightly different methods. Observer 1 rated 436 *B. indicus*-cross and *B. taurus* cattle via a temperament rating system similar to that used in Grandin (1993), assigning scores of 1 through 5. Each animal's temperament was assessed while the animal was in a non-restraining single-animal scale crate. Observer 2 rated 304 *B. indicus*-cross cattle in a hydraulic squeeze chute (crush) with a head stanchion. Observer 2 assigned scores of 1 through 4 designating behaviors similar to those denoted by the following five-point system:

- 1: calm, no movement
- 2: restless shifting
- 3: squirming, occasional shaking of device (squeeze chute or scale)
- 4: continuous vigorous movement and shaking of device
- 5(4): rearing, twisting, or violently struggling.

Restraint of animals in a hydraulic squeeze chute reduces the range of movement and therefore reduces the resolution of discrimination between categories on a rating scale; thus a four-point scale was used. No inter observer comparison can be made because of the differences in animal movement between the squeeze chute and scale and because of numerical differences in temperament rating scale. Due to these differences in method, the data sets have been analyzed separately and presented as two independent experiments. Experiments 1 and 2 will refer to data collected by observers 1 and 2, respectively.

Statistical analysis

Data were analyzed using the SAS GLM procedure (SAS, 1985). Average daily gain was analyzed with a model that included breed, gender (where appropriate), temperament, sire (breed) (as a random effect), and fat thickness. Temperament was analyzed using a model that included breed, gender (where appropriate), sire (breed), and fat thickness.

Pair wise comparisons were conducted between the means of each level of temperament score, breed, and gender.

RESULTS AND DISCUSSION

Table 1 lists the unshrunk on-test and off-test least squares mean weights, days on feed, and average daily gains for animals in the study.

TABLE 1: Least Squares Means for Growth Traits by Breeds [a]

Breed [b]	n	On-test wt, kg	Off-test wt, kg	Days on feed	Avg daily gain [c], kg/d
Braford	177	290	468	201	.95 .03
Red Brangus	70	308	507	206	.98 .04
Simbrah	65	320	552	212	1.10 .04
Angus	18	305	543	194	1.24 .06
Simmental/Red Angus	92	264	569	213	1.44 .02
Tarentaise/Angus	14	301	550	207	1.21 .09

a. Data listed are for all animals temperament scored by Observer 1.
b. Traits are adjusted to a constant fat thickness of 11 mm using analysis of covariance techniques. The model included breed, gender, (Brahman-cross only), sire(breed), and fat thickness.
c. Values are means SE. The error term for analysis of breed differences = sire (breed) (dfsOs *indicus*-cross = 73, dfsOS *taurus* = 64).

ANALYSIS OF BREED DIFFERENCES IN TEMPERAMENT

Experiment 1

Observer 1 collected data on the *Bos indicus* and *Bos taurus* cattle. Our analyses showed that temperament score differed between breed groups. No significant temperament score differences existed within *B. indicus*-cross cattle with respect to differing percentages of Brahman influence (1/4, 3/8, or 1/2 Brahman). Mean temperament scores of *B. indicus*-cross cattle were higher ($P < .001$) than those for *B. taurus* steers. This agrees with research that has shown that *B. indicus* cattle are more temperamental or excitable than *B. taurus* cattle. Because of these differences, weight gain data for *B. indicus*-cross and *B. taurus* breed groups were analyzed separately. Mean temperament scores by breed are presented in Table 2. Differences were present within the *B. indicus* cross breed group, with the Braford and Red Brangus cattle having more ($P < .05$) excitable temperaments than Simbrah cattle. Accurate representation of mean temperament score for individual *B. indicus*-cross breeds (Braford, 3.62; Red Brangus, 3.78; Simbrah, 2.89) and the *B. indicus* cross breed group (3.46) necessitated that heifers be omitted from this analysis because only steers were present in the *B. taurus* breeds.

Even though breed group differences were statistically significant, they may not represent true breed-based differences in temperament due to confounding by geographic origin. As was discussed in the Materials and Methods section, all *B. indicus*-cross breeds were obtained from a single location, Angus and Tarentaise x Angus cattle were obtained from a second location, and Simmental x Red Angus cattle were obtained from a third location.

TABLE 2: Least Squares Means for Temperament Score by Breed, Steers Only (Experiment 1)	
Breed[b]	Mean Temperature Ranking[b,c]
Braford	3.62 .15[d]
Red Brangus	3.78 .22[d]
Simbrah	2.89 .22[e]
Bos indicus cross	3.46 .09[g]
Angus	1.70 .19[f]
Simmental x Red Angus	1.77 .07[f]
Tarentaise x Angus	2.36 .31[e]
Bos taurus	1.80 .10[g]

a. Model included breed, sire (breed), and fat thickness. The error term for analysis of breed differences = sire (breed) (^{df}*Bos indicus*-cross individual breeds = 75; ^{df}*Bos taurus* individual breeds = 51; ^{df}allbreed means = 123).

b. 1 = calm, no movement; 2 = restless shifting; 3 = squirming occasional shaking of restraint device; 4 = continuous vigorous movement and shaking of restraint device; 5 = rearing, twisting or violently struggling.

c. Values are means ± SE.

d, e, f Means with different superscripts differ (P < .05).

g. Means differ (P < .001).

Experiment 2

No difference (P < .4) in temperament existed among any of the *B. indicus* breeds observed in the squeeze chute. Braford cattle had an average temperament score of 2.0 - .12, Red Brangus cattle had a score of 2.18 - .17, and Simbrah cattle had a score of 2.11 - .14, on the 1 to 4 rating system. No *B. taurus* cattle were included in this experiment (data not shown).

ANALYSES OF WEIGHT GAIN DIFFERENCES

Experiment 1

Our results show a significant effect of temperament ranking on average daily gain in *B. indicus*-cross and *B. taurus* cattle (Table 3). The *B. taurus* steers with the calmest temperaments had .19 kg/d greater ($P < .05$) mean average daily gain than the steers with the highest temperament scores or most excitable temperaments. With the exception of *B. indicus*-cross steers and heifers that had a temperament score of 1, average daily gains in both breed groups decreased as temperament scores increased. The *B. Indicus* cattle with calm temperaments (scores of 1) do not fit with this pattern, because they had the lowest average daily gains (.75 kg/d). We speculate, however, that the small number of animals (n = 4) and large standard error may have contributed to this apparently contradictory result.

Table 3: Least Squares Means for Average Daily Gain for Animals Temperament Ranked for Experiment 1

	Bos taurus[c]		**Bos indicus-cross**[d]	
Temperament ranking[a,b]	**n**	**Avg daily gain,e kg/d**	**n**	**Avg daily gain,e kg/d**
1	37	1.38 .05f	4	.75 .12h
2	70	1.29 .04g	40	1.07 .04f
3	17	1.19 .06g	94	1.02 .03fg
4	0		113	1.01 .03fg
5	0		61	.97 .04gh

a. Model included temperament, breed, gender (*B. indicus*-cross only), sire (breed), and fat thickness. The error term for analysis of temperament differences = residual (^{df}*Bos indicus*-cross = 274; ^{df}*taurus*-cross = 84).
b. 1 = calm, no movement; 2 = restless shifting; 3 = squirming, occasional shaking of restraint device; 4 = continuous vigorous movement and shaking of restraint device; 5 = rearing, twisting or violently struggling.
c. Steers only.
d. Steers and heifers.
e. Values are means ± SE.
f, g, h Within each main effect, means with different superscripts differ (P < .05).

Experiment 2

Observer 2 temperament ranked 304 *B. indicus*-cross cattle on the four-point system described previously (Table 4). Temperament score was a significant source of variation in average daily gain. Animals with temperament scores of 1 or 2 had higher (P < .05) average daily gains than animals with temperament scores of 3.

Table 4: Least Squares Means for Average Daily Gain for Animals Temperament Ranked for Experiment 2

Temperament ranking [a]	*Bos indicus*-cross [b]	
	n	Avg daily gain,[b] kg/d
1	89	1.04 .03[c]
2	119	1.05 .03[c]
3	76	.95 .03[d]
4	20	.94 .06[c,d]

a. Model included temperament, breed, gender, sire (breed), and fat thickness. The error term for analysis of temperament differences = residual (df = 267).
b. Values are means i SE.
c, d Means with different superscripts differ (P < .05).

The use of two observers and different experimental methods attests to the robustness of our results and the strength of the temperament effect on weight gain. Due to the lack of body restraint in the scale there was an increased ability for animal movement. As a result, observer 1 assigned more scores of 4 (25.9%) or 5 (14.0%) than observer 2 assigned scores of 4 (6.6%). Despite those differences, the results derived from the study remain consistent. We conclude from these results that the driving force behind average daily gain differences was primarily a product of calm temperaments, as opposed to excitable temperaments. Stated another way, calm cattle had increased average daily gains rather than excitable cattle having decreased average daily gains. More research, however, is necessary to confidently establish this.

ANALYSIS OF GENDER DIFFERENCES

Because heifers were present in *Bos indicus*-cross groups only, gender analyses were limited to the *B. indicus*-cross breed group. Gender was a significant source of variation, not only in average daily gain, as would be expected, but also in average temperament scores. Regardless of observer or temperament ranking system, heifers consistently had higher temperament scores than their male contemporaries (Table 5). In Experiment 1, heifers had a mean temperament score of 3.72, and steers had a mean temperament score of 3.39. In Experiment 2, the mean temperament score of heifers was 2.23 and that of steers was 1.97.

Table 5: Gender Differences in Mean Temperament Score in *Bos indicus*-Cross Cattle

	Mean temperament ranking [b]	
Gender [a]	Experiment 1	Experiment 2
Heifers	3.72 .11c	2.23 .10d
Steers	3.39 .11c	1.97 .10d

a. Model included breed, gender, sire (breed), and fat thickness. The error term for analysis of gender differences = residual (dfobserver 1 = 278; dfobserver 2 = 270).
b. Values are means - SE.
c. Means differ (P < .01).
d. Means differ (P < 05).

Similar gender differences in temperament have been found in British and European Continental (exotic) cattle (Stricklin et al., 1980). Other research, which focused on *B. taurus* breeds, found similar trends, but no significant differences in temperament due to gender were detected (Tulloh, 1961; Shrode and Hammack, 1971). We hypothesize that gender differences may be evident only in certain breeds. For example, due to calmer temperaments among *B. taurus* breeds, gender differences may not be as pronounced as the gender differences in *B. indicus* or *B. indicus*-cross breeds (Eld er et al., 1980; Fordyce et al., 1988).

Studies with rodents, which typically exhibit fear or anxiety (typically considered to be synonymous), have shown common, though inconsistent, gender differences in behavior (Gray, 1987; Johnston and File, 1991). Studies of fear may contribute to our knowledge of temperament by considering that fear, as a physiological state of the nervous system, ultimately results in certain behaviors (Gray, 1987). Additionally, Boissy (1995) defined fearfulness as a trait that determines the extent to which an individual becomes frightened in alarming situations.

The evolutionary and (or) adaptive mechanisms underlying gender differences in temperament are not fully understood. Practical experience on ranches has shown that heifers are more temperamental than cows. The fact that this calming of their disposition occurs just after parturition is verified by rodent experiments. Just after parturition and during lactation, rats exhibit a decrease in emotional reactivity or fearfulness (Hard and Hansen, 1985). Nulliparous rats were more fearful than parturient females in a variety of tests, including those that measured emergence latencies from a box into an open field test arena and the inclination to flee from an intruder (Fleming and Luebke, 1981). Reduced fearfulness of parturient female rats is most likely hormonally mediated (Fleming and Luebke, 1981).

In addition to genetically based differences in temperament, the possibility also exists for temperament to be influenced by growth-promotant implant protocols, which are completely confounded by gender; however, we found no research to support or refute this possibility in heifers. Two studies using steers and bulls have been conducted to examine behavioral effects of zeranol implants. Neither study showed a significant effect of implantation on agitation scores (Vanderwert et al., 1985; Baker and Gonyou, 1986).

Experience also affects reactions to handling and restraint. Crookshank et al. (1979) showed that agitation and cortisol levels in cattle were decreased over multiple handling experiences. Gentling of animals is at least somewhat successful at reducing aversion to restraint and handling, although not enough to overcome the effects of highly aversive procedures (Hargreaves and Hutson, 1990). European Continental cattle that were worked through a squeeze chute repeatedly in a single day became increasingly agitated (Grandin, 1993). Calm Angus bulls, however, did not become agitated with additional passes through working facilities (B. D. Voisinet, unpublished data). Other research, however, has shown that if given the opportunity to avoid highly aversive handling procedures, such as electro-immobilization, sheep will do so consistently over many trials (Grandin et al., 1986). Differences in the results between studies is likely due to differing levels of fear and how the animal perceives the aversiveness of a procedure. Animals are able to discriminate between different kinds of human interaction, aversive or non-aversive (Gonyou et al., 1986) and also between different areas of a restraint system where highly averse events occurred (Rushen, 1986). The levels of aversion expressed by an individual animal, however, are relatively persistent across multiple handling experiences (Fordyce and Goddard, 1984; Lyons, 1989; Grandin, 1993). Because of this and regardless of whether agitation in response to a particular handling event increases or decreases over time, one should expect agitation levels or temperament for an individual animal to remain relatively consistent with respect to its contemporaries. Heritability estimates of cattle temperament show that it is a moderately heritable trait (Shrode and Hammack, 1971; Stricklin et al., 1980; Fordyce et al., 1988).

Even though an economic analysis has not been completed at this time, the benefits of selecting for calmer or more docile animals may be more than enhanced animals and handler safety and decreased facility wear. Another advantage of selecting cattle with calmer temperaments would be increased welfare because injuries to the animal would be reduced.

Research is needed to determine the physiological mechanisms underlying the effect of temperament on average daily gain.

IMPLICATIONS

Selection for calm temperaments may become a key factor in maximizing production efficiency of cattle weight gains in feedlots. Cattle temperament is heritable, and temperament differences persist when animals are rated over a period of time. These two factors, considered together, imply that careful selection for a calm temperament may not only improve animal and handler safety but also increase economic returns via improved average daily gains.

REFERENCES

Baker, A. M., and H. W. Gonyou (1986) *J. Anim. Sci.* 62:1224

Boissy, A. (1995) Fear and fearfulness in animals. *Q. Rev. Biol.* 70: 165

Crookshank, H. R., M. H. Elissalde, R. G. White, D. C. Clanton, and H. E. Smalley (1979) Effect of transportation and handling of calves upon blood serum composition. *J. Anim. Sci.* 48:430

Drugociu, G., L. Runceanu, R. Nicorici, V. Hritcu, and S. Pascal (1977) Nervous typology of cows as a determining factor of gender and productive behaviour. *Anim. Breed. Abstr.* 45:1262 (Abstr.)

Elder, J. K., J. F. Kearnan, K. S. Waters, G. H. Dunwell, F. R. Emmerson, S. G. Knott, and R. S. Morris (1980) A survey concerning cattle tick control in Queensland. Use of resistant cattle and pasture spelling. *Aust. Vet. J.* 56:219

Fleming, A., and C. Luebke (1981) Timidity prevents the virgin female rat from being a good mother: Emotionality differences between nulliparous and parturient females. *Physiol. & Behav.* 27:863

Fordyce, G. E., R. M. Dodt, and J. R. Wythes (1988) Cattle temperaments in extensive beef herds in northern Queensland. Factors affecting temperament. *Aust. J. Exp. Agric.* 28:683

Fordyce, G. E., and M. E. Goddard (1984) Maternal influence on the temperament of *Bos indicus*-cross cows. *Proc. Aust. Soc. Anim. Prod.* 15:345

Gonyou, H. W., P. H. Hemsworth, and J. L. Barnett (1986) Effects of frequent interactions with humans in growing pigs. *Appl. Anim. Behav. Sci.* 16:269

Grandin, T. (1993) Behavioral agitation during handling of cattle is persistent over time. *Appl. Anim. Behav. Sci.* 36:1

Grandin, T. (1994) Solving livestock handling problems. *Vet. Med.* 89: 989

Grandin, T., S. E. Curtis, T. M. Widowski, and J. C. Thurmon. (1986) Electro-immobilization versus mechanical restraint in an avoid-avoid choice test for ewes. *J. Anim. Sci.* 62:1469

Gray, J. A. (1987) *The Psychology of Fear and Stress* (2nd Ed.) Cambridge University Press, Cambridge, U.K.

Hard, E., and S. Hansen (1985) Reduced fearfulness in the lactating rat. *Physiol. & Behav.* 35:641

Hargreaves, A. L., and G. D. Hutson (1990) The effect of gentling on heart rate, flight distance and aversion of sheep to a handling procedure. *Appl. Anim. Behav. Sci.* 26:243

Hearnshaw, H., and C. A. Morris (1984) Genetic and environmental effects on a temperament score in beef cattle. *Aust. J. Agric. Res.* 35:723

Johnston, A. L., and S. E. File (1991) Gender differences in animal tests of anxiety. *Physiol. & Behav.* 49:245

Lasater, L. M. (1972) *The Lasater Philosophy of Cattle Raising.* Texas Western Press, The University of Texas at El Paso

Lyons, D. M. (1989) Individual differences in temperament of dairy goats and the inhibition of milk ejection. *Appl. Anim. Behav. Sci.* 22:269

O'Connor, S. F., J. D. Tatum, D. M. Wulf, R. D. Green, and G. C. Smith (1997) Genetic effects on beef tenderness in *Bos indicus* composite and *Bos taurus* cattle. *J. Anim. Sci.* (In press)

Rushen, J. (1986) Aversion of sheep to electro-immobilization and physical restraint. *Appl. Anim. Behav. Sci.* 15:315

SAS (1985) *SAS User's Guide: Statistics* (Version 5 Ed.) SAS Inst. Inc., Cary, NC.

Shrode, R. R., and S. P. Hammack (1971) Chute behavior of yearling beef cattle. *J. Anim. Sci.* 33:193 (Abstr.)

Stricklin, W. R., C. E. Heisler, and L. L. Wilson. (1980) Heritability of temperament in beef cattle. *J. Anim. Sci.* 51(Suppl. 1):109 (Abstr.)

Tulloh, N. M. (1961) Behaviour in cattle yards. II. A study of temperament. *Anim. Behav.* 9:25

Vanderwert, W., L. L. Berger, F. K. McKeith, A. M. Baker, H. W. Gonyou, and P. J. Bechtel (1985) Influence of zeranol implants on growth, behavior and carcass traits in Angus and Limousin bulls and steers. *J. Anim. Sci.* 61:310

SIXTEEN

Objective Scoring of Animal Handling and Stunning Practices at Slaughter Plants

Originally published in (1998) Objective Scoring of Animal Handling and Stunning Practices at Slaughter Plants, *Journal of the American Veterinary Medical Association*, Vol. 212, pp. 36-39.

This paper is important because it contains the origin of the objective numerical scoring system that I developed for evaluating animal welfare at slaughter plants. The survey was funded by the United States Department of Agriculture. Bonnie Buntain at the USDA made this study possible. Before I did this survey of twenty-four federally inspected plants, Dr. Buntain had me visit with some of her colleagues to refine the methods. This paper is important because it formed the basis for the American Meat Institute Welfare Guidelines. Janet Riley invited me to write the AMI audit guide in 1997. This scoring tool was used in 1999 by McDonald's, Wendy's International, and Burger King to assess welfare in the plants that supplied them. During this time, I trained auditors from these major companies to assess animal welfare and enforce higher standards. The data obtained in this initial survey showed that stunning and handling practices were poor. After the restaurant audits started, both stunning and handling greatly improved.

ABSTRACT

Objective: To develop objective methods for monitoring animal welfare at slaughter plants to ensure compliance with the Humane Methods of Slaughter Act.

Design: Survey of existing procedures.

Sample Population: 24 federally inspected slaughter plants.

Procedure: 6 variables evaluated at each plant were stunning efficacy, insensibility of animals hanging on the bleeding rail, vocalization, electric prod use, number of animals slipping, and number of animals falling. Results of 11 beef plants, only 4 were able to render 95% of cattle insensible with a single shot from a captive-bolt stunner. Personnel at 7 of 11 plants placed the stunning wand correctly on 99% or more of pigs and sheep. At 4 beef plants, percentage of cattle prodded with an electric prod ranged from 5% at a plant at which handlers only prodded cattle that refused to move to 90% at another plant. Use of electric prods at 6 pork plants scored for prod use ranged from 15 to almost 100% of pigs. Percentage of cattle that vocalized during stunning and handling ranged from 1.1% at a plant at which electric prods were only used on cattle that refused to move to 32% at another plant at which electric prods were used on 90% of cattle and a restraint device was inappropriately used to apply excessive pressure.

Clinical Implications: To obtain the most accurate assessment of animal welfare at slaughter plants, it is important to score all of the aforementioned variables (*J Am Vet Med Assoc* 1998;212:36-39).

FULL ARTICLE

During the past 25 years, animal handling and stunning at more than 100 US slaughter plants have been increasingly scrutinized. During the past 10 years, handling practices have apparently improved, but there is still great variation in enforcement of the Humane Methods of Slaughter Act.[1] It is the responsibility of the USDA veterinarian in charge at each federally inspected slaughter plant to ensure that animals are handled humanely and are painlessly stunned. However, what one veterinarian may consider to be good handling practices, another may think is abusive cruelty. Objective methods of scoring and quantifying animal welfare are needed to ensure compliance with the Humane Methods of Slaughter Act.

For a scoring system to be effective, it should be simple to use under commercial conditions. Minimal training should be required to conduct an assessment of handling and stunning practices. It is important for scoring to be objective so that all observers may obtain similar results. For example, subjective evaluations of "rough handling" may vary greatly among observers, because it is not possible to define rough handling in specific and precise terms. However, an objective measure, such as tabulating the number of animals prodded with electric prods, is easy to define.

PROCEDURE

In 1996, a survey was conducted for the USDA on stunning and handling practices at federally inspected plants.[2] Objective scoring and subjective evaluations were used to assess animal welfare. Twenty-four plants in 10 states were visited. All visits were announced and scheduled. Sixteen plants in the vicinity of 4 randomly chosen cities were visited. To gain cooperation of the meat—packing industry confidentiality of survey data was ensured by surveying an additional 8 plants, the identities of which were not disclosed to the USDA. Yes or no scoring was used for measuring the following 6 variables: stunning efficacy insensibility on the bleeding rail, electric prod use, vocalization, number of animals slipping, and number of animals falling.

RESULTS

Stunning Scores: One hundred to 200 animals were scored at each of 11 pork or sheep plants to determine the percentage of animals on which stunning wands were correctly positioned. The electric stunning wand must be correctly positioned on the head to ensure that electricity passes through the brain to produce instant unconsciousness. Personnel at 7 plants positioned the stunning wand correctly on 99% of animals. At 2 plants, the electric stunning wand was placed in the wrong position on 10 to 35% of animals. The main cause of this problem was poor wand design. Of 11 pork and sheep plants, 10 used wands with sufficient current to ensure instant unconsciousness. Recommendations on parameters for electrical current for stunning are reported elsewhere.[3-8]

At beef and veal plants, the percentage of cattle or calves rendered instantly insensible with one use of a captive-bolt stunner was tabulated. Personnel at 4 of 11 plants were able to render 95 to 100% of cattle insensible with one use of the captive-bolt stunner. At 6 plants, more than 10% of cattle were insufficiently stunned during the first attempt. At 4 plants, the cause of an ineffective first attempt was poor stunner maintenance. At 2 other plants, the cause of ineffective use was poor ergonomic design of the pneumatic captive-bolt stunner, which was bulky and required excessive physical effort to aim. Thus, when operators became fatigued, animals were not appropriately stunned and additional stunning attempts were required.

Insensibility on the bleeding rail: At beef, pork, and sheep plants, animals hanging on the rail were classified as sensible or insensible. Animals that had partial signs of sensibility were classified as sensible. Animals hanging on the rail that had rhythmic breathing, vocalization, eye reflexes in response to touching, eye blinking, or arched back righting reflex were tabulated as sensible and not properly stunned. Animals should hang limply on the rail and have a floppy head.[4] Limb movements were ignored. In the survey more than 1,000 pigs and 1,000 cattle were observed. One of the pigs and one of the cattle were scored as sensible, displaying an arched back righting reflex and all 5 indicators of sensibility, respectively.

Slipping and falling: Evaluation of the number of animals slipping and falling is an important measure of welfare, because quiet handling is impossible when animals become agitated as a result of losing their footing. Slipping and falling were scored independently for each animal. The number of animals slipping or falling in the crowd pen, single-file chute, stunning box, or restrainer was recorded.

Of 11 beef plants evaluated, slick floors at 2 caused cattle or veal calves to fall. At these 2 plants, 8 and 12% of cattle fell. At 4 beef plants, 15 to 30% of cattle slipped in the stunning chute area, but at 6 plants, none of the cattle slipped. At 1 plant, 3% of cattle slipped, but none fell. Slipping in the stunning box Caused cattle to become agitated at 3 plants.

Slipping and falling were not detected at 2 sheep plants. Nine pork plants were surveyed for slipping and falling. None of the pigs slipped or fell in the stunning chute area. However, in the unloading area, which Was not formally scored, pigs fell while walking on a slick unloading ramp.

Electric prod use: Reducing use of electric prods will improve animal welfare, because repeated prodding often results in increased vocalization and behavioral agitation. Use of electric prods is easy to measure objectively. Twelve pork and beef plants were scored for electric prod use. None of the personnel at these plants used electric prods in the unloading area or stockyard pens. At 8 plants, well-trained employees only prodded animals that balked and refused to move. At 4 beef plants with well-trained employees, the percentage of cattle prodded with an electric prod was 5, 10, 10, and 64%. The plant with the 5% prod use score slaughtered more than 200 cattle/h. At the plant at which 64% of cattle had to be prodded, it was attributed to cattle being able to see a person's hand under the door of the stunning box. Electric prod use at these 4 plants was limited to the area of the single-file chute and stunning box or restrainer entrance. Prods were not used in crowd pens. Good animal welfare and a low percentage of animals prodded can be obtained in older facilities, as evidenced by a beef plant with old-fashioned facilities that had a low score of 10%.

At 4 pork plants at which an electric prod was used only on pigs that balked, the percentage of pigs prodded in crowd pens was 0, 18, 18, and 80%. At the plant with the highest use (80%), pigs balked because of shadows. All 4 pork plants had single-file chute systems and slaughter line speeds of more than 500 pigs/h. A fifth pork plant, which was not formally scored, had a state-of-the-art, double single-file chute system that was free of air drafts, shadows, and shiny reflections that would impede movement. All of its equipment had been engineered to produce little noise. Personnel responsible for moving pigs were able to keep up with a slaughter line speed of more than 800 pigs/h, despite using only plastic paddles and nonelectrified prods. Overall, pigs were prodded more than cattle. This may be attributable to the fact that modern swine hybrids bred to produce extremely lean pigs are extremely excitable and difficult to move.

At 4 plants, it appeared that employees handled animals roughly and used electric prods in an excessive manner. After baseline percentages were tabulated, employees were instructed to fill the crowd pen only half full and to attempt to move animals by tapping them on the rear before resorting to use of an electric prod. At the 2 beef plants with excessively rough handling, electric prod use in the single-file chute was reduced from 83 to 17%. At pork plants with excessively rough handling, electric prod use in the single-file chute decreased from 44 to 15%. Employees were able to move a sufficient number of animals to keep up with slaughter line speed when prod use was reduced.

Distractions such as sparkling reflections, air blowing in the faces of approaching animals, people visible in front of animals, hissing air, or shadows can make animals balk.[9] Distractions that cause balking, in turn, increase electric prod use. Nineteen beef and pork plants were evaluated for distractions. Twelve plants were subjectively rated excellent or acceptable for distractions, but 7 plants were rated not acceptable or had a serious problem with distractions that could impede animal movement. Most distractions could be fixed with minimal expense. At 1 plant, shadows made it impossible to reduce electric prod use. Four beef plants had air hissing from pneumatic valves, which could easily be remedied by use of an inexpensive muffler. At 2 other plants, animals balked because of noise from a ventilation fan and high-pitched whistling from a pump.

Vocalization: Vocalization scoring can be used as an indicator of animal discomfort. In cattle, an increase in vocalization during restraint is correlated with increases in cortisol concentration.[10] Restraining cattle in an inverted position causes a significantly greater number of cattle to vocalize, compared with restraint in an upright position. Vocalization (moos or bellows) were tabulated in 100 to 250 cattle at each plant. Each of the cattle was scored as a vocalizer or non-vocalizer. Attempts were not made to measure intensity of vocalization or to count the number of vocalizations per animal. Cattle were scored while they were passing through the crowd pen, single-file chute, stunning box, or restrainer. Vocalization was not scored in stockyard pens, because cattle standing undisturbed often vocalize to each other. Mean percentage of cattle vocalizing was 4.5% at 4 plants with careful, quiet handling and 22% at the 2 plants with rough handling. Vocalization scores for the 4 plants with careful, quiet handling were 1.1, 2.6, 6.6, and 7.5%. Vocalization scores for the 2 plants with rough handling were 12 and 32%. When excessive electric prodding in the 2 rough-handling plants was stopped, percentage of vocalizing cattle decreased significantly ($X^2 = 21.68$; $P < 0.01$) from a mean of 22% to a mean of 7%.

Prodding with an electric prod caused more than half of the vocalizations at all beef plants. Remaining vocalizations were caused by other aversive events, such as excessive pressure from a restraint device powered by pneumatic cylinders, ineffective stunning, or slipping on the floor of the stunning box. One hundred twelve of 1,125 cattle vocalized during handling in the stunning chute area. With the exception of 2 cattle, all vocalizations were caused by the aforementioned events. Because 98% of vocalizing cattle were responding to an easily observed aversive event, decreasing the percentage of cattle that vocalize at beef plants may contribute to improved welfare.

Since the survey was conducted, 4 large beef plants that had restrainer conveyors have been visited. At all of these plants, handlers moved cattle at a walk, and electric prods were only used on cattle that refused to move. At 3 of these 4 plants, 3% of cattle vocalized. At 1 plant, approximately 10% of cattle vocalized because of a slippery entrance ramp to the restrainer conveyor.

It was not possible to score the number of pigs that vocalized during handling in the crowd pen and single-file chute, because it was impossible to count specific squeals and accurately identify particular pigs squealing in the crowd pen. However, it was possible to count specific pig squeals in the conveyor restrainer, and the results were tabulated. Intensity of pig squeals is correlated with stress.[11] Researchers in England used a sound meter that recorded intensity of squeals throughout a 5-minute period.[12] They found that intensity of squealing was correlated with physiologic stress measurements. Plants that carefully handle pigs and use electric prods minimally have less squealing.

Vocalization scores for cattle and pigs in the restrainer or stunning box were tabulated for 14 plants, At 7 pork plants, the percentage of pigs that squealed in the restrainer was 0 to 2% for 6 plants. At the remaining plant, the percentage was 14%. At that plant, the restrainer had a missing part. At a large pork plant that was not included in the survey many pigs vocalized and struggled in a V-shaped conveyer restrainer when 1 side of the restrainer was moving faster than the other side.

At 3 of 7 beef plants, none of the cattle vocalized in the conveyer restrainer or stunning box. At the remaining 4 plants, the percentage of cattle that vocalized during restraint was 0.6, 0.6, 6.0, and 7.5%. The plant with the score of 6% used an electric device to paralyze cattle for stunning, and the plant with the vocalization score of 7.5% had cattle that slipped on the stunning-box floor and had excessive pressure exerted on them by a restraining gate. It has been reported in 3 studies[13-15] that restraining animals with electricity is highly aversive.

Vocalization scores were not tabulated for sheep. Whereas cattle in the stunning-chute area vocalize in direct response to an observed aversive event, sheep naturally vocalize to each other when they are being quietly moved through chutes. Therefore, vocalization would not be a reliable indicator of aversive events that might be detrimental to sheep welfare.

DISCUSSION

To assess animal welfare during handling, stunning, and slaughter, 6 variables should be evaluated (Appendix). These variables are easily quantifiable, objective measures of animal handling and stunning practices. Scoring of these variables can easily be done under commercial conditions. Scores more accurately reflect the true performance of a plant when a large number of animals are evaluated. It is recommended that investigators score a minimum of 50 animals/variable. At large plants, assessing 100 to 200 animals is recommended.

It is highly recommended to evaluate at the beginning and end of work shifts. Results of the survey reported here indicated that the number of ineffective stuns during the first attempt on an animal, using 2 captive-bolt stunner, increased late in the shift, when operators were fatigued.

It is important to assess all of the aforementioned variables. For example, if only stunning efficacy was measured, personnel at a plant might be tempted to use a restraint device that applied excessive pressure. Vocalization and electric prod scores would reveal the inadequacy of such procedures.

Slaughter plant management should work to continuously improve scores, which will help improve animal welfare. Use of objective scoring also helps to improve animal welfare, and cost to the industry will be minimal. Objective scoring also has the benefit of reducing variability in enforcement of the humane slaughter regulations.

APPENDIX

Variables that should be evaluated to assess animal welfare during handling, stunning, and slaughter:

- **Stunning efficacy**

  Percentage of animals rendered insensible by use of a captive-bolt stunner on the first attempt

  Percentage of animals on which electrodes are placed in the correct position so that electricity will pass through the brain during electric stunning

- **Insensibility of stunned animals hanging on the bleeding rail**

  Percentage of stunned animals that do not have signs of sensitivity or partial sensibility, such as rhythmic breathing, blinking, eye reflexes, righting reflexes, or vocalization.

- **Vocalization (should not be used when electricity is used to paralyze an animal because paralysis may prevent animals from vocalizing)**

  Percentage of cattle that vocalize in the stunning-chute area, which consists of the stunning box or restrainer, crowd pen, and single-file chute. Each animal is scored as a vocalizer or non-vocalizer.

  Percentage of pigs that vocalize in the retainer or stunning pen. Sound meters might be most efficacious in chutes and crowd pens.

  Not recommended for assessing sheep.

- **Electric prod use**

  Percentage of animals prodded with an electric prod in the single-file chute area and crowd pen. Because it is often difficult to observe both of these areas simultaneously, separate observations should be made in each area.

- **Slipping**

  Percentage of animals that slip in the stunning chute area. A slip is scored if it interferes with normal walking or causes an animal to become behaviorally agitated.

- **Falling down**

  Percentage of animals that fall down in the stunning chute area. A fall is recorded when any part of the animal's body other than the hooves touches the floor.

REFERENCES

1. Humane Methods of Slaughter Act of 1978. Public law 95445. Federal Register 1979, 44(232):68809—68817.
2. Grandin T. Survey of stunning and handling practices in federally inspected beef, veal, pork, and sheep slaughter plants. Project No. 36023200002086. Beltsville, Md: USDA—Agriculture Research Service, 1996.
3. Council of Europe. Council directive of 18 November on stunning of animals before slaughter (74/577/EEC). Off J Eur Commun 199; 316:10—11.
4. Grandin T. Euthanasia and slaughter of livestock. J Am Vet Med Assoc 1994, 204:1354-1360.
5. Gregory NG. Humane slaughter, in Proceedings. 34th Int Congr Meat Sci Technol 1988, 46-49.
6. Gregory NG, Wotton SB. Sheep slaughtering procedures. Ill. Head to back electric stunning. Br Vet J 1984, 140:570—575.
7. Hoenderken R. Electrical and carbon dioxide stunning of pigs for slaughter. In: Eikelenboom G, ed. Stunning of Animals for Slaughter. Boston: Martinus Nijhoff Publishing, 1983, 59-63.
8. Troeger K, Woltersdorf W. Measuring stress in pigs during slaughter. Fleischweirtsch 1989, 69:373—376.
9. Grandin T. Factors that impede animal movement at slaughter plants. J Am Vet Med Assoc 1996, 209:757—759.
10. Dunn CS. Stress reactions of cattle undergoing ritual slaughter using two methods of restraint. Vet Rec 1990, 126:522—525.
11. White RG, DeShazer JA, Tressler CJ, et al. Vocalizations and physiological response of pigs during castration with and without anesethetic. J Anim Sci 1995, 73:381—386.
12. Warris PD, Brown SN, Adams SjM. Relationships between subjective and objective assessments of stress at slaughter and meat quality in pigs. Meat Sci 1994, 38:329—340.
13. Grandin T, Curtis SE, Widowski TM, et al. Electro-immobilization versus mechanical restraint in an avoid-avoid choice test. J Anim Sci 1986, 62:1469-1480.
14. Pascoe PJ. Humaneness of electro-immobilization unit in cattle. J Wt Res 1986, 10:2252—2256.
15. Lambooy E. Electro anesthesia or electro-immobilization of calves, sheep and pigs by the Feenix Stockstill. Vet Q 1985, 7:120-126.

SEVENTEEN

Behavioral Genetics and Animal Science

Grandin, T. and Deesing, M. (2022) *Genetics and the Behavior of Domestic Animals,* 3rd Edition. pp. 1-47

The first edition of this book was published in 1998, Academic Press in San Diego, California.

The third edition was published by Elsevier, Amsterdam, The Netherlands.

This chapter is important because it illustrates the relationship between genetic factors and animal behavior. Old studies that are still relevant today arc reviewed in this chapter. In the early '90s, my co-author, Mark Deesing, spent hours in old library stacks finding these studies. The chapter also describes the founders of animal behavior research. There were two schools of thought. The Skinnerians believed that the environment totally shaped behavior, and the Ethologists studied animals in their natural environments. Both disciplines did not recognize animal emotions. Today, animal emotions are fully recognized, and the modern research is described.

ABSTRACT

The partnership between humans and domestic animals is natural. The human brain is hardwired to emotionally respond to animals. Beginning with the domestication of wolves, this chapter covers the process of domestication and reviews the early work of behaviorists and ethologists who refused to accept emotional states in animals. Modern behavior research employs methods developed by behaviorists and ethologists combined with neuroscience and genetics. Emotional systems in the brain drive behavior. Confusion between different emotional systems may explain conflicting findings in the behavior literature. Behavior in an open field test may be motivated either by fear, separation distress, or novelty seeking. Each emotion is controlled by separate subcortical systems. A novel open field arena can frighten a prey species, but it may activate seeking in a predator. Genetics affects the strength of fear, novelty seeking, and separation distress. Behavior is shaped by a complex interaction between genetics and experience.

INTRODUCTION

There have been huge advances in genetics since the publication of the first edition in 1998 and the second edition in 2014. This chapter contains many older studies that still provide valuable insights into genetic effects on animal behavior. It has been fully updated with information from the latest research. The original introduction from the first edition has been kept.

A bright orange sun is setting on a prehistoric horizon. A lone hunter is on his way home from a bad day at hunting. As he crosses the last ridge before home, a quick movement in the rocks, off to his right, catches his attention. Investigating, he discovers some wolf pups hiding in a shallow den. He exclaims, "Wow... cool! The predator... in infant form."

After a quick scan of the area for adult wolves, he cautiously approaches. The pups are all clearly frightened and huddle close together as he kneels in front of the den... all except one. The darkest colored pup shows no fear of the man's approach. "Come here you little predator! Let me take a look at you," he says. After a mutual bout of petting by the man and licking by the wolf, the man suddenly has an idea. "If I take you home with me tonight, maybe mom and the kids will forgive me for not catching dinner... again."

The opening paragraphs depict a hypothetical scenario of a man first taming the wolf. Although we have tried to make light of this event, the fact is no one knows exactly how or why this first encounter took place. More than likely, the "first" encounter between people and wolves occurred more than once. New technologies that have been recently developed enable the extraction of DNA from ancient fossils (Frantz et al., 2020; McHugo et al., 2019).

The earliest possible dates for domestication may have been 25,000–40,000 years ago (Pavlidis and Somel, 2020). Previous studies suggest that dogs were domesticated 14,000 years ago (Boessneck, 1985). However, Ovodov et al. (2011) reported finding dog fossils 33,000 years old in Siberia. Domestication of dogs may have begun before 35,000 years ago in what Galibert et al. (2011) described as a period of proto domestication. Early hunter-gatherers may have captured wolf pups that became tame and habituated to living with human groups. Some wolves may have become aggressive as they matured and were killed or chased away. Others remained submissive and bred with less fearful wolves scavenging around human settlements. Analysis of mitochondrial DNA of 67 dog breeds and wolves from 27 localities indicates that dogs may have diverged from wolves over 100,000 years ago (Vita et al., 1997). Other researchers question this finding and suggest that dogs were domesticated 5,400–16,300 years ago from many maternal lines (Pang et al., 2009; Savolainen et al., 2002). Other evidence suggests that domestication occurred in several different regions (Boyko, 2011). Ancient breeds such as the Australian Dingo, Basenjis, and New Guinea singing dogs, all originated in areas where there were no wolves (Larsen et al., 2012). Possibly, the scenario at the beginning of this chapter happened many times.

Another scenario is that wolves domesticated themselves. The presumption is that calm wolves with low levels of fear were more likely to scavenge near human settlements. Both Coppinger and Smith (1983) and Zenner (1963) suggest that wild species that later became domesticants started out as camp followers. Some wolves were believed to have scavenged near human settlements or followed hunting parties. Modern dog breeds probably were selectively bred from feral village dogs (Boyko, 2011).

The human brain is biologically programmed to pay attention to animals. Electrical recordings from the human amygdala, a brain structure involved with emotion, showed that pictures of animals caused a larger response than pictures of landmarks, people, or objects (Morman et al., 2011). Both threatening and nonthreatening animal pictures evoked the same response. Maybe this shows the importance of animals in our past.

GENETICS SHAPES BEHAVIOR

Genetic differences in animals affect behavior. A major goal of the third edition of this book is to review both old and new research on individual differences within a breed and behavioral differences between breeds. Well-done behavioral studies never become obsolete. Scientists may discover new ways to interpret "why" a certain behavior occurs, but well-done behavioral studies always retain their value. During our literature review, we found many new studies that verified older studies reviewed in the first edition. A good example is a classic work on fearfulness and social rein statement (separation distress) in quail (Faure and Mills, 1998). These emotional traits can be independently strengthened or weakened by selective breeding. Recent experiments with foxes showed distinct differences in the genome of foxes bred to be either tame or aggressive (Kukekova et al., 2018).

This book is aimed at students, animal breeders, researchers, and anyone who is interested in animal behavior. There has been a great increase in research on genetic mechanisms that affect both behavior and physical traits of animals. In this chapter, we review studies of behavioral differences with a focus on domestic animals such as dogs, cattle, horses, pigs, sheep, and poultry. Detailed reviews of genetic mechanisms and molecular biology are beyond the scope of this book.

GENETIC EFFECTS OF DOMESTICATION

Price (1984) defined *domestication* as follows:

> *a process by which a population of animals becomes adapted to man and the captive environment by some combination of genetic changes occurring over generations and environmentally induced developmental events recurring during each generation.*

Major behavioral differences exist between domesticated animals and their wild relatives. For example, the jungle fowl is much more fearful of novel objects and strange people compared to the domestic white Leghorn chicken (Campler et al., 2008). A strong genetic component underlies differences in fearfulness between jungle fowl and domestic chickens (Agnvall et al., 2012).

Domestication may have been based on selection for tameness. In long term selection experiments designed to study the consequences of selection for the "tame" domesticated type of behavior, Belyaev (1979) and Belyaev et al. (1981) studied foxes reared for their fur. The red fox (*Vulpes fulva*) has been raised on seminatural fur farms for over 100 years and was selected for fur traits and not behavioral traits. However, the foxes had three distinctly different behavioral responses to people. Thirty percent were extremely aggressive, 60% were either fearful or fearfully aggressive and 10% dis played a quiet exploratory reaction without either fear or aggression. The objective of this experiment was to breed animals similar in behavior to domestic dogs. By selecting and breeding the tamest individuals, 20 years later the experiment succeeded in turning wild foxes into tame, border collielike fox-dogs. The highly selected "tame" population of (fox-dog) foxes actively sought human contact and would whine and wag their tails when people approached (Belyaev, 1979). This behavior was in sharp contrast to wild foxes that showed extremely aggressive and fearful behavior toward men. Keeler et al. (1970) described this behavior:

Vulpes fulva (the wild fox) is a bundle of jangled nerves. We had observed that when first brought into captivity as an adult, the red fox displays a number of symptoms that are in many ways similar to those observed in psychosis. They resemble a wide variety of phobias, especially fear of open spaces, movement, white objects, sounds, eyes or lenses, large objects, and man, and they exhibit panic, anxiety, fear, apprehension, and a deep trust in the environment. They are (1) catalepsy-like frozen positions, accompanied by blank stares, (2) fear of sitting down, (3) withdrawal, (4) runaway flight reactions, and (5) aggressiveness. Sometimes the strain of captivity makes them deeply disturbed and confused or may produce a depression-like state. Extreme excitation and restlessness may also be observed in some individuals in response to many changes in the physical environment. Most adult red foxes soon after capture break off their canine teeth on the mesh of our expanded metal cage in their attempts to escape. A newly captured fox is known to have torn at the wooden door of his cage in a frenzy until he dropped dead from exhaustion.

Belyaev (1979) and Belyaev et al. (1981) concluded that selection for tameness was effective despite the many undesirable characteristics associated with it. For example, the tame foxes shed during the wrong season and developed black and white patterned fur. Changes were also found in their hormone profiles, and the monestrous (once a year) cycle of reproduction was disturbed. The tame foxes would breed at any time of the year. Furthermore, changes in behavior occurred simultaneously with changes in tail position and ear shape, and the appearance of a white muzzle, forehead blaze, and white shoulder hair. The white color pattern on the head is similar to many domestic animals (Belyaev, 1979) (Figs. 1.1 and 1.2). The most doglike foxes had white spots and patterns on their

Fig. 1.
Wildtype fox before Belyaev started selective breeding for tameness. Reprinted from the *Journal of Heredity* by permission of Oxford University Press.

Fig. 2.
Selecting for tameness for many generations altered the body shape and coat color pattern. Foxes selected for tameness resembled dogs.

heads, drooping ears, and curled tails and looked more like dogs than the foxes that avoided people. The behavioral and morphological (appearance) changes were also correlated with corresponding changes in the levels of sex hormones. The tame foxes had higher levels of the neurotransmitter serotonin (Popova et al., 1975). Serotonin is known to inhibit certain kinds of aggression (Belyaev, 1979). Serotonin levels are increased in the brains of people who take Prozac (fluoxetine).

The fox experiments started by Dmitry Belyaev and Lyudmila Trut in the late 1950s are still ongoing. Researchers have bred both tame and aggressive foxes. Genomic sequencing indicates that they have distinct genetic profiles for genes involved in brain development (Rosenfield et al., 2019; Kukekova et al., 2018). Lord et al. (2020) claim that the fox experiments are not relevant for studying domestication because the original fur farm foxes had already been selected to breed in captivity and for coat color. They were not truly wild foxes. One can argue about this, but the original Belyaev (1979) experiments clearly show that selection for the single behavioral trait of tameness resulted in physical changes in the animal that were not related to behavior.

BASIC GENETIC MECHANISMS

Since the first and second editions, there has been a huge increase in research on genetic mechanisms. By the time this book is published, some of the material on genetic mechanisms may be obsolete. However, behavior studies reviewed in this book will remain useful to scientists working to discover new genetic mechanisms. The classical genetic concepts of recessive and dominant traits discovered by Gregor Mendel explain only a small fraction of the genetic factors affecting inheritance. It is beyond the scope of this book to provide an in-depth review of genetic mechanisms, instead, we describe key principles that will make it easier for the non-geneticist to read and understand the latest papers. Life is complicated. Increasingly complex non-Mendelian genetic mechanisms are being discovered (Schoenfelder and Fraser, 2019; Harich et al., 2020). They are best viewed as networks of information (Hayden, 2012). Below is an outline of some basic genetic mechanisms that produce changes in the appearance and behavior of animals.

Single-nucleotide polymorphisms

A single-nucleotide polymorphism (SNP) is a single code change in a single base pair of DNA. Guryev et al. (2004) state that SNPs are a major factor in genetic variation. In some Mendelian diseases, single SNPs or multiple SNPs are involved in disease inheritance (Kong et al., 2009; Shastry, 2002).

Repeats

Repeats are also called tandem repeats, single sequence repeats, or genetic stutters. Repeats are sequences of DNA code repeated more than once, and the number of repeats can vary within a gene. The number of repeats can determine many traits ranging from the length of a dog's nose to variations in brain development (Fondon and Garner, 2004; Fondon et al., 2008). Marshall et al. (2021) state that repeats are involved in both normal function and disease. Newer genomic sequencing methods will facilitate counting repeats.

Copy number variations

Copy number variations (CNVs) are rearrangements of genetic code. It is likely CNVs contribute greatly to genetic variation by modifying genetic expression (Chaignat et al., 2011; Henrichsen et al., 2009). CNVs are often spontaneous (de novo) mutations not inherited from the parents. The many types of CNVs range from translocation of pieces of genetic code to deletions of genetic code or to extra copies of code. Twelve percent of the human genome is in copy number variable regions of the genome (Redon, 2006). CNVs are very numerous in the brain and the immune system (Harich et al., 2020).

Jumping genes

Also called transposable elements (transposons), jumping genes are short segments of genetic material that transport themselves throughout the genome in a "cut and paste," or a "copy and paste" manner. Mikkelson et al. (2007) state that transposons are a "creative force" in the evolution of mammalian gene regulation. Jumping genes are more numerous in the brain than in liver or heart cells (Vogel, 2011).

Coding DNA

The very small percentage of the genome that specifically codes for proteins is used in the development of the animal. Until recently, only coding DNA was sequenced. Coding DNA is only 2% of the human genome (Marshall et al., 2021).

Noncoding DNA, also called regulatory DNA

In the 1980s, this was called junk DNA because it does not code for proteins. Researchers have discovered that noncoding DNA has a regulatory function and approximately 80% of the noncoding DNA is transcribed by RNA and has biochemical functions (ENCODE Project Consortium, 2012). Noncoding DNA is the "computer operating system" that directs the coding DNA. Noncoding DNA may be the gene's "project managers" that orchestrate and direct the sequence of building proteins (Saey, 2011). Chakravarti and Kapoor (2012) state that to understand the genes that code for proteins, the regulatory noncoding DNA needs to be understood. Some portions of non-coding DNA are highly conserved, and similar sequences occur in many different animals. Other portions of non-coding genome may rapidly evolve (Maher, 2012). Coding regions of DNA that direct the development of basic patterning of the body (fox genes) are highly conserved across many species from arthropods to mammals (Linn et al., 2008). Early pattern formation of the notochord and neural tube is also highly conserved (Richardson, 2012). In both plants and animals, the embryos of many species look similar during the mid-embryonic stage of development. The mid-stage of development is "dominated by ancient genes" (Quint et al., 2012). A basic principle is that similar traits in a species originate from a highly conserved genetic code. Traits that have recently changed originate from newer code. Research shows that changes in noncoding DNA are drivers of evolutionary change. The human neocortex has a higher percentage of young genes expressed during fetal development compared to mice (Zhang et al., 2011). Transposable bits of code can make changes in regulatory DNA (Mikkelson et al., 2007). The fact that similar sections of noncoding DNA occur in many species indicates its important function. Changes in noncoding DNA during environmental changes cause stickleback fish to adapt by changing traits, such as body shape, skeletal armor, or the ability to live in salt or freshwater (Hockstra, 2012). Sections of noncoding regulatory DNA evolve and change along with the coding DNA (Jones et al., 2012). Studies in other animals also show a role for noncoding DNA mutations in the development of domestic animals (Anderson, 2012).

Exome

The exome is the DNA sequencing of all the protein coding regions of the genome. The noncoding DNA is left out of the exome.

RNA transcriptome

The RNA transcriptome is the DNA code that is read and transcribed by RNA. Sequencing the genome indicates that the animal's genome contains a particular piece of genetic code. The transcriptome indicates whether or not the code was transcribed by RNA and expressed as either a protein or involved in regulatory functions.

De novo mutations

De novo mutations are random mutations that are not inherited. Common de novo mutations are CNVs, SNPs, and other changes in DNA code.

Quantitative trait loci

Quantitative trait loci (QTLs) are regions of DNA containing many nucleotide base pairs associated with continuous traits such as height or temperament. QTLs are not associated with simple discreet Mendelian traits such as hair and eye color. QTLs are associated with phenotypic traits influenced by many genes (polygenic). They have limitations and QTLs may fail to detect genetic code that has the opposite effect on the trait being studied (York, 2018).

Haplotypes

Haplotypes are a group of genes linked together and inherited as a group.

Epigenetics

An animal's DNA may contain a certain sequence of genetic code, but it may be locked out by epigenetic mechanisms. Environmental influences can either lock out sections of code or unlock them. For example, epigenetic mechanisms either upregulate (make more anxious) or downregulate the nervous system of rodents depending on how much the pregnant mother was stressed, or how often a rodent is attacked by another rodent (Nestler, 2011, 2012).

Lamarckism

Jean-Baptiste Lamarck (17441829) was a French naturalist remembered for a theory of inheritance of acquired characteristics, more commonly referred to as soft inheritance, Lamarckism, or the theory of use/disuse. Lamarck

believed that animals could acquire a certain trait during their lifespan and that the trait could be passed to the next generation. Part of Lamarck's argument is actually supported by the field of epigenetics and other parts of his theories were wrong. One part of this branch of science is based on the proteins (histones) binding to the DNA, winding it into a small enough shape to fit into the cell (Probst et al., 2009). Chemicals cause histones to bind either tighter or looser. Certain influences during life can influence the tightening or loosening of histones, these changes are then passed on (Sarma and Reinberg, 2005). If a piece of genetic code is loose, it will be easy to read, if it is tighter, it will not be read. This changes the expression of the genes and to a large extent explains the difference between twins at older ages, even when their genes are exactly the same (Fraga, 2005; Poulsen et al., 2007). Another epigenetic modification that serves to regulate gene expression without altering the underlying DNA sequence is DNA methylation. In simple terms, DNA methylation acts to "turn on" or "turn off" a gene. The methylation "lock" prevents a section of the DNA code gene from being read and expressed. For example, maternal obesity before and during pregnancy in mice affects the establishment of body weight regulatory mechanisms in her baby. This is caused by methylation locks that lock out a section of the genetic code. Overweight mothers give birth to offspring who become even heavier, resulting in amplification of obesity across generations (Champagne and Curley, 2009).

Some evidence suggests that the environment can make lasting changes to the expression of genes via epigenetic mechanisms—changes that may be passed on to future generations (Crews, 2010). Studies in rats show that epigenetics influences maternal behavior and the effect can be passed on from one generation to the next (Cameron, 2008). The offspring of rat mothers who display high levels of nurturing behavior such as licking and grooming are less anxious and produce less stress hormones, compared to the offspring of less nurturing mothers. In turn, the female offspring of nurturing mothers become nurturing mothers themselves. The effects of maternal behavior are mediated in part through epigenetic mechanisms. In people, if the father has early life stress as a child, he may pass on brain changes to his child (Karlsson et al., 2020). This may be via an epigenetic mechanism.

BRAIN GENETICS MORE COMPLEX THAN OTHER TRAITS

For centuries, intensive selection in dogs has narrowed the gene pool for traits such as body shape and type of coat. Researchers have discovered that only a few genomic regions control many dog appearance traits (Boyko et al., 2010). This is not true for behavior. The genetics that controls brain development is much more complex. Selective breeding experiments in foxes and other experiments involving selection for appearance traits show that those traits are sometimes linked to behavioral traits. Why does selecting for a calm temperament produce a black and white fox? When the first edition of this book was written, these unusually linked traits were unexplained. The long running ENCODE project (2012), which is mapping the noncoding regions of DNA, may help provide answers. Regions of noncoding DNA are not always located adjacent to the piece of code it regulates. There are long range interactions. Sanyal et al. (2012) and Giammartino et al. (2020) state that regulatory elements and coding DNA are in complex three-dimensional networks. Maybe when long strands of DNA are folded up, the temperament and coat color regions are folded up beside each other. Schoenfelder and

Fraser (2019) state that regulatory information gets transcribed when it is in physical contact with target genes.

A BRIEF HISTORICAL REVIEW OF ANIMAL BEHAVIOR STUDY

This historical review is not intended to be completely comprehensive. Our objective is to discuss some of the early discoveries important for our current understanding of animal behavior, with particular emphasis on genetic influences on behavior in domestic animals.

Early in the 17th century, Descartes came to the conclusion "that the bodies of animals and men act wholly like machines and move in accordance with purely mechanical laws (in Huxley, 1874)." After Descartes, others undertook the task of explaining behavior as reactions to purely physical, chemical, or mechanical events. For the next three centuries, scientific thought on behavior oscillated between a mechanistic view that animals are "automatons" moving through life without consciousness or self-awareness and an opposing view that animals had thoughts and feelings similar to those of humans.

In "On the Origin of the Species" (1859), Darwin's ideas about evolution began to raise serious doubts about the mechanistic view of animal behavior. He noticed animals share many physical characteristics and were one of the first to discuss variation within a species, both in behavior and in physical appearance. Darwin believed that artificial selection and natural selection were intimately associated. Darwin (1868) cleverly outlined the theory of evolution without any knowledge of genetics. In "The Descent of Man" (1871) Darwin concluded that temperament traits in domestic animals are inherited. He also believed, as did many other scientists of his time, that animals had subjective sensations and could think. Darwin wrote: "The differences in mind between man and the higher animals, great as it is, is certainly one of degree and not of kind."

Other scientists realized the implications of Darwin's theory on animal behavior and conducted experiments investigating instinct. Herrick (1908) observed the behavior of wild birds to determine, first, how their instincts are modified by their ability to learn, and second, the degree of intelligence they attain. On the issue of thinking in animals, Schroeder (1914) concluded: "The solution, if it ever comes, can scarcely fail to illuminate, if not the animal mind, at least that of man." It is evident that by the end of the 19th century, scientists studying animal behavior in natural environments learned that the mechanical approach could not explain all behavior.

BEHAVIORISM

During the middle of the 20th century, scientific thought again reverted to the mechanical approach and behaviorism reigned throughout America. The behaviorists ignored both genetic effects on behavior and the ability of animals to engage in flexible problem-solving. The founder of behaviorism—Watson—stated, "differences in the environment can explain all differences in behavior (1930)." He did not believe that genetics had any effect on behavior. In "The Behavior of Organisms," the psychologist Skinner (1958) wrote that all behavior could be explained by the principles of stimulus-response and operant conditioning.

The first author visited with Dr Skinner at Harvard University in 1968. Skinner responded to a question about the need for brain research by saying, "We don't need to know about the brain because we have operant conditioning (Grandin and Johnson, 2005)." Operant conditioning uses food rewards and punishments to train animals and shape their behavior. In a simple Skinner box experiment, a rat can be trained to push a lever to obtain food when a green light turns on, or to push a lever very quickly to avoid a shock when a red light appears. The signal light is the "conditioned stimulus." Rats and other animals can be trained to perform a complex sequence of behaviors by chaining together a series of simple operant responses. Skinner believed that even the most complex behaviors can be explained as a series of conditioned responses.

However, in a Skinner box a rat's behavior is very limited. It's a world with very little variation, and the rat has little opportunity to use its natural behaviors. It simply learns to push a lever to obtain food or prevent a shock. Skinnerian principles explain why a rat behaves a certain way in the sterile confines of a 30x3x30 cm Plexiglas box, but they don't reveal much about the behavior of a rat in the local dump. Outside of the laboratory, a rat's behavior is more complex.

INSTINCTS VERSUS LEARNING

Skinner's influence on scientific thinking slowed a bit in 1961 following the publication of "The Misbehavior of Organisms" by Brelands and Brelands. Their paper described how Skinnerian behavioral principles collided with instincts. The Brelands were trained Skinnerian behaviorists who attempted to apply strict principles of operant conditioning to animals trained at fairs and carnivals. Ten years before this classic paper, Brelands and Brelands (1951) wrote, "we are wholly affirmative and optimistic that principles derived from the laboratory can be applied to the extensive control of animal behavior under non laboratory condition." However, by 1961, after training more than 6000 animals as diverse as reindeers, cockatoos, raccoons, porpoises, and whales for exhibition in zoos, natural history museums, department store displays, fairgrounds, trade convention exhibits, and television, the Brelands wrote a second article featured in the Breland and Breland (1961), which stated, "our background in behaviorism had not prepared us for the shock of some of our failures."

One of the failures occurred when the Brelands tried to teach chickens to stand quietly on a platform for 1012 seconds before they received a food reward. The chickens would stand quietly on a platform in the beginning of training. However, once they learned to associate the platform with a food reward, half (50%) started scratching the platform, and another 25% developed other behaviors, such as pecking the platform. The Brelands salvaged this disaster by developing a wholly unplanned exhibit involving a chicken that turned on a jukebox and danced. They first trained chickens to pull a rubber loop that turned on some music. When the music started, the chickens would jump on the platform and start scratching and pecking until the food reward was delivered. This exhibit made use of the chicken's instinctive food-getting behavior. The first author remembers as a young adult seeing a similar exhibit at the Arizona State Fair of a piano-playing chicken in a little red barn. The hen would peck the keys of a toy piano when a quarter was put in the slot and would stop when the food came down the chute. This exhibit also worked because it was similar to a Skinner box in the laboratory. It also utilized the natural pecking behavior of the chicken.

The Brelands experienced another classic failure when they tried to teach raccoons to put coins in a piggy bank. Because raccoons are adept at manipulating objects with their hands, this task was initially easy. As training progressed, however, the raccoons began to rub the coins before depositing them in the bank. This behavior was similar to the washing behavior raccoons do as instinctive food-getting behavior. The raccoons at first had difficulty letting go of the coin and would hold and rub it. However, when the Brelands introduced a second coin, the raccoons became almost impossible to train. Rubbing the coins together "in a most miserly fashion," the raccoons got worse and worse as time went on. The Brelands concluded that the innate behaviors were suppressed during the early stages of training and sometimes for long into the training, but as training progressed, instinctive food-getting behaviors gradually replaced the conditioned behavior. The animals were unable to override their instincts and thus a conflict between conditioned and instinctive behaviors occurred.

ETHOLOGY

While Skinner and his fellow Americans were refining the principles of operant conditioning on thousands of rats and mice, ethology was being developed in Europe. Ethology is the study of animal behavior in natural environments. The primary concern of ethologists is instinctive or innate behavior (Eibi-Eibesfeldt and Kramer, 1958). Essentially, ethologists believed that the secrets to behavior are found in the animal's genes, and the way genes were modified during evolution to deal with particular environments. The ethological trend originated with Whitman (1898), who regarded behavioral reactions to be so constant and characteristic for each species that, like morphological structures, they may be of taxonomic significance. A similar opinion was held by Heinroth (1918, 1938). He trained newly hatched fledglings in isolation from adults of their own species and discovered that instinctive movements, such as preening, shaking, and scratching, were performed by young birds without observing other birds.

Understanding the mechanisms and programming of innate behavioral patterns and the motivation underlying behavior is the primary focus of ethologists. Konrad

Lorenz (1939, 1965, 1981) and Niko Tinbergen (1948, 1951) cataloged the behavior of many animals in natural environments. Together they developed the ethogram. An ethogram is a complete listing of the behaviors an animal performs in its natural environment. The ethogram includes both innate and learned behaviors.

An interesting contribution to ethology came from studies on egg rolling behavior in the graylag goose (Lorenz, 1965, 1981). When a brooding goose notices an egg outside her nest, Lorenz observed that an instinctive program triggers the goose to retrieve it. The goose fixates on the egg, rises to extend her neck and bill out over it, then gently rolls it back to the nest. This behavior is performed in a highly mechanical way. If the egg is removed as the goose begins to extend her neck, she still completes the pattern of rolling the nonexistent egg back to the nest. Lorenz (1939) and Tinbergen (1948) termed this a "fixed action pattern." Remarkably, Tinbergen also discovered that brooding geese can be stimulated to perform egg rolling on such items as beer cans and baseballs. The fixed action pattern of rolling the egg back to the nest can be triggered by anything outside the nest that even marginally resembles an egg. Tinbergen realized that geese possess a genetic releasing mechanism for this fixed action pattern. Lorenz and Tinbergen called the object that triggers the release of a fixed action pattern "sign stimuli." When a mother bird sees the gaping mouth of her young, it triggers the maternal feeding behavior and the mother feeds her young. The gaping mouth is another example of sign stimuli acting as a switch that turns on the genetically determined program (Herrick, 1908; Tinbergen, 1951).

Ethologists also explained the innate escape response of newly hatched goslings. When goslings are tested with a cardboard silhouette in the shape of a hawk moving overhead, it triggers a characteristic escape response. The goslings will crouch or run. However, when the silhouette is reversed to look like a goose, there is no effect (Tinbergen, 1951). Several members of the research community doubted the existence of such a hardwired instinct because other scientists failed to repeat these experiments (Hirsch et al., 1955). Canty and Gould (1995) repeated the classic experiments and explained why the other experiments failed. First, only goslings under 7 days old respond to the silhouette. Second, a large silhouette, which casts a shadow, must be used. Third, goslings respond to the perceived predator differently depending upon the circumstances. For example, birds tested alone try to run away from the hawk silhouette and birds reared and tested in groups tend to crouch (Canty and Gould, 1995). Nevertheless, fear is likely to be the basis of the response. Ducklings were shown to have higher heart rate variability when they saw the hawk silhouette (Mueller and Parker, 1980). Research by Balaban (1997) indicates that species-specific vocalizations and head movements in chickens and quail are controlled by distinct cell groups in the brain. To prove this, Balaban transplanted neural tube cells from developing quail embryos into chicken embryos. Chickens hatched from the transplanted eggs exhibited species-specific quail songs and bobbing head movements.

Do similar fixed action patterns occur in mammals? Fentress (1973) conducted an experiment on mice that clearly showed that animals have instinctive species-specific behavior patterns that do not require learning. Day old baby mice were anesthetized and had a portion of their front legs amputated. Enough of the leg remained that the mice could easily walk. The operations were performed before the baby mice had fully coordinated movements so there was no opportunity for learning. When the mice became adults, they still performed the species-specific face-washing behavior; normal mice close their eyes just before the foreleg passing over the face, and in the amputees, the eye still closed before the nonexistent paw hit it. The amputees performed the face-washing routine as if they still had their paws. Fentress (1973) concluded that the experiment proved the existence of instincts in mammals.

Two years after Breland's article, Jerry Hirsh (1963) at the University of Illinois, wrote a paper emphasizing the importance of studying individual differences. In it he wrote, "Individual differences are no accident. They are generated by properties of organisms as fundamental to behavior science as thermodynamic properties are to physical science."

ETHOLOGY AND BEHAVIORISM PROVIDE TOOLS TO STUDY EMOTIONS AND BEHAVIORS

Both the behaviorists and the ethologists avoided the question of whether or not animals had emotions. They both developed a strictly functional approach to the motivations of behavior (DeWaal, 2011). Until relatively recently, most behaviorists and ethologists did not get involved with neuroscience. A review of the neuroscience literature makes it clear that emotional systems in the brain drive behavior. The research tools provided by the disciplines of both ethology and behaviorism are essential to further our understanding of animal behavior. Ethology provides the methodology for studying animals in complex environments. Bateson (2012) discusses the need to study freely moving animals. Animal behavior is more complex in natural settings or on a farm. Lawrence (2008) reviewed the behavior literature and determined that domestic animal research was changing from studying the basic biology of domestic animal behavior to studying animal behavior related to specific animal welfare issues. Lawrence (2008) warns that too narrow a focus on specific welfare concerns may be detrimental to answering broader welfare issues such as the subjective state of animals.

NEUROSCIENCE AND BEHAVIOR

Modern neuroscience supports Darwin's view on emotions in animals. All mammal brains are constructed with the same basic design. They all have a brainstem, limbic system, cerebellum, and cerebral cortex. The cerebral cortex is the part of the brain used for thinking and flexible problem-solving. The major difference between the brains of people and animals is in the size and complexity of the cortex. The emotional systems serving as drivers for behavior are located in the subcortex and are similar in all mammals (Panksepp, 2011; Montag and Panksepp, 2017). Primates have a larger and more complex cortex than a dog or a pig. Pigs have a more complex cortex than a rat or a mouse. Furthermore, all animals possess innate species-specific motor patterns that interact with experience and learning in deter mining behavior. Certain behaviors in both wild and domestic animals are governed largely by innate (hardwired) programs.

Behaviors for copulation, killing prey, nursing young, and nest building tend to be more instinctual and hardwired. Experience and learning play a larger role in behaviors that require more flexibility such as finding food, social interactions, and hunting.

Another basic principle to remember is that animals with large, complex brains are less governed by innate behavior patterns. For example, bird behavior is governed more by instinct than that of a dog, whereas an insect would have more hardwired behavior patterns than a bird. This principle was clear to Yerkes (1905) who wrote:

Certain animals are markedly plastic or voluntary in their behavior, others are as markedly fixed or instinctive. In the primates, plasticity has reached its highest known stage of development; in the insects, fixity has triumphed, instinctive action is predominant. The ant has apparently sacrificed adaptability to the development of the ability to react quickly, accurately, and uniformly in a certain way. Roughly, animals might be separated into two classes: those that are in a high degree capable of immediate adaptation to their conditions, and those that are apparently automatic as they depend upon instinct tendencies to action instead of upon rapid adaptation. Since the publication of the second edition, researchers have embraced the concept of personality in animals (Cabrera et al., 2021; Finkelmeier et al., 2018). Emotions in animals are now being researched and discussed (Kremer et al., 2020; Burghardt, 2019). This is a major shift compared to the early 1990s. At this time, journal article reviewers forced the first author to remove the word *fear* from a cattle behavior paper. It had to be replaced with *agitated* (Grandin, 1993a,b).

EMOTIONAL SYSTEMS MOTIVATE BEHAVIOR

Great strides have been made in understanding how genetic factors affect behavior when scientists started by looking at brain systems that control emotions. This is the starting point for making it possible to sort out many conflicting results in behavioral studies. One big problem is that different words are used by different researchers to describe the same emotions (O'Malley et al., 2019).

The neuroscientist Jaak Panksepp outlined the major emotional systems located in the subcortical areas of the brain. The four main emotions are FEAR, RAGE, PANIC (separation distress), and SEEKING (novelty seeking) (Panksepp, 2005, 1998; Morris et al., 2011). He also listed three additional emotional systems of LUST, CARE (mother-young nurturing behavior), and PLAY. Each primary system is associated with a genetically based subcortical brain network. Panksepp (2011) and Montag and Panksepp (2017) defined the basic emotional circuits of mammalian brains:

Fear

An emotion induced by a perceived threat that causes animals to move quickly away from the location of the perceived threat, and sometimes hide. Fear should be distinguished from anxiety, which typically occurs without any certain or immediate external threat. Some examples of fear are reactions to exposure to sudden novelty, startle responses, and hiding from predators. Fear is sometimes referred to as behavioral reactivity, behavioral agitation, or a highly reactive temperament. This emotion is often referred to as Bold (low fear) and shy (high fear) (O'Malley et al., 2019; Laine and Oers, 2017).

Panic

Separation distress is an emotional condition in which an individual experiences excessive anxiety regarding separation from either home or from other animals that the individual has a strong emotional attachment to. One example of separation distress is a puppy or lamb vocalizing when it is separated from its mother. The PANIC system may also be activated when a single cow is separated from her herd. Sometimes referred to as social isolation stress or high social reinstatement behavior.

Rage

A feeling of intense anger. Rage is associated with the fight-or flight response and is activated in response to an external cue such as frustration or attempts to curtail an animal's activity. RAGE is the emotion that enables an animal to escape when it is in the jaws of a predator. In the research literature, it is also called aggression.

Seeking

The seeking system (novelty seeking) in the brain motivates animals to become extremely energized to explore the world but is not restricted to the narrow behavioristic concept of approach, or the pleasure/ reinforcement system. Seeking is a broad action system in the brain that helps coordinate feelings of anticipation, eagerness, purpose and persistence, wanting, and desire. This system promotes learning by urging animals to explore and to find resources needed for survival. A dog that excitedly sniffs and explores every room when turned loose in a strange house is an example of high SEEKING in an animal. In many behavior studies, this emotion is referred to as exploration (O'Malley et al., 2019). Open-field tests where an animal is placed along in an arena may be difficult to interpret from an emotional standpoint. Is the behavior in the open field motivated by PANIC (separation stress), FEAR, or SEEK? Perals et al. (2017) warn that other behavioral tests may be required to separate the traits of exploration and shyness (Fear).

Exciting new research has separated the variable of true novelty seeking from activity or separation distress. Mice were exposed in an open field to both a familiar and a novel object (Farahbakhsh and Siciliano, 2021; Ahmadlou et al., 2021). This makes it possible to separate the attraction to novelty from the other variables. These researchers may have discovered a new circuit in the brain that is associated with novelty seeking.

Lust

The lust system in the brain controls sexual desire or appetite. Sexual urges are mediated by specific brain circuits and chemistries that overlap but are distinct between males and females and are aroused by male and female hormones.

Care

The maternal nurturing system that assures that parents take care of their offspring. Hormonal changes at the end of pregnancy activate maternal urges that promote social bonding with the offspring.

Play

A key function of the play system is to help young animals acquire social knowledge and refine subtle social interactions needed to thrive. One motivation for PLAY is the dopamine energized SEEKING system.

The existence of these emotional structures is well documented, and review articles on this research can be found in Morris et al. (2011), Burgdorf and Panksepp (2006), and Panksepp (2011). Direct electrical or chemical stimulation of specific subcortical structures elicits emotional responses. The brain circuits controlling fear and seeking in animals have been extensively mapped (LeDoux, 2000; Reynolds and Berridge, 2008).

CONFUSION OF EMOTIONAL SYSTEMS MAY CONFOUND STUDIES

In the behavior literature, many inconsistencies exist in papers on novelty seeking and fear. In many studies, novelty seeking may be confused with other emotional systems when terms such as activity level or emotional reactivity are used. Confusion may also exist between FEAR and PANIC when the term reactivity is used. The FEAR and PANIC (separation distress) systems have totally different functions. Fear keeps an animal away from danger and the PANIC system prevents the offspring from getting separated from its mother and helps to keep social groups together. Some scientists assume that an open field test only evaluates fearfulness in animals. An animal alone in an open field may be reacting to separation from its mother or the social group. Confusing fear and separation distress is more likely when herding and flocking animals are tested in an open field, compared to animals that live a more solitary life. A term that is often used for describing behavior in an open field test is activity. It is likely that the motivation for activity is misinterpreted in many studies. Another big confusion in the scientific literature is behavior labeled personality. Zadar et al. (2017) state that tests for coping style did not correlate with personality assays. It is likely that the emotional systems are being mixed up. Panksepp's framework of the seven emotional systems may help sort out conflicting results in the scientific literature.

Research by Reynolds and Berridge (2008) has shown that the emotional traits of seeking novelty and fear are both controlled in a structure in the brain called the nucleus accumbens. When one end of the nucleus accumbens is stimulated, the animal becomes fearful. Stimulating the other end turns on seeking (Faure et al., 2008). There is a mixture of fear and seeking receptors in the middle portion of the nucleus accumbens. The discovery of this function for the nucleus accumbens may explain the "curiously afraid" behavior we observed in cattle. Cattle curiously approach a novel clipboard laid on the ground, but when the wind flaps the paper, they fearfully jump back. When the paper stops moving, the cattle approach it again. The nucleus accumbens may be in SEEK mode when cattle voluntarily approach and switches into a fear mode when the paper suddenly moves.

GENETICS AND EMOTIONAL SYSTEMS

Research clearly shows that genetic factors have a very strong effect on both fearfulness and novelty seeking (Campler et al., 2008; Clinton et al., 2007; Stead et al., 2006). Maternal factors had little effect on novelty seeking in rats (Stead et al., 2006). Cross-fostering also had little effect, which shows that novelty seeking is highly heritable. Animals that are high seekers have more dopaminergic activity in the nucleus accumbens (Dellu et al., 1996). In addition to the heritable component of fearfulness, environmental influences also have an effect on fearfulness. Stressful treatment of either a pregnant mother or her offspring can upregulate the fear system. Lemos et al. (2012) propose that severe stress can disable the appetitive system in the nucleus accumbens. To state it more simply, stress can break the animal's SEEK function and it will no longer explore. Other emotional systems are also subject to both genetic and environmental influences. The PANIC (separation distress) system is highly heritable. In sheep, separation distress measured by isolating a single animal and measuring how many times it bleats (vocalizes) shows that vocalization is highly heritable (Boissy et al., 2005). Strength of the sex drive is also heritable. When the first Chinese pig breeds were imported into the United States, caretakers observed that boars were more highly motivated to mate compared to the United

States and European commercial pigs. The Chinese pig was bred for large litter size and low levels of meat production.

The brain circuits that drive behavior are complicated. Bendesky et al. (2017) report that parental care behavior in mice has four separate brain circuits for licking their pups, retrieving their pups, huddling, and nest building. There is also increasing evidence that genetic factors that influence behavior in animals are also present in people. There is an interesting paper titled *Solitary Mammals as a Model for Autism* that reviewed genetic similarities between people with autism and animals that lead a more solitary life such as panthers or leopards (Resar, 2013). Two other studies showed that cattle temperament measured with a flight speed test was associated with genes for susceptibility of autism (Costilla et al., 2020; Chen et al., 2019). An extensive review article by Amanda et al. (2021) reported that half of the genes associated with farm animal behavior have also been identified as related to behavioral and neuronal disorders in people. The same genes that make dogs friendly are also associated with Williams-Beuren syndrome in people (Von Holdt, et al., 2017). These studies clearly show that the same genetic codes are associated with emotions in both people and animals. Why would genes that are associated with variations in animal behavior be associated with brain disorders in humans? Maybe it is due to having a more complex brain. Sikela and Searles (2018) have a paper that may help explain this. It is titled *Genomic Tradeoffs: Are Autism and Schizophrenia the Steep Price for a Human Brain?* Building a huge human brain from stem cells that rapidly multiply may create more opportunities for mistakes during development. This may explain why genes associated with animal behavior would also be associated with brain disorders in humans.

INTERACTIONS BETWEEN GENETICS AND EXPERIENCE

Some behavior patterns are similar between different species, and some are found only in a particular species. For example, the neural programs that enable animals to walk are similar in most mammals (Melton, 1991; Grillner, 2011). On the other hand, courtship rituals in birds are very species-specific (Nottebohm, 1977). Some innate behavior patterns are very rigid and experience has little effect on them. Other instinctive behaviors can be modified by learning and experience. The flehmann, or lip curl response of a bull when he smells a cow in estrus, and the kneel down (lordois) posture of a rat in estrus are examples of rigid behaviors. Suckling by newborn mammals is another example of a hardwired behavioral system. Suckling behavior does not vary. Newborn mammals suckle almost everything put in their mouth.

An example of an innate behavior affected by learning is burrowing behavior in rats. Boice (1977) found that wild Norway rats and albino laboratory rats both dig elaborate burrows. Learning has some effect on the efficiency of burrowing, but the configuration of the burrows was the same for both the wild and domestic rats. Albino laboratory rats dug excellent burrows the first time they were exposed to an outdoor pen. Nest building in sows is another example of the interaction between instinct and learning. When a sow is having her first litter, she has an uncontrollable urge to build a nest. Nest building is hardwired and hormonally driven. Widowski and Curtis (1989) showed that injections of prostaglandin F_{2a} induce nest building in sows. However, sows learn from experience how to build a better nest with each successful litter.

Other behaviors are almost entirely learned. Seagulls are known to drop shellfish on rocks to break them open, while others drop them on the road and let cars break them open (Grandin, 1995). Many animals ranging from apes to birds use tools to obtain food. New Calendonian crows use complicated tools to obtain food and can solve problems other birds cannot (Weir and Kacelnik, 2006). Griffin (1994) and Dawkins (1993) provide many examples of complex learned behaviors and flexible problem solving in animals.

Innate behaviors used for finding food, such as grazing, scavenging, or hunting, are more dependent on learning than behaviors used to consume food. Sexual behavior, nesting, eating, and prey killing behaviors tend to be governed more by instinct (Gould, 1977). The greater dependence on learning to find food makes animals in the wild more flexible and able to adapt to a variety of environments. Behaviors used to kill or consume food can be the same in any environment. Mayr (1974) called these different behavioral systems "open" or "closed" to the effects of experience. A lion hunting her prey is an example of an open system. The hunting female lion recognizes her prey from a distance and carefully stalks her approach. Herrick (1910) wrote, "the details of the hunt vary every time she hunts. Therefore, no combination of simple reflex arcs laid down in the nervous system will be adequate to meet the infinite variations of the requirements for obtaining food."

INTERACTIONS BETWEEN INSTINCTUAL HARDWIRED BEHAVIOR AND EXPERIENCE

Some of the interactions between genetics and experience have very complex effects on behavior. In birds, the chaffinch learns to sing its species-specific song even when reared in a soundproof box where it is unable to hear other birds (Nottebohm, 1970, 1979). However, when chaffinches are allowed to hear other birds sing, they develop a more complex song. The basic pattern of the canary song emerges even in the absence of conspecific (flock mate) auditory models (Metfessel, 1935; Poulsen, 1959). Young canaries imitate the song of adult canaries they can hear, and when reared in groups they develop song patterns that they all share (Nottebohm, 1977). Many birds, such as the white crowned sparrow, chaffinch, and parrot, can develop local song dialects (Nottebohm et al., 1976; Adler, 1996). Sparrows are able to learn songs by listening to recordings of songs with either pure tones or harmonic overtones. Birds trained with harmonic overtones learned to sing songs with harmonic overtones, but 1 year later, 85% of their songs reverted back to innate pure tone patterns (Nowicki and Marler, 1988). Experiments by Mundinger (1995) attempted to determine the relative contribution of genetics and learning in bird songs. Inbred lines of roller and border canaries were used in this study along with a hybrid cross of the two. The rollers were cross fostered to border hens and vice versa to control for the effects of maternal behavior. The roller and border males preferred to sing innate song patterns instead of copying their tutors. The hybrids preferred to learn some of both songs. Furthermore, canaries are capable of learning parts of an alien song but have a definite preference for their own songs. Comparing these animals to those in Brelands and Brelands (1961) exhibits, birds can be trained to sing a different song, but genetically determined patterns have a strong tendency to override learning. In reviewing all this literature, it became clear that innate patterns in mammals can be overridden. Unfortunately, the animals tend to revert back to innate behavior patterns.

THE PARADOX OF NOVELTY

Novelty is anything new or strange in an animal's environment. Novelty is a paradox because it is both fear-provoking and attractive. Paradoxically, it is most fear-provoking and attractive to animals with a nervous, excitable temperament. Skinner (1922) observed that pronghorn antelope, a very flighty animal, will approach a person lying on the ground waving a red flag. Kruuk (1972) further observed attraction and reaction to novelty in Thompson's gazelles in Africa. In small groups, Thompson's gazelles are most watchful for predators (Elgar, 1989). Animals that survive in the wild by flight are more attentive to novelty than more placid animals. Gazelles can also distinguish between a dangerous hunting predator and one that is not hunting. In Thompson's gazelles, the most dangerous predators attract the highest degrees of attraction. They often move close to a cheetah when the cheetah is not hunting. Furthermore, when predators walk through a herd of Thompson's gazelles, the size of the flight zone varies depending on the species of predator.

More recent research since the second edition may provide insights into why novelty is both scary and attractive. Christensen et al. (2020, 2021) found that horses that were more likely to explore a novel object could learn a discrimination task more quickly. In both birds and cattle, genetics has an effect on both the tendency to approach a novel object or avoid it (Hirata and Arimoto, 2018; Rozempolska-Rucinska et al., 2017). The explore (SEEK) trait and the FEAR treat are separate. Depending on the environment, the balance between SEEK or FEAR can be adjusted. Exploration facilitates learning but it may make an animal more vulnerable to predators. Fig. 3 shows the reaction of cattle to a novel box. Some animals approach it more quickly than others.

Fig. 3.
The reaction of these cattle to a novel box placed in their pen clearly shows differences in their behavior. One animal immediately investigates the box, and the others are more reluctant to approach. Many of the other cattle are watching. It is likely that future research will find that the variation in their behavior can be explained by inherited differences in fearfulness and exploration (SEEK) traits.

REACTION TO NOVELTY

Highly reactive animals are more likely to have a major fear reaction when confronted with sudden novelty. In domestic animals, examples of sudden novelty include being placed in a new cage, transport in a strange vehicle, an unexpected loud noise, or being placed in an open field. Using various experimental environments, Hennessy and Levine (1978) found that rats show varying degrees of stress and stress hormone levels proportional to the degree of novelty of the environment they are placed in. A glass jar is totally novel in appearance compared to a lab cage the animal is accustomed to.

Being placed in a glass jar was more stressful for rats than a familiar lab cage with no bedding.

Studies of reactions to novelty in farm animals have been conducted by Moberg and Wood (1982), Stephens and Toner (1975), and Dantzer and Mormede (1983). Calves show the highest degrees of stress when placed in an open-field test arena very dissimilar from their home pen (Dantzer and Mormede, 1983). Calves raised indoors were more stressed by an outdoor arena and calves raised outdoors were more stressed by an indoor arena. The second author is familiar with similar responses in horses. When horses are taken to the mountains for the first time, a well-trained riding horse accustomed to different show rings may panic when it sees a butterfly or hears a twig snapping on a mountain trail. Recent research with horses shows that when a familiar large complex object is rotated ninety degrees, a horse will react as if it is something new (Corgan et al., 2020). A child's brightly colored playset with a slide and a swing was used.

GENETIC FACTORS AND THE NEED FOR NOVELTY

In mammals and birds, normal development of the brain and sense organs requires novelty and varied sensory input. Nobel prizewinning research of Hubel and Wiesel (1970) showed that the visual system is permanently damaged if kittens do not receive varied visual input during development. Dogs are more excitable when raised in barren and non-stimulating environments (Melzack and Burns, 1965; Walsh and Cummins, 1975). Schultz (1965) stated, "when stimulus variation is restricted central regulation of threshold sensitivities will function to lower sensory thresholds." Krushinski (1960) studied the influence of isolated conditions of rearing on the development of passive defense reactions (fearful aggression) in dogs and found that the expression of a well-marked fear reaction depends on the genotype of the animal. In this experiment, Airedales and German shepherds were reared under conditions of freedom (in homes) and in isolation (in kennels). Krushinski (1960) found that the passive defense reaction developed more acutely and reached a greater degree in the German shepherds kept in isolation compared to the Airedales. In general, animals reared in isolation become more sensitive to sensory stimulation because the nervous system attempts to readjust for the previous lack of stimulation.

In an experiment with chickens, Murphy (1977) found that chicks from a flighty genetic line were more likely to become highly agitated when a novel ball was placed in their pen, but were more attracted to a novel food than birds from a calm line. Cooper and Zubek (1958) and Henderson (1968) found that rats bred to be dull greatly improved in maze learning when housed in a cage with many different objects. However, enriched environments had little effect on the rats bred for high intelligence. Greenough and Juraska (1979) found that rearing rats in an environment with many novel objects improves learning and results in increased growth of dendrites (nerve endings in the brain).

Pigs raised in barren concrete pens also actively seek stimulation (Grandin, 1989a,b; Wood-Gush and Beilharz, 1983; Wood-Gush and Vestergaard, 1991). Piglets allowed to choose between a familiar object and a novel object prefer the novel object (Wood-Gush and Vestergaard, 1991). Pigs raised on concrete are strongly attracted to objects to chew on and manipulate. The first author has observed that nervous, excitable hybrid pigs often chew and bite vigorously on boots or coveralls. This behavior is less common in the placid genetic lines of pigs. Although hybrid pigs are highly attracted to novelty, tossing a novel object into their pen will initially cause a strong flight response. Compared to calm genetic lines, nervous hybrid pigs pile up and squeal more when startled. Pork producers report that nervous, fast-growing, lean hybrid pigs also tail bite other pigs more often than calmer genetic lines of pigs. Tail biting occurs more often when pigs are housed on a concrete slatted floor that provides no opportunity for rooting.

Practical experience by both authors suggests that highly reactive horses are more likely to engage in vices such as cribbing or stall weaving when housed in stalls or runs where they receive little exercise. Denied variety and novelty in their environments, highly reactive animals adapt poorly compared to animals from calmer genetic lines (Huck and Price, 1975).

In summary, in both wild and domestic animals, novelty is both highly feared and necessary. Novelty is more desirable when animals can approach it voluntarily. Unfortunately, novelty is also fear-provoking when animals are suddenly confronted with it.

TEMPERAMENT IS NOT JUST ABOUT FEAR

Early research showed that animals as diverse as rats, chickens, cattle, pigs, and humans, genetic factors influence differences in temperament (fearfulness) (Royce et al., 1970; Blizard, 1971; Broadhurst, 1975; Fordyce et al., 1988; Fujita et al., 1994; Grandin, 1993b; Hemsworth et al., 1990; Kagan et al., 1988; Murphy, 1977; Murphey et al., 1980a; Reese et al., 1983; Tulloh, 1961). Trillmich and Hudson (2011) suggest that a broader area of animal personality needs to be studied. Today, the concept of temperament includes all the neurotransmitter systems that affect the strength of different emotional systems. Barr (2012) contains an outline of how behaviors such as fear and exploring are affected by different neurotransmitter systems. Some individuals are wary and fearful, and others are calm and placid. Boissy (1995) stated, "fearfulness is a basic psychological characteristic of the individual than predisposes it to perceive and react in a similar manner to a wide range of potentially frightening events." In all animals, genetic factors influence reactions to situations that cause fear (Boissy and Bouissou, 1995; Davis, 1992; Kagan et al., 1988; Murphey et al., 1980b). Therefore, temperament is partially determined by an individual animal's fear response. Rogan and LeDoux (1996) suggest that fear is the product of a neural system that evolved to detect danger and causes an animal to make a response to protect itself. Plomin and Daniels (1987) found a substantial genetic influence on shyness (fearfulness) in human children. Shy behavior in novel situations is considered a stable psychological characteristic of certain individuals. Shyness is also suggested to be among the most heritable dimensions of human temperament throughout the lifespan.

In an experiment designed to control for maternal effects on temperament and emotionality, Broadhurst (1960) conducted cross-fostering experiments on Maudsley reactive and nonreactive rats. These lines of rats are genetically selected for high or low levels of emotional reactivity. The results showed that maternal effects were not great enough to completely mask the temperament differences between the two lines (Broadhurst, 1960; Eysenck and Broadhurst, 1964). Maternal effects can affect temperament, but they are not great enough to completely change the temperament of a cross fostered animal that has a temperament very different from that of the foster mother. In an extensive review of the literature, Broadhurst (1975) examined the role of heredity in the formation of behavior and found that differences in temperament between rats persist when the animals are all raised in the same environment.

In their study of genetic effects on dog behavior in a large sample, Fuller and Thompson (1978) found that "simply providing the same defined controlled environment for each genetic group is not enough. Conditions must not only be uniform for all groups, but also favorable to the development of the behavior of interest." In wartime Russia, Krushinski (1960) investigated the ability of dogs to be trained for the antitank service or as trail dogs trained to track human scent. The dogs were tied to a spike driven into the ground and the person who regularly looked after them let them lick from a bowl of food for a few moments then summoned the dog to follow the man as he moved 1015 m away. The activity of each dog was measured with a pedometer for the next 2 minutes. The most active dogs were found to be the best antitank dogs. They were also fearless. In the antitank service, dogs were trained to run up to a tank and either run alongside of it or penetrate under the caterpillars of the tank. To do this, the dogs had to overcome their natural fear of a tank moving toward them at high speed. The less active dogs (as measured by the pedometer) were found to make the best trailer dogs. They slowly followed a trail and kept their noses carefully to the scent while negotiating the corners and turns on the trail. The more active dogs trailed at too high a speed and often jumped the corners and turn in the trail, which sometimes resulted in switching to another trail.

Mahut (1958) demonstrated an example of differences in fear responses between beagles and terriers. When frightened, beagles freeze, and terriers run around frantically. In domestic livestock, measuring fear reactions during restraint or in an open-field test reveals differences in temperament both between breeds and between individuals within a breed (Dantzer and Mormede, 1983; Grandin, 1993a; Murphey et al., 1980b, 1981; Tulloh, 1961). Fearful, flighty animals become more agitated and struggle more violently when restrained for vaccinations and other procedures (Fordyce et al., 1988; Grandin, 1993a). Fear is likely to be the main cause of agitation during restraint in cattle, horses, pigs, and chickens.

SPECIES DIFFERENCES IN EMOTIONAL REACTIONS TO SIMILAR TESTS

In an open-field test, a single animal is placed in an arena (Hall, 1934). Rodents often stay close to the arena walls whereas cattle may run around wildly and attempt to escape. Possibly the reaction of the rodent is motivated by fear and the reaction of a single bovine may be PANIC (separation distress). The rodent has an instinctual fear of open spaces and stays close to the walls. The motivation is to avoid open spaces where it is likely to be seen by a predator. The bovine is motivated to rejoin its herd-mates. The open-field test may be measuring different emotional systems in these two species. When fear is the motivator, a frightened animal may react in two different ways. It may run around frantically and try to escape or in another situation, it may freeze and stay still. Chickens often freeze when handled by humans. Jones (1984) called this "tonic immobility." The chickens become so frightened they cannot move. In cattle, brahman *Bos indicus* cattle are more likely to go into tonic immobility than *Bos taurus* breeds such as Hereford. In wildlife, forceful capture can cause enough fear to sometimes inflict fatal heart damage. Wildlife biologists call this capture myopathy. In summary, much is known about the complex phenomenon of fear, but many questions still remain.

BIOLOGICAL BASIS OF FEAR

Genetic factors influence the intensity of fear reactions. Genetic factors can also greatly reduce or increase fear reaction in domestic animals (Flint et al., 1995; Parsons, 1988; Price, 1984). Research in humans has clearly revealed some of the genetic mechanisms governing the inheritance of anxiety (Lesch et al., 1996). LeDoux (1992) and Rogan and LeDoux (1996) state that all vertebrates can be fear-conditioned. Davis (1992) recently reviewed studies on the biological basis of fear. Overwhelming evidence points to the amygdala as the fear center in the brain. A small bilateral structure located in the limbic system, the amygdala is where the triggers for "flight or fight" are located. Electrical stimulation of the amygdala is known to increase stress hormones in rats and cats (Matheson et al., 1971; Setckliev et al., 1961). Destroying the amygdala can make a wild rat tame and reduce its emotionality (Kemble et al., 1984). Destroying the amygdala also makes it impossible to provoke a fear response in animals (Davis, 1992). Blanchard and Blanchard (1972) showed that rats lose all of their fear of cats when the amygdala is lesioned. Furthermore, when a rat learns that a signal light means an impending electric shock, a normal response is to freeze. Destroying the amygdala eliminates this response (Blanchard and Blanchard, 1972; LeDoux et al., 1988, 1990). Finally, electrical stimulation of the amygdala makes humans fearful (Gloor et al., 1981). Animal studies also show that stimulation of the amygdala triggers a pattern of responses from the autonomic nervous system similar to that found in humans when they feel fear (Davis, 1992). Different types of fear are processed in separate neural circuits. Fear of predators, learned fear, and fear of conspecifics are some examples (Gross and Canteras, 2012). In birds, fear of a novel object is different from fear during tonic immobility. Fear can either elicit a response to freeze or flight to escape. Selection of quail for either high or low tonic immobility had little effect on reaction to a novel object (Saint-Dizier et al., 2008). Since the publication of the first edition of this book, scientists have learned that fear is not a single emotional trait. Further research shows that adding a third variable of human handling indicates that fear has more than one independent dimension (Richard et al., 2010).

Heart rate, blood pressure, and respiration also change in animals when the flight or fight response is activated (Manuck and Schaefer, 1978). All these autonomic functions have neural circuits to the amygdala. Fear can be measured in animals by recording changes in autonomic activity. In humans, Manuck and Schaefer (1978) found tremendous differences in cardiovascular reactivity in response to stress, reflecting a stable genetic characteristic of individuals.

FEARFULNESS AND THE MOTHERING INSTINCT CONFLICT

Fearfulness and the mothering instinct can conflict. This principle was observed first-hand by the second author during his experience raising Queensland Blue Heeler dogs. Some researchers may say that this observation is just anecdotal and has no value. This observation is being kept in the third edition because observations are essential for forming a hypothesis for future research studies. Annie's first litter was a completely novel experience because she had never observed another dog giving birth or nursing pups. She was clearly frightened when the first pup was born. It was obvious she did not know what the pup was. However, as soon as she smelled it, her maternal instinct took over and a constant uncontrollable licking began. Two years later, Annie's daughter Kay had her first litter. Kay was more fearful than her mother and her highly nervous temperament overrode her innate licking program. When each pup was born, Kay ran wildly around the room and would not go near them. The second author had to intervene and place the pups under Kay's nose. Otherwise, they may have died, Kay's nervous temperament and fearfulness were a stronger motivation than her motherly instinct.

NERVOUS SYSTEM REACTIVITY CHANGED BY THE ENVIRONMENT

Raising young animals in barren environments devoid of variety and sensory stimulation will have a detrimental effect on the development of the nervous system. Both the older research reviewed below, and the most recent research clearly show the benefits of an enriched environment for young animals. Young mice that are socially isolated for 2 weeks after weaning have social impairments and defects in the circuits in the prefrontal cortex of the brain (Yamamuro et al., 2020). Juvenile mice that have been bred to exhibit autistic behavior will have less anxiety and increased social behavior when they are raised in an enriched environment (Queen et al., 2020). A deprived environment can cause animals to be more reactive and excitable as adults. This is a long-lasting, environment induced change in how the nervous system reacts to various stimuli. Effects of deprivation during early development are also relatively permanent. Melzack and Burns (1965) found that puppies raised in barren kennels developed into hyperexcitable adults. In one experiment, deprived dogs reacted with "diffuse excitement" and ran around a room more than control dogs raised in homes by people. Presenting novel objects to the deprived dogs also results in "diffuse excitement." Furthermore, the EEGs of the kennel raised dogs remained abnormal even after they were removed from the kennel (Melzack and Burns, 1965). Research by Simons and Land (1987) shows that the somatosensory cortex in the brains of baby rats does not develop normally if sensory input is eliminated by trimming their whiskers. A lack of sensory input made the brain hypersensitive to stimulation. The effects persisted even after the whiskers had grown back.

Emotional reactivity develops in the nervous system during early gestation. Denenberg and Whimbey (1968) showed that handling a pregnant rat can cause her offspring to be more emotional and explore less in an open field compared to control animals. This experiment is significant because it shows that handling the pregnant mother had the opposite effect on the behavior of the infant pups. Handling and stressing pregnant mothers changed the gestational environment of the fetus resulting in nervous offspring. However, handling newborn rats by briefly picking them up and setting them in a container reduced emotional reactivity when the rats became adults (Denenberg and Whimbey, 1968). The handled rats developed a calmer temperament.

The adrenal glands are known to have an effect on behavior (Fuller and Thompson, 1978). The inner portions of the adrenals secrete the hormones, adrenaline, and noradrenaline, while the outer cortex secretes the sex hormones androgens and estrogens (reproductive hormones); and various corticosteroids (stress hormones). Yeakel and Rhoades (1941) found that Hall's (1938) emotional rats had larger adrenals and thyroids compared to the non-emotional rats. Richter (1952, 1954) found a decrease in the size of the adrenal glands in Norway rats accompanied by domestication. Several line and strain differences have been found since these early reports. Furthermore, Levine (1968) and Levine et al. (1967) showed that brief holding of baby rats reduces the response of the adrenal gland to stress. Denenberg et al. (1967) concluded that early handling may lead to major changes in the neuroendocrine system.

TAMING DOES NOT CHANGE NERVOUS SYSTEM REACTIVITY

Adult wild rats can be tamed and become accustomed to handling by people (Galef, 1970). This is strictly learned behavior. Taming full-grown wild animals to become accustomed to handling by people will not diminish their response to a sudden novel stimulus. This principle was demonstrated by Grandin et al. (1995) in training wild antelope at the Denver Zoo for low stress blood testing. Nyala are African antelope with a hair-trigger flight response used to escape from predators. During handling in zoos for veterinary treatments, nyala are often highly stressed and sometimes panic and injure themselves. Over a period of three months, Grandin et al. (1995) trained nyala to enter a box and stand quietly for blood tests while being fed treats. Each new step in the training had to be done slowly and carefully. Ten days were required to habituate the nyala to the sound of the doors on the box being closed.

All the training and petting by zookeepers did not change the nyala's response to a sudden, novel stimulus. When the nyala saw repairmen on the barn roof, they suddenly reacted with a powerful fear response and crashed into a fence. They had become accustomed to seeing people standing at the perimeter of the exhibit, but the sight of people on the roof was novel and very frightening. Sudden movements, such as raising a camera up for a picture, also caused the nyala to flee.

DOMESTIC VERSUS WILD AND FEAR RESPONSES

Wild herding species show much stronger fear responses to sudden novelty compared to domestic ruminants such as cattle and sheep. Domestic ruminants have attenuated flight responses due to years of selective breeding (Price, 1984). Wild ruminants learn to adapt in captivity and associate people with food but are more likely to become agitated and injure themselves when frightened by novel stimuli (Grandin, 1993b, 1997). Fear is more likely when they are prevented from fleeing by a fence or other barrier. Principles for training and handling all herding animals are basically similar. Training procedures used on flighty antelope or placid domestic sheep are the same. The only difference is the amount of time required. Grandin (1989) demonstrated this by training placid Suffolk sheep to voluntarily enter a tilting restraining device in one afternoon. The nyala at the Denver Zoo took 3 months to train. Each new procedure had to be introduced in smaller increments to prevent an explosive flight reaction.

NEOTENY

Neoteny is the retention of the juvenile features in an adult animal. Genetic factors influence the degree of neoteny in individuals. Neoteny is manifested both behaviorally and physically. In the foreword to *The Wild Canids* (Fox, 1975), Conrad Lorenz adds a few of his observations on neoteny and the problems of domestication:

> The problems of domestication have been an obsession with me for many years. On the one hand, I am convinced that man owes the lifelong persistence of his constitutive curiosity and explorative playfulness to partial neoteny that is indubitably a consequence of domestication. In a curiously analogous manner does the domestic dog owe its permanent attachment to its master to a behavioral neoteny that prevents it from ever wanting to be a pack leader? On the other hand, domestication is apt to cause an equally alarming disintegration of valuable behavioral traits and an equally alarming exaggeration of less desirable ones.

Infantile characteristics in domestic animals are discussed by Price (1984), Lambooij and van Putten (1993), Coppinger et al. (1993), Coppinger et al. (1993), and Coppinger et al. (1987). The shortened muzzle in dogs and pigs is an example. Domestic animals have been selected for a juvenile head shape, shortened muzzles, and other features (Coppinger and Smith, 1983). Furthermore, retaining juvenile traits makes animals more tractable and easy to handle. The physical changes are also related to changes in behavior.

Genetic studies point to the wolf as the ancestor of domestic dogs (Isaac, 1970). During domestication, dogs retained many infant wolf behaviors. For example, wolf pups bark and yap frequently, but adult wolves rarely bark. Domestic dogs bark frequently (Fox, 1975; Scott and Fuller, 1965). Wolves have hardwired instinctive behavior patterns that determine dominance or submission in social relationships. In domestic dogs, the ancestral social behavior patterns of the wolf are fragmented and incomplete. Frank and Frank (1982) observed that the rigid social behavior of the wolf has disintegrated into "an assortment of independent behavioral fragments." Malamutes raised with wolf pups fail to read the social behavior signals of the wolf pups. Further comparisons found that the physical development of motor skills is slower in the malamute. Goodwin et al. (1997) studied 10 different dog breeds, ranging from German shepherds and Siberian huskies to bulldogs, cocker spaniels, and terriers. They found that breeds that retained the greatest repertoire of wolflike social behaviors were breeds that physically resembled wolves, such as German shepherds and huskies. Barnett et al. (1979) and Price (1984) both conclude that experience may also cause animals to retain juvenile traits. Gould (1977) found that the effects of neoteny are determined by changes in a few genes that determine the timing of different developmental stages. On average, wolves are smarter than dogs on a test involving spatial orientation and pulling rope. All the adult wolves passed the most complex version of this test, but only five out of 40 German shepherds could do it (Hiestand, 2011).

OVER-SELECTION FOR SPECIFIC TRAITS

There are countless examples in the medical literature of serious problems caused by continuous selection for a single trait (Dykman et al., 1969; Steinberg et al., 1994). People experienced in animal husbandry know that over-selection for single traits can ruin animals. Good dog breeders have known this for centuries. Some traits that appear unrelated are in fact linked. Wright (1922, 1978) demonstrated this clearly by continuous selection for hair color and hair patterns in inbred strains of guinea pigs. Selection for hair color and patterns resulted in decreased reproduction in all the strains. Furthermore, differences in temperament, body conformation, and the size and shape of internal organs were found. Belyaev (1979) showed that continuous selection for tameness in foxes reduced maternal behavior and caused neurological problems. Graded changes occurred in many traits over several years of continuous selection for tame behavior. Physiological and behavioral problems increased with each successive generation. In fact, some of the tamest foxes developed abnormal maternal behavior and cannibalized their pups. Belyaev et al. (1981) called this "destabilizing selection," in contrast to "stabilizing selection," found in nature (Dobzhansky, 1970; Gould, 1977).

There are countless examples in the veterinary medical literature of abnormal bone structure and other physiological defects caused by over-selecting for appearance traits in dog breeds (Ott, 1996). The abnormalities range from bulldogs with breathing problems to German shepherds with hip problems. Scott and Fuller (1965) reported the negative effects of continuous selection for a certain head shape in cocker spaniels:

> In our experiments, we began with what were considered good breeding stocks, with a fair number of champions in their ancestry. When we bred these animals to their close relatives for even one or two generations, we uncovered serious defects in every breed... Cocker spaniels are selected for a broad forehead with prominent eyes and a pronounced "stop" or angle, between the nose and forehead. When we examined the brains of some of these animals during autopsy, we found that they showed a mild degree of hydroencephaly, that is, in selecting for skull shape, the breeders accidentally selected for a brain defect in some individuals. Besides all this, in most of our strains, only about 50% of the females were capable of rearing normal, healthy litters, even under nearly ideal conditions of care.

OVER-SELECTION IN LIVESTOCK

In pigs and cattle, indiscriminate selection for production traits such as rapid gain and leanness result in more excitable temperaments in pigs (Grandin, 1994). The first author has observed thousands of modern lean hybrid pigs. Lean hybrids have a greater startle response and are more excitable and difficult to drive through races compared to older genetic lines with more back fat. Separating single animals from the group is also more difficult. Research shows that cattle with an excitable temperament have lower weight gains and more meat quality problems (Cafe´ et al., 2011; Silveira et al., 2012; Voisinet et al., 1997a,b). This research illustrates that selection away from a very excitable temperament would be beneficial. However, ranchers report that mothering ability is reduced by over-selecting for extremely calm temperaments. There are always tradeoffs.

LINKS BETWEEN DIFFERENT TRAITS

Casual observations by the first author indicate that the most excitable pigs and cattle have long slender bodies and fine bones. Some lean hybrid pigs have weak legs and normally brown eyed pigs now have blue eyes. Blue eyes are often associated with neurological problems (Bergsma and Brown, 1971; Schaible, 1963). Furthermore, pigs and cattle with large bulging muscles often have calmer temperaments compared to lean animals with less muscle definition. However, animals with double muscling (hypertrophy) have more excitable temperaments (Holmes et al., 1972). Double muscling is abnormal and may have opposite effects on temperament compared to normal muscling. Deafness in pointer dogs selected for nervousness is another example of apparently unrelated traits being linked (Klein et al., 1988). There appears to be a relationship between thermoregulation and aggressiveness. Sluyter et al. (1995) found that wild mice selected for aggressiveness used larger amounts of cotton to build their nests than mice selected for low aggression. This effect occurred in both laboratory and wild strains of mice.

The complexity of genetic interactions continues to frustrate researchers using high-tech "knockout" gene procedures. Genes are knocked out in a gene targeting procedure and prevented from performing their normal functions. Knockout experiments show that blocking different genes can have unexpected effects on behavior. In one experiment, super-aggressive mice were created when genes involved with learning were inactivated (Chen et al., 1994). The mutant mice had little or no fear and fought until they broke their backs. In another experiment, knockout mutants demonstrated normal behavior until they had pups, which they failed to care for (Brown et al., 1996). In another experiment, Konig et al. (1996) disabled the gene that produces enkephalin (a brain opioid substance) and found unexpected results. Enkephalin is a substance normally involved in pain perception. Mice deficient in this substance were very nervous and anxious. They ran frantically around their cages in response to noise. Traits are linked in unexpected ways and completely isolating single gene effects may be impossible. Researchers suggest using caution and being careful not to jump to conclusions about claims by those who say they found an "aggression gene" or a "maternal gene" or an "anxiety gene." To use an engineering analogy, one cannot conclude the "picture center" in a television set was found after cutting one circuit inside the set used to create the pictures. Gerlai (1996) and Crawley (1996) also warn that knocking out the same gene in two different species may have different effects on behavior. This is due to the complex interactions between many different genes.

Thirty-five years ago, behavior geneticists found that inheritance of behavior is complex. Fuller and Thompson (1978) concluded, "It has been found repeatedly that no one genetic mechanism accounts exclusively for a particular kind of behavior."

TRANSGENIC MICE

Since the first edition of this book was published, hundreds of papers have been published on mouse models of conditions such as autism and various behavioral extremes (Roullet and Crawley, 2011; Silverman et al., 2010). Models of high-novelty-seeking mice, timid mice, fearful mice, and obsessive-compulsive mice are used to study these conditions in humans. There are different behavioral tests for the hundreds of types of special mice. Crawley (2007, 2008) devoted an entire book to behavioral testing of transgenic mice. Eisener-Dorman et al. (2009) stated that knockout procedures do not always work correctly to produce the desired effects and are concerned that transgenic mice are not always what they were originally bred to be.

RANDOM FACTORS AND NON-INHERITED EFFECTS ON VARIATION

Behavioral geneticists find it impossible to completely control variation in some traits. Gartner (1990) found that breeding genetically similar inbred lines of rats failed to stop weight fluctuations. Even under highly standardized laboratory conditions, body weights continued to fluctuate between animals. Pig breeders also observe that commercially bred hybrid lines of pigs do not gain weight at the same rate. Even in genetically identical animals, random unknown factors affect variability. In utero factors may be one cause, other causes are unknown. Dr. Daryl Tatum and students at Colorado State University found both body conformation and meat quality variation in 50% English (*Bos taurus*) and 50%

Brahman (*Bos indicus*) cattle. Some animals showed more Brahman characteristics, with larger humps and longer ears. However, the body conformation was not half English and half Brahman, and meat characteristics varied as well. Cattle that looked more Brahman had tougher meat. The animals had about 10% variation from the body shape and meat characteristics of Brahman half-bloods.

Two new studies clearly show two other non-inherited causes of variation. Linneweber et al. (2020) studied wiring variations in the visual system of insects. These variations can have a profound effect on behavior. Even if the genetics is the same, the neurons will always grow slightly differently. In people, monozygotic identical twins may have differences. These differences may be due to mutations that occur right before the developing embryo splits in two (Jonsson et al., 2021).

CONCLUDING THOUGHTS

Many animal breeders are moving away from selecting for a narrow range of production traits. There are always tradeoffs. When decisions are made from the genetic selection. This analogy might be helpful when making decisions. When selecting for different traits, imagine that an animal is country, with a financial budget that can be divided between the economy (production) infrastructure (reproduction and bone structure) and military (immune function). If a large percentage of the money is allocated to the economy (production), then the infrastructure sector and the military may get less. This may result in poorer reproduction bone structure or immune function. Poor reproduction in a high-producing dairy cow is a good example.

There is a complex interaction between genetic and environmental factors which determine how an animal will behave. The animal's temperament is influenced by both genetics and learning. Another principle to remember is that changes in one trait, such as temperament, can have unexpected effects on other unrelated traits. Over-selection for a single trait may result in undesirable changes in both behavioral and physical traits.

REFERENCES

Adler, A., 1996. How songbirds get their tunes. *Sci. News* 149, 280-281.

Agnvall, B., Jongren, M., Strandberg, E., Jensen, P., 2012. Heritability and genetic correlations of fear related behavior in Red Jungle fowl—possible implications for early domestication. *PLoS ONE* 7 (4), 35162.

Ahmadlou, M., Houba, J.H.W., van Vierbergen, J.F.W., Giannouli, M., et al., 2021. A cell type specific cortico-subcortical brain circuit for investigatory and novelty seeking behavior. *Science* 372 (6543), 704.

Amanda, B., et al., 2021. A systematic review of genomic regions and candidate genes underlying behavioral traits in farmed mammals and their link with human disorders. *Animals* 11 (3), 715. Available from: https://doi.org/10.3390/ani11030715.

Anderson, L., 2012. How selective sweeps in domestic animals provide new insight into biological mechanisms. *J. Intern. Med.* 271, 114.

Balaban, E., 1997. Changes in multiple brain regions underlie species differences in complex, congenital behavior. *Proc. Natl. Acad. Sci. USA* 94, 2001-2006.

Barnett, S.A., Dickson, R.G., Hocking, W.E., 1979. Genotype and environment in the social interactions of wild and domestic "Norway" rats. *Aggress. Behav.* 5, 105-119.

Barr, C.S., 2012. Temperament in animals. In: Zentner, M., Shiner, R.L. (Eds.), *Handbook of Temperament*. Guildford Press, pp. 251-272.

Bateson, P., 2012. Behavioral biology: past and a future. *Ethology* 118, 216-221.

Belyaev, D.K., 1979. Destabilizing selection as a factor in domestication. *J. Hered.* 70, 301-308.

Belyaev, D.K., Ruvinsky, A.O., Trut, L.N., 1981. Inherited activation inactivation of the star gene in foxes. *J. Hered.* 72, 267-274.

Bendesky, A., Kwon, Y.M., Lassance, J.M., Lewarch, C.L., Yaa, S., Peterson, B.K., et al., 2017. The genetic basis of parental care evolution in monogamous mice. *Nature* 544, 434-438.

Bergsma, D.R., Brown, K.S., 1971. White fur, blue eyes, and deafness in the domestic cat. *J. Hered.* 62, 171-185.

Blanchard, D.C., Blanchard, R.J., 1972. Innate and conditioned reactions to threat in rats with amygdaloid lesions. *J. Comp. Physiol. Psychol.* 81, 281-290.

Blizard, D.A., 1971. Autonomic reactivity in the rat: effects of genetic selection for emotionality. *J. Comp. Physiol. Psychol.* 76, 282-289.

Boessneck, J., 1985. (Domestication and its sequelae) Die Domestikation und irhe Folgen Tieraerzil. *Prax.* 13 (4), 479-497.

Boice, R., 1977. Burrows of wild and albino rats: effects of domestication, outdoor raising, age, experience and maternal state. *J. Comp. Physiol. Psych.* 91, 649-661.

Boissy, A., 1995. Fear and fearfulness in animals. *Q. Rev. Biol.* 70 (2), 165-191.

Boissy, A., Bouissou, M.F., 1995. Assessment of individual differences in behavioral reactions of heifers exposed to various feareliciting situations. *App. Aim. Behav. Sci.* 46, 1731.

Boissy, A., Bouix, J., Orgeur, P., Paindron, P., Bibe, B., LeNeindre, P., 2005. Genetic analysis of emotional reactivitiy of sheep, effects of genotypes of lambs and their dams. *Genet. Sel. Evol.* 37, 381-401.

Boyko, A.R., 2011. The domestic dog: man's best friend in the genomic era. *Genome. Biol.* 12, 216-226.

Boyko, A.R., Quignon, P., Li, L., Schoenebeck, J.J., Degenhardt, J.D., Lohmueller, K.E., 2010. A simple genetic architecture underlies morphological variation in dogs. *PLoS Biol.*

Brelands, K., Brelands, M., 1951. A field of applied animal psychology. *Am. Psychol.* 6, 202204.

Brelands, K., Brelands, M., 1961. The misbehavior of organisms. *Am. Psychol.* 16, 681684.

Broadhurst, P.L., 1960. Analysis of maternal effects in the inheritance of behavior. *Anim. Behav.* 9, 129-141.

Broadhurst, P.L., 1975. The Maudsley reactive and nonreactive strains of rats: a survey. *Behav. Genet.* 5, 299-319.

Brown, J.R., Ye, H., Bronson, R.T., Dikkes, P., Greenberg, M.E., 1996. A defect in nurturing in mice lacking the immediate early gene for *B. Cell* 86, 297-309.

Burgdorf, J., Panksepp, J., 2006. The neurology of positive emotions. *Neurosci. Biobehav. Rev.* 30, 173-187.

Burghardt, G.M., 2019. The place for emotions in behavior systems research. *Behav. Process.* 166, 103881. Available from: https://doi.org/10.1016/j.beproc.2019.06004.

Cabrera, D., Nilsson, J.R., Griffen, B.D., 2021. The development of animal personality across ontogeny: a cross species review. *Anim. Behav.* 173, 137144. Available from: https://doi. org/10.1016/j.anbehav.2021.01.003.

Cafe´, L.M., Robinson, D.L., Ferguson, D.M., McIntyre, B., Beesink, G.H., Greenwood, P.L., 2011. Cattle temperament: persistence of assessments and associations with productivity, carcass and meat quality traits. *J. Anim. Sci.* 89, 1452-1465.

Cameron, N.M., 2008. Epigenetic programming of phenotypic variations in reproductive strategies in the rat through maternal care. *J. Neuroendocrinol.* 20, 795-801.

Campler, M., Jongren, N., Jensen, P., 2008. Fearfulness in red jungle fowl and domesticated white leghorn chickens. *Behav. Process.* 61, 39-43.

Canty, N., Gould, J., 1995. The Hawk/Goose experiment: sources of variability. *Anim. Behav.* 50, 1091-1095.

Chaignat, E., YahyaGraison, E., Henrichsen, C.N., Chrast, J., Schultz, F., Pradervand, S., 2011. Copy number variation modifies expression time courses. *Genome. Res.* 21, 106-113.

Chakravarti, A., Kapoor, A., 2012. Mendelian puzzles. *Science* 335, 930-931.

Champagne, F.A., Curley, J.P., 2009. Epigenetic mechanisms mediating the long-term effects of maternal care on development. *Neurosci. Biobehav. Rev.* 33 (4), 593-600.

Chen, C., Rainnie, D.G., Greene, R.W., Tonegawa, S., 1994. Abnormal fear response and aggressive behavior n mutant mice deficient for a calcium calmodulin kinase II. *Science* 266, 291-294.

Chen, Q., Zhang, F., Qu, K., Hanif, Q., Shen, J., Peng, J., et al., 2019. Genome wide association study identifies genomic loci associated with flight reaction in cattle. *J. Anim. Breed. Genet.* 137 (5), 477-485.

Christensen, J.W., Beblein, C., Malmkrist, J., 2020. Development and consistency of fearfulness in horses from foal to adult. *Appl. Anim. Behav. Sci.* 232, 105-106.

Christensen, J.W., Ahrendt, L.P., Malmkrist, J., Nicol, C., 2021. Exploratory behavior towards novel objects is associated with enhanced learning in young horses. *Sci. Rep.* 11 (1), 1428. Available from: https://doi.org/10.1038/S4159802080833W.

Clinton, S.M., Vazquez, D.M., Kabbaj, M., Kabbaj, M.H., Watson, S.J., Akil, H., 2007. Individual differences in novelty seeking and emotional reactivity correlate with variation in maternal behavior. *Horm. Behav.* 51, 655-664.

Cooper, R.M., Zubek, J.P., 1958. Effects of enriched and restricted early environments on learning ability of bright and dull rats. *Can. J. Psychol.* 12, 159-164.

Coppinger, R.P., Smith, C.K., 1983. The domestication of evolution. *Environ. Conserv.* 10, 283-292.

Coppinger, R., Glendinning, J., Torop, E., Mattthay, C., Sutherland, M., Smith, C., 1987. Degree of behavioral neoteny differentiates canid polymorphy. *Ethology* 75, 85-108.

Coppinger, et al., 1993. Coppinger, L., Coppinger, R., 1993. Dogs for herding livestock. In: Grandin, T. (Ed.), *Livestock Handling and Transport.* CAB International, Wallingford, UK, pp. 179-196.

Corgan, M., Grandin, T., Matlock, S., 2020. Evaluating the recognition of a large rotated object in domestic horses (Equus cabellos). *J. Anim. Sci.* 98 (Suppl. 4). Available from: https://doi. org/10.1093/jas/skaa278.002.

Costilla, R., Kemper, K.E., Byrne, E.M., PortoNeto, L.R., Carvalheiro, R., Purfield, D.C., et al., 2020. Genetic control of temperament traits across species: association of autism disorder risk genes with cattle temperament. *Genetic. Sel. Evol.* Available from: https://doi.org/ 10.1186/s1271102000569Z.

Crawley, J.N., 1996. Unusual behavioral phenotypes of inbred mouse strains. *Trends. Neurosci.* 19 (5), 181-182.

Crawley, J.N., 2007. *What's Wrong with My Mouse?* Wiley and Sons, New Jersey.

Crawley, J.N., 2008. Behavioral phenotyping strategies for mutant mice. *Neuron.* 57, 809818.

Crews, D., 2010. Epigenetics, brain, behavior, and the environment. *Hormones.* 9, 40-50.

Dantzer, R., Mormede, P., 1983. Stress in farm animals: a need for reevaluation. *J. Anim. Sci.* 57, 618.

Darwin, C.R., 1868. *The Variation of Animals and Plants under Domestication.* John Murray, London.

Davis, M., 1992. The role of the amygdala in fear and anxiety. *Annu. Rev. Neurosci.* 15, 353-375.

Dawkins, M.S., 1993. *Through Our Eyes Only: The Search for Animal Consciousness.* Freeman, New York.

Dellu, F., Piazza, P.V., Mayo, W., LeMoal, M., Simm, H., 1996. Novelty seeking in rats, behavioral characteristics and possible relationships with sensation seeking in man. *Neuropsycholobiology* 34, 136-145.

Denenberg, V.H., Whimbey, A.E., 1968. Experimental programming of life histories: towards an experimental science of individual differences. *Dev. Psychobiol.* 1 (1), 55-59.

Denenberg, V.H., Brumaghim, J.T., Haltmeyer, G.C., Zarrow, M.X., 1967. Increased adrenocortical activity in the neonatal rat following handling. *Endocrinol. (Baltim.)* 81, 1047-1052.

DeWaal, F.B., 2011. What is animal emotion. Ann. *N. Y. Acad. Sci.* 122, 191-206.

Dobzhansky, T., 1970. *Genetics of the Evolutionary Process.* Columbia University Press, New York.

Dykman, R.A., Morphee, O.D., Peters, J.E., 1969. Like begets like: behavior tests, classical auto nomic and motor conditioning in two strains of pointer dogs. *Ann. N. Y. Acad. Sci.* 159, 976-1007.

Eibi-Eibesfeldt, I., Kramer, S., 1958. Ethology, the comparative study of animal behavior. *Q. Rev. Biol.* 33, 181-211.

Eisener-Dorman, A.F., Lawrence, D.A., Bolivar, V.J., 2009. Cautionary insights on knockout mouse studies: the gene or not gene. *Brain Behav. Immun.* 23, 318-324.

Elgar, M.A., 1989. Predator vigilance and group size in mammals and birds, A critical review of the empirical evidence. *Biol. Rev. Camb. Philos. Soc.* 64, 13-33.

ENCODE Project Consortium, 2012. An integrated encyclopedia of DNA elements in the human genome. *Nature* 489, 57-74.

Eysenck, H.J., Broadhurst, P.L., 1964. Experiments with animals: introduction. In: Eysenck, H.J. (Ed.), *Experiments in Motivation*. Macmillan, New York, pp. 285-291.

Farahbakhsh, Z.Z., Siciliano, C.A., 2021. Neurobiology of novelty seeking. *Science* 372 (6543), 684-685.

Faure, Mills, 1998. Faure, J.M., Mills, AD, 1998. Improving the adaptability of animals by selection. In: Grandin, T. (Ed.), *Genetics and the Behavior of Domestic Animals*. Academic Press.

Faure, A.S.M., Reynolds, J.M., Richard, Berridge, C., 2008. Mesolimbic dopamine in desire and dread enabling motivation to be generated by localized glutamate disruption of the nucleus accumbens. *J. Neurosci.* 28, 7184-7192.

Fentress, J.C., 1973. Development of grooming in mice with amputated forelimbs. *Science* 179, 204-205.

Finkelmeier, M.A., Langbein, J., Puppe, B., 2018. Personality research in mammalian farm animals: concepts, measures and relationship to welfare. *Front. Vet. Sci.* Available from: https://doi.org/10.3389/fvets.2018.00131.

Flint, J., Corley, R., DeFries, J.C., Fulker, D.W., Gray, J.A., Miller, S., 1995. A simple genetic basis for a complex physiological trait in laboratory mice. *Science* 269, 1432-1435.

Fondon III, J.W., Garner, H.R., 2004. Molecular origins of rapid and continuous morphological evolution. *Proc. Natl. Acad. Sci.* 101, 18058-18063.

Fondon III, J.W., Hammack, E.A.D., Hannak, A.J., King, D.G., 2008. Simple sequence repeats: genetic modulators of brain function and behavior. *Trends. Neurosci.* 31, 328-334.

Fordyce, G., Dodt, R.M., Wythes, J.R., 1988. Cattle temperaments in extensive herds in northern Queensland. *Aust. J. Exp. Agric.* 28, 683-687.

Fox, M.W. (Ed.), 1975. *The Wild Canids: Their Systematics, Behavioral Ecology and Evolution*. Van Nostrand-Reinhold, New York.

Fraga, M.F., 2005. Epigenetic differences arise during the lifetime of monozygotic twins. *PNAS* 102, 10604-10609.

Frank, H., Frank, M.G., 1982. On the effects of domestication on canine social development and behavior. *Appl. Anim. Ethol.* 8, 507-525.

Frantz, L.A.F., bailey, D.G., Larson, G., Orlando, L., 2020. Animal domestication in the era of ancient genomics. *Nat. Rev. Genet.* 21, 449-460.

Fujita, O., Annen, Y., Kitaoka, A., 1994. Tsukuba high and low emotional strains of rats (*Rattus norvegicus*): an overview. *Behav. Genet.* 24, 389-415.

Fuller, J.L., Thompson, W.R., 1978. *Foundations of Behavior Genetics*. Mosby, St. Louis, MO.

Galef, B.G., 1970. Aggression and timidity: responses to novelty in feral Norway rats. *J. Comp. Physiol. Psychol.* 70, 370-381.

Galibert, F., Quignon, P., Hitte, C., Andrew, C., 2011. Towards understanding dog evolutionary and domestication history. *C. R. Biol.* 334, 190-196.

Gartner, K., 1990. A third component causing random variability besides environment and genotype. A reason for the limited success of a thirty-year-long effort to standardize laboratory animals. *Lab. Anim.* 24, 71-77.

Gerlai, R., 1996. Gene-targeting studies of mammalian behavior: is it the mutation or the back ground genotype? *Trends. Neurosci.* 19 (5), 177-181.

Gloor, P., Oliver, A., Quesney, L.F., 1981. The role of the amygdala in the expression of psychic phenomenon in temporal lobe seizures. In: BenAri, Y. (Ed.), *The Amygdaloid Complex*. Elsevier/NorthHolland, New York, pp. 489-507.

Goodwin, D., Bradshaw, J.W.S., Wickens, S.M., 1997. Paedomorphosis affects visual signals of domestic dogs. *Anim. Behav.* 53, 297-304.

Gould, S.J., 1977. *Ontogeny and Phylogeny*. Harvard University Press (Belknap Press), Cambridge, MA and London.

Grandin, T., 1989. Environmental causes of abnormal behavior. *Large. Anim. Vet.* 4 (3), 13-16.

Grandin, T. 1989a. Grandin, 1989a Effects of Rearing Environment and Environmental Enrichment on Behavior and Neural Development in Young Pigs (Ph.D. dissertation). University of Illinois, Urbana.

Grandin, T., 1989b. Voluntary acceptance of restraint by sheep. *Appl. Anim. Behav. Sci.* 23, 257-261.

Grandin, T., 1993a. Behavioral agitation during handling of cattle is persistent over time. *Appl. Anim. Behav. Sci.* 36, 19.

Grandin, T., 1993b. Behavioral principles of cattle handling under extensive conditions. In: Grandin, T. (Ed.), *Livestock Handling and Transport*. CAB International, Wallingford, UK, pp. 43-57.

Grandin, T., 1994. Solving livestock handling problems. *Vet. Med.* 89, 989-998.

Grandin, T., 1995. *Thinking in Pictures*. Doubleday, New York.

Grandin, T., 1997. Assessment of stress during handling and transport. *J. Anim. Sci.* 75, 249-257.

Grandin, T., Johnson, C., 2005. *Animals in Translation.* Scribner, New York, NY.

Grandin, T., Rooney, M.B., Phillips, M., Canibre, R.C., Irlbeck, N.A., Graffam, W., 1995. Conditioning of nyala (*Tragelaphus angasi*) to blood sampling in a crate with positive reinforcement. *Zoo. Biol.* 14, 261-273.

Greenough, W.T., Juraska, J.M., 1979. Experience induced changes in fine brain structure: their behavioral implications. In: Hahn, M.E., Jensen, C., Dudek, B.C., (Eds.), *Development and Evolution of Brain Size: Behavioral Implications.* Academic Press, New York, pp. 295-320.

Griffin, D., 1994. Animal Minds. University of Chicago Press, Chicago, IL. Grillner, S., 2011. Human locomotion circuits. *Science* 334, 912-913.

Gross, C.T., Canteras, N.S., 2012. The many paths of fear. *Nat. Rev. Neurosci.* 13, 651.

Guryev, V., Berezikov, E.,malik, R., Plasterk, R.H.A., Cuppen, E., 2004. Single nucleotide polymorphisms associated with rat expressed sequences. *Genome. Res.* 14, 1438-1443.

Hall, C.S., 1934. Emotional behavior in the rat, I. Defecation and urination as measures of individual differences in emotionality. *J. Comp. Psychol.* 18, 385-403.

Hall, C.S., 1938. The inheritance of emotionality. *Sigma. Xi. Q.* 26, 17-27.

Harich, M., van der Voet, M., Klein, M., Cizek, P., Fenckova, M., Schenck, A., et al., 2020. From rare copy number variants to biological processes in ADHD. *Am. J. Psychiatry.* Available from: https://doi.org/10.1176/appi.ajp.2020.19090923.

Hayden, E.C., 2012. Life is complicated. *Nature* 464, 664-667.

Heinroth, O., 1918. Reflektorische Bewegungen bei Voegeln. *J. Ornithol.* 66 (1 and 2).

Heinroth, O., 1938. *Aus dem Leben der Vogel.* Springer, Berlin.

Hemsworth, P.H., Barnett, J.L., Treacy, D., Madgwick, P., 1990. The heritability of the trait fear of humans and their association between this trait and subsequent reproductive performance of gilts. *Appl. Anim. Behav. Sci.* 25, 85-95.

Henderson, H.D., 1968. The confounding effects of genetic variables in early experience research: can we ignore them? *Dev. Psychobiol.* 1, 146-152.

Hennessy, M.G., Levine, S., 1978. Sensitive pituitary-adrenal responsiveness to varying intensities of psychological stimulation. *Physiol. Behav.* 21, 295297.

Henrichsen, C.N., Chaignet, E., Reymond, A., 2009. Copy number variants diseases and gene expression. *Hum. Mol. Genet.* 18, R1R8.

Herrick, F.H., 1908. The relation of instinct to intelligence in birds. *Science* 27, 847-850.

Herrick, C.J., 1910. The evolution of intelligence and its organs. *Science* 31, 718.

Hiestand, L., 2011. A comparison of problem solving and spatial orientation in the wolf (*Canis lupis*) and dog (Canis familiaris). *Behav. Genet.* 41, 840-857.

Hirata, M., Arimoto, C., 2018. Novel object response in beef cattle grazing a pasture as a group. *Behav. Process.* 157, 315-319.

Hirsch, J., Lindley, R.H., Tolman, E.C., 1955. An experimental test of an alleged sign stimulus. *J. Comp. Physiol. Psychol.* 48, 278-280.

Hirsh, J., 1963. Behavior genetics and individuality understood. Science 142, 14361442. Hockstra, H.E., 2012. Stickleback catch of the day. *Nature* 484, 46-47.

Holmes, J.H.G., Robinson, D.W., Ashmore, C.R., 1972. Blood lactic acid and behavior of cattle with hereditary muscular hypertrophy. *J. Anim. Sci.* 55, 10111014.

Hubel, D.H., Wiesel, T.N., 1970. The period of susceptibility to the physiological effects of unilateral eye closure in kittens. *J. Physiol.* London 206, 419.

Huck, U.W., Price, E.O., 1975. Differential effects of environmental enrichment on the open field behavior of wild and domestic Norway rats. *J. Comp. Physiol. Psychol.* 89, 892-898.

Huxley, T.H., 1874. On the hypothesis that animals are automata and its history. *Collected Essays, vol. 1, Methods and Results: Essays.* Published in 1901 by MacMillan, London, p. 218.

Isaac, E., 1970. *Geography and Domestication.* Prentice-Hall, Englewood Cliffs, NJ.

Jones, F.C., Grabberr, M.G., Chan, Y.F., Russell, P., 2012. The genomic basis of adoptive evolution in three spine stick backs. *Nature* 484, 55-62.

Jones, R.B., 1984. Experimental novelty and tonic immobility in chickens (*Gallus domesticus*). *Behav. Processes.* 9, 155-260.

Jonsson, H., Magnusdottir, E., Eggertsson, H.P., et al., 2021. Differences between germlines genomes of monozygotic twins. *Nat. Genet.* 53, 27-34.

Kagan, J., Reznick, J.S., Snidman, N., 1988. Biological bases of childhood shyness. *Science* 240, 167-171.

Karlsson, H., Merisaari, H., Karlsson, L., et al., 2020. Association of cumulative paternal early life stress with white matter in newborns. *J. Am. Med. Assoc.* 3 (11), e2024832.

Keeler, C., Mellinger, T., Fromm, E., Eade, I., 1970. Melanin, adrenalin and the legacy of fear. *J. Hered.* 61, 81-88.

Kemble, E.D., Blanchard, D.C, Blanchard, R.J., Takushi, R., 1984. Taming in wild rats following medial amygdaloid lesions. *Physiol. Behav.* 32, 131-134.

Klein, E., Steinberg, S.A., Weiss, S.R.B., Matthews, D.M., Uhde, T.W., 1988. The relationship between genetic deafness and fearrelated behaviors in nervous pointer dogs. *Physiol. Behav.* 43, 307-312.

Kong, A., Steinthorsdottir, V., Masson, G., Thorleifsson, G., 2009. Parental origin of sequence variants associated with complex diseases. *Nature* 462, 868-874.

Konig, M., Zimmer, A.M., Steiner, H., Holmes, P.V., Crawley, J.M., Brownstein, M.J., 1996. Pain responses, anxiety and aggression in mice deficient in preproenkephalin. *Nat. (Lond.)* 383, 535-538.

Kremer, L., Holkenborg, K., Reimert, I., Bolhuis, J.E., Webb, L.E., 2020. The nuts and bolts of animal emotion. *Neurosci. Biobehav. Rev.* 113, 273-289.

Krushinski, L.V., 1960. Animal behavior—its normal and abnormal development. In: Wortis, J. (Ed.), *International Behavioural Sciences Service*. Consultants Bureau, New York.

Kruuk, H., 1972. *The Spotted Hyena*. University of Chicago Press, Chicago, IL.

Kukekova, A.V., Johnson, J., Xiang, X., Feng, S., Liu, S., Rando, H.M., et al., 2018. Red fox genome assembly identifies genomic regions associated with tame and aggressive behaviors. *Nat. Ecol. Evol.* 2, 1479-1491.

Laine, V.N., Oers, K., 2017. The quantitative and molecular genetics of individual differences in animal personality. In: Vonk, J., Weiss, A., Kucza, S. (Eds.), *Personality in Non-Human Animals*. Springer Publishing, pp. 55-72.

Lambooij, E., van Putten, G., 1993. Transport of pigs. In: Grandin, T. (Ed.), *Livestock Handling and Transport*. CAB International, Wallingford, UK.

Larsen, G., Karlsson, E.K., Perri, A., Webster, M.T., Ho, S.Y.W., Peters, J., 2012. Rethinking dog domestication by integrating genetics, archeology, and biogeography. *Proc. Natl. Acad. Sci.* 109, 8878-8883.

Lawrence, A.B., 2008. Applied animal behavior science: past, present, and future prospects. *Appl. Anim. Behav. Sci.* 115, 124.

LeDoux, J.E., 1992. Brain mechanisms of emotion and emotional learning. *Curr. Opin. Neurobiol.* 2 (2), 191-197.

LeDoux, J.E., 2000. Emotion circuits in the brain. *Annu. Rev. Neurosci.* 23, 155-184.

LeDoux, J.E., Iwata, J., Ciccheti, P., Reis, D.J., 1988. Different projections of the central amygdaloid nucleus mediate autonomic and behavioral correlates of conditioned fear. *J. Neurosci.* 8, 2517-2529.

LeDoux, J.E., Ciccheti, P., Nagoraris, A., Romanski, L.M., 1990. The lateral amygdaloid nucleus: sensory interface of the amygdala in fear conditioning. *J. Neurosci.* 10, 1062-1069.

Lemos, J.C., Wanata, M.J., Smith, J.S., Reyes, B.A.S., Hollon, N.G., VanBockstaele, E.J., 2012. Severe stress switches CRF action in the nucleus accumbens from appetitive to aversive. *Nature* 490, 402-406.

Lesch, K.P., Bengel, D., Heils, A., Sabol, D.Z., Greenburg, B.D., Perri, S., 1996. Association of anxiety-related traits with a polymorphism in the serotonin transporter gene regulatory region. *Science* 274, 1527-1531.

Levine, S., 1968. Influence of infantile stimulation on the response to stress during preweaning development. *Dev. Psychobiol.* 1 (1), 67-70.

Levine, S., Haltmeyer, G.C., Karas, G.G., Denenburg, V.H., 1967. Physiological and behavioral effects of infantile stimulation. *Physiol. Behav.* 2, 55-59.

Linn, Z., Ma, H., Nei, M., 2008. Ultra conserved coding regions outside the homeobox in mammalian box genes. *Evol. Biol.* 8, 260.

Linneweber, G.A., et al., 2020. A neural developmental origin of behavioral individuality in the Drosophila visual system. *Science* 367 (6482), 1112-1119.

Lipkind, D., Sakov, A., Kafkafi, N., Elmer, G.I., Benjamin, Y., Golani, 2004. New replicable anxiety related measures of wall vs. center behavior of mice in the open field. *J. Appl. Physiol.* 97, 347-359.

Lord, K.A., Larson, G., Coppinger, R.P., Karlsson, E.K., 2020. The history of farm foxes undermines domestication syndrome. *Trends Ecol. Evol.* 35 (2), 125136. Available from: https://doi.org/10.1016/j.tree.2019.10.011.

Lorenz, K.S., 1939. Vergleichende Verhaltensforschung. *Zool. Anz.* 12, 69109.

Lorenz, K.S., 1965. *Evolution and Modification of Behavior*. University of Chicago Press, Chicago.

Lorenz, K.Z., 1981. The Foundations of Ethology. Springer Verlag, New York. Maher, B., 2012. The human encyclopedia. *Nature* 489, 46-48.

Mahut, H., 1958. Breed differences in the dog's emotional behavior. *Can. H. Psycho.* 12 (1), 35-44.

Manuck, S.B., Schaefer, D.C., 1978. Stability of individual differences in cardiovascular reactivity. *Physiol. Behav.* 21, 675-678.

Marshall, J.N.G., Lopez, A.I., Pfaff, A.L., 2021. Variable number tandem repeats their emerging role in sickness and health. *Exper. Bio. Med.* 2021. April 1.

Matheson, B.K., Branch, B.J., Taylor, A.N., 1971. Effects of amygdaloid stimulation on primary adrenal activity in conscious cats. *Brain. Res.* 32, 151-167.

Mayr, E., 1974. Behavioral programs and revolutionary strategies. *Am. Sci.* 62, 650-659.

McHugo, G.P., Dover, M.J., MacHugh, D.E., 2019. Unlocking the origins and biology of domestic animals using ancient DNA and paleo genomics. *BMC Bol.* 17, 98. Available from: https://doi.org/10.1186/S.1291501907247.

Melton, D.A., 1991. Pattern formation during animal development. *Science* 252, 234-241.

Melzack, R., Burns, S.K., 1965. Neurophysiological effects of early sensory restriction. *Exp. Neurol.* 13, 163-175.

Metfessel, M., 1935. Roller canary song produced without learning from external source. *Science* 81, 470.

Mikkelson, T.S., Wakefield, M.J., Aken, B., 2007. Genome of the marsupial Monodephis domestica reveals innovation in noncoding sequence. *Nature* 447, 167-177.

Moberg, G.P., Wood, V.A., 1982. Effects of differential rearing on the behavioral adrenocortical response of lambs to a novel environment. *Appl. Anim. Ethol.* 8, 269-279.

Montag, C., Panksepp, J., 2017. Primary emotional systems and personality: An evolutionary perspective. *Front. Psychol.* 11. Available from: https://doi.org/10.3369/fpsyg.2017.00464.

Morman, F., Dubois, J., Korblith, S., Milosovljevic, M., Cerf, M., 2011. A category specific response to animals in the human right amygdala. *Nat. Neurosci.* 14, 1247-1249.

Morris, C.L., Grandin, T., Irlbeck, N.A., 2011. Companion animal symposium: environmental enrichment for companion exotic and laboratory animals. *J. Anim. Sci.* 89, 4227-4238.

Mueller, H.C., Parker, P., 1980. Cardiac responses of domestic chickens to hawk and goose models. *Behav. Processes.* 7, 255-258.

Mundinger, P.C., 1995. Behavior analysis of canary song inter-strain differences in sensory learning and epigenetic rules. *Anim. Behav.* 50, 1491-1511.

Murphey, R.M., Moura Duarte, F.A., Torres Penedo, M.C., 1980a. Approachability of bovine cattle in pastures: breed comparisons and a breed X treatment analysis. *Behav. Genet.* 10, 171-181.

Murphey, R.M., Moura Durate, F.A., Coelho, Novaes, W., Torres Penedo, M.C., 1980b. Age group differences in bovine investigatory behavior. *Dev. Psychobiol.* 14 (2), 117-125.

Murphey, R.M., Moura Duarte, F.A., Torres Penedo, M.C., 1981. Responses of cattle to humans in open spaces: breed comparisons and approach avoidance relationships. *Behav. Genet.* 11 (1), 37-48.

Murphy, L.B., 1977. Responses of domestic fowl to novel food and objects. *Appl. Anim. Ethol.* 3, 335-349.

Nestler, E., 2011. Hidden switches in the mind. Sci. Am. Dec. 7783. Nestler, E.J., 2012. Stress leaves its molecular mark. *Nature* 490, 171172.

Nottebohm, F., 1970. Ontogeny of bird song. *Science* 1678, 950-956.

Nottebohm, F., 1977. Asymmetries in neural control of vocalization in the canary. In: Harnard, S. (Ed.), *Lateralization of the Nervous System*. Academic Press, New York and London.

Nottebohm, F., 1979. Origins and mechanisms in the establishment of cerebral dominance. In: Gazzaniza, M. (Ed.), *Handbook of Behavioral Neurobiology*. Plenum, New York.

Nottebohm, F., Stokes, T.M., Leonard, C.M., 1976. Central control of song in the canary (*Serinus canaria*). *J. Comp. Neurol.* 165, 457-486.

Nowicki, S., Marler, P., 1988. How do birds sing? *Music. Percept.* 5, 391-421.

O'Malley, C.I., Turner, S.P., D'Eath, R.B., Steibel, J.P., Bates, R.O., Ernst, C.W., et al., 2019. Animal personality in management and welfare of pigs. *Appl. Anim. Behav. Sci.* 2018, 104821. Available from: https://doi.org/10.1016/j.applanim.2019.06002.

Ott, R.S., 1996. Animal selection and breeding techniques that create diseased populations and compromise welfare. *J. Am. Vet. Med.* 208, 1969-1974.

Ovodov, et al., 2011. Ovodov, N.D., Crockford, S.J., Kuzmin, Y.V., Higham, T.F.G., Hodgins, G.W.L. and der Plicht, J. (2011) *PLOS ONE* 6 (7), e22821. Available from: https://doi.org/ 10.1371/Journal.pone.0022821.

Pang, J.F., Kleutsch, C., Zau, X., Zhang, A.B., Luo, L.Y., 2009. mtDNA data indicated a single origin of dogs south of the Yangtzee River less than 16,300 years ago from numerous wolves. *Mol. Biol. Evol.* 26, 2849-2864.

Panksepp, J., 1998. *Affective Neuroscience*. Oxford University Press, New York, NY.

Panksepp, J., 2005. Affective consciousness: core emotional feelings in animals and humans. *Conscious. Cogn.* 14, 3080.

Panksepp, J., 2011. The basic emotional circuits of mammalian brains: do animals have inner lives? *Neurosci. Biobehav. Rev.* 35, 1791-1804.

Parsons, P.A., 1988. Behavior, stress and variability. *Behav. Genet.* 18 (3), 293-308.

Pavlidis, P., Somel, M., 2020. Of dogs and men. *Science* 370 (6516), 522-523.

Perals, D., Griffin, A.S., Bartomeus, I., Sol, D., 2017. Revisiting the open field test: what does it really tell us about animal behavior? *Anim. Behav.* 123, 6979. Available from: https://doi. org/10.1016/janbehav.2016.10.006.

Plomin, R., Daniels, D., 1987. Why are children in the same family so different from one another? *Behav. Brain. Sci.* 10, 160.

Popova, N.K., Voitenko, N.N., Trut, L.N., 1975. Changes in serotonin and 5hydroindoleacetic acid content in the brain of silver foxes under selection for behavior. *Proc. Acad. Sci. USSR* 233, 1498-1500.

Poulsen, H., 1959. Song learning in the domestic canary. *Z. Tierpsychol.* 16, 173178.

Poulsen, P., Esteller, M., Vaag, A., Fraga, M.F., 2007. The epigenetic basis of twin discordance in age-related diseases. *Pediatr. Res.* 61, 38R42R.

Price, E.O., 1984. Behavioral aspects of animal domestication. *Q. Rev. Biol.* 59, 132.

Probst, A.V., Dunleavy, E., Almouzni, G., 2009. Epigenetic inheritance during the cell cycle. *Nat. Rev. Mol. Cell. Biol.* 10, 192-206.

Queen, N.J., Boardman, A.A., Patel, R.S., Sice, J.J., Mo, X., Cao, L., 2020. Environmental enrichment improves metabolic and behavioral health in BTBR mouse model of autism. *Psycho-neuro-endocrinology* 111, 104476. Available from: https://doi.org/10.1016/ jpsyneuen2019.104476.

Quint, M., Drost, H.G., bagel, A., Ullrich, K.K., Bonn, M., Gross, I., 2012. A transcript? hour glass in plant embryogenesis. *Nature* 490, 98-108.

Redon, R., 2006. Global variation in copy number in the human genome. *Nature.* 444, 444-454.

Reese, W.G., Newton, J.E.O., Angel, C., 1983. A canine model of psychopathology. In: Krakowski, A.J., Kimball, C.P. (Eds.), *Psychosomatic Medicine.* Plenum, New York, pp. 25-31.

Resar, J., 2013. Solitary Mammals as a model for autism. *J. Comp. Psychol.* 128 (1). Available from: https://doi.org/10.1037/90034519.

Reynolds, S.M., Berridge, K.C., 2008. Emotional environmental retune the balance between appetitive vs fearful functions in nucleus accumbens. *Nat. Neurosci.* 11, 423-425.

Richard, S., Lend, N., SaintDizier, H., Letemer, C., Faure, J.M., 2010. Human handling and presentation of novel objects evoke independent dimensions of fear in quail. *Behav. Processes.* 85, 18-23.

Richardson, M.K., 2012. A phylotypic stage for all animals? *Dev. Cell.* 22, 903-904.

Richter, C.P., 1952. Domestication of the Norway rat and its implications for the study of genetics of man. *Am. J. Hum. Genet.* 4, 273-285.

Richter, C.P., 1954. The effects of domestication and selection on the behavior of Norway rat. *J. Nat. Cancer Inst.* (U.S.). 15, 727-738.

Rogan, M.T., LeDoux, J.E., 1996. Emotion: systems cells, and synaptic plasticity. *Cell (Camb. Mass.)* 83, 369-475.

Rosenfield, C.S., Hekman, J.P., Johnson, J.L., Lyu, Z., Orgega, M.T., Joshi, T., et al., 2019. Hypothalamic transcriptions of tame and aggressive silver foxes (Vulpes vulpes) identifies gene expression differences shared across brain regions, *Gene. Brain Behav.* 19 (1). Available from: https://doi.org/10.1111/gbb.12614.

Roullet, F.I., Crawley, J.N., 2011. Mouse models of autism: Testing hypotheses about molecular mechanisms Roulette, F.I., Crawley, J.N., 2011 *Curr. Top. Behav. Neurosci.* 7, 187-212.

Royce, J.R., Carran, A., Howarth, E., 1970. Factor analysis of emotionality in ten strains of inbred mice. *Multidiscip. Res.* 5, 1948.

RozempolskaRucinska, I., Kibala, L., Prochniak, T., Zieba, G., Lukaszewicz, M., 2017. Genetics of the novel object test outcome in laying hens. *Appl. Anim. Behav. Sci.* 139, 73-76.

Saey, T.M., 2011. Missing link. *Sci. News* 22-25.

SaintDizier, H., Letemier, C., Levy, F., Richard, S., 2008. Selection for tonic immobility duration does not affect response to novelty. *Appl. Anim. Behav. Sci.* 112, 297-306.

Sanyal, A., Lajole, B.R., Jain, G., Dekker, J., 2012. The long-range interaction landscape of gene promoters. *Nature* 489, 109-113.

Sarma, K., Reinberg, D., 2005. Histone variants meet their match. *Nat. Rev. Mol. Cell Biol.* 6, 139-149.

Savolainen, P., Zhang, Y.P., Lup, J., Lundeberg, J., Leitner, T., 2002. Genetic evidence for an East Asian origin of domestic dogs. *Science* 298, 610-613.

Schaible, R., 1963. Clonal distribution of melanocytes n piebald spotting and variegated mice. *J. Explor. Zool.* 172, 181-200.

Schoenfelder, S., Fraser, P., 2019. Long-range enhances contacts in gene expression control. *Nat. Rev. Genet.* 20, 437-455.

Schroeder, C., 1914. Thinking animals. *Nat. (Lond.)* 94, 426-427.

Schultz, D., 1965. *Sensory Restriction.* Academic Press, New York.

Scott, J.P., Fuller, J.L., 1965. *Genetics and the Social Behavior of the Dog.* University of Chicago Press, Chicago.

Setckliev, J., Skaug, O.E., Kaada, B.R., 1961. Increase of plasma 17hygroxy corticosteroids by cerebral cortisol and amagdaloid stimulation in the cat. *J. Endocrinol.* 22, 119-129.

Shastry, B.S., 2002. SNP alleles in human disease and evolution. *J. Hum. Genet.* 47 (11), 0561-0566.

Sikela, J.M., Searles, V.B., 2018. Genomic tradeoffs: are autism and schizophrenia, the steep price for a human brain. *Hum. Genet.* 137 (1), 113.

Silveira, I.D.B., Fischer, J., Farinatti, H.E., Restle, J., Filho, D.C.A., deMenezes, L.F.G., 2012. Relationship between temperament with performance and meat quality of feedlot steers with predominantly Charolais or Nellose breed. *Rev. Bras. Zootech.* 41.

Silverman, J.L., Yang, M., Lord, C., Crawley, J.N., 2010. Behavioral phenotyping assays for mouse models of autism. *Nat. Rev. Neurosci.* 11, 490-502.

Simons, D., Land, P., 1987. Dearly experience of tactile stimulation influences organization of somatic sensory cortex. *Nat. (Lond.)* 326, 694-697.

Skinner, M.P., 1922. The pronghorn. *J. Mammal.* 3, 82-106.

Skinner, F.F., 1958. *Behavior of Organisms*. AppletonCenturyCrofts, New York.

Sluyter, F., Bult, A., Lynch, C.B., VanOortmerssen, G.A., Kookhaus, J.S., 1995. Comparison between house mouse lines selected for attack latencies or nest building. Evidence for a genetic basis of alternative behavioral strategies. *Behav. Genet.* 25 (3), 247-252.

Stead, J.D.H., Clinton, S., Neal, C., Schneider, J., 2006. Selective breeding for divergence in novelty seeking traits: heritability and enrichment in spontaneous, anxiety related behaviors. *Behav. Genet.* 36, 697-712.

Steinberg, S.A., Klein, E., Killens, R.L., Udhe, T.W., 1994. Inherited deafness among nervous pointer dogs. *J. Hered.* 85, 56-59.

Stephens, D.E., Toner, J.M., 1975. Husbandry influences on some physiological parameters of emotional responses in calves. *Appl. Anim. Ethol.* 1, 233-243.

Tinbergen, N., 1948. Social releasers and the experimental method required for their study. *Wilson. Bull.* 60, 652.

Tinbergen, N., 1951. *The Study of Instinct*. Clarendon Press, Oxford University Press, New York, Tinbergen, N., 1951.

Trillmich, F., Hudson, R., 2011. The emergence of personality in animals: the need for a developmental approach. *Dev. Psychobiol.* 53, 505-509.

Tulloh, N.M., 1961. Behavior of cattle in yards, II. A study of temperament. *Anim. Behav.* 9, 25-30.

Vita, C., Savolainen, P., Maldoado, J.E., Amorim, I.R., Rice, J.E., Honeycutt, R.L., 1997. Multiple and ancient origins of the domestic dog. *Science* 276, 1687-1689.

Vogel, G., 2011. Do jumping genes spawn diversity. *Science* 332, 300-301.

Voisinet, B.D., Grandin, T., O'Connor, S.F., Tatum, J.D., Deesing, M.J., 1997a. Bosindicus cross feedlot cattle with excitable temperaments have tougher meat and a higher incidence of borderline dark cutters. *Meat. Sci.*

Voisinet, B.D., Grandin, T., Tatum, J.D., O'Conner, S.F., Struthers, J., 1997b. Feedlot cattle with calm temperaments have higher average daily gains than cattle with excitable temperaments. *J. Anim. Sci.* 75, 892-896.

Von Holdt, B.M., Shuldner, E., Koch, I.J., Kartinel, R.Y., Hogan, A., Bruebaker, L., 2017. Structural variants in genes associated with human Williams-Beuren Syndrome underlie stereotypical hyper sociality in domestic dogs. *Sci. Adv.* 3 (7), e17000398.

Walsh, R.N., Cummins, R.A., 1975. Mechanisms mediating the production of environmental induced brain changes. *Psychol. Bull.* 82, 986-1000.

Watson, J.B., 1930. *Behaviorism*. W.W. Norton, New York.

Weir, A.A.S., Kacelnik, A., 2006. New Caledonian crows (Corvus moneduloides) creatively redesign tools by bending or unbending metal strips according to needs. *Anim. Cogn.* 9 (4), 317-334.

Whitman, C.O., 1898. "Animal behavior". Biol. Lec. Marine Biological Laboratory, Woods Hole, MA.

Widowski, T.M., Curtis, S.E., 1989. Behavioral response of per parturient sows and juvenile pigs to prostaglandin F2a. *J. Anim. Sci.* 67, 3266-3276.

WoodGush, D.G.M., Beilharz, R.G., 1983. The enrichment of a bare environment for animals in confined conditions. *Appl. Anim. Ethol.* 10, 209.

WoodGush, D.F.M., Vestergaard, K., 1991. The seeking of novelty and its relation to play. *Anim. Behav.* 42, 599-606.

Wright, S., 1922. The effects of inbreeding and crossbreeding on guinea pigs. *U.S. Dep. Agric. Bull.* 1090.

Wright, S., 1978. The relation of livestock breeding to theories of evolution. *J. Anim. Sci.* 46 (5), 11921200.

Yamamuro, K., Bicks, L.K., Levanthal, M.B., Kato, D., Im, S., et al., 2020. A prefrontal paraventricular thalamus circuit requires juvenile social experiences to regulate adult sociability in mice. *Nat. Neurosci.* 23, 1240-1252.

Yeakel, E.H., Rhoades, R.P., 1941. A comparison of the body and endocrine gland (adrenal, thy roid and pituitary) weights of emotional and non-emotional rats. *Endocrinol. (Baltim.)* 28, 337-340.

Yerkes, R.M., 1905. Animal psychology and criteria of the psychic Yerkes, 1905 *J. Comp. Neurol. Psychol.* 15, 137.

York, R.A., 2018. Assessing the genetic landscape of animal behavior. *Genetics* 209 (1), 223232. Available from: https://doi.org/10.1534/genetics.118.300712.

Zadar, J., Balogh, A., Favati, A., Jensen, P., Leimer, O., Loulie, H., 2017. A comparison of animal personality and coping styles in red jungle fowl. *Anim. Behav.* 130, 209220. Available from: https://doi.org/10.1016/j.anbehav.2017.06.024.

Zenner, F.E., 1963. *A History of Domesticated Animals*. Harper & Row, New York.

Zhang, Y.E., Landback, P., Vibranovski, M.D., Long, M., 2011. Accelerated recruitment of new brain development genes into the human genome. *PLoS Biol.* 9 (10), e1001179.

FURTHER READING

Coppinger, R., Schneider, R., 1993. Evolution of working dog behavior. In: Serpell, J. (Ed.), *The Domestic Dog: Its Evolution, Behavior and Interactions with People*. Cambridge University Press, Cambridge, UK.

Darwin, C.R., 1859. *On the Origin of Species*. Oxford University Press.

Darwin, C., 1871. *The Descent of Man and Selection in Relation to Sex*. Modern Library, New York.

Giammortino, D.C.D., Polyzos, A., Apostolou, E., 2020. Transcription factors: building hubs in 3D space. *Cell Cycle* 19 (9), 23952410. Available from: https://doi.org/10.1080/ 15384101.2020.1805238.

Momozawa, Y., Yuari, Y., Kusunose, R., Kikusui, T., Mori, Y., 2005. Association of equine temperament and polymorphism in dopamine D4 gene. Mamm. *Genome*. 16, 538544.

Wilson, V., Guenther, A., Overli, O., Saltmann, M.W., Attachul, D., 2019. Future directions for personality research: contributing new insights to understanding animal behavior. *Animals* 9 (5), 240. Available from: https://doi.org/10.3390/ani9050240.

EIGHTEEN

The Relationship Between Reaction to Sudden Intermittent Movements and Sounds and Temperament

Originally published in (2000) Lanier, J.L., Grandin, T., Green, R.D., Avery, D. and McGee, K. The Relationship Between Reaction to Sudden Intermittent Movements and Sounds and Temperament, *Journal of Animal Science*, Vol. 76, No. 6, pp. 1467-1474.

The temperament of an animal is related to its reactivity to sudden sounds and motions. Cattle that become more agitated in an auction ring were likely to be startled. Old breeding bulls were less reactive than younger steers and heifers. Cattle with a hair whorl high on the forehead were more reactive than cattle with a hair whorl low on the forehead.

ABSTRACT

Casual observations indicated that some cattle are more sensitive to sudden movement or intermittent sound than other cattle. Six commercial livestock auctions in two states and a total of 1,636 cattle were observed to assess the relationship between breed, gender, and temperament score on the response to sudden, intermittent visual and sound stimuli, such as the ring-man swinging his arm for a bid and the sound of him briefly yelling a bid. A 4-point temperament score was used to score each animal while it was in the ring.

The scores used were 1) walks and(or) stands still, with slow, smooth body movements; 2) continuously walks or trots, and vigilant; 3) gait is faster than a trot (runs even a couple of steps), with fast, abrupt, jerky movements, and very vigilant; and 4) hits the ring fence, walls, partitions, or people with its head. Animals were observed for flinches, startle responses, or orientation toward sudden, intermittent sounds, motions, and tactile stimulation, such as being touched with a cane or plastic paddle. The cattle observed were mostly *Bos taurus* beef breeds and Holstein dairy cattle. Holsteins were more sound-sensitive ($P = .02$) and touch-sensitive ($P < .01$) than beef cattle. Sensitivity to sudden, intermittent stimuli (e.g., sound, motion, and touch) increased as temperament score (excitability) increased. Cattle with a temperament score of 1 were the least sensitive to sudden, intermittent movement and sound and those with a temperament score of 4 were the most sensitive ($P < .01$). This same relationship was sometimes observed for touch but was not statistically significant. Motion-sensitive cattle were more likely than non-sensitive cattle to score a temperament rating of 3 or 4 ($P < .01$). Steers and heifers were more motion-sensitive than the older bulls and cows ($P = .03$). Beef cattle urinated ($P < .01$, $n = 1{,}581$) and defecated ($P < .01$, $n = 1{,}582$) more often in the ring than did dairy cattle. Cattle that became agitated during handling in an auction ring were the individuals that were most likely to be startled by sudden, intermittent sounds and movements. Reactivity to sudden, intermittent stimuli may be an indicator of an excitable temperament.

INTRODUCTION

There is an increasing interest in management options for improving animal welfare. There has been extensive research conducted to assess stress associated with handling and husbandry procedures (Mitchell et al., 1988; Lay et al., 1992; Zavy et al., 1992; Rushen et al., 1998). There is a significant relationship in cattle between temperament and productivity. Cattle that became agitated during restraint in a squeeze chute had lower weight gains and tougher meat (Voisinet et al. 1997a,b). Burrow and Dillon (1997) found that cattle that exited more slowly from a squeeze chute had greater weight gains than those that exited the squeeze chute quickly. Drugociu et al. (1977) reported that dairy cows with calm temperaments had increased milk production. Stressful treatment during growth can have adverse effects on meat quality in lambs (Bramblett et al., 1963). Producers are becoming increasingly interested in assessing temperament because excitable animals have reduced weight gain (R. D. Green, unpublished data). Temperament is definitely heritable (Shrode and Hammack, 1971; Hearnshaw and Morris 1984; Fordyce et al. 1988).

Casual observations at auctions indicated that cattle in the auction ring are most likely to flinch and startle in response to sudden, intermittent stimuli such as a ring-man waving his arm or yelling for a bid and children running near the ring. The purpose of the study was to determine whether the reaction of cattle to sudden, intermittent motions, sounds, and touch in an auction ring is related to their overall temperament. This could be useful to producers for temperament-testing cattle.

MATERIALS AND METHODS

Animals

Two observers collected data in six different commercial auctions during the summer of 1998. Five auctions were located in Colorado and the sixth was in Texas. All five auction houses in Colorado were east of the Rocky Mountains. Two were located in the north, one in central Colorado, and the last two in southern Colorado. The Texas auction house was 161 km from Fort Worth. A total of 1,636 beef cattle were observed. They were 74.4% British and European breeds (*Bos taurus*) and 21.4% Holsteins (*Bos taurus*). Ninety-three *Bos indicus* cattle consisted of Brahman, Watusi, and crosses with *Bos taurus* breeds. Most of the European and British breeds were Angus, Hereford, Charolais, Simmental, and their crosses. Breeds were categorized based on the auctioneer's announcement of cattle breeds. Cattle with longer ears, loose dewlap, and a dorsal hump were classified as Brahman crosses. Single cattle (n = 1,543, 94.3%) and cow/calf pairs (n = 93, 5.7%) weighing 182 kg or greater were recorded. Only 3.2% of the total sample were Brahman or Brahman cross. Interviews with ranchers indicated that they were not selling their heat-tolerant cattle (Brahman or Brahman crosses) due to drought conditions. Data were collected only on the cow when a cow-calf pair was sold. Cattle weighing less than 182 kg were considered juveniles and were not part of the study. Data were collected while each animal was in the auction ring, while the gates were closed and the auctioneer was soliciting bids. Animals that initially entered the ring alone or with a calf were scored. Cattle that were taken out of the ring and later brought back through were not scored.

Observers

Observers were always centered in the first to third rows of seats nearest the auction ring or 3 m from the center, toward the gate from which the cattle entered the sale ring. Seating near the entrance gate was ideal for the collection of data because the majority of the cattle remained in this area of the ring during bidding. Inter and intra rater reliability tests conducted at two of the auctions used in the study, with two separate observers, demonstrated reliability between the two experienced observers ($P > .05$). However, the reliability does decrease if the observer is unfamiliar with cattle behavior, cattle flight zones, or has not previously practiced scoring cattle behavior in an auction ring.

Prior to the collection of data, the observers practiced the recording of data at three different auctions (140 cattle). These data were used to refine methodology and were not included in the study. During the study, the first 10 animals observed at each new auction were used for practice and were not included in the analysis of data.

Scoring temperament

The first observer collected data on animal weight, breed, color, and gender (bull, steer, cow, or heifer). A second observer collected behavioral data. Both observers sat in the spectator area of the auction barn, and both had full view of the animals and the auction ring. Reactivity to external stimuli (e.g., noise, being touched) was not used to determine temperament score. Activity level of each animal was the primary scoring criterion, followed next by the head and neck position of the animal. All scoring was done while the amplified auctioneer chanted. Each animal was in the auction ring for a period of approximately 15 to 30 s.

Temperament rating: activity level in the auction ring

The following scoring system was used to rate cattle behavior in the auction ring:

1 = Walks and (or) stands still. Slow smooth body movements. Head and neck in a lowered, relaxed position. The head and neck may be thrust forward.

2 = Continuously walks or trots. Vigilant. Head and neck are slightly raised above back, slightly lowered below back, or level with back.

3 = Gait is faster than a trot (runs even a couple of steps). Fast, abrupt, jerky movements. Very vigilant.

4 = Hits the ring fence, walls, partitions, or people with its head. Contact with the ring fence, walls, partitions, or people due to licking, smelling, or bumping into or brushing up against with its body were not considered as a rating of 4. A 4 was given if the animal attempted to go under, through, or jump or climb over a barrier, regardless of activity level (i.e., standing, walking, or running).

Behavior rating: aggression or escape

Animals that were rated as either a 3 or a 4 were further rated for aggressive behavior (A), or escape behavior (B):

A = Aggression behavior = Pawing the ground while the head is lowered, lunging forward at a person or object with the head slightly lowered, lowering and shaking head at a person or object, or charging a person or object. Aggressive behavior head position: Head and neck were held high above back held close to the ground, or slightly raised above back.

B = Escape behavior = Head and neck were stretched forward and either slightly raised above back, slightly lowered, or level with back.

For example, a cow that was walking continuously around the ring with its head held slightly raised above its back would rate a 2. However, if the cow then attempted to climb out of the auction ring, the rating of 2 would be void and she would be recorded as an "escape" 4.

Scoring animal response to sudden, intermittent environmental stimuli

Animals that flinched and (or) oriented immediately toward a sudden sound, motion, combination of sound and motion, or touch were scored as sensitive to those particular stimuli. A flinch was scored if the animal gave a startle response, or its skin quivered immediately after the stimulus. The auction houses would not allow the use of a controlled movement or sound stimulus to test each animal's startle reaction. One hundred forty cattle were observed in three different practice auctions to determine the naturally occurring intermittent movements, sounds, and touches that were most likely to cause cattle to flinch, jump, quiver, or orient.

Stimuli scoring

The following naturally occurring intermittent movements, sounds, and touch stimuli were used. Movements: 1) ring-man swinging an arm to take a bid; 2) audience deliberately waving at an animal; or 3) young children running within 2 m of the ring. Sounds: 1) ring-man briefly yelling a one syllable bid without the aid of a microphone; 2) air hoses used to move cattle outside of the auction ring but audible in the spectator seating; 3) children yelling; or 4) a "rattle paddle" shaken or hit on a wall or fence. Touch stimuli: 1) hit with a paddle, cane, or whip by the ring-man; or 2) poked with a cane by the audience.

Movements related to startle response and not flight zone were recorded. To avoid being confounded by an animal reacting to a movement made directly in front of its face, motions that were close to the animal's face were not scored. All other occurrences of the above movements, touches, and sounds were scored. The observer must have been able to discern between an animal reacting to a movement that applies pressure to the flight zone and causes the animal to move away, and a movement that does not affect the flight zone and causes a startle response. Response of an animal to a stimulus was not used in determining the animal's temperament. For example, if the ring-man touched a cow with a cane, and the cow jumped and flicked her ears, a temperament rating based on this response was not given.

Animals exposed to the above movements, sounds, and touches were scored. Animals were either scored as "yes," sensitive, or "no," not sensitive. Reactions of an animal to sudden environmental stimuli used to score an individual as "yes," included flinching, jumping, whole body quivers, and ear and (or) head orientation toward the stimulus. Only one of the criteria was needed in order for an animal to be scored as sensitive to sudden environmental stimuli. Reactions to motion, tactile stimulation, and combinations of auditory and visual (motion) stimulation were scored as discrete binomial variables.

Scoring of an animal's reactivity to sudden sounds, motion, or being touched was recorded for those animals for which the stimulus occurred while the animal was in the ring and the auctioneer was chanting. The first sound, motion, and touch detected by the observer were used for scoring sensitivity. Interobserver reliability tests demonstrated that neither all behaviors nor reactions to all behaviors could be reliably observed and recorded, due to the speed of the auction. It was found that interobserver reliability was very high (92%) if each observer recorded the first behavior that he or she observed, rather than attempting to record the first behavior that occurred.

If a sudden stimulus occurred while the auctioneer was silent, the response was not scored. This was done for consistency for the type of background noises all animals would receive, and to control variance. In addition, during the practice recording of cattle behavior in the auction ring, it was observed that the constant chant of the auctioneer appeared to separate out those cattle that had become accustomed to a low volume of noise and stress but that were actually reactive under extreme conditions. Reactivity to external stimuli was not used to determine temperament score.

Animals that urinated and(or) defecated in the auction ring were recorded as either "yes," they did, or "no," they did not.

STATISTICAL ANALYSIS

Data analyses were conducted with the use of chi-square (SAS, 1991). The effect of breed and gender were controlled for by chi-square and the logistic regression gen-mod procedure (SAS, 1995). The GLM procedure (SAS, 1985) controlled for auction, breed, and gender. Results and conclusions were identical from both analytical methods. Intra and interobserver reliability were verified using a paired T-test (SAS, 1991).

RESULTS

The breakdown of gender within Holsteins was 4.4%, 8.5%. 6.8%, 80.3% for steers, bulls, heifers, and cows, respectively. Genders of the beef type cattle were 7.2%, 21.4%, 15.8%, and 55.6% for steers, bulls, heifers, and cows, respectively.

SOUND SENSITIVITY

There were differences in responses to sudden, intermittent sounds among the temperament score groups (Table 1) (chi-square 51.31, P <.01, n = 928; GLM P < .01, F = 18.03). Analysis by least significant differences found the same effect. Holstein cattle were significantly more sound-sensitive (P = .02, n = 918) than beef cattle. Of those individuals scored for sound sensitivity, 34.9% were Holstein and 27.4% were beef cattle. The percentage of bulls and steers that were aggressive in the sale ring and were sensitive to sound was 14.6% (P = .01, n = 64).

TABLE 1: Percentage and Fraction of Each Temperament Score Group That Was Sensitive to an Environmental Stimulus [a]

Temperament Score	Intermittent Motion	Intermittent Sound	Sound and Motion	Touch
1	20.43% 38/186	13.07% 26/199	43.33% 13/30	29.41% 10/34
2	38.54 % 227/589	29.58 % 147/497	74.58 % 88/118	47.62 % 60/126
3	61.02 % 180/295	42.34 % 94/222	82.5 % 33/40	52.17 % 24/46
4	66.67 % 8/12	70.0 % 7/10	100 % 2/2	33.3 % 1/3

a. All are significant at the P .05 level (GLM and chi-square; SAS, 1985) except for touch sensitivity. Fractions are actual numbers. The numerator is the number of animals sensitive. The denominator is the total number of animals with that particular ring score that were scored for sensitivity. Ring scores ranged from 1 for a calm animal to 4 for a highly agitated animal.

MOTION SENSITIVITY

Motion-sensitive cattle were more likely to score a temperament rating of 3 or 4 than were non-sensitive cattle (chi-square = 85.27 P < .01, n = 1,082; GLM P < .01, F = 30.74) (Table 1). There was no significant difference between Holstein (38%) and beef cattle (44%) that were motion-sensitive (P = .13). The percentages of motion-sensitive cattle were 50.65% of the heifers, 38.26 % of cows, 43.27% of the bulls, and 46.91% of the steers (Table 2).

SOUND AND MOTION SENSITIVITY

Sensitivity to sudden sound and motion, (e.g., a ring-man swinging his arm upward as he called out a bid) increased as overall temperament score in the auction ring increased (chi-square = 15.42, P <.01, n = 190 GLM; P < .01, F = 5.48) (Table 1). There was no difference in reaction between Holstein and beef type cattle for combinations of sound and motion sensitivity.

All are significant at the P .05 level. Fractions are actual numbers. The numerator is the number of animals sensitive. The denominator is the total number of animals that were scored for sensitivity to the stimulus.

TOUCH SENSITIVITY

Holsteins were significantly more touch-sensitive (P <.01, n = 208) than beef cattle (Table 2). Heifers (63.6%) were the most touch-sensitive, followed by bulls (55.4%), steers (50.0%), and cows (38.5%) (P = .05, n = 214) (Table 2).

Table 2: Percentage and Fraction of Cattle That Were Sensitive to an Environmental Stimulus

Animal	Intermittent Motion	Intermittent Sound	Sound and Motion	Touch
Holstein and Beef Type Cattle				
Heifers	50.65 % 77/152	34.09 % 45/132	82.61% 19/23	63.64 % 14/22
Cows	38.26 % 251/656	31.94 % 183/573	70.53 % 79/112	38.52 % 47/122
Steers	46.91 % 38/81	29.69 % 19/45	41.17 % 5/12	50.00 % 7/14
Bulls	43.27 % 90/208	20.57 % 36/175	75.00 % 36/48	55.36 % 31/56
Holstein Cattle Only				
Heifers	31.58 % 6/19	58.82 % 10/17	100.00 % 2/2	80.00 % 4/5
Cows	35.15 % 84/239	33.33 % 71/213	82.35 % 14/17	57.90 % 11/19
Steers	50.00 % 7/14	55.56 % 5/9	0.00 % 0/2	66.67 % 2/3
Bulls	60.00 % 12/20	23.53 % 4/17	100.00 % 10/10	81.82 % 9/11
Beef Type Cattle Only				
Heifers	53.23 % 66/124	58.82 % 10/17	84.21 % 16/19	62.50 % 10/16
Cows	40 % 162/405	33.33 % 71/213	68.89 % 62/90	35.35 % 35/99
Steers	46.15 % 30/65	55.56 % 5/9	50.00 % 5/10	45.45 % 5/11
Bulls	41.53 % 76/183	23.53 % 4/17	67.57 % 25/37	50.00 % 22/44

COMBINED STIMULI EFFECTS

Ninety-one percent of the motion-sensitive bulls and steers and 89% of the cows and heifers were sensitive to combinations of sound and motion (P <.01, n = 42, and P> .01, n = 72, respectively). Sixty-nine percent of the cows and heifers that were sound-sensitive were also sensitive to being touched (P .01, n 32).

GENDER DIFFERENCES

Differences between temperament score and ~ were found. Bulls were the calmest in the auction ring, followed by cows. Steers and heifers were the most agitated in the ring (P <.01, n = 1,614).

URINATION AND DEFECATION

Beef cattle urinated (P <.01, n = 1,581) and defecated (P < .01, n = 1,582) more often in the ring than dairy cattle (95% of beef cattle vs 5% of Holstcins and 85% of beef cattle vs 15% of Holsteins, respectively). Bulls and steers defecated in the ring more often than females (P <.01, n = 1,635). Cattle with a temperament rating of 3 or 4 were less likely to defecate in the auction ring (P <.01, n = 1,613). These highly excitable cattle probably defecated before reaching the auction ring.

AUCTION EFFECT ON TEMPERAMENT

The effect of auction on all measured behaviors except vocalization and motion sensitivity was significant (P < .05). Motion sensitivity was not affected by auction (location). Temperament scores at some auctions were significantly higher (P <.05) than at other auctions.

DISCUSSION

Our results indicate that reactivity to intermittent sounds and sudden movements is significantly related to numerical ranking of cattle temperament during handling in a commercial auction ring. One of the advantages of observing cattle in commercial auction houses was that it made it possible to observe very large numbers of cattle. The disadvantage of commercial auctions was that it was not possible to control all the variables. Conducting observations in six different auction houses and visiting the auctions more than once (except for the Texas auction) helped prevent variables that were unique to one auction from confounding the results. Eight auction houses were visited but two were not used in our observations because rough handling and overuse of electrical prods made a very high percentage of animals become extremely agitated with a temperament score of 3 or 4. Differentiation of the temperament scores and observing the animal's reaction to intermittent stimuli would have been impossible. Correlation between methods used to move animals and temperament scores were not evaluated.

The sounds that were most effective for eliciting a response were often accompanied with sudden movement (e.g., the ring-man shouting while swinging an arm into a raised position). Stimuli that were most effective for eliciting a startle response were intermittent, high-pitched sounds and sudden movements. In rats, sound pulses of 3,000 to 7,000 Hz elicited less of a startle response than sound pulses of 15,000 to 23,000 Hz (Blaszczyk and Tajchert, 1997).

The intermittent stimuli chosen in our observations were based on observations made at the three different auctions used for practice. The stimuli chosen were the ones that were most effective for eliciting a startle reaction. We noticed that high-pitched, intermittent sounds of the ring-man yelling "hey" or a young child yelling had a greater effect on the cattle than the amplified auctioneer's chant, gates slamming, or phones ringing. Waynert et al. (1999) found that sounds made by people while handling cattle had a greater effect on heart rate and reactivity than equipment sounds such as gates banging. Pajor et al. (1999) reported that shouting at cows was very aversive. Our own observations indicated that the constant sound of the auctioneer's chant did not directly elicit a startle response compared to sudden, intermittent stimuli. However, the background noise of the chant may sensitize the animal to intermittent stimuli. Research with rats shows that a constant background noise enhances an acoustic startle response (Schanbacker et al., 1996). High-pitched sounds have a greater effect on an animal's heart rate than low pitched sounds (Tailing et al., 1996). High-pitched sounds with a rising pitch are used in dog training to signal an animal to do something. For example, a whistle signals an animal to come. A low-pitched sound is used to inhibit an activity (McConnell, 1990).

Talling et al. (1996) reported that piglets had increased heart rates when they were exposed to high frequency and high-intensity (sound pressure) sounds, whereas piglets' movement was associated only with loudness. In another experiment (Talling et al., 1998), swine exposed to intermittent, sudden sounds were more reactive than when they were exposed to a constant sound. This study is of particular interest because it showed that intermittent sounds had a greater effect.

Cattle and horses have ears that are more sensitive than human ears. They are especially sensitive to high-frequency sounds (Heffner and Heffner, 1983; Grandin, 1996; Smith, 1998). Therefore, noises that are a whisper to humans are quite audible to cattle. Trnka (1977) reported an inverse relationship between level of sound and abnormal behavior in dairy cattle. Noises in auction houses are diverse in frequency and source, so auction houses provide a good setting for observing cattle's reaction to intermittent sound.

The physiology of the eye and how that relates to instinctual behavior may explain the results found for reaction to a sudden motion. Prey species have visual adaptations for survival in the wild (Craig, 1981). In general, these adaptations are wide field of vision (especially while the head is lowered) (Prince, 1970; Coulter and Schmidt, 1993) and bulbous eyes on the side of the head. They also have slit-shaped pupils, whereas most predatory animals have round pupils (Smith, 1998). Grazing animals have a smaller binocular field of vision than predatory animals and a reduced ability to see objects above them compared to humans (Prince, 1970; Lynch et al., 1992). Prey animals have relatively weak eye muscles, which inhibits the ability to quickly focus on nearby objects; this may explain the tendency of horses to shy from nearby, sudden movement (Prince, 1970; Coulter and Schmidt, 1993). While grazing, the visual system of a prey animal has an increased ability to detect movement, which helps protect the animal from predators. The latest research indicates that cattle, sheep, and goats are dichromats with cones that are most sensitive to yellowish-green (552 to 555 nm) and blue-purple (444 to 445 nm) light (Jacobs et al., 1998). Dichromatic vision may provide an animal with better vision for detecting motion than full color vision (Pick et al., 1994; Miller and Murphy, 1995). LeDoux (1996) states that sudden movements have the greatest activating effect in the amygdala. The amygdala is a region in the brain that controls fearfulness (LeDoux, 1996; Rogan and LeDoux, 1996).

It is possible that motion-sensitive cattle are simply ineffective at visual search (Humphreys, 1996) and have a greater desire to orient to an object (e.g., the exit) than their conspecifics that are not motion-sensitive. Like horses, cattle may have the tendency to shy from sudden motion because of the morphology of their eyes.

There was no difference in temperament between single animals alone in the ring and cows with calves at their sides. It was not within the scope of this study to investigate the behavior of larger groups of animals. Grouped cattle tend to be less behaviorally agitated during routine handling (Ewbank, 1968; Grandin, 1987).

The differences found between genders were also expected. Voisinet et al. (1997b) found that heifers were more excitable than steers. Fleming and Luebke demonstrated that virgin female rats were more excitable than mature male rats. Hard and Hansen (1985) found that female rats became less fearful after parturition and the onset of lactation. This may explain why cows had lower temperament scores than heifers.

Predictions of cattle temperament in unfamiliar environments are becoming increasingly important in today's cattle industry. Animals that are calm and placid on their ranch may become agitated and stressed when they are confronted with a novel situation such as the fairgrounds, feedlots, auctions, and slaughterhouses (Grandin, 1997; Grandin and Deesing, 1998). This is especially a problem in cattle that have an excitable, nervous temperament. Visual stimuli can disrupt handling (Grandin, 1996, 1980). Both cattle and deer orient and face a moving person in a field (Grandin and Deesing, 1998; Hodgett et al., 1998). On detection of motion, prey species visually orient to the source of the movement and watch until they determine that the stimulus is or is not a danger. After such a determination, the animal either returns to its previous activity or takes appropriate evasive action (B. J. Smith, personal communication, 1999). This reaction to visual stimuli can adversely affect smooth animal handling. For example, cattle that are going down an alley may balk at seeing a hat blowing in the wind. After the cattle have determined that the hat is not a danger, they will proceed calmly down the alley.

Temperament scores at some auctions were significantly higher ($P < .05$) than at other auctions. This may be due to differences in animal handling before the cattle entered the auction ring. No data regarding animal handling outside of the auction ring were collected. Electrical prods were used extensively and indiscriminately in the two auctions in which data were not collected. Use of electrical prods in this manner caused normally calm cattle to become agitated and aggressive and (or) to injure themselves during the auction. The relationship between overall behavior in the auction ring and reactivity to sudden, intermittent stimuli was significant in all six auction houses. The differences in animal handling between auctions may have had an effect on cattle temperament. Two auctions in Texas were excluded from the study because the extremely rough handling and excessive electrical prodding caused all animals that entered the auction ring to run (ring score of 3). All other auctions surveyed had a consistent percentage of animals in each ring score and therefore the effect of auction handling on temperament was thought to be minimal. No data on individual auction handling practices, other than brief notes, were collected.

A survey conducted by R. D. Green (unpublished data) found commercial cow/calf producers ranked disposition, after birth weight, as their second most important selection trait in bulls. Their top three reasons for desiring bulls with calm dispositions were 1) excitable bulls lose weight, 2) temperament is heritable, and 3) there is a high labor cost associated with wilder cattle. Producers know that calm handling of cattle (Stricklin and Kautz-Scanavy, 1984) and calm cattle (Burrow and Dillon, 1997; Voisinet et al., 1997a,b; Smith, 1998) can increase productivity.

IMPLICATIONS

Cattle that become agitated during handling in an auction ring are more sensitive to sudden touches and sudden, intermittent movements and sounds, such as the ring-man yelling and waving his arm, a plastic "rattle paddle" slapping a fence, or children yelling or running. Reactivity to intermittent stimuli may be useful for predicting which cattle would be more likely to become agitated when exposed to a new place such as an auction, feedlot, or meat packing plant.

REFERENCES

Blaszczyk, J. W., and K. Taichert. 1997. Effect of acoustic stimulus characteristics on the startle response in hooded rats. *Acta Neurobiol. Exp.* (Wars.) 57:315-321.

Bramblett, V. D., M. D. Judge, and G.E. Vail. 1963. Stress during growth: II. Effects on palatability and cooking characteristics of lamb meat. *J. Anim. Sci.* 22:1064-1067.

Burrow, H. M., and R. D. Dillon. 1997. Relationship between temperament and growth in a feedlot and commercial carcass traits of *Bos indicus* crossbreds. *Aust. J. Exp. Agric.* 37:407-411.

Coulter, D. B., and G. M. Schmidt. 1993. Special senses 1: Vision. In: M. J. Swenson and W. O. Reece (Ed.) Duke's Physiology of Domestic *Animals* (11th Ed.). pp 803815. Comstock Publishing Associates, Ithaca, NY.

Craig, J. V.1981. *Domestic Animal Behavior: Causes and Implications for Animal Care and Management.* Prentice Hall, Englewood Cliffs, NJ.

Drugociu, G., L. Runceanu, R. Nicorici, V. Hritcu, and S. Pascal. 1977. Nervous typology of cows as a determining factor of reproductive and productive behaviour. *Anim. Breed.* 45:1262 (Abstr.).

Ewbank, R. 1968. The behavior of animals in restraint. In: M. W. Fox (Ed.) *Abnormal Behavior in Animals.* pp.159-178. W. B. Saunders Co., Philadelphia, PA.

Fleming, A., and C. Luebke. 1981. Timidity prevents virgin female rat from being a good mother: Emotionality differences between nulliparous and parturient females. *Physiol. Behav.* 27:863-868.

Fordyce, G., R. Dodt, and J. Wythes. 1988. Cattle temperaments in extensive beef herds in northern Queensland. 1. Factors affecting temperament. *Aust. J. Exp. Agric.* 28:683-687.

Grandin, T. 1980. Observations of cattle behavior applied to the design of handling facilities. *Appl. Anim. Ethol.* 6:1933.

Grandin T. 1987. Animal handling. *Vet. Clin.* 3:323-338.

Grandin, T. 1996. Factors which impede animal movement in slaughter plants. *J. Am. Vet. Med. Assoc.* 209:757-759.

Grandin T. 1997. Assessment of stress during handling and transport. *J. Anim. Sci.* 75:249-257.

Grandin, T., and M. J. Deesing. 1998. Genetics and behavior during handling, restraint, and herding. In: T. Grandin (Ed.) *Genetics and the Behavior of Domestic Animals.* pp.113144. Academic Press, San Diego, CA.

Hard, E., and S. Hansen. 1985. Reduced fearfulness in the lactating rat. *Physiol. Behav.* 35:641-643.

Hearnshaw, H., and C. Morris. 1984. Genetic and environment effects on a temperament score in beef cattle. *Aust. J. Agric. Res.* 35:723-727.

Heffner, R. S., and H. E. Heffner. 1983. Hearing in large mammals: Horses (*Equus caballus*) and cattle (*Bos taurus*). *Behav. Neurosci.* 97:299-309.

Hodgett, B. V., J. R. Waas, and L. R. Matthews. 1998. The effects of visual and auditory disturbance on the behavior of red deer (*Cervus elaphus*) at pasture with and without shelter. *Appi. Anim. Behav.* Sci. 55:337-351.

Humphreys, G. W. 1996. Neuropsychological aspects of visual attention and eye movements: A synopsis. In: W. H. Zangemeister (Ed.) *Visual Attention and Cognition.* pp.73-78. Elsevier, Amsterdam.

Jacobs, G. H., J. F. Deegan, and J. Neitz. 1998. Photopigment basis for dichromatic color vision in cows, goats and sheep. *Visual Neurosci.* 15:581-584.

Lay, D. C., Jr., T. H. Friend, R. D. Randel, C. L. Bowers, K. K. Grissom, and O. C. Jenkins. 1992. Behavioral and physiological effects of freeze or hotiron branding on crossbred cattle. *J. Anim. Sci.* 70:330-336.

LeDoux, J. 1996. *The Emotional Brain.* Simon and Schuster, New York.

Lynch, J. J., G. N. Hinch, and D. B. Adams. 1992. *The Behaviour of Sheep: Biological Principles and Implications for Production.* CAB International, Wallingford, Oxon, U.K.

McConnell, J. C. 1990. Acoustic structure a receiver response in domestic dogs (canis familiaris). *Anim. Behav.* 39:897-904.

Miller, P. E., and C. J. Murphy. 1995. Vision in dogs. *J. Am. Vet. Med. Assoc.* 12:1623-1634.

Mitchell, G., J. Hattingh, and M. Ganhao. 1988. Stress in cattle assessed after handling, transport and slaughter. *Vet. Rec.* 123:201-205.

Pajor, E. A., J. Rushen, and A. M. de Pasille. 1999. Aversion learning techniques to evaluate dairy cow handling practices. *J. Anim. Sci.* 77(Suppl. 1):149 (Abstr.).

Pick, D. F., G. Lovell, S. Brown, and D. Dail. 1994. Equine color perception revisited. *Appl. Anim. Behav. Sci.* 42:61-65.

Prince, J. H. 1970. The eye and vision. In: M. J. Swenson (Ed.) *Duke's Physiology of Domestic Animals* (8th Ed.). pp 1135-1159. Comstock Publishing Associates, Ithaca, NY.

Rogan, M. T., and J. E. LeDoux. 1996. Emotion: Systems, cells and synaptic plasticity. *Cell* 83:369-375.

Rushen, J., Munksgaard, L. de Passille, M. B. Jensen, and K. Thodberg. 1998. Location of handling and dairy cows' response to people. *Appi. Anim. Behav. Sci.* 55:259-267.

SAS. 1985. *SAS User's Guide: Statistics* (5th Ed.). SAS Inst. Inc., Cary, NC.

SAS. 1991. *Applied Statistics and the SAS Programming Language* (3rd Ed.). SAS Inst. Inc., Cary, NC.

SAS. 1995. *Logistic Regression Examples Using the SAS System* (Version 6, 1st Ed.). SAS Inst. Inc., Cary, NC.

Schanbacher, A., M. Koch, K. D. Pilz, and H. C. Schnitzlen. 1996. Lesions of the amygdala do not affect enhancement of an acoustic startle response by background noise. *Physiol. Behav.* 60:13411346.

Shrode, R. R., and S. P. Hammack. 1971. Chute behavior of yearling beef cattle. *J. Anim. Sci.* 33:193 (Abstr.).

Smith, B. 1998. *Moving 'Em: A Guide to Low Stress Animal Handling.* University of Hawaii, Graziers Hui, Kamuela, HI.

Stricklin, W. R., and C. C. Kautz-Scanavy. 1984. The role of behavior in cattle production: A review of research. *Appi. Anim. Ethol.* 11:359-390.

Talling, J. C., N. K. Waran, and C. M. Wathes. 1996. Behavioural and physiological responses of pigs to sound. *Appi. Anim. Behav. Sci.* 48:187-202.

Talling, J. C., N. K. Waran, C. M. Wathes, and J. A. Lines. 1998. Sound avoidance by domestic pigs depends upon characteristics of the signal. *Appi. Anim. Behav. Sci.* 58:255-266.

Trnka, J. 1977. The effect of an increased acoustic noise level on the behavior of dairy cows of the Danish Red breed./ Vliv zvysene hiadiny akustickeho tlaku na zivotni projevy dojnic plemene danske cervene. *Zivocisna Vyroba* 22:665-671.

Voisinet, B. D., T. Grandin, S. F. O'Connor, J. D. Tatum, and M. J. Deesing. 1997a. *Bos indicus*-cross feedlot cattle with excitable temperaments have tougher meat and a higher incidence of borderline dark cutters. *Meat Sci.* 46:367-377

Voisinet, B. D., T. Grandin, J. D. Tatum, S. F. O'Connor, and J. J. Struthers. 1997b. Feedlot cattle with calm temperaments have higher average daily gains than cattle with excitable temperaments. *J. Anim. Sci.* 75:892-896.

Waynert, D. F., J. M. Stookey, K. S. Schwartzkopf-Genwein, J. M. Watts, and C. S. Waltz. 1999. Response of beef cattle to noise during handling. *Appi. Anim. Behav. Sci.* 62:27-42.

Zavy, M. T., P. E. Juniewicz, W. A. Phillips, and D. L. Von Tungein. 1992. Effects of initial restraint, weaning and transport stress on baseline and ACTH stimulated cortisol responses in beef calves of different genotypes. *Am. J. Vet. Res.* 53:552-557.

NINETEEN

Cattle Vocalizations Are Associated with Handling and Equipment Problems at Beef Slaughter Plants

T. Grandin (2001) Cattle vocalizations are associated with handling and equipment problems at beef slaughter plants, *Applied Animal Behavior Science*, Vol. 71, pp. 191-201.

This paper was further work to validate vocalization scoring as an indicator of stress in cattle. A previous paper showed that 99 percent of cattle vocalizations (bellowing) were associated with something that was either scary or frightening. When cattle bellow loudly during handling or restraint, there is a serious welfare issue that must be corrected. Some of the things that often need to be corrected are reducing electric prod use and pressure applied by a restraint device. This paper clearly shows how making simple equipment improvements reduced vocalization scores.

ABSTRACT

Vocalization of cattle in commercial slaughter plants is associated with observable aversive events such as prodding with electric prods, slipping in the stunning box, missed stuns, sharp edges on equipment or excessive pressure form a restraint device. A total of 5806 cattle were observed during handling and stunning in 48 commercial slaughter plants in the United States, Canada and Australia during the calendar year of 1999. Each animal was scored as either a vocalizer or a non- vocalizer. In 20 plants (42%), 0-1% of the cattle vocalized, in 12 plants (25%) 2-3% vocalized, in 12 plants (25%) 4-10% vocalized and in four plants (8%) more than 10% vocalized. In three plants repeated use of an electric prod on 95% or more of the cattle that balked and refused to move was associated with vocalization percentages of 17, 16 and 12%. In five plants, the percentage of cattle that vocalized was reduced by making modifications to plant equipment. Reducing the voltage on a rheostat controlled electric prod reduced the vocalization percentage from 7 to 2% in the first plant. In three other plants, the incidence of cattle backing up and balking was reduced by illuminating a dark entrance or adding a false floor to a conveyor restrainer. A false floor eliminates the visual cliff effect. The percentage of cattle that vocalized was reduced from 8 to 0%, 9 to 0% and 17 to 2%. Since balking was reduced, electric prod use was also reduced. In the fifth plant, reduction of the pressure exerted by a neck restraint reduced the percentage of cattle that vocalized from 23 to 0%. In the five plants where modifications were made, a total of 379 cattle were observed prior to equipment modifications and 342 after modification. The mean percentage of cattle that vocalized was 12.8% before the modifications and 0.8% after the modifications ($P < 0.001$). Vocalization scoring can be used to identify handling and equipment problems that may compromise animal welfare.

INTRODUCTION

Previous research by Grandin (Jonsson et al., 2021) indicated that 99% of the cattle vocalizations during handling and stunning in a commercial slaughter plant were associated with observable aversive events such as prodding cattle with an electric prod (goad), missed stuns, slipping in the stun box or excessive pressure from a restraint device. Studies of both cattle and pigs indicate that increased vocalization is correlated with physiological indicators of stress (Dunn, 1990; Warriss et al., 1994; White et al., 1995). Watts and Stookey (1998) found that hot iron branded cattle had greater rates of vocalization than freeze branded or sham branded cattle.

Due to increasing public concerns about animal welfare, large restaurant companies (e.g., McDonald's Corp. and Wendy's International) have added animal handling and stunning audits to their food safety audits of slaughter plants. There is a need for a simple scoring system to identify cattle handling problems that would be detrimental to animal welfare. The scoring system must be simple enough to quickly teach to many food safety auditors. The auditors also have time constraints because they have to audit more than one plant in a day.

In a previous study (Grandin, 1998a), ill-trained employees in some plants repeatedly used an electric prod on cattle that were moving quietly and easily through the races. When the percentage of cattle prodded with an electric prod was reduced, the percentage of cattle that vocalized was significantly reduced. The previous study (Grandin, 1998a) was conducted before major customers of the slaughter plants started auditing animal welfare.

A unique feature of the data which will be presented in this paper is that industry awareness about the importance of animal welfare has greatly increased. Plant management is now more motivated to train employees to minimize electric prod use and to handle the cattle quietly. The data on vocalization which will be presented in this paper was collected during announced welfare audits by a major customer.

The variable of poor employee behavior had largely been removed and electric prods were only used on cattle that refused to move. This made it possible to use vocalization scoring to identify problems with equipment which impeded easy orderly movement of the cattle. If cattle frequently backed up or balked, an electric prod had to be used to keep the processing line full. Increased electric prod was associated with increased vocalization (Grandin, 1998a).

Previous observations indicate that distractions such as seeing moving people up ahead, sparkling reflections, shiny metal that jiggles or air blowing into the faces of approaching cattle or pigs will often cause them to balk and back up (Grandin, 1982, 1996, 1998b). Animals also have a tendency to move from a darker place to a more brightly illuminated place (Grandin, 1982; Van Putten and Elshof, 1978; Tanida et al., 1996).

Cattle often balk when they enter a dark race, stunning box or restrainer. Installing a lamp to illuminate the entrance of conveyor restrainer will often facilitate cattle movement into the restrainer (Grandin, 1996, 1998b). The light must illuminate the entrance of the restrainer or stunning box and it must not shine into the eyes of the approaching cattle.

Another equipment problem which often causes cattle to balk is the visual cliff effect in conveyor restrainers. In large plants, cattle ride on a conveyor during stunning. There are two types of conveyor restrainers for cattle, the V-conveyor restrainer which holds the cattle between two moving conveyers that form a V and the center track (double rail) conveyor.

The V-conveyor restrainer used for cattle is a larger version of the system used for pigs (Regensburger, 1940). In the center track, the animals straddle a moving conveyor (Giger et al., 1977; Grandin, 1988, 1991). Conveyor restrainer systems are mounted more than 2 m above the floor of the plant. Observations indicate that cattle entering the conveyor restrainer often balk if they see the "visual cliff" effect through the open bottom of the conveyor restrainer. Grandin (1991) installed a false floor located slightly below the animal's dangling feet to provide the illusion of a solid floor to walk on. The false floor improved cattle movement into the restrainer. Ruminants can see depth and they are sensitive to the "visual cliff" effect (Lemmon and Patterson, 1964).

The first objective of this study was to provide a survey of the percentage of cattle that vocalized in plants that had made some efforts to reduce electric prod use and improve handling. A second objective was to use vocalization scoring to identify equipment problems associated with vocalizations and to compare different plants based on the type of equipment they had. A third objective was to correct equipment problems in plants where this was possible and to compare the percentage of cattle that vocalized before and after modification.

METHODS

Slaughter plant samples

Data were collected in 48 commercial beef slaughter plants in the United States, Canada and Australia during regular operations. A total of 5806 cattle were observed in Federally inspected plants. The plant visits were announced and scheduled with plant management several weeks before the audit. The audits on animal handling and stunning were conducted by McDonald's Corp. (43 plants), Wendy's International (two plants) or by Agriculture Canada (three plants) during the calendar year of 1999. Line speeds among plants varied from 5 to 390 cattle per hour. There were eight plants under 50 cattle per hour, nine between 51 and 100 per hour, six between 101 to 200 per hour, 11 between 201 to 300 per hour and 14 over 300 per hour. The sample was very representative of the large and medium sized plants in the United States and Canada. In Canada plants in three different provinces were included. In the United States plants in 11 different states were visited. In Australia, one of the restaurant companies chose two of their major suppliers for auditing.

Audit methods

The author collected data in 27 plants and an auditor trained by the author collected data in the rest. When the author or the auditor arrived in the stunning and handling area, the line was running. Since the audits were both scheduled and announced in advance, the employees in the plants had been informed by management to minimize electric prod use. To pass the audit, the handlers had to be able to move cattle with a minimum of vocalization. If more than 3% of the cattle vocalized (moo or bellow) the plant failed the audit. A percentage of 3.51 would fail. This is the criterion in the American Meat Institute Guidelines. To pass, each plant was also required to keep the slaughter line full. If cattle balked and refused to move, the employees sometimes had to use electric prods to keep the line full.

Vocalization scoring

At each plant, 100-500 cattle were scored in plants with line speeds of over 100 cattle per hour. Fifty cattle were scored in plants slaughtering under 100 cattle per hour due to time constraints. In one plant, 66 cattle were scored due to a breakdown and 25 and 26 cattle were scored in two small plants with chain speeds of 5 and 20 per hour. No attempt was made to pick certain groups of cattle. Animals moving through the system at the time of the visit were scored. A variety of cattle types were observed ranging from Holstein and beef breed cows to fed steers and heifers. All the plants operating over 300 per hour processed fattened feedlot cattle.

The scoring method for vocalization was the same as Grandin (1998a). Each animal that passed through the single file race, stunning box or restrainer conveyor was scored as either a vocalizer or a non-vocalizer. The smallest vocal sounds were recorded as a vocalizer. No attempt was made to assess the intensity of the vocalizations. The person doing the scoring stood near the entrance of the stunning box or restrainer conveyor. Vocalizations from cattle standing undisturbed in the lairage were not scored. Scoring vocalization was only done during actual moving of the cattle through the race into the stun box or restrainer and during stunning.

Electric prod use and equipment observations

Electric prod use was not numerically scored in most plants, but plants were classified as either (1) no electric prods; (2) electric prods used; (3) electric prods used on 95% or more of the cattle. Auditors also took notes on aversive events associated with vocalization, and on handling facility problems which caused cattle to balk and refuse to move, which led to increased electric prod usage.

Vocalization survey

The percentage of cattle that vocalized were calculated for each plant and grouped into five groups of 0-1% of the cattle vocalizing, 2-3%, 4-10%, 10-15% and over 15%. Percentages were rounded off to the nearest whole number where 1.51 would be 2% and 1.49 would be 1%.

Comparisons between plants

Comparisons were made between plants on various factors which may affect vocalization percentages. The factors compared were restraint equipment type, electric prod use, line speed and the presence or absence of observable equipment problems which caused cattle to frequently balk and back up and refuse to move. Other equipment problems which were tabulated were slipping in the stunning box and excessive pressure applied by a restraining device. Each plant was classified for stunning restraint type: V-conveyor restrainer, center track conveyor restrainer or stunning box.

Equipment modifications to reduce vocalizations

In five plants equipment problems associated with vocalization percentages of 7% or more of the animals were corrected. Four plants were chosen because the modifications could be easily made while the author was present. At many plants there was insufficient time to make modifications and then rescore vocalizations. In four plants, the modifications were made the same day of the audit. The modifications in these four plants were made in the middle of the same group of cattle, so that cattle from the same group were scored for the percentage of cattle that vocalized both before and after the modification. The following modifications were made:

Plant 1 - reduce electric prod voltage;
Plant 2 - install spotlight to illuminate a dark center track conveyor entrance (Grandin, 1998b);
Plant 3 - install a false floor in the conveyor restrainer (Grandin, 1991);
Plant 4 - reduce pressure applied by a head restraint.

The author scored vocalizations both before and after the modifications. In a fifth plant a V-conveyor restrainer without a false floor was replaced by a center track conveyor restrainer that had a false floor. The percentage of cattle that vocalized was scored on two different days.

Statistical methods

Chi square was used to determine the effect of equipment modifications on vocalization and a *t*-test for a difference between two independent means was used for comparisons made between different types of plants (Bruning and Kintz, 1968).

TABLE 1: Percentage of Cattle That Vocalized during Handling and Stunning in 48 Beef Slaughter Plants

Percentage of cattle vocalizing (%)	Number of plants	Percentage of plants (%)
0-1	20	42
2-3	12	25
4-10	12	25
10-15	1	2
Over 15	3	6

RESULTS

Vocalization survey

In 42% (20) of the plants, 0-1% of the cattle vocalized (Table 1). All of the plants in the 0-1% vocalization group had cattle that moved easily through the race and into the conveyor restrainer or stunning box with a minimum of balking. The line speed varied from 5 to 390 cattle per hour. No animal in the 0-1% vocalization group was touched more than once with an electric prod. Six of these plants (30%) did not use electric prods. In one plant that operated at a speed of 113 cattle per hour, 96% of the cattle entered the stunning box when the door was opened without being touched. This plant had done extensive experimentation with lighting to improve cattle movement into the stunning box.

Twenty five percent (12) of the plants had vocalization percentages ranging from 2 to 3% of the cattle. Electric prods were used on some of these cattle, but few animals balked to the point where multiple pokes with an electric prod were required to move them. There were four plants (8%) that had vocalization percentages of 10% or more of the cattle. The percentage of cattle that vocalized during handling in these plants was 12, 16, 17 and 23%. In two of these plants, cattle balked and refused to enter the conveyor restrainer. To keep the line supplied with cattle, the employees repeatedly prodded 95% or more of the cattle with an electric prod. The line speeds were 200 and 390 per hour. At a third plant, a head restraint powered by a pneumatic cylinder applied excessive pressure to the animal's neck, causing it to vocalize. In the fourth plant, electric prods had to be used on 95% or more of the cattle because they balked and refused to enter the stunning box.

Comparisons between the plants

Twenty-nine plants (60%) had stunning boxes and the range in line speeds was 5-390 per hour. Thirteen plants (27%) had a center track conveyor restrainer and six (13%) had a V-conveyor restrainer. The line speed in plants with conveyors ranged from 150 to 390 per hour. The average percentages of cattle that vocalized was 3.87% in a stunning box, 2.35% in a center track conveyor restrainer and 7.16% in a V-conveyor restrainer. Vocalization was significantly elevated in the V-conveyor restrainer compared to the center track conveyor restrainer ($T = 2.53$, $P > 0.05$).

Ten plants (21%) used no electric prods. There was no significant difference in vocalization percentages in plants that used electric prods versus plants that did not ($T = 1.39$). The means were 1.5% of the cattle vocalizing in plants with no prods and 3.77% vocalizing in plants that used electric prods. However, when plants that used, no electric prods were compared to the three plants that used electric prods on 95% of more of the cattle, there was a highly significant difference ($T = 3.64$, $P > 0.01$). The means were 1.5% for the plants with no prods and 15% in plants which used prods on 95% or more of the cattle.

The mean percentages of cattle that vocalized at different line speeds were: 0-50 cattle per hour, 4.31%; 51-100 cattle per hour, 2.88%; 101-200 cattle per hour, 3.83%; 201-300 per hour, 3.72% and over 300 per hour, 2.14%.

TABLE 2: Vocalization Reduction after Modification of Cattle Handling Equipment in Plants where Electric Prods Were Used Only on Cattle That Balked[a]

Plant	Number of Animals	Initial vocalization (%)	Number of Animals	After Modification (%)	X^2 (P)	Equipment modifications
1	100	7	100	2	0.10	Reduce electric prod voltage
2	100	8	100	0	0.01	Install light at entrance to center track conveyor restrainer to reduce balking
3	66[b]	9	29[b]	0	0.10	Install false floor in center track conveyor restrainer to eliminate visual cliff effect
4	13[c]	23	13	0	Sample too small	Reduce pressure applied to neck by a head restraint
5	100	17	100	2	0.001	Replace V-conveyor restrainer with a center track conveyor restrainer which has a false floor, a well-lighted entrance, and a conveyor belting curtain on the discharge end to prevent incoming cattle from seeing into the plant
Combined	379	12.8	342	0.8	X^2 = 28.6P < 0.001	

a. An electric prod was only used on cattle which balked and refused to move into the conveyor restrainer or stunning box.
b. Small number due to stop for lunch. False floor installed during lunch and had breakdown after lunch.
c. Small numbers due to time constraints.

There were 32 plants (67%) that had 3% or less of the cattle vocalizing. Out of these 32 plants, 29 plants (91%) had no observable equipment problem which caused cattle to balk and refuse to move. In these 32 plants there was no slipping in the stunning box or excessive pressure applied by restraint equipment. There were 16 plants where more than 3% of the cattle vocalized. Eleven out of 16 of these plants (68%) had an observable equipment problem which was associated with vocalization. Some of the observed equipment problems that caused cattle to balk and back up were moving equipment, which was visible to the cattle, no false floor in the conveyor restrainer, and a dark conveyor restrainer entrance. Increased balking resulted in increased electric prod use. Other problems which were observed in these plants were slipping in the stunning box and excessive pressure exerted by a head restraint device. Plants where more than 3% of the cattle vocalized had significantly more observable equipment problems that were associated with vocalization ($X^2 = 18.19 > 0.001$). Therefore, plants that passed the audit and had 3% or less of the cattle vocalizing had fewer observable handling and restraint equipment problems.

Equipment modifications to reduce vocalizations

Equipment modifications made in five plants significantly reduced vocalizations (Table 2, $X^2 = 28.6$, $P < 0.001$). A total of 379 cattle in five plants were observed before equipment modification and 342 after equipment modification. For the five plants combined, the average vocalization percentage was reduced from 12.8 to 0.8%. The modifications made in three of the plants reduced balking and cattle backing up. This resulted in less electric prod use. In one plant, the voltage on an electric prod was reduced and, in another plant, pressure applied to the animal's neck by a head restrainer was reduced.

DISCUSSION

Variation in electric prods and driving methods

A large variability in types and uses of electric prods may explain why the absence or the presence of an electric prod had no significant effect on vocalization. However, when 95% or more of the cattle had to be prodded with an electric prod due to backing up, balking or refusing to move forward, electric prod use significantly increased vocalization. In an earlier study stated, the percentage of cattle prodded with an electric prod was scored

(Grandin, 1998a). Further observations indicated that using this method to compare handling in different plants had problems. The problem with electric prod use scoring is that when animals balked and refused to move, employees sometimes abusively hit or poked them in sensitive areas to obtain a favorable score on electric prod use. In some plants it is very difficult to determine if an animal was actually shocked with an electric prod. Hutson et al. (2000) observed that when a battery-operated electric prod was used as an aversive stimulus the animal did not always receive a shock. In the US, the plants use a variety of different electric prods which range from a severe shock which causes most cattle to vocalize to a 15 V shock which seldom causes vocalization. When battery operated electric prod use was observed it was difficult to determine whether the button was pushed to deliver a shock. It makes no sense for an auditor to penalize a plant for touching cattle with an electric prod that is so weak that they may not feel it. The data collected during the study indicated that the tendency of cattle to vocalize was less variable than the variability of electric prod types.

In many of the plants, electric prods were no longer used as the primary driving tool for moving cattle. An electric prod was only picked up and used when a non-electrified driving aid failed to move the cattle. Some common non electrified driving aids were: flags, a plastic bag on a stick or a plastic paddle stick that looks like a small ore. Most of the plants had eliminated electric prods in the crowd pen and electric prod use was confined to the stunning box or restrainer entrance. Many of the plants that had cattle moving easily through the race and stunning box or restrainer had quiet employees who seldom whistled or yelled. Waynert et al. (1999) found that yelling and whistling caused a greater increase in cattle heart rate than the sound of a gate slamming.

The effect of a false floor in the conveyor restrainer on cattle movement

A significantly higher percentage of cattle vocalized in the V-conveyor restrainer compared to the center track restrainer. Unfortunately, the variable of the type of conveyor restrainer was confounded by the presence or absence of a false floor which prevented the incoming animal from seeing the 2 m drop off under the conveyor. Only one V-conveyor restrainer out of six had a false floor and all but one of the center track restrainers had a false floor. The one V-conveyor restrainer which had a false floor had a vocalization percentage of 1% and the one center track restrainer, which did not have a false floor had a vocalization percentage of 9%. Elimination of the "visual cliff" effect by a false floor is an important design feature which will facilitate animal movement into a conveyor restrainer.

Reasons for vocalization reduction after modification

In the first plant, vocalizations were reduced because the intensity of the shock delivered by an electric prod was reduced. Prior to lowering the voltage, many cattle vocalized when the prod was held on them. In three plants, the modifications reduced balking and backing up which reduced electric prod use. In the first of the three plants, a light was installed on the entrance of a dark restrainer conveyor. Previous observations have shown that animals tend to move from a darker area to a more brightly illuminated area (Grandin, 1996; Van Putten and Elshof, 1978; Grandin, 1996). In another plant, the installation of a false floor reduced balking and refusal to enter a center track restrainer. Replacement of a V-conveyor restrainer that did not have a false floor, with a center track conveyer restrainer that had both a false floor and a metal shielding to prevent the incoming animals from seeing people moving in front of the restrainer; greatly reduced balking and the need to use an electric prod to induce the cattle to enter the restrainer. Grandin (1991, 1998b, 2000) states that cattle will enter a center track restrainer more easily and remain calmer, if in addition to a false floor the conveyor restrainer is equipped with a solid metal cover located above the animal's back. This cover must be long enough to block the animal's vision of activities outside the restrainer race until its back feet are completely off the entrance ramp. In other words, the animal must be fully restrained and riding on the conveyor before it is permitted to see out.

In the last plant, the most likely reason for reduced vocalizations was that loosening the pressure applied to the animal's neck by a head restraint reduced pain or discomfort.

Effect of line speed

Line speed appeared to have little effect on the percentage of cattle that vocalized because even at 390 cattle per hour, cattle were able to walk into the restrainer at their normal slow walking speed.

Additional factors that may affect vocalization

Persons who are auditing slaughter plants should make a checklist of equipment problems that can cause cattle to balk and refuse to move. In addition to the items discussed previously, other problems that may cause balking are, air drafts blowing in an approaching animal's face, sparkling reflections off of wet floors, dangling chain ends and moving reflections (Grandin, 1996, 1998b, 2000). Restrainer systems should be inspected for broken sharp edges and excessive pressure. Auditors must also be aware of situations that may prevent vocalization such as an extremely tight restraint device that restricted breathing or electrical immobilization. Electrical immobilization paralyzes the animal (Grandin et al., 1986; Lambooy, 1985; Pascoe, 1986; Grandin, 1998c). Vocalization may also occur if an animal is isolated alone too long in the stunning box. There are also definite genetic effects on the tendency of Angus and Holstein cattle to vocalize (Stookey, 2000, personal communication; Jane Morigan, 1999, personal communication). Watts and Stookey (2000) are concerned that a plant could have an acceptable low percentage of cattle vocalizing one day and a higher unacceptable percentage on another day. To address this concern, additional cattle from another group should be scored if animals vocalize in the absence of an aversive event.

CONCLUSION

Vocalization scoring in commercial beef slaughter plants can be used to locate handling and equipment problems in specific plants. Plant management can use it to monitor handling detect problems in their own plants. The author also views it as a method for objectively identifying plants that have severe animal welfare problems associated with excessive electric prod use due to cattle either balking or refusing to move or excessive pressure from a restraint device. All 20 plants with vocalization percentages of 0-1% were free of distractions, lighting or visual cliff problems which make animals balk. All 15 plants with vocalization percentages of 5% or more had easily identifiable equipment problems. Sixty-nine percent of the plants had vocalization percentages of 3% or less of the cattle. This is an easily attainable minimum standard of performance that can be attained by both large and small plants.

ACKNOWLEDGMENTS

The author wishes to thank Gary Platt and all of the other auditors in the McDonald's Corp. system who participated in the audits. The audits were funded by McDonald's Corp., Wendy's International, and Agriculture, Canada.

REFERENCES

Bruning, J.L., Kintz, B.L., 1968. *Computational Handbook of Statistics*, Scott. Forestman, Glenview, IL.

Dunn, C.S., 1990. Stress reactions of cattle undergoing ritual slaughter using two methods of restraint. *Vet. Rec.* 126, 522-525.

Giger, W., Prince, R.P., Kinsman, D.V., 1977. Equipment for low stress animal slaughter. *Trans. Am. Soc. Agric. Eng.* 20, 571-578.

Grandin, T., 1982. Pig behavior studies applied to slaughter plant design. *Appl. Anim. Ethol.* 9, 141-151.

Grandin, T., 1988. Double rail restrainer conveyor for livestock handling. *J. Agric. Eng. Res.* 41, 327-338.

Grandin, T., 1991. Double rail restrainer for handling beef cattle. Technical Paper No. 915004. *Am. Soc. Agric. Eng.*

Grandin, T., 1996. Factors that impede animal movement at slaughter plants. *J. Am. Vet. Med. Assoc.* 209, 757- 759.

Grandin, T., 1997. Good management practices for animal handling and stunning. *Am. Meat Inst.*, Washington, DC.

Grandin, T., 1998a. The feasibility of using vocalization scoring as an indicator of poor welfare during cattle slaughter. *Appl. Anim. Behav. Sci.* 56, 121-128.

Grandin, T., 1998b. Solving livestock handling problems in slaughter plants. In: Gregory, N.G. (Ed.), *Animal Welfare and Meat Science*. CAB International, Wallingford, UK, pp. 42-63.

Grandin, T., 1998c. Objective scoring of animal handling and stunning practices in slaughter plants. *J. Am. Vet. Med. Assoc.* 212, 36-43.

Grandin, T., Curtis, S.E., Widowski, T.M., Thurmon, J.C., 1986. Electro-immobilization versus mechanical restraint in an avoid-avoid choice test. *J. Anim. Sci.* 62, 1469-1480.

Grandin, T., 2000. Handling and welfare of livestock at slaughter plants. In: Grandin, T. (Ed.), *Livestock Handling and Transport*, 2nd Edition. CAB International, pp. 409-439.

Hutson, G.D., Ambrose, T.J., Barnett, J.L., Tilbrook, A.J., 2000. Development of a behavioural test of sensory responsiveness in the growing pig. *Appl. Anim. Behav.* Sci. 66, 187-202.

Lambooy, E., 1985. Electro-anesthesia or electro immobilization of calves, sheep and pigs with Feenix Stockstill. *Vet. Quart.* 7, 120-126.

Lemmon, W.B., Patterson, G.H., 1964. Depth perception in sheep: effects of interrupting the mother-neonate bond. *Science* 145, 835-836.

Pascoe, P.J., 1986. Humaneness of electro-immobilization unit for cattle. Am. J. Vet. Res. 10, 2252-2256. Regensburger, R.W., 1940. Hog stunning pen. US Patent 2185949. United States Patent Office, Washington, DC.

Tanida, H., Mivra, A., Tanaka, T., Yoshimoto, T., 1996. Behavioral responses of piglets to darkness and shadows. *Appl. Anim. Behav. Sci.* 49, 173-183.

Van Putten, G., Elshof, W.J., 1978. Observations of the effects of transport on the well-being and lean quality of slaughter pigs. *Anim. Reg. Stud.* 1, 247-271.

Warriss, P.D., Brown, S.N., Adams, S.J.M., 1994. Relationship between subjective and objective assessments of slaughter and meat quality in pigs. *Meat Sci.* 38, 329-340.

Watts, J.M., Stookey, J.M., 1998. Effects of restraint and branding on rates and acoustic parameters of vocalization in beef cattle. *Appl. Anim. Behav. Sci.* 62, 125-135.

Watts, J.M., Stookey, J.M., 2000. Vocal behaviour in cattle: the animal's commentary on its biological processes and welfare. *Appl. Anim. Behav. Sci.* 3, 5-22.

Waynert, D.E., Stookey, J.M., Schwartzkopf-Genwein, J.M., Watts, C.S., 1999. Response of beef cattle to noise during handling. *Appl. Anim. Behav. Sci.* 62, 27-42.

White, R.G., DeShazer, J.A., Tressler, C.J., Borcher, G.M., Davey, S., Waninge, A., Parkhurst, A.M., Milanuk, M.J., Clems, E.T., 1995. Vocalizations and physiological responses of pigs during castration with and without an anesthetic. *J. Anim. Sci.* 73, 381-386.

TWENTY

Transferring the Results of Behavioral Research to Industry to Improve Animal Welfare on the Farm, Ranch, and Slaughter Plant

T. Grandin (2003) Transferring the Results of Behavioral Research to Industry to Improve Animal Welfare on the Farm, Ranch, and Slaughter Plant, *Applied Animal Behavior Science*, Vol. 81, pp. 215-228.

There is lots of valuable research that never gets transferred out of the academic world to the people in the livestock industry who really need it. In this paper, I explain how I successfully transferred the center track restrainer from a research lab at the University of Connecticut to the meat industry. Writing was an important method for getting research results used by the livestock industry. I wrote about my projects and did not try to hold on to my intellectual property. Writing for many diverse audiences gets new research results adopted. I wrote numerous articles in the livestock industry trade press. Today it is blogs and videos. Successful transfer of research often requires more time than doing the research.

ABSTRACT

Knowledge obtained from research has been effectively transferred to the agricultural industry in some areas and poorly transferred in others. Knowledge that has been used to create a product such as a pharmaceutical or a device is more likely to be adopted by industry than a behavioral management technique. During my career, I have observed that some people will purchase a new cattle-handling system, which is designed with animal behavioral principles, but they will continue to handle cattle roughly. People are more willing to purchase new equipment than they are to use easy to learn, low stress handling techniques. Even when financial benefits are clear, some people find it difficult to believe that a behavioral management method really works.

From my experience, I have learned that successful transfer of knowledge and technology to industry often requires more work than doing the research. For an effective transfer of technology to take place, the method or equipment must be used successfully by the people who initially adopt it. If the new piece of equipment fails on the first or second place that attempts to adopt it, transfer to the industry may fail. In this paper, I describe a successful case study of transfer of a conveyor restrainer system, based on behavioral principles, from the research lab to US and Canadian beef slaughter plants. I also describe the successful implementation of a measurement system for auditing animal handling in slaughter plants. Based on my experience, the following steps for successful transfer of behavior research to the industry are: (1) Communicate your results outside the research community. Write articles in popular and industry magazines. Speak at producer meetings and develop websites that can be used to transfer research results into practice. (2) Choose places (e.g., farms or plants) that have managers who believe in your research and be prepared to spend a lot of time with the first place that uses your findings. (3) Closely supervise other early adopters to prevent mistakes which could cause the method or technology to fail. (4) Do not allow your technology to get tied up in patent disputes.

INTRODUCTION

There is a large amount of behavioral knowledge that has not been successfully transferred to the industry. Based on my experience in the United States, many people in the pork industry do not know about the extensive research that has been done on pig behavior. Some upper-level managers of large corporate pig farms are not even aware that behavior is a field of research. There is a need to do a much better job of transferring research results into industry language and practices. This paper is divided into five sections: (1) How to get people to recognize the importance of behavioral knowledge and behaviorally-based management methods, and addressing the problem that some people in the industry do not want to discover that a common agricultural practice may be stressful or painful; (2) How to encourage the successful transfer of behaviorally based technology to the industry; (3) How to avoid failure to transfer a good piece of equipment that is based on behavioral research to the industry; (4) How to maintain and motivate excellent stockmanship and animal handling; and (5) Conclusions.

RECOGNIZING THE IMPORTANCE OF BEHAVIORAL MANAGEMENT OF ANIMALS

Years ago, W.D. Hoard, the founder of *Hoard's Dairyman* magazine recognized the importance of good stockmanship (Rankin, 1925). Since that time, many research studies have shown that good stockmanship improves animal productivity. Seabrook's (1972) work was some of the first to show the importance of good stockmanship on productivity of dairy farms. Albright (1978) reported that cows with small flight zones that allowed people to approach them were more productive and gave more milk. Hemsworth et al. (1981) reported that farms on which sows were more willing to approach a person had greater productivity and more piglets per sow per year. In another study, milk production was lowered when a handler who had previously treated cows in a severely aversive manner was present (Rushen et al., 1999). A recent study by Munksgaard et al. (2001) showed that the presence of an additional person who had treated cows in a mildly aversive manner did not affect milk yield. In this study, the cows were milked by the regular dairy farm staff. In most studies linking stockmanship with productivity, the most significant variable is the behavior of the regular milker or caretaker. It is likely, that if the cows in Munksgaard et al.'s (2001) study had associated the regular milker with aversive acts, there would have been a more detrimental effect on milk production than that found when the cows associated a bystander with something aversive.

Further research by Voisinet et al. (1997) and Fell et al. (1999) has shown that cattle that became agitated during handling in a squeeze chute had lower weight gains and more sickness. In their book, Hemsworth and Coleman (1998) review a number of studies indicating that people who have a good attitude towards animals have more productive animals.

Even though numerous studies conducted on almost every livestock species show the advantages of good stockmanship, adoption of good stockmanship principles has been slower than adoption of behaviorally-based facility designs. During the last 20 years, I have written many articles on behaviorally-based cattle-handling principles (Grandin, 1980, 1987, 1996, 1998a). However, many people still handle beef cattle roughly and have poor stockmanship skills. On some large dailies, the level of stockmanship is still poor. Why is all this research on good stockmanship ignored by a large portion of the industry? I have observed that people will adopt new handling equipment based on behavioral principles much more quickly than they will adopt behavioral principles that they have to learn and practice. In my business, I sell twice as many books on how to build corrals and races based on behavioral principles than I sell videotapes that can be used to train stockpersons, even though the costs of the books and tapes are nearly the same. Many people want the "thing" rather than learning better management practices. They think buying the technology is all that they need to do.

Even when a behavioral method has been documented to make money, people have been slow to adopt it. In one slaughter plant I documented a US $500 to US $1000 savings per day by training people to handle cattle quietly. After I left, they quickly reverted back to their old rough ways.

Why are some people so reluctant to learn and adopt behavioral principles of quiet animal handling? I have observed many managers that want a quick, easy fix and learning good stockmanship skills requires time and effort. My own visits to several hundred farms and meat plants indicate that the operations that maintain good stockmanship have either an owner or a high-level manager who insists upon it. I would also like to speculate that one reason may be that to be a good stockman one must recognize that the animal is a conscious being that has feelings. It is not a machine or just an economic entity. Hemsworth and Coleman (1998) have done extensive work on the attitude of the stockperson and its effect on productivity in pigs and dairy cattle. Hemsworth et al. (2002) found that stockpersons who have a positive attitude towards their cows have higher milk yields and fewer negative tactile interactions with cows, such as hitting. Hemsworth et al. (2000) further reported that stockperson training improved the stockperson's attitude towards the cows and subsequently improved productivity. The materials that these researchers have used to train stockpersons teach a lot of basic information about cow behavior. It is likely that learning more about how cows behave would make a person less likely to view an animal as a machine. From my own work on teaching people basic principles of beef handling, I have found that many people are surprised to learn that "dumb" cattle have so many different behaviors.

I have also observed that some management people in the livestock industry do not want to find out that a commonly used agricultural practice is either stressful or painful to an animal. For example, several studies have clearly shown that castrating piglets without an anesthetic is stressful and painful (McGlone and Hellman, 1988; McGlone et al., 1993; White et al., 1995). Some industry people have criticized researchers for doing studies on the use of anesthetics and analgesics for dehorning and castration because it would increase costs. My discussions with different researchers suggest that funding for research in areas where the results may force the change of a common agricultural practice is often difficult to get.

I have also observed that there is a certain percentage of physiologists and veterinarians who do not recognize the importance of behavior in assessing the pain or stress associated with certain procedures. The animal may be violently struggling and vocalizing but the physiologist will say, "It is not distressed because its physiological measures are low."

Would the physiologist say the same thing if a person was screaming in pain when a dentist drill hit a nerve? To address this issue, more research is needed on brain neurotransmitter systems so that behavior can be correlated with activity in specific brain systems, which are known to be associated with distress in humans.

While discussing distress in animals, it is important to separate the variables of pain and fear. Brain research has shown that the mechanisms and circuits that process pain and those that process fear are different. Fear is processed in more primitive lower brain areas than pain. Fear conditioning takes place in a subcortical circuit through the amygdala (LeDoux, 1996; Rogan and LeDoux, 1996). It can take place when the entire cortex is removed (Medina et al., 2002). The role of the frontal cortex is different for fear and pain. Activation of the frontal cortex is required to extinguish a conditioned fear response (Rogan and LeDoux, 1996). However, just the opposite happens for pain (Woolf, 1983). When the cortex is removed, a rat can no longer suffer from pain. Brain scan studies in humans show that activation of the frontal cortex makes pain worse and reduces fear. Much more research is needed to differentiate animal responses to pain or fear (Fischer et al., 1998; Apkarian et al., 2001; Fulbright et al., 2001). Both of these variables either singly or combined would create distress.

Another component of the broad term distress would be physical stresses such as fatigue or heat stress during a long truck ride. All of these variables would interact together to cause distress. For example, a tame animal that is accustomed to handling would have less fear during loading and unloading from a truck than an animal that is not accustomed to being handled (Grandin, 1997a). Antelopes that had been trained to voluntarily enter a crate for blood sampling had almost baseline cortisol levels and significantly lower glucose levels compared to animals immobilized with a dart (Phillips et al., 1998).

HOW TO ENCOURAGE SUCCESSFUL TRANSFER OF BEHAVIORAL BASED TECHNOLOGY TO THE INDUSTRY

Research results on animal handling, transport and stunning methods have been successfully transferred to the livestock industry in some areas. Hoenderken's (1982) and Gregory and Wotton's (1984) work on the electrical parameters for stunning pigs and sheep are used worldwide. Transportation guidelines for stocking density in trucks are also widely used (Warris, 1998; Knowles, 1998; Tarrant and Grandin, 2000). My own work on the design of cattle and pig handling systems for ranches, feedlots and slaughter plants is used worldwide (Grandin, 1980, 1982, 1992, 1997b, 199.8a, 2000a).

One of the reasons I was able to successfully transfer cattle-handling facility designs to the industry is that I wrote over a hundred articles in the livestock industry press on behaviorally based cattle-handling. I also posted the designs on my website http://www.grandin.com and gave talks at cattle producer meetings. I gave away the designs and made a living by charging for custom designs and consulting. People are often too reluctant to give information away. I discovered that when I gave out a lot of information, I got more consulting jobs than I could handle. However, it may be advantageous to keep quiet about new ideas when they are in the early development stage.

CASE HISTORY OF A SUCCESSFUL BEHAVIORALLY-BASED TECHNOLOGY TRANSFER

The center track or double rail conveyor restrainer for handling cattle in slaughter plants is now being successfully used in 26 plants in the United States, Canada and Australia. Half of the cattle in the US and Canada are handled in this system when they go to slaughter. The case history of this system is a good example of a technology that started in the research laboratory and was adopted by many of the world's largest beef plants.

The Council for Livestock Protection funded the original project during the early 1970s. The council was a consortium of US animal welfare groups, which included the Humane Society of the United States and the American Society of Prevention of Cruelty to Animals. In the early 1970s, the council gave a grant of US $60,000 to researchers at the University of Connecticut to develop a method to replace cruel shackling and hoisting of conscious calves and sheep by one rear leg, which was commonly used prior to kosher slaughter. The Connecticut researchers began with a complete search of all patents and literature to determine the "state of the art" prior to inventing new designs. A complete review of the literature is important to prevent "reinventing the wheel." The researchers invented the idea of having the calf or lamb straddle a moving conveyor and they were able to demonstrate that this method of restraint was low stress (Westervelt et al., 1976; Giger et al., 1977). The Council patented the system so that no one else could patent it and block transmission of the invention to the industry.

The Connecticut researchers initially developed a laboratory prototype, but many more components had to be invented to make a commercially usable system. In 1985, the Council for Livestock Protection gave another US $100,000 grant to design and build a system for a commercial veal slaughter plant. I was hired to do this job. I invented a new entrance design to. facilitate calf entry onto the conveyor and I added adjustable sides for different sized animals (Grandin, 1988). To make my design available, I published the drawings and placed them in the public domain. This prevented people outside the US from patenting it. The system was then installed in two other veal plants.

I knew that this system would also work really well for large cattle, but plant managers were reluctant to try the new design until I obtained a second grant from an independent animal welfare advocate in Florida. This made it possible to give a restrainer to a plant that was willing to pay for building remodeling costs. Since larger cattle are wilder and more difficult to handle than tame milk formula fed veal calves, I had to modify the system even further to keep the large cattle calm. I added a solid roof (hold down) over the head of the cattle to block their vision and to prevent them from seeing an escape route until they were fully restrained (Grandin, 1991). This roof did not press down and physically hold the animal down (Fig. 1), but simply blocked the animal's view of an escape route until the animal's back feet were off the entrance

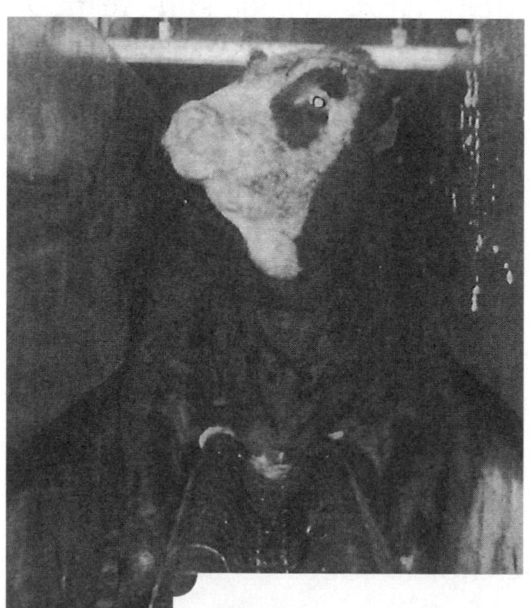

Fig. 1.
A steer sits quietly on the center track conveyor which is shaped to fit its brisket. The solid roof over the steer's head must be long enough to block his vision until his back feet are off the entrance ramp. The maximum width of the conveyor is 30 cm (reprinted from Grandin, 2000a).

ramp and he was completely settled down on the conveyor restrainer and supported by his brisket and belly. A solid false floor was also added to prevent animals from seeing the "visual cliff" effect as they entered the restrainer (Fig. 2). The conveyor was mounted 2 m

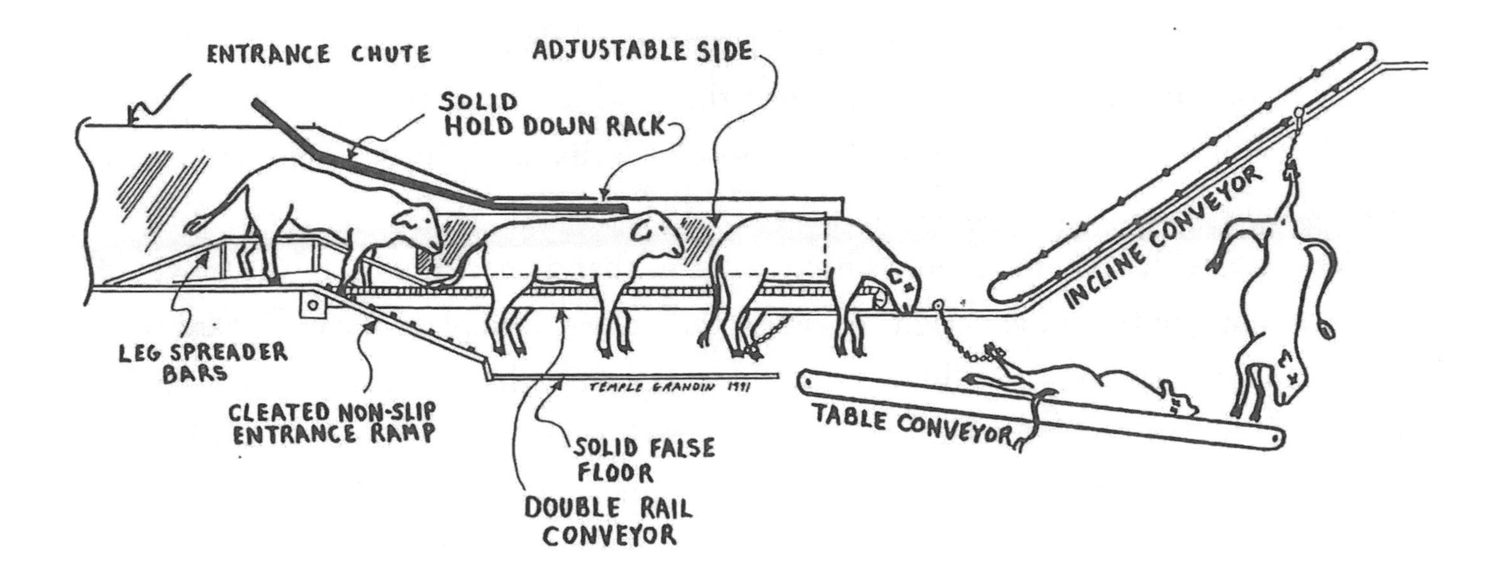

Fig. 2.
Diagram of the center track (double rail) conveyor restrainer system. A false floor below the animal's feet prevents incoming cattle from seeing a 2 m high drop off under the conveyor. A nonslip entrance ramp is essential. Cattle back out if they slip (reprinted from Grandin, 2000a).

off the floor and animals would often refuse to enter if they could see the steep drop off (Grandin, 2001). The false floor provided the optical illusion of a solid floor to walk on. As the animal entered, it was centered high on the moving conveyor and its feet were 20 cm above the false floor. These extra pieces of metal that controlled what the cattle could see were essential. Cattle remained calm and entered the restrainer easily when the vision blocking panels were installed. Prior to installation of the false floor and solid roof the cattle often became extremely agitated. Details of design are important.

Transferring this technology out of the first successful beef plant to the rest of the industry was very time consuming. I visited the next seven plants that installed the equipment to make sure it was installed correctly. Improper installation and bad modifications made by steel welding companies caused big problems. I had to be there at startup to correct these problems. Equipment installers often failed to install the false floor or shortened the overhead roof that blocked the cattle's vision. They did not understand a purely behavioral reason for having extra metal that would need to be cleaned. At the second plant, the welders shortened the solid roof, and the cattle often struggled. I demonstrated the need for a longer roof by laying a 70 cm x 60 cm wide piece of cardboard across the system to lengthen it. That piece of cardboard instantly made 450 kg cattle with large flight zones become calm. That demonstration of the power of behavioral principles in the design convinced the welders to replace the cardboard with additional metal sheets. Three different equipment construction companies did not build the false floor and I had to go to the plants and put it back on, because the cattle balked and refused to enter the restrainer and ride on the moving conveyor. The false floor was clearly marked on all the drawings provided to the equipment companies. Transfer of the restrainer technology to the industry could have failed if I had not been there to correct mistakes made by the early adopters.

One month before this lecture, I visited a recent restrainer installation. They had built the false floor and the overhead roof correctly, but they had omitted two belly rails that keep the cattle centered as they enter the restrainer. This is a third important behavioral component of the design. To induce the large cattle to straddle the leg spreader bar, the entrance race must be made narrower at the animal's belly height; and be wider at the floor to provide the animal with adequate space for walking (Fig. 3). When the belly rail was missing, the cattle had a tendency to walk on one side of the leg spreader bar instead of straddling it.

Working on the center track conveyor restrainer system convinced me of the power of using behavior principles for handling large cattle that were not accustomed to close contact with people. Some of the range cows successfully handled in this system were wild animals. The behavioral principles necessary for keeping the animals calm in this system are:

1. Block vision of an escape route until the animal is completely restrained.
2. Do not allow incoming animals to see the visual cliff effect.
3. Equipment must move with steady motion and have no jerky motion.
4. Slipping frightens animals and the system must have a nonslip entrance ramp.
5. Optimal pressure: The adjustable sides had to hold the animal snugly enough, so it felt held but not so tightly as to cause pain or discomfort.

HOW TO SUCCESSFULLY TRANSFER KNOWLEDGE

Scientists need to spend more time communicating with the public and the livestock community about the importance of their work. Doing this often requires writing many similar articles for the livestock magazines. The research must also be published in the scientific literature to provide a permanent, accessible record of the knowledge. This prevents knowledge from being lost.

Scientists also need to learn how to write without jargon. Scientific writing needs to be precise, but sometimes a simple word is just as precise. One of the things that I have learned during my career of successfully transferring behavioral knowledge is that transfer of the knowledge is more difficult and time consuming than doing the research that generates the knowledge. It is also extremely important that the first places that attempt to adopt the technology are successful. If the early adoptees of a technology fail, the word will get out and convincing more people to adopt the technology will be more difficult. To

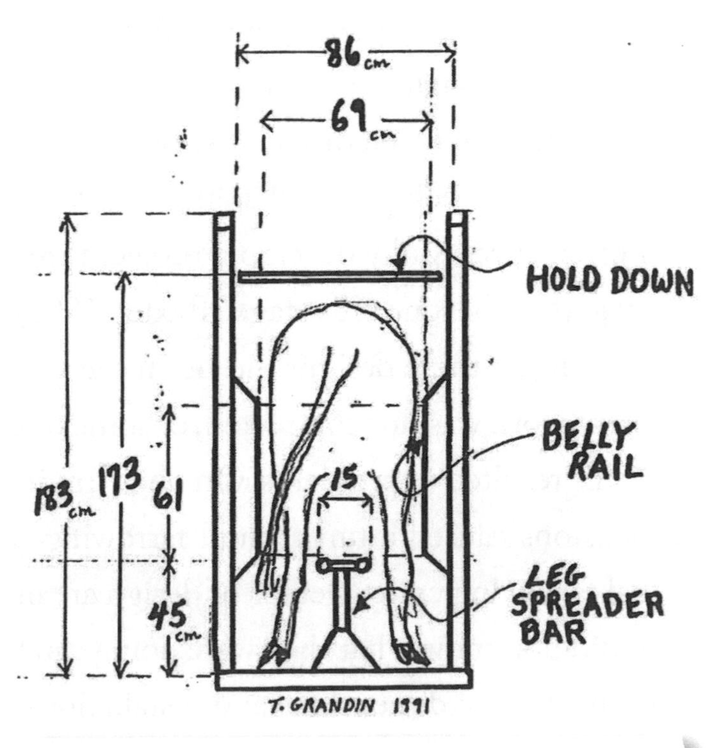

Fig. 3.
Belly rail that keeps the animal centered as it enters the center track conveyor.

ensure that the early adopters are successful requires a great deal of time working to ensure that the method or equipment is used correctly.

HOW TO AVOID FAILURE TO TRANSFER RESEARCH RESULTS TO INDUSTRY

Researchers and producers have developed a number of behaviorally-based designs for equipment that have not been adopted by the industry. These include several excellent farrowing stalls for sows that allow the sow to turn around. These stalls could be cost competitive, and they could be easily adopted. Another excellent behaviorally-based design that has not been adopted by the largest US pork companies is a feeder design based on the ergonomics of pig eating behavior that prevents the pig from wasting feed. A kinematic analysis of the saw's eating movements was used to identify the spatial requirements for optimal feeder design and reduced feed wastage (Taylor et al., 1986, 1987, 1989ab, and Taylor, 1990). Many commonly used feeder designs have feed wastage of up to 20%.

Why are these excellent and cost effective behaviorally-based designs not being widely used? Baxter states that one difficulty is that there is a fundamental difference between the approach of scientists and designers to a problem. Scientists seek knowledge and designers seek practical solutions. My own observations have shown that scientists can often successfully develop a project through the prototype stage, and they have difficulty getting beyond this stage (Baxter, 1995).

Maybe these designs did not receive enough promotion, or some of the final development work was not completed, or some of the early adopters tried to modify the design and this resulted in problems. On one farm in the US, I know that changes in some of the dimensions caused a turnaround farrowing stall to perform more poorly than a conventional farrowing crate. Details of design are important, and modifications made by people who did not know what they were doing can block the transfer of promising technologies. Modification of dimensions and installation mistakes can cause systems installed by early adopters to fail.

Even today I have to constantly check up on clients to make sure they build my cattle systems correctly. In June 2001, I checked up on a feedlot that was building a new curved race system. A draftsman had made a drawing mistake that would have mined the system. Successful transfer of technology requires constant checking of what people are doing in the field.

Some technologies do not get adopted due to patent fights when one company buys up the patent rights to block adoption of a technology. Another business entity that has impeded adoption of some improved designs is building contractors. Building contractors design things for ease of construction. For example, ventilation in many animal buildings in the US has taken some steps backwards. They are designed for ease of construction and to sell fans instead of proper ventilation. For example, a naturally ventilated building with a pitched roof and a wide ridge vent will stay cooler in the summer than a building with a flatter roof and a small ridge vent. Contractors prefer to build the flatter roof and a small ridge vent because it is easier to build and is more profitable for them. To compensate for the poorer air movement, they have to install fans. Contractors will often design things to benefit contractors instead of benefiting the animals. I could provide countless other examples. The solution is for the owner to insist that the contractor build from designs that the owner supplies. My best jobs were built by owners who insisted that the contractor not deviate from my specifications.

HOW TO MAINTAIN EXCELLENT STOCKMANSHIP AND ANIMAL HANDLING

As stated previously, people are often more willing to purchase new technology than to make a sustained commitment to better behavioral management of animals. Good stockmanship and handling is impossible if a facility is understaffed, and the people are overworked. I have also observed that the method of payment affects the quality of handling. Cattle have significantly fewer bruises when the producer is held financially accountable for bruises (Grandin, 1981). Handling animals on a "piece work" basis where speed is rewarded encourages rough handling. I have seen many cattle severely injured when operation of a squeeze chute for vaccinating cattle is based solely on the numbers of cattle vaccinated per hour.

I have trained many people to handle cattle and pigs in a careful, quiet manner. For a few weeks following training, the handling practices would be good but in many cases the people reverted back to rough practices such as excessive electric prod (goad) use. I call this "bad becoming normal." I have observed that people often lapse back into old bad habits, and they do not even realize it. One of the reasons that this occurs is because the handlers have no standard to compare their performance against. To combat this problem, I developed a simple scoring system for animal handling at the slaughter plant (Grandin 1998b,c, 2001).

The scoring system enables management to quantify variables such as the percentage of animals shocked with an electric prod (goad), the number of animals that fall down, the percentage of cattle that vocalize during handling and the percentage of animals stunned correctly. These variables were chosen after I collected data on over 1000 cattle at slaughter plants throughout the US (Grandin, 1998b). Cattle vocalizations during movement through the chutes and during stunning are highly correlated with aversive events. Out of 1125 cattle, 98% of the vocalizations during active handling were associated with an aversive event such as an electric goad, missed stuns, slipping and falling or excessive pressure from a restraint device (Grandin, 1998c). A further study of over 5000 cattle in 40 commercial slaughter plants indicated that cattle vocalizations were associated with equipment problems (Grandin, 2001). Vocalizations during handling or during surgical procedures are highly correlated with physiological measures of stress (Dunn, 1990; Warriss et al., 1994; White et al., 1995). People manage the things that they measure. Measurement also provides a benchmark that enables people to see if their performance has improved or become worse.

Other potential methods for objectively scoring cattle handling on farms and ranches are a police radar speed camera or constructing a squeeze chute with sensors to monitor how hard an animal struggles (Burrow and Dillon, 1997; Schwartzkoph-Genswein et al., 1998). Burrow and Dillon (1997) found that cattle that ran more quickly out of a squeeze chute in which they were restrained for vaccination had lower weight gains. Animals that have been shocked repeatedly will run faster out of a squeeze chute and are more likely to struggle. Exit speed from the squeeze chute is already being used on farm audits. To simplify the audit, each bovine is scored as exiting at either a walk, trot or canter.

Animal welfare legislation and requirements specified by large customers such as supermarkets and restaurants serve as powerful motivators to use behavioral methods. In Europe, legislation has prompted the use of research results on welfare friendly animal housing. In the United States, welfare requirements of the McDonald's Corporation and Wendy's International have greatly improved handling and stunning of cattle and pigs at slaughter plants (Grandin, 2000b). These restaurant companies are using the scoring system I developed for monitoring animal welfare in slaughter plants (Grandin, 1998b). When a restaurant auditor visits a plant, 100 cattle or pigs are scored. For cattle, the minimum passing scores are 95% of the cattle stunned with one shot and 100% rendered insensible on the bleed rail. Seventy-five percent of the cattle must be moved through the entire system without an electric goad and only 3% can vocalize. During the year 1999, a total of 48 plants were audited. Three percent or less of the cattle vocalized in 67% of the plants. Plants that repeatedly failed audits have been suspended for several months or removed from the approved supplier list. This provides a strong economic incentive to handle animals carefully. In 1996, only 30% of the plants stunned 95% or more of the cattle with one shot, and after McDonald's started these audits, 90% of the plants passed (Grandin, 1998b, 2000b). For the last 3 years this level of performance has been maintained through the use of yearly audits. Implementing the scoring system required many days of work. I visited over 30 beef and pork slaughter plants to teach restaurant auditors how to use the scoring system. Audits by McDonald's and Wendy's have greatly improved animal handling and stunning in US slaughter plants (Grandin, 2000b). To achieve compliance, most plants have implemented employee training and their quality assurance department conducts weekly audits using the same scoring system.

Plant managers had to make sure their employees used behavior principles in order to reduce or eliminate the use of electric prods so that they were in compliance with the welfare guidelines of their customers (i.e. Grandin, 1980, 1982, 1992). Making several simple changes in the design of the handling area such as lighting or plant ventilation made it possible to reduce or eliminate electric prod usage (Grandin, 1996, 2001). Some of the changes that improved animal movement and made it possible to reduce electric prod use were, installing solid sides on races, lighting a dark restrainer conveyor entrance, moving lamps to eliminate reflections on a wet floor, eliminating air drafts that blew in the faces of approaching animals and filling the crowd pen that leads to the single file race only half full instead of stuffing it full. With the exception of two plants, all of the 46 plants were able to comply with the standards without making large capital investments. The purchase of lamps, inexpensive nonslip floor gratings or shields to block animal vision were usually the only things they had to buy.

Another good motivator for good stockmanship is financial incentive resulting from improvements in animal productivity or from a reduction in bruises or broken bones. I have been shown internal company figures that indicate that in the poultry industry, paying catchers an incentive for reducing wing breakage resulted in a reduction in broken wings of over 100%. Poultry companies in both the US and England have implemented incentive pay based on reducing broken wings and dead birds. At one company, broken wings were reduced from 3 to 41 % on a per bird basis. Similar economic incentives could be applied on farms. A farrowing manager will work extra hard if he/she receives extra money for weaning more piglets. Milkers could be paid incentives for increased milk production.

CONCLUSIONS

Scientists need to take more initiative to get their research results adopted by the industry. From my experience, the necessary steps for successful transfer of behavioral research results to the industry are:

1. Communicate your research results by speaking to producer groups, writing in producer magazines and making websites. Results should be published in refereed scientific journals because these papers are easier to access in libraries.

2. Find a place (farm or plant) that will try your research findings and be prepared to spend a lot of time there. This place must have cooperative management. The manager of the place must believe in what you are attempting to accomplish.
3. You must supervise installation and correct mistakes made by the early adopters of your methods or technology.
4. Do not allow your research findings to get tied up in a patent dispute because this often blocks adoption of your research findings. This is especially important for behavioral research. Make your money with consulting fees to help people to use your research.

REFERENCES

Albright, J.L., 1978. The behavior and management of high yielding dairy cows. In: *Proceedings of the British Oil and Cake Mills (BOCM)*, Silcock Conference, Silcock, London, UK, p. 31.

Apkarian, A.V., Thomas, P.S., Krauss, B.R., Szeverenyi, N.M., 2001., Prefrontal cortical hyperactivity in patients with sympathetically mediated chronic pain. *Neurosci. Lett.* 311, 193-197.

Baxter, M., 1995. There is more to design than behavior. In: *Proceedings of the Animal Behavior and Design of Livestock and Poultry Systems.* Northeast Regional Agricultural Engineering Service, Cooperative Extension, Cornell University, Ithaca, NY, pp. 73-82.

Burrow, H.M., Dillon, R.D., 1997. Relationship between temperament and growth in a feedlot and commercial carcass traits in *Bos indicus* crossbreds. *Aust. J. Exp. Agric.* 37, 407-411.

Dunn, C.S., 1990. Stress reactions of cattle undergoing ritual slaughter using two methods of restraint. *Vet. Rec.* 126, 522-525.

Fell, L.R., Colditz, LG., Walker, K.H., Watson, D.L., 1999. Associations between temperament, performance and immune function in cattle entering a commercial feedlot. *Aust. J. Exp. Agric.* 39, 795-802.

Fischer, H., Andersson, J.L., Furmark, T., Fredrikson, M., 1998. Brain correlates of an unexpected panic attack: a human positron emission tomographic study. *J. Neurosci. Lett.* 251, 137-140.

Fraser, D., Phillips, P.A., Thompson, B.K., 1988. Initial test of a farrowing crate with inward sloping sides. *Livestock Prod. Sci.* 20, 249-256.

Fulbright, R.K., Troche, C.J., Skudlarski, P., Gore, J.C., Wexler, B.E., 2001. Functional MRI imaging of regional brain activation associated with the effective experience of pain. *Am. J. Roetgenol.* 177, 1205-1210.

Giger, W., Prince, R.P., Westervelt, R.G., Kinsman, D.M., 1977. Equipment for low stress animal slaughter. Trans. *Am. Soc. Agric. Eng.* 20, 571-578.

Grandin, T., 1980. Observations of cattle behavior applied to the design of cattle handling facilities. *Appl. Anim. Ethol.* 6, 1931.

Grandin, T., 1981. Bruises on southwestern feedlot cattle. *J. Anim. Sci.* 53 (1), 213 (Abstract).

Grandin, T., 1982. Pig behavior studies applied to slaughter plant design. *Appl. Anim. Ethol.* 9, 141-151.

Grandin, T., 1987. Animal handling. *Vet. Clin. North Am. Food Anim. Pract.* 3, 323-328.

Grandin, T., 1988. Double rail restrainer conveyor for livestock handling. *J. Agric. Eng. Res.* 41, 327-338.

Grandin, T., 1991. Double rail restrainer for handling beef cattle. Paper No. 915004. American Society of Agricultural Engineers, St. Joseph, MI.

Grandin, T., 1992. Observations of cattle restraint devices for stunning and slaughtering. *Anim. Welfare* 1, 8591.

Grandin, T., 1996. Factors that impede animal movement at slaughter plants. *J. Am. Vet. Med. Assoc.* 209, 757759.

Grandin, T., 1997a. Assessment of stress during handling and transport. *J. Anim. Sci.* 75, 249-257.

Grandin, T., 19976. The design and construction of facilities for handling cattle. *Livestock Prod. Sci.* 49, 119.

Grandin, T., 1998a. Handling methods and facilities to reduce stress in cattle. *Vet. Clin. North Am. Food Anim. Pract.* 14, 325-341.

Grandin, T., 19986. Objective scoring for animal handling and stunning practices in slaughter plants. *J. Am. Vet. Med. Assoc.* 212, 3693.

Grandin, T., 1998c. The feasibility of using vocalization scming as an indicator of poor welfare during slaughter. *Appl. Anim. Behav. Sci.* 56, 121-128.

Grandin, T., 2000a. *Livestock Handling and Transport*, second ed. CAB International, Wallingford, Oxon, UK.

Grandin, T., 2006. Effect of animal welfare audit of slaughter plants by a major fast-food company in cattle handling and stunning practices. *J. Am. Vet. Med. Assoc.* 216, 848-851.

Grandin, T., 2001. Cattle vocalizations are associated with handling and equipment problems at beef slaughter plants. *Appl. Anim. Behav. Sci.* 71, 191-201.

Gregory, N.G., Wotton, S.B., 1984. Sheep slaughtering procedures. III. Head to back electrical stunning. *Br. Vet. J.* 140, 570-575.

Hemsworth, P.H., Coleman, G.J., 1998. *Human Livestock Interactions: The Stock Person and the Productivity and Welfare of Intensely Fanned Animals.* CAB International, Wallingford, Oxon, UK.

Hemsworth, P.H., Barnett, J.L., Hensen, C., 1981. The influence of handling by humans on the behavior, growth and carticostenoids in the juvenile female pig. *Horm. Behav.* 15, 396-403.

Hemsworth, P.H., Coleman, G.J., Barnett, J.L., Borg, S., 2000. Relationships between human animal interactions and productivity of commercial dairy cows. *J. Anim. Sci.* 78, 2821-2831.

Hemsworth, P.H., Coleman, G.L., Barnett, J.L., Borg, S., Dowling, S., 2002. The effects of cognitive behavioral intervention on the attitude and behavior of stockpersons and the behavior and productivity of commercial dairy cows. *J. Anim. Sci.* 80, 68-78.

Hoenderken, R., 1982. Electrical and carbon dioxide stunning of pigs for slaughter. In: Eikelenboom, G. (Ed.), *Stunning Animals for Slaughter.* Martinus-Nijhoff, Dorderecht, pp. 59-63.

Knowles, T.G., 1998. A review of road transport of sheep. *Vet. Rec.* 144, 197201.

LeDoux, J.E., 1996. *The Emotional Brain.* Simon & Schuster, New York.

Lou, Z., Hurnik, J.P., 1994. An ellipsoid furrowing crate: its ergonomical design and effects on pig productivity. *J. Anim. Sci.* 72, 26102616.

McGlone, J.J., Hellman, J.M., 1988. Local and general anesthetic effects on behavior and performance of 2- and 7-week-old castrated and non-castrated piglets. *J. Anim. Sci.* 66, 3049-3058.

McGlone, J.J., Nicholson, RI., Hellman, J.M., Herzog, D.N., 1993. The development of pain associated with castration and attempts to prevent castration induced behavioral changes. *J. Anim. Sci.* 71, 1441-1446.

Medina, J.P., Repa, J.C., Mauk, M.D., LeDoux, J.E., 2002. Parallels between cerebellar and amygdala dependent conditioning. *Nat. Rev. Neurosci.* 3, 122-131.

Munksgaard, L., dePasille, A.M., Rushen, J., Harkin, M.S., Kristensen, A.M., 2001. Dairy cows fear of people. Social learning, milk yield and behavior at milking. *Appl. Anim. Behav. Sci.* 73, 15-26.

Phillips, M., Grandin, T., Graffam, W., Irlbeck, N.A., Cambre, RC., 1998. Crate conditioning of bongo (*Tragelaphus euycerus*) for veterinary and husbandry procedures at Denver Zoological Garden. *Zoo Biol.* 17, 25-32.

Rankin, G.W., 1925. *The Life of William Dempster Hoard.* W.D. Hoard and Sons Press, Fort Atkinson, WI.

Rogan, M.T., LeDoux, J.E., 1996. Emotion: systems, cells, synaptic plasticity. *Cell* 85, 469-475.

Rushen, J., dePasille, A.M.B., Munksgaard, L., 1999. Fear of people by cows and effects on milk yield behavior and heart rate at milking. *J. Dairy Sci.* 82, 720-727.

Schwartzkoph-Genswein, K.S., Stookey, J.M., Crowe, J.M., Genswein, RM., 1998. Comparison of image analysis, exertion force and behavior measurements for assessment of beef cattle response to hot iron and freeze branding. *J. Anim. Sci.* 76, 972-979.

Seabrook, M.F., 1972. A study to determine the herdsman's personality in milk yield. *J. Agric. Labor Sci.* 1, 45-59.

Tanant, V., Grandin, T., 2000. Cattle transport. In: Grandin, T. (Ed.), *Livestock Handling and Transport.* CAB International, Wallingford, Oxon, UK, pp. 151-173.

Taylor, I.A., 1990. Design of the sow feeder: a systems approach. University Microfilms International, Ann Arbor, MI (Ph.D. Dissertation, University of Illinois, Urbana Champaign, IL).

Taylor, I.A., Curtis, S.E., 1989a. Grower Feeders: 11 Designs Reviewed. *National Hog Farmer.* December 1989.

Taylor, I.A., Curtis, S.E., 1989b. Nursery Feeders: Researchers Find Feed Waste Ranges from 2% to 11% in Tests. *National Hog Fanner.* May 1989.

Taylor, I.A., Curtis, S.E., Feinmehl, RI., 1986. Amounts of feed mated gilts spilled and left with six feeder models. *J. Anim. Sci.* 63 (1), 96 (Abstract).

Taylor, I.A., Groppel, J.L., Curtis, S.E., 1987. Kinematic analysis of eating movements in sows. *J. Anim. Sci.* 65 (1), 233 (Abstract).

Taylor, I.A., Curtis, S.E., Taylor, K., Russell, K., 1989. Feed wastage from eleven commercial models of nursery feeder. *J. Anim. Sci.* 67 (Suppl. 1), 94 (Abstract).

Voisinet, B.D., Grandin, T., Tatum, J.D., O'Connor, S.F., Struthers, J.J., 1997. Feedlot cattle with calm temperaments have higher average daily gains than cattle with excitable temperaments. *J. Anim. Sci.* 75, 892-896.

Warris, P.D., 1998. Choosing appropriate space allowances for slaughter pigs transported by road: a review. *Vet. Rec.* 142, 449-454.

Warriss, P.D., Brown, S.N., Adams, S.J.M., 1994. Relationship between subjective and objective assessment of stress at slaughter and meat quality of pigs. *Meat Sci.* 38, 329-340.

Westervelt, R.G., Kinsman, D., Prince, R.P., Giger, W., 1976. Physiological stress measurement during slaughter of calves and lambs. *J. Anim. Sci.* 42, 831-834.

White, R.C., DeShazer, J.A., Tressler, C.J., Borches, G.M., Davey, S., Waninge, A., Parkhust, A.M., Milanuk, M.J., Clems, E.I., 1995. Vocalizations and physiological response of pigs during castration with and without anesthetic. *J. Anim. Sci.* 73, 381-386.

Woolf, C.J., 1983. Evidence for central component of postinjury pain hypersensitivity. *Nature* 306, 686.

TWENTY-ONE

Special Report Maintenance of Good Animal Welfare at Beef Slaughter Plants by Use of Auditing Programs

T. Grandin (2005) Special Report Maintenance of Good Animal Welfare at Beef Slaughter Plants by Use of Auditing Programs, *Journal of the American Veterinary Medical Association*, Vol. 226, No. 3, pp. 370-373.

In this paper I summarize the results of the first four years of the welfare audits done by restaurant companies. There were huge improvements compared to my USDA survey that was published before the restaurant audits started. Today I have been in beef plants where there have been further improvements in the percentage of cattle rendered unconscious with one application of a captive bolt. Many plants are at 99 percent and above. The few animals that are missed are immediately re-stunned. Continuing audits by major buyers and increased enforcement by the USDA are helping to keep standards high.

The public has become increasingly concerned about poor farm animal welfare. Although all veterinarians do not work with farm animals, it is important that all can answer questions on that subject from their clients or others. One area of concern is how animals are treated at slaughter plants. Many people are not aware that restaurant companies have been auditing animal welfare in slaughter plants since 1999. This has resulted in great improvements in cattle handling and stunning. The McDonald's Corporation started their animal welfare auditing program in 1999. Wendy's International and Burger King Corporation began their programs in 2000 and 2001, respectively. The author worked with all 3 companies on the implementation of their programs.

Restaurant companies use their own auditors (who visit the plants routinely for food safety audits) to perform their welfare audits. During the last 5 years, the author has worked with McDonald's Corporation, Burger King Corporation, and Wendy's International to train food-safety auditors to score each slaughter plant for animal welfare by use of an objective numerical scoring system.[2,3] Instead of relying on a subjective assessment, the auditor counts the percentage of cattle that was correctly stunned with 1 shot, percentage moved with an electric prod, and percentage that vocalized (moo or bellow) during handling and stunning. Vocalization is associated with blood cortisol concentrations in cattle.[4] A previous study[5] revealed that 98% of the cattle that vocalized during handling and stunning in a slaughter plant had vocalizations that were associated with an obvious aversive event, such as nonfatal attempts at stunning, slipping, use of electric prods, contact with sharp metal edges, or excessive pressure from a restraint device.

Prior to the start of the audits in 1999, conditions in beef slaughter plants were poor. A survey conducted for the USDA in 1996 indicated that only 3 of 10 plants were able to correctly stun 95% of the cattle with a single shot,[3] which is an industry guideline for animal welfare.[2] In 1999, the percentage of beef plants that could attain this score rose to 90%.[1] The main cause of poor stunning prior to 1999 was failure of slaughter plants to perform factory-specified maintenance on stunners. Only approximately half of the plants maintained and operated their equipment correctly. In some plants, there was little or no supervision of the employees who stunned cattle.

When the first audits started in 1999, they resulted in plant management personnel making cattle welfare a priority. Many plant managers have now implemented their own internal self-audits, which occurred after McDonald's Corporation removed a major beef plant from its approved suppliers list in 1999. The purpose of this report was to review 4 years of animal welfare audit data to determine whether improvements in cattle welfare have been maintained since 1999.

MATERIALS AND METHODS

Beef slaughter-plant audit data from McDonald's Corp, Burger King Corp, Wendy's International, and authors were compiled. All auditors were trained by the author at a minimum of 3 slaughter plants each. The auditors had academic degrees in either food science or animal science and were all trained and experienced food safety auditors in beef and pork slaughter plants. The data was collected during 1999, 2000, 2001, and 2002. During the first 2 years, data were collected by McDonald's auditors in 49 beef plants in 12 states. Forty-one plants were audited during 1999 and 49 plants during 2000. During each of the next 2 years, data were collected in 57 beef plants in 17 states during visits by auditors from McDonald's Corporation, Wendy's International, Burger King Corporation, or the author. Overall, 66 different plants were audited during the 4-year period. Some plants were not audited every year because of plant closings, removal from approved suppliers lists, or scheduling problems. There were 28 (42%) beef-fed slaughter plants and 38 (58%) plants that processed cull beef cows or old Holstein dairy cows and bulls. The audited plants constitute approximately 90% of the large and medium-sized slaughter plants in the United States.

Data collected during 1999 (41 audits) and 2000 (49 audits) were from a single audit in each slaughter plant. Data collected during 2001 (79 audits) and 2002 (97 audits) consisted of a single audit or multiple audits of each slaughter plant. The slaughter plants were scored on certain variables. The percentage of cattle stunned correctly (i.e., no signs of return to sensibility) with the first attempt was recorded.[1] The percentage of cattle moved with an electric prod and the percentage of cattle that vocalized during handling and stunning were also recorded. In plants that had >1 audit/year, mean data were calculated so that each plant would be evenly represented. During each visit, each plant was also classified as either passing the insensibility audit (i.e., 100% of the cattle rendered completely insensible prior to hoisting on the bleed rail) or failing the insensibility audit.

A plant failed on insensibility if any sign of return to sensibility after hoisting, such as eye reflexes, blinking, breathing, or a righting reflex, was observed. Stunning scores and insensibility data were not recorded in 3 kosher plants, although the kosher plants were evaluated for electric prod use and vocalization. Each animal was scored on a yes or no basis for each variable. Stunning and insensibility data were not collected in the kosher plants because slaughter was performed without stunning. Insensibility is not instantaneous after kosher slaughter.

An animal was scored as a vocalizer if it made any vocal sound when it was in the stunning box, restrainer, or lead-up chute leading to the stunning box. Vocalizations of cattle standing undisturbed in the yards were not counted. The animal was scored as moved by an electric prod if it was touched with an electric prod because it was difficult to determine when a shock was given. Balking and cattle refusing to enter the stunning box or restrainer were subjectively evaluated by noting animals that backed up frequently in the chute.

Each slaughter plant was classified as either a beef-fed plant or a cow-bull plant and placed in 1 of 5 line-speed categories of < 50 cattle/h, 51 to 100 cattle/h, 101 to 200 cattle/h, 201 to 300 cattle/h, and > 300 cattle/h. Slaughter plants were also classified into 3 equipment types regarding the manner of holding the animal during stunning, including conventional stun boxes that held only 1 animal, a center track conveyor restrainer, and a V-conveyor restrainer. In the older V-conveyor restrainer system, the cattle are held between 2 moving conveyors that form a V, and in the more modern center-track restrainer, the cattle straddle a moving conveyor.[6,7]

None of the non-kosher beef slaughter plants used a head restraint device to hold the head during stunning. In the 3 kosher plants, a head restrainer was used, and the animal was held in an upright position during kosher slaughter. In plants with line speeds of > 100 cattle/h, 100 cattle were observed for the audit; in plants with lower line speeds, either 50 animals or 1 hour of production was observed. These sample sizes were chosen on the basis of time constraints of the auditors; the auditors had to perform the animal welfare audit and a food-safety audit in 1 day Descriptive statistics were calculated for all variables.

RESULTS

During the 4-year auditing period, a mean SD of 97.2 ± 6.21% (mode, 99%; median, 98%) of the cattle were correctly stunned with the first attempt. Animals that were not correctly stunned were immediately re-stunned before hoisting or bleeding. By year, the mean percentages of cattle that were insensible after the first stunning attempt were 96.2% for 1999, 98.9% for 2000, 97.4% for 2001, and 96.7% for 2002. Thus, the initial improvements noted during the audits in 1999 had been maintained. On the basis of a voluntary industry standard that was adopted by the restaurant companies, a satisfactory score is considered to be 95%. The line speed varied from < 50 to 390 cattle/h. Line speed was < 50 cattle/h in 16 (24%) of the slaughter plants, 51 to 100 cattle/h in 13 (20%) plants, 101 to 200 cattle/h in 10 (15%) plants, 201 to 300 cattle/h in 21 (32%) plants, and > 300 cattle/h in 6 (9%) plants. The mean stunning scores for each line-speed category were 96.2%, 98.9%, 97.4%, and 96.7%, respectively.

The percentages of cattle stunned correctly with 1 attempt in each type of restraint equipment were 97.2% for conventional stunning boxes, 97.1 % for center track conveyor restrainers, and 97.0% for V-conveyor restrainers. Forty (61%) plants had conventional stunning boxes that held I animal, 20 (30%) plants had a center track conveyor restrainer, and 6 (9%) plants had a V-conveyor restrainer. All plants with a line speed of < 100 cattle/h had conventional single animal stun boxes.

All plants used either a cartridge-fired, hand-held stunner or a pneumatic stunner. Both types of stunners had a penetrating captive bolt. None of the plants used a nonpenetrating mushroom head stunner. In the year 2000, the use of pneumatic stunners that injected air into the brain was discontinued because of concerns about bovine spongiform encephalopathy and the possibility that such stunners would cause contamination of meat with brain tissue. During 2001 and 2002, specific data on stunner type and brand were not collected. However, observations by the auditors indicated that between 2001 and 2002, many plant maintenance departments started using a testing device to measure bolt velocity to ensure the stunners were operating correctly.

Among all slaughter-plant audits during the 4-year period (n = 266), the insensibility audit failed during 9 (3%) of the audits and passed during 257 (97%) of the audits. In the plants that failed the insensibility audit, only 1 animal with signs of sensibility was found in each slaughter plant. Among the 66 slaughter plants, 60 (91%) passed all of their insensibility audits. The 6 plants that failed consisted of 2 beef-fed plants and 4 cow-bull plants. Three of 4 of these cow-bull plants failed more than 1 audit. Stunning of bulls was the cause of failure in 2 of 4 cow-bull plants. By year, the number of audits in which slaughter plants passed the audit was 40 of 41 (98%) during 1999, 47 of 49 (96%) during 2000, 75 of 79 (95%) during 2001, and 94 of 97 (97%) during 2002.

Mean ± SD vocalization score for all 4 years was 2.14 ± 2.63% of the cattle vocalizing during stunning and handling (mode, 0%; median, 2%). By year, mean vocalization scores were 2.4% for 1999, 1.8% for 2000, 2.2% for 2001, and 1.7% for 2002. On the basis of the accepted industry standard, a vocalization score of < 3% was considered acceptable. The mean vocalization score in 1996 before the audits began was 7.7%. The worst slaughter plant in 1996 had a vocalization score of 32%. The worst vocalization score in 1999 was 17% and in 2002 was 6%. Mean vocalization scores for the different types of cattle holding equipment were 2.5% for conventional stun boxes, 1.1% for center track restrainers, and 2.5% for V-conveyor restrainers.

No data were collected on electric prod type or voltage. During the 4-year period, a mean SD of 17.2 25.7% of the cattle were moved with an electric prod (mode, 0%; median, 6%). By year, 28.6% of cattle were prodded during 1999, 19.9% during 2000, 17.1% during 2001, and 21.3% during 2002. The percentages of slaughter plants that met the industry standard of < 25% of cattle prodded were 76% in 1999, 67% in 2000, 76% in 2001, and 82% in 2002. Electric prods were used on 15.2% of the beef-fed cattle and 29% of the cows and bulls. Observations by auditors suggested that Holstein dairy cows balked more than beef-fed cattle (steers and heifers). The percentage of cattle moved with an electric prod for each line-speed category was 19.8%, 27.0%, 12.5%, 24.1%, and 25.1%, respectively. Mean percentage of cattle moved with an electric prod in each restrainer type was 10.1% for conventional stun boxes, 16.0% for center track conveyor restrainers, and 39.1% for V-conveyor restrainers.

DISCUSSION

Evaluation of animal welfare audit data suggested that improvements in stunning and handling of cattle that started in 1999 were maintained through 2002. Plant management personnel and equipment suppliers have made many additional improvements that could not be numerically quantified. Some slaughter plants now conduct their own internal weekly or monthly animal welfare audits with the numerical scoring system. The manufacturers of stunners have responded to the increasing concern about animal welfare by developing equipment to help slaughter-plant maintenance departments do a better job of maintaining stunners. At the start of the audits in 1999, only I manufacturer of stunners marketed a test stand used to measure bolt velocity. Measuring bolt velocity enables maintenance personnel to determine if a stunner is functioning correctly. Today, test stands are commercially available for all 4 of the commonly used stunners. The maintenance of stunners has improved, and many plant managers have implemented documented maintenance programs. Prior to 1999, inadequate stunner maintenance was a major cause of poor stunning. It is the author's opinion that during the last 4 years, stunning of bulls and cows with heavy skulls is still a problem area. Methods to correct stunning problems are outlined in other publications.[8,9] During 2003, many slaughter plants installed a new, more powerful pneumatic stunner that may reduce problems with stunning bulls and cows with heavy skulls.

Maintenance of chutes, stunning boxes, and restrainers has also improved during the last 4 years. Many slaughter plants have relocated switches and valves that operate gates or restrainers in more convenient locations. Ergonomic handles have been installed on bulky pneumatic stunners. These improvements were made when management personnel began to give higher priority to animal welfare.

Many small changes can yield big improvements in the movement of cattle in slaughter plants. Cattle that move easily without balking are easier to stun because they are less likely to be agitated. Almost every slaughter plant has removed distractions that make cattle balk and refuse to move. They have moved lights to eliminate shiny reflections on wet floors. Lamps have been placed on dark restrainer entrances to facilitate cattle entry because cattle will often refuse to enter a dark place.[9] In plants in which approaching cattle could see people ahead, shields have been installed to block the cattle's view Some plants have modified their ventilation systems to prevent air drafts from blowing into the faces of approaching cattle because this may cause cattle to balk. Another modification made on several conveyor restrainer systems was to add a false floor to prevent incoming cattle from experiencing the visual cliff effect.[7] These systems are located 2 to 3 m off the floor. If cattle look down and see a steep drop-off, they often refuse to enter.[10] Ruminants have depth perception.[11]

In 2 slaughter plants, data were collected before and after some simple, inexpensive improvements were made. At the first plant, 4 modifications of the stunning box were made, including nonslip flooring and a light over the stun box entrance. The lamp was positioned to provide indirect illumination of the stun box. The third and fourth modifications were rubber stops to prevent gates from banging and a small (2.5 X 15-cm) hole cut at the cow's eye level in the solid stun box entry door to promote following (the hole must not provide a view into the slaughter plant). The plant's scores improved; stunning score was 93% before the changes and 97% after the changes, vocalization caused by electrical prodding was reduced from 4% to 1% of the cattle, and electric prod use decreased from 31% to 7% of the cattle. In the second plant, air blowing toward the cattle through the center track conveyor restrainer entrance was stopped. This resulted in less electric prodding, and vocalization decreased from 4.5% to 1% of the cattle.

Slaughter-plant management personnel have also improved employee training, and the American Meat Institute (the meat packer trade association) sponsors a 3-day animal welfare conference to train managers. Most plant personnel now use alternative nonelectric aids to reduce the use of electric prods. Flags, inflated plastic bags, and plastic paddle sticks are the 3 most popular aids. These should be used quietly to guide cattle and be the primary aids for moving cattle. In most plants, electric prod use is restricted to the stunning box or restrainer entrance. The electric prod is used only when it is needed to move a stubborn animal. One plant maintenance department invented an effective driving aid that produces a strong vibration that is not painful.

Some problem areas were recognized. A few slaughter plants had chronic problems and either barely passed an audit or failed by 1 or 2 percentage points. The main problem in these plants was that management personnel did not give animal welfare high priority. During the last 4 years, 4 plant managers were replaced after audits failed- problems in those plants were eliminated with new management. In 2002, 4 new beef plants entered the restaurant audit system and 3 of 4 failed the audit. One plant had a 19% stunning score; that slaughter plant had an untrained operator, a broken stunner, and no knowledge of what was expected during an audit.

The USDA has also increased enforcement of the Humane Slaughter Act. In 2002, the USDA hired 18 veterinarians whose main responsibility is to travel to slaughter plants to enforce humane slaughter regulations. Discussion with those veterinarians and telephone calls to the author from slaughter plants inspected by those veterinarians indicated that many of the problems occurred in smaller plants that were not audited by a restaurant company. In 2002 and 2003, the USDA increased enforcement of humane handling practices during truck unloading. Until that time, truck unloading at a slaughter plant was not monitored by the USDA.

Audits performed by the restaurant companies have maintained the improvements that were started in 1999. Basing the audits on easily measured performance standards has helped maintain the improvement. Slaughter-plant managers know exactly what is expected and can monitor their own performance and determine whether their operation is improving or getting worse. Some people have criticized numerical scoring because it allows a plant to make a few mistakes-absolute perfection is impossible, but high standards can be maintained.

REFERENCES

1. Grandin T. Effect of animal welfare audits of slaughter plants by a major fast-food company on cattle handling and stunning practices. J. Am. Vet. Med. Assoc. 2000. 216:848-851.
2. Grandin T. Good management practices for animal handling and stunning. Washington, DC: American Meat Institute, 1997.
3. Grandin T. Objective scoring of animal handling and stunning practices at slaughter plants. J. Am. Vet. Med. Assoc. 1998. 212:36-39.
4. Dunn CS. Stress reactions of cattle undergoing ritual slaughter using two methods of restraint. Vet. Rec. 1990. 126:522-525.
5. Grandin T. The feasibility of using vocalization scoring as an indicator of poor welfare during slaughter. Appl. Anim. Behav. Sci. 1998. 56:121-128.

6. Grandin T. Double rail restrainer for handling beef cattle. Technical paper No. 915004. St Joseph, Mich: American Society of Agricultural Engineers, 1991. 1-14.
7. Grandin T. Transferring results of behavioral research to industry to improve animal welfare on the farm, ranch and slaughter plant. Appl. Anim. Behav. Sci. 2003. 81:215-228.
8. Grandin T. Return to sensibility problems after penetrating captive bolt stunning of cattle in commercial beef slaughter plants. J. Am. Vet. Med. Assoc. 2002. 221:1258-1261.
9. Daly CC, Gregory NG, Wotton SB. Captive bolt stunning of cattle-effects on brain function and the role of bolt velocity Br. Vet. J. 1987. 143:574-580.
10. Grandin T. Factors that impede animal movement at slaughter plants. J. Am. Vet. Med. Assoc. 1996. 209:757-759.
11. Lemmon WB, Patterson GH. Depth perception in sheep: effects of interrupting the mother-neonate bond. Science. 1964. 145:835-836.

TWENTY-TWO

Implementing Effective Animal-Based Measurements for Assessing Animal Welfare on Farm and Slaughter Plants

T. Grandin (2021) Chapter 4: Implementing Effective Animal Based Measurements for Assessing Animal Welfare on Farm and Slaughter Plants, In: T. Grandin (Editor) *Improving Animal Welfare: A Practical Approach*, 3rd Edition, CABI International, Wallingford Oxfordshire, UK, pp. 60-83. First edition published in 2010.

This chapter provides instructions for people who will be implementing animal welfare auditing. Large buyers of meat need simple, clear guidance. I included this chapter so that readers of this collection will learn how effectively auditing programs work. Vague wording does not work. One person's definition of proper animal handling or adequate space will be different than another person's definition.

SUMMARY

When standards and guidelines are being written for animal welfare, vague wording should be avoided. Avoid terminology such as "proper handling" or "adequate veterinary care." What one person thinks is proper handling, another person may consider the handling practice as abusive. Both the World Organisation for Animal Health (OIE) and many animal specialists recommend the use of Animal based outcome measures.

Some of the commonly used outcome measures are body condition score, percentage of lame animals that have difficulty in walking, eye damage from high ammonia levels, animal cleanliness, sores/lesions, thermal stress (panting, shivering) and feather or coat condition. There is less emphasis on Resource based input standards that specify how to build animal housing. Resource based standards are still required for feeder space, water access space, and truck loading densities. Both legislative and private standards in either a particular country or particular private program may have types of housing systems that are prohibited. All types of legislative and private standards usually ban abusive handling practices. Two examples are breaking tails or dragging non-ambulatory downed animals. Major stakeholders in both developing and developed countries can all agree that abusive practices cause suffering. The most effective audit and inspection programs have three components: (i) internal (first party) audits done by a farm's own stock-people or veterinarian; (ii) third-party audits done by an independent auditing organization; and (iii) managers or inspectors (second party audits) from either the corporate or government agencies who check random farms to ensure that the auditors are doing their job.

LEARNING OBJECTIVES

- Learn how to write clear standards and guidelines that will be interpreted the same way by different people
- Know the difference between animal-based outcome measures and input Resource based standards
- How to determine the most important core criteria or critical points to prevent abuse or neglect
- Provide easy-to-use measures for assessing body condition, lameness, injuries, condition of haircoat/feathers, animal handling, hygiene, heat and cold stress, and presence of abnormal behavior
- How to set up effective animal welfare auditing programs

INTRODUCTION: THE NEED FOR CLEAR STANDARDS AND METHODS TO IMPROVE CONSISTENCY BETWEEN DIFFERENT AUDITORS AND INSPECTORS

Many standards and regulations for animal welfare, disease control, food safety, and many other important areas are too vague and subjective. When vague standards are used, there will be great variation in how they are interpreted. One inspector may interpret the guidelines in a very strict manner, and another may allow abuse to occur and have a lax interpretation. The author has trained many auditors and inspectors to assess animal welfare in slaughter plants and on farms. These people need specific information on what conditions are acceptable and what conditions would either be a failed audit or a violation of the law. They ask very specific questions. For example, what is the size of a bruise that counts as a bruised carcass? When does nudging an animal with your foot become kicking the animal, which would be considered an act of abuse? When does tapping an animal with a driving aid progress to beating? A video titled "Proper Use of Livestock Driving Tools" demonstrates when tapping has definitely progressed to beating. An empty corrugated cardboard box is used. When it starts to be crushed by being hit by a driving aid, tapping has progressed to prohibited beating. In the poultry barn, they asked for a very specific description of acceptable and nonacceptable litter condition. A good example of a clear description of good litter condition would be: the litter does not transfer soil on to the animals and there is some loose, dry material for chickens to scratch in. Vague terminology, such as "handle animals properly" or "provide adequate veterinary care," should be avoided. What one person may consider as proper handling; another person may think are abusive methods.

IMPROVING INTEROBSERVER RELIABILITY

A good auditor training program is essential to make an auditing program really effective. Good training will improve consistency between different auditors. Variation in judgements between different auditors can be reduced through good training (Webster, 2005). When assessors are trained and everybody uses the same interobserver reliability than others. This may be due to the design of the different scoring tools. It is also essential that the same sampling strategy is used. Van Os et al. (2018) found that sampling strategy had an effect on interobserver reliability. Some statistically based sampling strategies are too complex for on-farm use. Another factor that can make comparison of scores difficult is differences in numbering systems. Some systems use a numerical 1 for the normal condition and others use a zero. When comparing scores, make sure that the numbering system is the same.

Interobserver reliability of a welfare assessment can be improved greatly when the wording of the standard is clear. A measure with good interobserver reliability will produce similar scores when it is done by different people. A study by Phythian et al. (2013) reported good interobserver agreement for body condition, lameness, and eye damage (Phythian et al., 2019). High interobserver agreement and repeatability are possible for the body condition of dairy cows and lameness (difficulty in walking) in dairy cows (Thomsen et al., 2008; D'Eath, 2012; Vasseur et al., 2013).

SIMPLE YES/NO SCORING WILL IMPROVE INTEROBSERVER RELIABILITY

Data collected by the author in 66 US slaughter plants indicated that there were no significant differences between three different restaurant auditors when they had to count the percentage of cattle rendered insensible with a single shot from a captive bolt gun ($P = 0.529$) and count the percentage of cattle that vocalized (moo or bellow) in the race and stun box ($P = 0.22$). A possible reason for high interobserver agreement was the use of simple yes/no scoring. For example, each animal was scored as silent or vocal. Unfortunately, there were significant differences between auditors on the percentage of cattle moved with an electric goad ($P = 0.004$). The reason for this was that the standards for stunning and vocalization were clearly written and the standards for electric goad use were not clear. Some auditors counted all touches with the electric goad and others did not. Some auditors did not know that all touches with an electric goad should count, because it is impossible to determine accurately if the electric button has been pushed.

A comparison of three commonly used dairy welfare evaluation assessment tools indicated that they all picked out the 20% of the worst dairies accurately, but they differed greatly on other measures (Stull et al., 2005). Nicole (2014) compared the Welfare Quality system with two industry assessments. All three assessments identified the farms with the poorest welfare.

Research done by Smulders et al. (2006) with pigs showed that high interobserver reliability could be attained for evaluating welfare and behavior on the farm when simple yes/no scoring was used. These researchers developed an easy to administer behavior test, and the results correlated highly with physiological measures of stress, such as salivary cortisol, urinary epinephrine, norepinephrine, and production traits. On the startle test, a pen of pigs was scored as either fearful or not fearful. A yellow ball 21 cm in diameter was tossed into the pen. The pen of animals was rated as fearful if more than half the animals initially ran away. Startling is influenced by both environmental enrichment and the attitude of the stockperson (Grandin et al., 1987; Hemsworth et al., 1989; Beattie et al., 2000). Animals that have both environmental enrichment and a good stockperson caring for them will have a weaker, less fearful startle response. One of the reasons this test had such good inter observer reliability was because it was simple.

YES/NO SCORING MAY UNDERESTIMATE SOME SEVERE WELFARE PROBLEMS

Yes/no scoring works really well for animal handling slaughter practices and certain behaviors. Yes/no scoring is not recommended for welfare issues such as foot pad lesions on poultry. If the birds are scored as normal or lesioned, both severe and minor lesions would be combined. Many birds with small nonpainful lesions may have acceptable welfare, but birds with large lesions would suffer. Therefore, a three-point scoring system should be used of normal, minor, and severe. It is important to identify birds with severe lesions.

ASSESSMENT TOOLS FOR COMMERCIAL USE SHOULD BE SIMPLER THAN ASSESSMENT USED FOR RESEARCH

It is the author's opinion, based on the field training of over 600 auditors and inspectors, that a student auditor needs to be accompanied by a highly experienced person on visits to five operations before they carry out audits on their own. If they are going to audit or inspect dairies, they need to visit five dairies, and if they are going to assess poultry slaughter, they need to visit five poultry plants.

There is a difference between an audit or assessment tool that is used commercially to assess animal welfare and measurements used for detailed scientific study or veterinary diagnosis. A screening tool used for commercial certifications or legislative compliance has to be much simpler, to make the training of large numbers of auditors easier. The purpose of a screening tool is to determine that there is a problem. The purpose of more sophisticated measures is to diagnose and fix a problem or to carry out scientific research. An example of a good research tool is the Welfare Quality Network (2009) dairy assessment. Unfortunately, it is too time consuming for routine use as a certification tool (Heath et al., 2014). The Welfare Quality Network contains good assessment tools for pigs, poultry, and beef. Parts of these assessments can be incorporated into simpler audits. Andreasen et al. (2014) developed a simpler dairy cow audit that correlated well with Welfare Quality. De Jong et al. (2016) also found that the Welfare Quality protocol for broiler chickens could be simplified to make it more practical.

VAGUE WORDING SHOULD BE ELIMINATED TO IMPROVE CONSISTENCY

Several vague words that should be eliminated from all standards and guidelines are: "adequate," "proper," and "sufficient." One person's interpretation of proper handling may be totally different from another person's. For example, if a standard states that the handler should "minimize the use of electric goads," one person may interpret this as almost never using a goad and another may think that using it once on every animal is in compliance with this standard.

In the USA, some examples of vague wording in United States Department of Agriculture (USDA) standards are to avoid "excessive electric goad use" or avoid "unnecessary pain and suffering." It is impossible to train an auditor or an inspector on what is excessive use of an electric goad and have good interobserver reliability between different inspectors. One inspector may suspend meat inspection and shut down a plant for using an electric goad once on every animal, and another may think this is perfectly normal and does not penalize the plant. The author has repeatedly observed great inconsistency in the enforcement of vague USDA directives and standards. The use of objective numerical scoring can improve agreement greatly between different auditors or government inspectors.

In another example, a standard states that pigs should have adequate space on a truck or in a housing system. This is too vague. An example of a more effective way to write the standard is that the pigs must have enough space so they can all lie down at the same time without being on top of each other. Two other examples of clearly written standards are: "ammonia levels in an animal house must not exceed 25 ppm, and 10 ppm is the goal"; and "tail docking of dairy cows is prohibited." (Kristensen and Wathes, 2000; Kristensen et al., 2000; Jones et al., 2005) Some welfare standards require that animals must be housed on pasture. There should be a good definition of the minimum requirements for pasture. At what point does barren, overgrazed ground switch from being a pasture to a dirt lot? The author suggests that to qualify as pasture, 75% or more of the area occupied by the animals must have vegetation with a root system (Fig. 1).

Fig. 1.
Pigs housed outdoors on pasture. For the ground to qualify as being pasture, 75% or more of the paddock where the animals are housed must have vegetation with a root system. The pasture in this photograph has some bare spots, but it would definitely be in compliance for the entire pasture. There is a bare spot where the pigs congregated. Bare areas can be reduced if feeders are moved on a regular basis.

Traffic safety standards are a good example of clearly written standards that work effectively in many countries. People who write welfare standards should use traffic laws as a model. When a car is speeding, the police officer pulls it over after he has measured its speed. He does not stop cars that he thinks are speeding, he measures the car's speed. Both the drivers and the police officers know the numerical speed limits and that rules such as a stop sign mean stop. In most countries, enforcement of traffic laws is effective and fairly uniform.

Unfortunately, there are some politicians and policy makers who deliberately make standards vague. A political appointee who was in charge of food inspection in a major developed country resisted the author's suggestions to make standards less vague. He admitted that his agency wanted enforcement flexibility. The reason he wanted this was so enforcement could be strengthened or weakened depending on the political conditions. His vague standards resulted in one inspector being super strict and others very lax. There was no consistency between inspectors.

COMPUTERIZED ARTIFICIAL INTELLIGENCE (AI) SYSTEM MAY IMPROVE ACCURACY OF ASSESSMENTS

The field of using computers to assess animal welfare indicators is rapidly advancing. Computer systems can use machine learning or artificial intelligence to do some of the same evaluations that people do now. Some of these systems have the ability to recognize individual animals with facial recognition software. To calibrate these systems, their performance is compared with visual scoring by highly trained people (Alsaaod et al., 2019). At the time of updating this book for the third edition, some of these systems are now in commercial use. Systems that use artificial intelligence *must* be trained by experts. To train a system to detect either lame animals, lesions on animals or body condition, highly qualified human evaluators must tell the computer which animals have different scores. The AI program then compares new animals it is observing with the acceptable and not acceptable animals that it was trained with. The computer system is dependent on the quality of its training. Many of the most practical systems use sophisticated cameras.

A commercial camera system from Delavel was as accurate as manual scoring of dairy cow body condition (Mullins et al., 2019). Another camera system can often outperform human observers on lameness scoring (Schlageter-Tello et al., 2018). At the slaughter plant, a camera system can automatically evaluate ear and tail lesions on pigs (Brünger et al., 2019; Blomke et al., 2020) and foot pad lesions on broiler poultry. Systems for on-farm use to monitor behaviour and lameness of poultry and pigs are also available (Silvera et al., 2017; Benjamin and Yiks, 2019).

THREE MAJOR TYPES OF ANIMAL WELFARE STANDARDS: ANIMAL-BASED MEASURES, PROHIBITED PRACTICES, AND RESOURCE-BASED STANDARDS

Table 1 outlines different types of welfare standards. Animal based measures of welfare problems that can be directly observed by an auditor when they visit a farm can be very effective for improving welfare. Some common examples are body condition score (BCS), lameness (difficulty in walking), coat or feather condition, animal cleanliness, and sores, swellings, and injuries. They are outcomes of poor management practices. The World Organisation for Animal Health (OIE) has moved towards the use of more animal-based

TABLE 1: Different Types of Welfare Standards

There are five basic types of standards for assessing animal welfare:

1. Animal based outcome measures, also called performance standards or outcome criteria. Animal based measures can be used both as assessment tools and for setting standards. When they are used as a standard, a certain level of performance is required. An example of "excellent" would be a farm with only 5% lame, lactating cows (most emphasis).
2. Practices that are prohibited (most emphasis).
3. Resource based requirements, also called engineering input, or design standards (less emphasis).
4. Subjective evaluations (less emphasis) for improving actual conditions in the field.
5. Recordkeeping, stockperson training documents and paperwork requirements. Documentation of management procedures and standard operating procedures (SOPs) (less emphasis) for improving actual conditions in the field. Regulators may require more emphasis.

outcome standards (OIE, 2019a,b,c). The use of outcome based standards is recommended by many animal welfare researchers (Hewson, 2003; Whay et al., 2003, 2007; Webster, 2005; Velarde and Dalmau, 2012; Grandin, 2017). The large European animal welfare assessment project is also emphasizing outcome-based measures (Welfare Quality Network, 2009; EFSA, 2012, 2019; Linda Keeling, 2008, personal communication). Keeling states that measures should be: (i) science based; (ii) reliable and repeatable; and (iii) feasible and practical to be implemented in the field. Work on animal based assessment has been done by Whay et al. (2003, 2007) and LayWel (2006). Directly observable conditions are easy for auditors to score and quantify. Some important examples are BCS, lameness scoring, falling during handling, and animal cleanliness scoring. There are many published scoring systems that have pictures and diagrams that make it easy to train auditors (LayWel, 2006; Welfare Quality Network, 2009). Edmonson et al. (1989) is one example. For a welfare audit, scoring will be easier because the assessor only has to identify animals that are too skinny.

For assessing animal handling, scoring systems where handling faults are scored numerically are easy to implement and are very effective (Maria et al., 2004; Grandin, 2005, 2010; Edge and Barnett, 2008; Losada-Espinosa et al., 2017). Some of the items that are measured are the percentage of animals that fall, the percentage poked with an electric goad, the percentage that vocalize (moo, bellow, squeal), or the percentage that move faster than a trot or walk. Lameness scoring of dairy cows has high interobserver repeatability of locomotion scores of individual cows, and the variation between observers is low (Winckler and Willen, 2001). This indicates that a five-point lameness scale will provide reliable data. A study with broiler chickens indicated that a three-point lameness scoring system had better inter observer agreement than a six-point system (Webster et al., 2008). To improve inter observer reliability, scoring systems with more than five categories should be avoided. An exception to this recommendation may be sophisticated computerized AI systems. In another study, observers who visited seven dairy farms on three occasions had high interobserver repeatability on the variables of lameness scoring, kicking and stepping during milking, cow cleanliness, and avoidance (flight) distance

(De Rosa et al., 2003). They had similar results in water buffalo for kicking and stepping and avoidance distance (De Rosa et al., 2003). Lameness was nonexistent in buffalo, and cleanliness scores are meaningless because buffalo wallow.

At the slaughter plant, directly observable Animal based conditions that would be detrimental to welfare can be assessed easily. A few examples are bruises, death losses during transport, broken wings on poultry, hock burn on poultry, disease conditions, poor body condition, lameness, injuries, and animals covered with manure. On the farm, abnormal behavior such as stereotypical pacing, bar biting in sows, cannibalism in poultry, excessive startle response, and tail biting are also easy to quantify and observe numerically.

ANIMAL BASED WELFARE CRITERIA THAT ARE NOT DIRECTLY OBSERVABLE

Animal health measures that can be obtained from producers' records are important indicators of welfare problems. Some common examples are death losses, culling rates, animal treatment records, and health records. These measures are useful, but directly observable measures should be weighted more heavily, because records can be falsified. Paperwork and records are more important for tracing disease outbreaks than for assessing welfare.

ANIMAL BASED MEASURES ARE CONTINUOUS

It is impossible to never have a sick animal or never have a lame animal. When handling animals, doing it perfectly is not attainable. All of the Animal based measures are *continuous variables*. When a standard is being written, a decision has to be made on the acceptable level of faults. These decisions should be based on a combination of scientific data, ethical concerns, and field data on levels that can actually be attained.

The percentage of faults that is considered acceptable may vary between different countries or customer specifications. Many people who are concerned about animal welfare have difficulty writing a standard that allows for some faults to occur. Unless a numerical limit is placed on an Animal based measure, it is impossible to enforce it in an objective manner. An example would be 5% as the maximum acceptable percentage of lame animals. Vague terms such as "minimizing lameness" should not be used, because one auditor may think 50% is acceptable and another may consider 5% lame animals as a failed audit.

For example, the scoring system used in slaughter plants (Grandin, 1998a, 2019) allows 1% of the animals to fall during handling. Some people thought that allowing this level of falling was abusive. Data from audits done in many beef and pork plants indicated that for a plant to reliably pass an audit of 100 animals at the 1% level, their actual falling percentages dropped to less than one in 1000. The reason why the standard has remained at 1% is that during an audit of 100 animals, a farm or plant should not be penalized for a single animal that may have jumped and fallen or was spooked by additional people getting too close to it. Putting a hard number on the allowable percentage of animals that fall has brought about remarkable improvements. The use of hard numbers also prevents the gradual deterioration of practices.

Many people are reluctant to assign hard numbers, or they make the allowable numbers of bruised, injured, or lame animals or birds so high that the worst operations can pass. For example, the National Chicken Council in the USA originally set the limit for broken wings during catching and transport at 5% of the chickens. When progressive managers improved their catching practices, broken wings dropped to 1% or less. The standard is now much stricter.

PROHIBITED PRACTICES THAT SHOULD BE BANNED TO MAINTAIN MINIMUM WELFARE STANDARDS

These standards are much easier to write because they are discrete variables that leave no room for interpretation. A banned practice is either occurring or not occurring. OIE (2019a,b,c,d,e), NAMI (2019), and AVMA (2016) all contain lists and descriptions of abusive practices that should never be used. Some examples are beating animals, poking sensitive areas, such as the eyes, mouth or anus, breaking tails, throwing or dropping animals or lifting animals by their tail or head, hooves, or ears. Other acts of abuse are deliberately driving livestock over the top of downed animals or deliberately slamming gates on animals.

PROHIBITED PRACTICES THAT ARE BANNED IN SPECIFIC COUNTRIES OR RETAIL HIGH WELFARE PROGRAMS

Intensive housing such as sow gestation stalls or small laying hen cages are permitted in many countries. However, they are prohibited in many high welfare retail programs and in a few countries. Gestation stalls are permitted in the OIE guidelines (OIE, 2019e). Other practices that are permitted in many countries are castration, dehorning, beak trimming of chickens or tail docking. In high welfare programs, some of the above practices may either be prohibited or have additional specifications such as requiring the use of analgesics for pain. When a farm is being assessed, the assessor must be knowledgeable of the practices that are either banned or have specific requirements.

RESOURCE BASED ENGINEERING RESOURCE OR DESIGN CRITERIA STANDARDS

Input based standards may also be called engineering resource or design standards. These standards tell producers how to build housing or specify space requirements.

Input based standards are easy to write clearly. They may work well with one breed of animal and poorly for another. For example, a small hybrid hen needs less space than a large hybrid hen. A single space guideline would not work for both large and small chickens. In many cases, Resource based standards should be replaced with Animal based ones. However, there are other situations where input-based standards are recommended.

USEFUL RESOURCE INPUT STANDARDS

Resource based standards work well for specifying baseline minimum conditions for acceptable levels of welfare. Some examples are minimum space requirements on transport vehicles, space in housing, feed troughs, and water trough space. For nipple waterer systems, specify the number of animals per nipple. Another important input standard is limiting ammonia levels in animal houses to 25 ppm, with 10 ppm being the goal. The above items can all be quantified with numbers. For these standards, charts with animal species and weights can easily be made to fit local conditions. Animal weight is often a better variable to use because breed characteristics change. For example, Holstein and Angus cattle were much smaller in the 1960s and 1970s compared with 2015. Resourced based standards work poorly when they specify exactly how specific details of animal housing should be designed. For example, detailed specifications on how to build dividers and neck rails on dairy cow stalls (cubicles) is not recommended because an outdated specification may block innovation and new designs. To detect a problem with poorly designed or poorly maintained stalls, Animal based outcome measures such as injuries, swellings, sleeping posture, or cow cleanliness should be used. If the stalls are either badly designed or poorly bedded, the cows will have higher percentages of swollen hocks and lameness. Figure 2 shows a dairy cow cubicle that is sufficient width for comfortable lying for a small cow, but the stalls are too short for a large cow.

Fig. 2.

Dairy cow cubicle (free stall) that is of sufficient width for a cow to lie in the natural position of having her head curled back against her body. For the large cow on the left, the stall is too short.

To have a bare minimum acceptable level of welfare, animals must *never* be jammed into a crate or pen so tightly that they have to sleep on top of each other. The author has observed some caged layer farms where hens have had to walk on top of each other to reach the feeder. An input standard should be implemented to ban this type of abuse. However, on broiler chicken farms, both Dawkins et al. (2004) and Meluzzi et al. (2008) found that stocking density was not a straightforward indicator of welfare. Severe welfare problems such as footpad lesions or mortality were more related to the poor condition of the litter (Meluzzi et al., 2008). The problems were worse in the winter when air-flow through the building was reduced. There have been serious problems with conducting audits when a resource standard is vague. One example was a standard for pasture; it stated that animals must have access to pasture. There was no stipulation on how much time the cows had to be on pasture.

TRAINING OF AUDITORS AND ASSESSORS TO IMPROVE CONSISTENCY OF AUDITS

Unfortunately, it is impossible to eliminate all subjectivity from an animal welfare audit. Some variables that will require subjectivity to evaluate are the overall maintenance of a facility and the attitude of the staff. One of the best ways for an auditor or inspector to become skilled in subjective measures is to visit many different places. This will provide a range of farms or plants ranging from excellent to bad.

Photographs and videos of abusive practices on animals in poor conditions or poorly maintained equipment should be used for training auditors and inspectors. Photographs of both poorly maintained and well-maintained equipment can be used. Common faults in equipment such as broken gates, dirty ventilation fans, dirty water troughs, and worn-out milking machine equipment are shown alongside properly maintained equipment.

CLEAR COMMENTS

Auditors and inspectors must write clear comments. The author has reviewed many audits and inspection reports, and there are too many vague comments, or not enough comments. There should be well written comments to describe conditions on failed audit items and areas of noncompliance. Observations of both really good and really bad practices should be recorded. Well written comments will help both customers and regulatory officials to make wise decisions after an audit or inspection is failed. Some examples of vague and well written comments are shown in Table 2.

TABLE 2: Examples of Poor and Well-Written Comments

Vague comments	Well-written comments that explain the actual problem
Rough handling of pigs	The handler kicked piglets and threw them across the room
Poor stunning	The broken stun gun misfired and failed to render the animal insensible about one-third of the time
Poor litter in poultry barn	The litter was clumped, chopped-up newspaper. It was wet and transferred soil on to the birds

RECORDKEEPING AND PAPERWORK REQUIREMENTS

Well-kept records are essential for an efficient farm that has high standards. They are also essential for identification and traceback for disease control purposes. However, there is an unfortunate tendency for some regulatory agencies to turn the entire audit into a paperwork audit. This may create a situation where the paperwork is correct but the conditions out on the farm may be poor. Records are valuable for looking at culling rates and the longevity of animals such as dairy cows and breeding sows. The longevity of an animal is one important indicator of animal welfare (Barnett et al., 2001; Engblom et al., 2007). On some farms, a dairy cow lasts for only two lactations.

CORE CRITERIA AND CRITICAL CONTROL POINTS TO PREVENT POOR WELFARE

Some items on an audit are much more important than others. The situation must be avoided where the paperwork has been inspected and it passes but the farm is full of skinny, emaciated, lame animals. There are certain major core standards or critical non-compliance that must be met in order to pass an audit. Many of the most important core standards are directly observable Animal based measures that are the outcomes of many bad practices or conditions. Another term used for critical control points is Key Welfare Indicators. Some examples are moving animals by poking sensitive areas such as the rectum, high ammonia levels in a building that cause eye damage, or dirty animals covered with manure with no dry place to lie down, or emaciated animals. The principle of Hazard Analysis Critical Control Points (HACCP) is used in many audit systems. The principle is to have relatively few critical control points (CCPs) or core standards that must *all* be in compliance to pass an audit. An effectivc CCP is an outcome measure of many poor practices. HACCP systems were originally used for food safety. When the first HACCP programs were developed, they were often too complicated, with too many CCPs. Later versions of the programs were simplified. The numerical outcome based scoring system the author developed for slaughter plants has five core standards or CCPs that are continuous outcome based measures that are scored numerically, one engineering or Resource based standard on watering animals, and one discreet core criterion that prohibits any act of abusive handling (Grandin, 1998a, 2005, 2010; NAMI, 2019). A plant has to pass on *all* seven core standards to pass a welfare audit. The HACCP principles are being applied increasingly to animal welfare audits (von Borell et al., 2001; Edge and Barnett, 2008; Losada-Espinosa et al., 2017). There are a number of different ways that core standards or CCPs can be classified. The rationale behind the author's classification is for easier implementation in the field.

For each species of animal, training materials will have to be developed for local conditions. The reason for placing less emphasis on examining records is because the author has observed that some records are faked. It is the author's opinion that auditing paperwork is an important part of a welfare audit, but it should be weighted less heavily than the directly observable items in Tables 3, 4, 5, 6, and 7.

TABLE 3: Body Condition Scoring System

1. Emaciated—ribs and spine showing. Not acceptable
2. Thin
3. Normal
4. Over condition
5. Obese—extremely fat—sometimes is a welfare problem

Conditions vary greatly in different parts of the world. Body condition scoring charts should be made to fit local conditions.

TABLE 4: Simple Four-Point Lameness (Difficulty in Walking) Scoring of Cattle, Sheep, Pigs, and Other Mammals

1. Walks normally with smooth, even steps.
2. Walks with a limp or has a stiff gait, with head down or bobbing head. The animal keeps up when a group of animals are walking (classify as mild lameness).
3. Walks with difficulty but still fully mobile. It cannot keep up and is left behind when a group of animals is walking (classify as severe lameness). Lame sheep will also walk further back in a group.
4. Can barely stand and walk and may become non-ambulatory. Same as a 5 on the Zinpro dairy cow scales (classify as severe lameness and not fit for transport). On the New Zealand four-point scale, the most severe score is assigned to a fully mobile animal that cannot keep up with the walking herd. This animal would be fit for transport.

Note: Some guidelines will label normal walking as 0 and renumber degrees of lameness with 1, 2, 3. A detailed photographic guide to identify hoof lesions that cause lameness can be found in Mulling et al. (2014).

PROBLEMS WITH CONVERTING MULTIPLE MEASURES TO A SINGLE SCORE

Busy managers want data that are easy to evaluate, and they like to have all the welfare and food safety data combined into single scores. The danger of converting multiple measures to a single score is that serious problems may be masked. For example, when multiple welfare indicators on a European Welfare Quality Network dairy audit were combined into a single score, a dairy with 47% lame cows received an acceptable rating. A dairy with 47% lame cows would have very poor welfare. Additional studies have also shown problems with converting welfare audit data to a single score (Graaf et al.). On Animal based measures, such as lameness, a few percentage points over the 5% level would be points off, but if 25% of the cows were lame, the audit would be failed. Therefore, each animal-based measure could have two levels of failure—points off and audit failure.

THE MOST IMPORTANT ANIMAL BASED MEASURES (CCPS) THAT ASSESS MANY WELFARE PROBLEMS

Body Condition Scoring (BCS)

There are many different published scoring systems. It is often best to use charts that have been developed in your country. For American Holstein and European dairy cows, there are good pictures and charts in Wildman et al. (1982), Ferguson et al. (1994), University of Wisconsin (2005), and Welfare Quality Network (2009). All the articles are free on the Internet. The BCS pictures should be on a plastic laminated card that assessors can always have with them. Body condition scoring systems range from three scores to nine scores. The most practical systems use four or five scores. It is recommended to use a scoring system that producers in the local area are familiar with. For example, cattle that had starved to death in extremely cold conditions, such as –18°C, often did not have an emaciated body condition score of 1. Autopsy revealed that there was no fat around the heart and kidney. To have adequate welfare, cattle that live in very cold climates must have better (fatter) body condition than cattle that live in warmer climates. Laminated cards with photographs of a rear view and side view of animals that are acceptable and are not acceptable are recommended. Poor body condition can be caused by a lack of food, parasites, disease, or poor management of rBST (rBST is a growth hormone given to dairy cows to increase milk production). It is an important outcome measure of poor management, nutrition, and health. Lactating animals will usually be thinner, with a poorer BCS than nonlactating animals.

Lameness (difficulty in walking mobility) scoring

Lameness is caused by many different conditions. Measuring the percentage of lame animals is an outcome of a variety of factors that cause animals to become lame. Some examples of factors that may increase the percentage of lame animals are:

- rapid growth in poultry and pigs genetically selected for growth
- wet bedding or muddy lots
- standing on wet concrete
- lack of hoof trimming or foot care
- disease conditions—for example, hoof or digital dermatitis
- poor leg conformation
- founder (laminitis) due to feeding high levels of concentrates
- lameness due to feeding high levels of beta agonists, such as ractopamine or zilpaterol (this is more likely to be a problem during hot weather)
- poor handling that causes animals to slip and fall
- poorly designed cubicles (free stalls) in dairies
- genetic selection for high milk production.

Lameness in chickens can be measured with a simple three-point scoring system (Dawkins et al., 2004; Webster et al., 2008). The scores are: 1– Walks normally for ten paces; 2– Walks crookedly and obviously lame for ten paces; 3– Downer or not able to walk ten paces.

Some auditing systems use numbers 0, 1, 2. Knowles et al. (2008) refers to videos for training auditors that can be accessed online. The website Zinpro.com/lameness (accessed 14 August 2020) has excellent videos for scoring lameness in cattle with a five-point system. Cattle with scores of 3, 4, or 5 on a five-point scale would be classified as lame. On this scale, a 1 is completely normal and a 5 can barely stand and walk.

Since the second edition of this book, a simpler four-point scoring system has become popular (Edwards-Callaway et al., 2017). Lameness is classified as mild or severe, depending on whether or not the animal can keep up with its herd-mates when the group is walking. A three-point lameness scoring system works well in poultry, but it may underestimate the percentage of lame cattle. There are many variations of lameness scoring systems with slight differences in wording. When scoring systems are used for research, it is important to specify in the methods section the lameness scoring tool that has been used. If only part of the herd is scored, the sampling method must also be described. There has also been great progress in the development of automated lameness detection (Leary et al., 2020).

Ideally, all lactating cows should be scored for lameness. On very large dairies where this is not possible, research by Hoffman et al. (2013) indicated that a sample of lactating cows should be taken in the middle of each group of cows exiting the milking parlour. A study done in 50 pasture-based dairies indicated that scoring the last 30% of the cows could be used as a screening test to identify dairies with lameness problems (Beggs et al., 2019). New methods for measuring lameness with electronic devices are also being developed. In dairies and other operations where animals are kept tied, lameness can be assessed by watching the cow to see if she shifts weight from side to side or shows uneven weight bearing (Gibbons et al., 2014). There are differences in numbering in some systems. One popular four point lameness scoring system uses a zero for normal and then 1, 2, 3 (Edwards-Callaway et al., 2017). When lameness scores are being compared between farms, *all* data *must* use the same numbering system. Observing animals when they are resting can provide clues about possible lameness. Figures 2 and 3 show cattle in normal lying positions. Figure 4 shows a sheep sitting like a dog. This posture is often due to either sore joints or sore feet.

Fig. 3.
Normal lying position of cattle. The front legs are tucked up under the animal. If one front leg is extended out straight, the animal may have a sore leg.

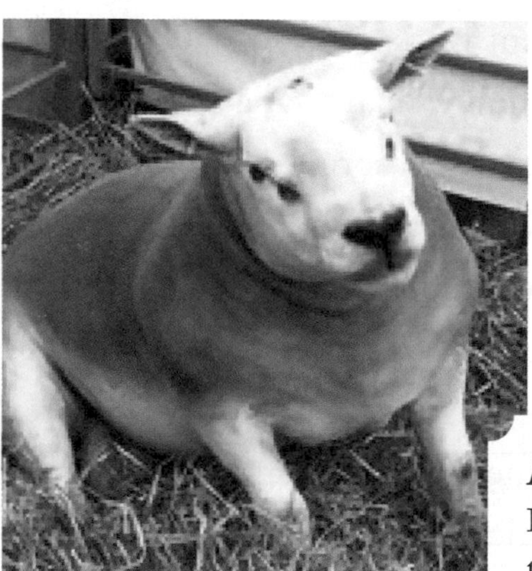

Fig. 4.
If sheep or cattle sit in a dog-sitting position, this is an indicator that the front legs may be sore. Dog-sitting is not a normal position.

Cleanliness

Table 5 suggests a scoring system for fecal matter and mud on animals.

TABLE 5: Scoring Fecal Matter and Mud on Animals

For all species, a simple 1, 2, 3, 4 scoring system can be used. This score is used for fecal matter, fecal matter mixed with earth, and fecal matter mixed with bedding that has adhered to the animal or bird. This scoring system can be used for dairy cows, feedlot cattle, pigs, sheep, and poultry. Specify whether or not the soil is a thin layer or large chunks of manure or mud.

1. Completely clean legs, belly/breast, and body. Birds must be completely clean.
2. Legs are soiled but belly/breast and body are clean (Fig. 5).
3. Legs and belly/breast are soiled.
4. Legs, belly/breast, and the sides of the body are soiled (Fig. 6).

Soil on animals that is caused by an animal's attempts to cool itself should not be confused with fecal matter on animals that is due to a failure to provide a dry place for the animals to lie down. Some systems will label clean as 0 and renumber degrees of dirtiness as 1, 2, 3. Recent research indicates that deep mud in feedlots can reduce weight gain by 14% to 38%.

Fig. 5.
These dairy cows have been grazed on lush green pastures. Their bellies, udder, and upper legs are clean. They would score a 2 for cleanliness in a scoring system where 1 is clean.

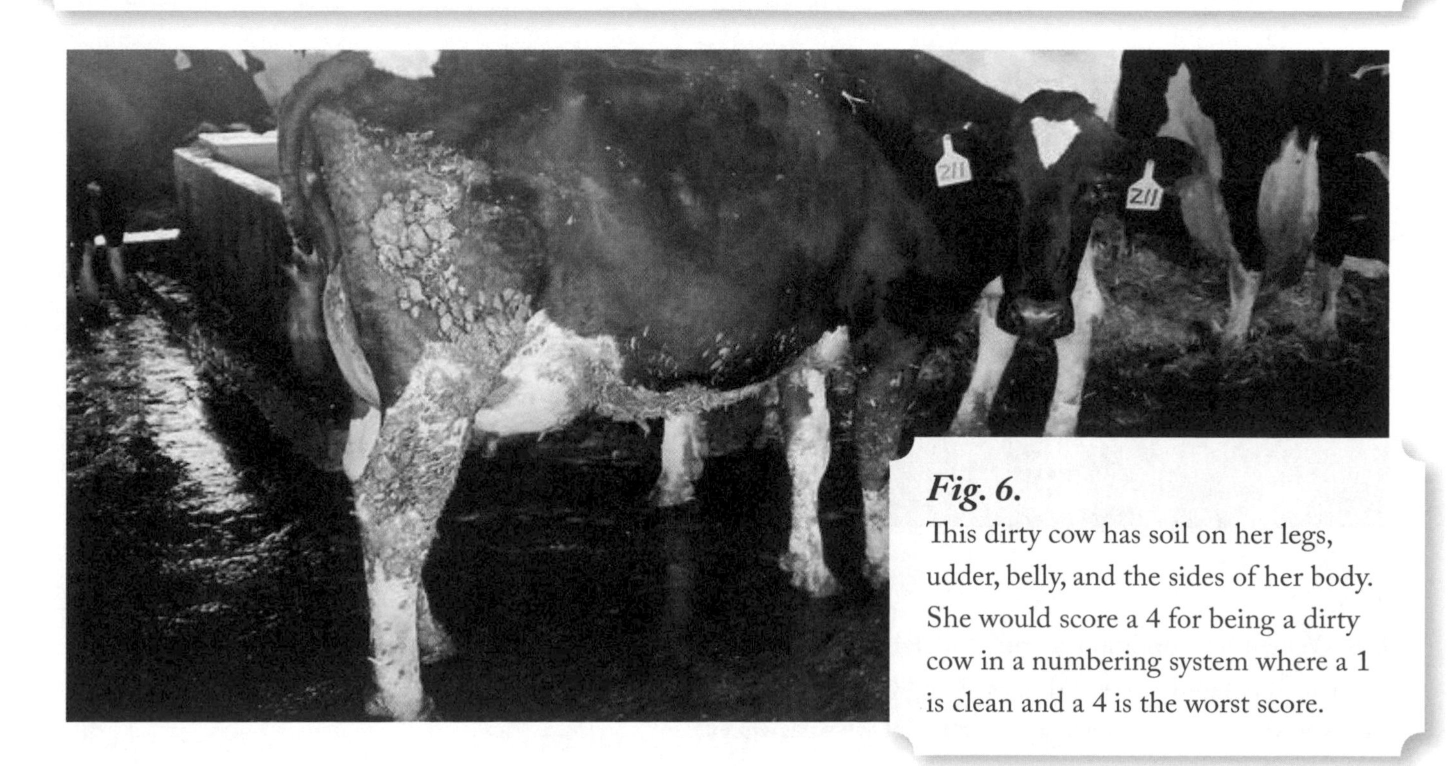

Fig. 6.
This dirty cow has soil on her legs, udder, belly, and the sides of her body. She would score a 4 for being a dirty cow in a numbering system where a 1 is clean and a 4 is the worst score.

Condition of coat or feathers

When coats and feathers are being evaluated, it is important to determine the cause of the damage. The three major causes are: abrasion against a cage or feeder; damage inflicted by another bird; or internal and external parasites.

There are published scoring systems for feather condition in poultry. The best systems have separate scores for feather wear and damage caused by other birds. The LayWel website (www.laywel.eu, accessed 14 August 2020) has good pictures for scoring feather condition in poultry (Fig. 7). The type of housing system will have an effect on the pattern of feather damage. Hens on deep litter have greater damage on the head and neck, and caged birds have more wing and breast damage (Bilcík and Keeling, 1999; Mollenhorst et al., 2005). Damage on the head is most likely due to pecking from other chickens, and wing damage is more likely to be caused by abrasion against equipment.

For livestock, poor coat condition is an indicator of problems such as untreated external parasites or mineral deficiencies. These can occur in organic systems where drugs for treating parasites are not allowed. In cattle, bald spots are an indicator of untreated external parasites. This would be a severe welfare problem.

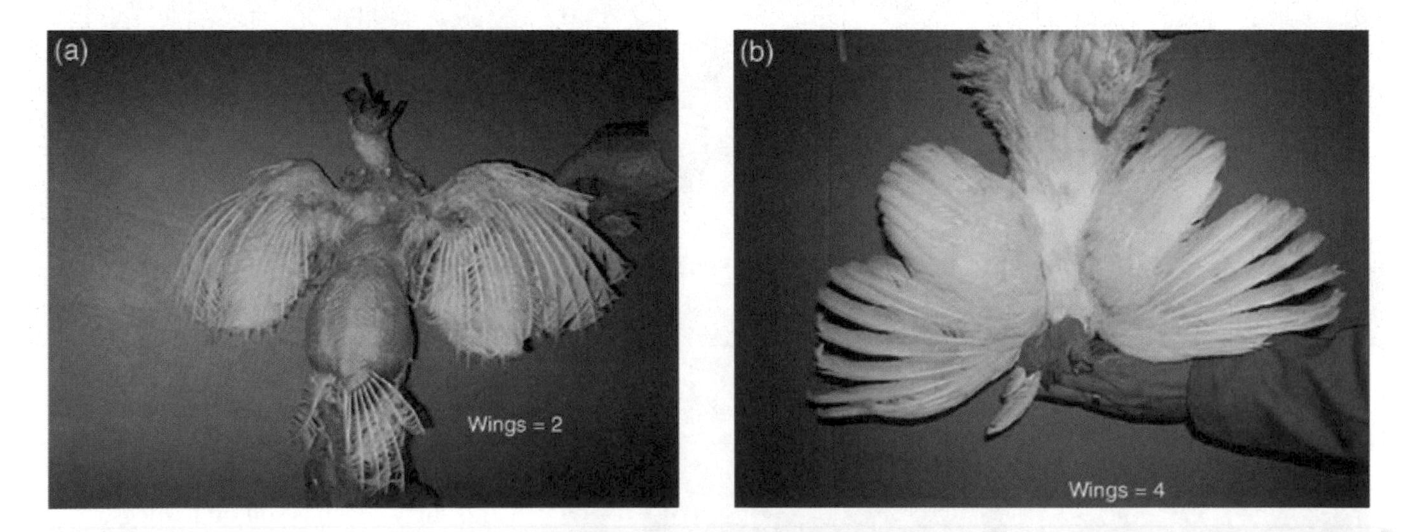

Fig. 7.

Lay Wel feather condition scoring for laying hens. It is a four-point scale, where a 1 is almost bald and a 4 is normal. This figure shows (a) score 2 and (b) score 4. The Lay Wel website also has pictures for scoring breasts, backs, and necks. (Photographs courtesy of www.laywel.eu).

Sores and injuries

Pictures and charts for scoring will need to be provided for each species. In pigs, pressure sores on the shoulder are an indicator of poor management in gestation stalls, and bites and wounds would be scored in group housing. For dairy cows living in free stall (cubicle) housing, lesions and swellings on the legs are an important measure. Cows in cubicles that are frequently bedded have fewer lesions (Fulwider et al., 2007). Improving cow comfort and reducing knee lesions, lameness, and dirty animals increased milk production (Robichaud et al., 2019). Free stalls that are too small also increase lesions. Leg lesions on dairy cows are an excellent outcome measure of both poor management and poor stall design. On birds, injuries caused by pecking, cannibalism, or aggressive roosters can be quantified easily. In pigs, injuries such as bitten tails and ears can be quantified. In broiler chickens, lesions and burns on the feet can be scored with a chart showing different levels of severity (Fig. 8). Wet litter is one major cause of footpad lesions.

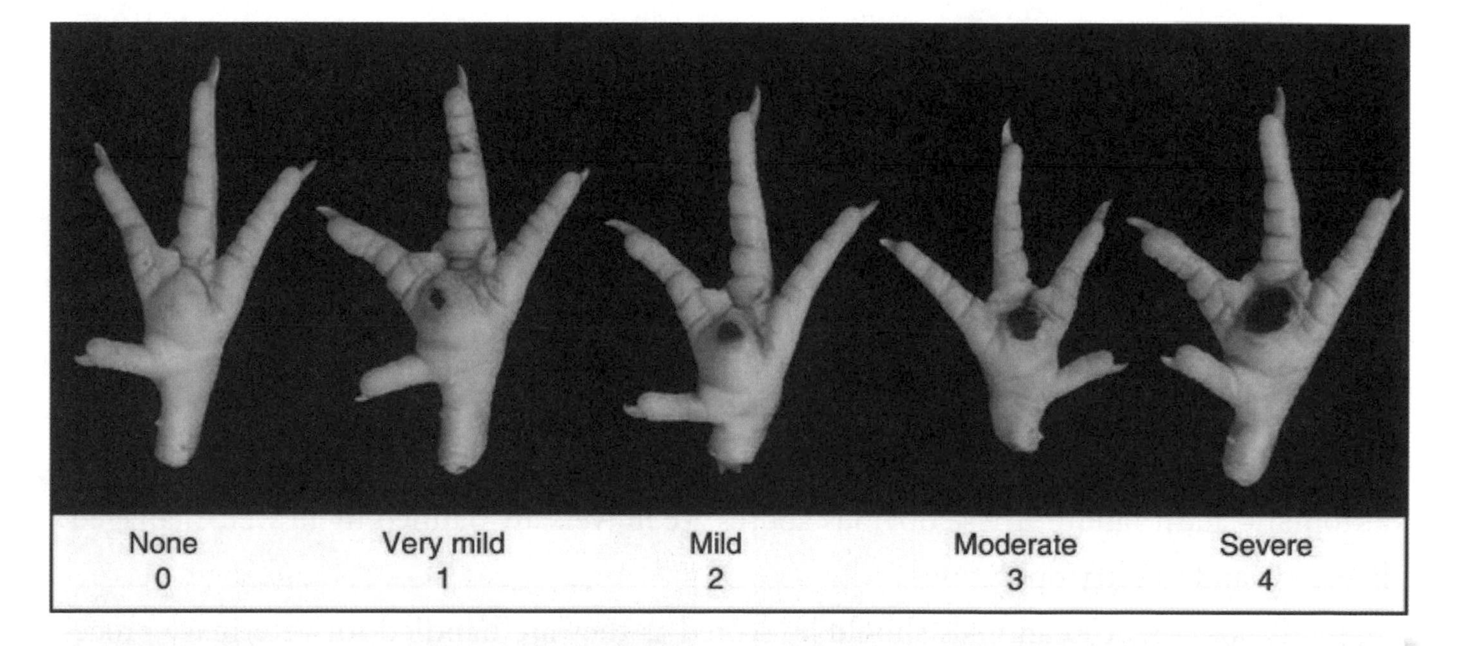

Fig. 8.
Broiler chicken foot burn chart. Footpad lesions are caused by poor litter conditions. Lesions that are brown or otherwise discolored usually occur during a later stage of growth. Lesions that are white and not discolored at slaughter are more likely to have occurred early in the growing period.

Animal-based outcome measures for broiler chickens

The eight critical outcome-based measures for broiler chickens are: broken wings due to handling, breast blisters, hock burn, foot pad lesions, cleanliness, dead on arrival, condemnations and ascites (damaged hearts). Figure 8 shows a five-point scoring system for foot pad lesions. If yes/no scoring is used, the none and very mild lesions are combined into pass (acceptable) and the mild, moderate, and severe are all fail. However, a three-point scoring system is recommended for hock burn and foot pad lesions. All eight of these measures can be easily assessed at the abattoir (Saraiva et al., 2016; Grandin, 2017). Stocking density is another important variable and it must be measured at the farm. A simple method for determining if a farm is overstocked is conducted by walking through the barn on the day of catching. If it is impossible for the chickens to move 1 m (3 ft) away without piling, the farm is definitely overstocked. On-farm assessments should also include lameness (gait score), 7-day mortalities, and total mortalities.

Obvious neglected health problems

Table 6 outlines other important CCP that should be scored. Some examples of neglected health problems would be advanced necrotic cancer eye in cattle, bald spots due to untreated parasites, large infected untreated injuries, large untreated abscesses, necrotic prolapses, or other obvious neglected conditions.

Handling scoring

The acts of abuse in Table 7 that can occur during handling should be grounds for an automatic audit failure. These obvious abuses are universally banned in all well-managed livestock and poultry operations.

If possible, 100 animals should be scored as they are handled for veterinary procedures during truck loading/unloading or moving to slaughter at an abattoir. Each animal is scored on a per animal yes/no basis. For vocalization and electric prod use, each animal is scored as either silent or vocalizing, or moved with an electric prod or not moved with one. The main continuous variables that are scored are the percentage of animals. Further information on handling scoring is in Welfare Quality Network (2009), Grandin (2010, 2014, 2019), OIE (2019a) beef cattle standards, and OIE (2019d).

TABLE 6: Other Important Critical Control Points That Should Be Scored with Animal-Based Outcome Measures

- Thermal stress—open-mouth breathing in cattle indicates severe heat stress, or huddling or piling in pigs indicates cold stress.
- Condition of haircoat on mammals.
- Condition of feathers on poultry.
- Condition of feet, all species.
- Sores and injuries on the body.
- Swellings and injuries on the legs.
- Neglected health problems.
- Eye damage or lung lesions due to high ammonia or dust level.
- Size of flight zone. Good stockmanship: animals approach people. Poor stockmanship: animals run away.

TABLE 7: Acts of Abuse That Would Result in a Failed Animal Welfare Audit

1. Beating, kicking, or throwing animals or birds.
2. Poking out eyes or cutting tendons.
3. Dragging live animals.
4. Deliberately driving animals over the top of other animals.
5. Breaking tails.
6. Poking animals in sensitive areas such as the rectum, eyes, nose, ears, or mouth to move them.
7. Deliberately slamming gates on animals.
8. Jumping or throwing animals off a truck that has no drop-down tailgate or unloading ramp.

All of the handling measures in Table 8 are outcomes of either poorly trained people or deficiencies in the facilities. For example, falling can be caused by either rough handling methods, slick flooring (if the floor is too slick, the animals will be more likely to slip and fall), or the animal not being fit for handling and transport. Vocalization scoring is a sensitive indicator of both equipment design problems and poorly trained people. Vocalizations during the handling and restraint of cattle and pigs are associated with the following bad practices (Grandin, 1998b, 2001):

- excessive use of electric goads
- slipping on the floor
- sharp edge in a restrainer
- malfunctioning stunning equipment
- slamming gates on animals
- excessive pressure applied by a restraint device
- being left alone in a race or stun box.

Vocalization scoring should not be used for sheep, because severely abused sheep often remain silent. Many large slaughter plants have been able to easily achieve vocalization percentages of 3% or less of the cattle (Grandin, 2005, 2019). Industry data also indicate that in many plants, only 5% or fewer of the pigs squeal in the restrainer.

TABLE 8: Handling Scoring—Score When Animals Are Being Moved by a Stockperson

- Percentage falling.
- Percentage moved with an electric goad.
- Percentage vocalizing (moo, bellow, squeal) (do not use vocalization scoring for sheep).
- Percentage moved faster than a walk or trot.
- Percentage that run.
- Percentage that strike fences or gates.

Heat and cold thermal stress

Many animals have died or have had their welfare severely compromised by either heat or cold stress. The thermal neutral zone of an animal is very variable and it depends on genetics, length of hair/coat, access to shade, and many other factors. If animals or birds are panting, they are heat stressed, and shivering or huddling animals are cold stressed. Open-mouth breathing in cattle is a sign of severe heat stress (Gaughan et al., 2008; Gaughan and Mader, 2014) and either shade or water sprays should be provided to relieve it. Auditors and inspectors can count the number of panting animals easily. For use under commercial conditions, the scoring system should be simplified to open-mouth breathing, severe heat stress, or closed-mouth breathing. It is beyond the scope of this book to discuss all aspects of thermal comfort, but conditions that have caused death from thermal stress must be corrected. For heat stress, some of the corrective actions would be fans, sprinklers, shades, extra water, or changing animal genetics (Fig. 9). For cold stress,

Fig. 9.
Well-designed shades and sprinklers keep cattle cool in this feedlot located in the arid, dry southwestern USA. The cattle stay clean and dry in the arid climate. Feedlot shades must be installed north and south so that the shadow will move and there is sufficient space for all animals to lie under the shade. This prevents mud buildup under the shade. Shades should be 3.5 m (12 ft) high. Places with rainfall under 50 cm/year (20 in/year) are the best locations for feedlots. Pens should have 2–3% slope away from the feed trough to provide drainage and prevent mud buildup.

some possible corrective actions would be shelter, heat, additional bedding, or changing animal genetics. Some of the worst problems with thermal stress occur when animals are moved between different climatic regions. It takes several weeks for an animal's body to acclimatize to either hot or cold conditions. When animals have to move from a cold area to a hotter area, it is best to transport them to the hotter region during its cool season. Transporting cold-acclimatized cattle to a hot desert region may result in higher death losses. In extremely cold climates under –15°C, supplying bedding for feedlot cattle will improve weight gain (Smerchek and Smith, 2020).

Abnormal behavior associated with poor welfare and assessment of behaviors associated with positive emotions

Feather and coat condition scoring can be used to detect damage that is caused by other animals. Cannibalism in hens and wool pulling in sheep are two forms of abnormal behavior that can be detected by examination for injuries. In pigs, abnormal behavior such as biting tails and ears can be quantified by counting the percentage of pigs with

TABLE 9: Abnormal Animal Behaviors That Are Not Acceptable

Abnormal animal behavior should be prevented either with environmental enrichment, feeding more roughages, or changing animal genetics. Count the percentage of animals performing these abnormal behaviors. Use scoring to determine if interventions are effective for reducing abnormal behaviors.

- Tongue rolling in beef and dairy cattle.
- Tail biting in pigs.
- Urine sucking.
- Pacing—all species.
- Repetitive behaviors – all species.
- Feather pecking in poultry.
- Pulling out hair or wool.
- Barbiting sows.

injuries. The author has observed that some genetic lines of rapidly growing lean pigs have a higher percentage of tail biting compared with others. In bovines, tongue rolling and urine sucking are abnormal behaviors that can be prevented by feeding roughage (Montoro et al., 2013; Webb et al., 2013). Providing straw to finishing pigs reduced tail damage (Wallgren et al., 2019). Animals housed in barren intensive environments may be more likely to engage in stereotypies such as bar biting or pacing. Most animal welfare specialists agree that lameness, disease, or injury is bad for welfare, but there is likely to be more disagreement on how heavily behavioral measures should be weighted. Bracke et al. (2007) developed a technique where a number of different animal welfare scientists were polled about behavioral needs. Statistics were used to determine the behavioral needs that the specialists ranked as most important. The highest ranking was given to materials for pigs to root and explore, and prevention of tail biting. This group consensus approach will be important for writing guidelines in areas where the science may be less clear and decisions are based partially on ethical concepts. In 2001, the EU passed a directive that pigs must have access to fibrous materials such as straw to root and chew.

Positive behaviors and Qualitative Behavioral Assessments (QBA)

To achieve a minimum level of animal welfare will require the prevention of injuries, health problems, and other problems discussed in this chapter. More and more welfare scientists see the importance of using indicators for truly positive animal welfare (Boissy et al., 2007; Edgar et al., 2013; Mellor, 2017). Having a barely acceptable level of welfare may not be adequate.

Some examples of behaviors that indicate a positive emotional state are: animals that have a small flight zone and approach people, play behavior, dust bathing in chickens, and normal grooming behaviors.

Qualitative behavioral assessment (QBA) is a method for assessing animal emotionality by having people assign words to animal behavior such as calm, curious, nervous, agitated, or angry. There is a definite statistical relationship between words assigned by observers and the physiological state of the animal (Rutherford et al., 2012; Stockman et al., 2012; Wickham et al., 2015). QBA may be more effective for assessing short-term stressors such as transport or handling. Kniese et al. (2016) reported that QBA was not consistent between multiple farm visits. This method is not sufficiently developed for routine use on commercial farms. In the future, powerful computer programs may make it easy to use.

HOW TO SET CRITICAL LIMITS FOR ANIMAL BASED STANDARDS (ALSO CALLED BENCHMARKING)

To avoid vague guidelines, numerical limits have to be put on Animal based standards to determine passing and failing scores. The limits have to be set high enough to make the industry improve, but not so high that people will either say it is impossible or fight conforming to the standard. When the author worked on auditing slaughter plants, three out of four plants failed their first audit (Fig. 10). They were then given time to improve their practices, with no penalties. When an Animal based outcome scoring system is first introduced, baseline data will need to be taken and the limit for an acceptable rating should be set so that the best 25–30% of the plants or farms will pass. To motivate farms that are still below the threshold, they should be given recognition for making improvements. The others have to be given a period of time to attain the standard. For example, slaughter plants should be required to correct minor problems within 30–60 days and major problems within 6 months. On farms, producers should be given up to 2 years to correct problems such as a high percentage of lame animals. This will make it easier to implement the system, because the places with excellent scores can be used as examples to show the bad operators that they need to improve. In parts of the world where practices are really poor, the standards may need to be raised again after the system is implemented and the farms have improved. Some people who are concerned about animal welfare have objected to having a threshold for allowable problems. In the practical world of real farms, perfection is impossible.

Continuous improvement approach

Another approach is not to set strict numerical limits but to use the ISO 9001 method of continually working to improve (Main et al., 2014). It is the author's opinion that a combination of thresholds (critical limits) for minimum numerical standards and continuous improvement should be used. Setting a minimum acceptable critical limit for a continuous measure such as lameness or body condition score prevents poor welfare. This approach is recommended for legislative enforcement or retailer auditing programs where a farm or slaughter plant is removed from an approved supplier list. Continuous improvement approaches can be used to improve welfare further. A list ranking farms on different measures can also be used to motivate continuous improvement.

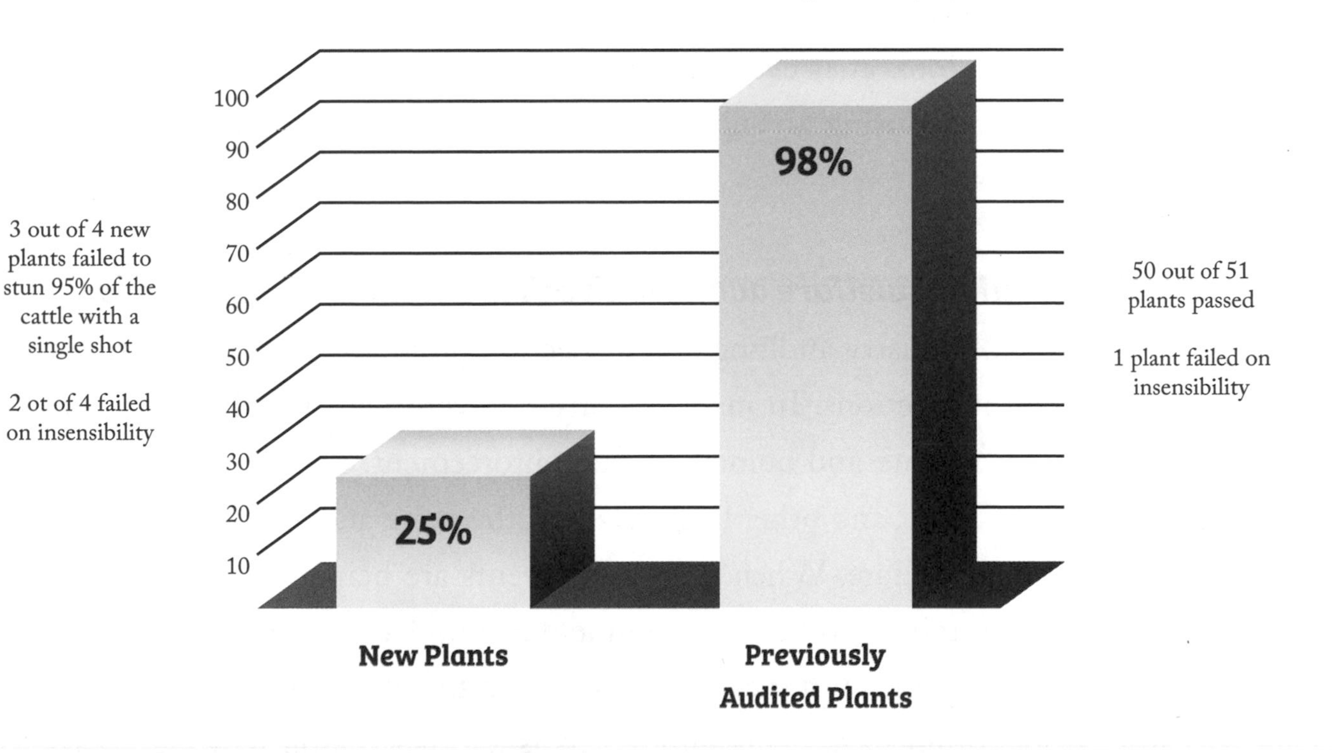

Fig. 10.
Comparison of beef plants audited for 4 years versus new plants audited for the first time. Only 25% of the slaughter plants passed their initial animal welfare and humane slaughter audit. Plant managers did not know what to expect, and they often engaged in bad practices because they were their normal practices. An initial assessment should be considered as a training session to educate both the management and the employees.

Implementation of successful auditing and assessment programs

When the author first started doing audits of slaughter plants, many people did really bad things in front of her because they did not know any better. In the early years, there was no difference between announced and unannounced audits. After people learned that certain practices were wrong, some managers became very skilled at acting "good" during an audit and then bringing back the electric goads when the audit was over. Audits and assessments of handling and transport practices are more likely to have different results between those obtained on an announced visit and those gained on an unannounced audit, whereas measures of lameness, body condition, dirty animals, or maintenance of facilities are more likely to be consistent between the two types of audit. Some slaughter plants in the USA have installed video auditing that can be viewed unannounced over a secure internet link. Video auditing is most effective when it is monitored by people outside the plant, who look at it every day at different random times. This solves the problem of acting "good" during an audit and then reverting back to old, bad practices after the auditor leaves.

Combining animal welfare audits with other audits and inspections

Most government and industry auditing and inspection programs have people that conduct several types of inspections. In many countries, government veterinary inspectors carry out both meat hygiene and animal welfare enforcement. For most of the audits done by McDonald's, Tesco, and other large retailers, the same auditor has to audit both food safety and animal welfare. When farm assessments are being done, the same person may have to audit animal welfare, environmental compliance, and animal drug use. Bundling auditor and inspection duties is often required to cut down on many duplicate trips. Because people often have many duties, this is another reason why clear, easy to implement standards and guidelines are essential. To comply with EU standards, a person has to be designated as that animal welfare specialist.

Structure of effective audit and inspection systems

The most effective audit and inspection systems have a combination of internal and external audits. The best systems that have been implemented by private industry have three parties:

- Internal audits (first party) done on a daily or weekly basis either by employees at the plant or by a farm's local veterinarian.
- Audits by people employed by the corporate office of a meat-buying customer (second party). They should visit a certain percentage of their suppliers each year.
- Third-party audits done by an independent auditing company. They are conducted once or twice a year on every plant and farm.

One advantage of using a third-party independent auditing company is to prevent a conflict of interest. For example, a meat buyer in the company corporate office might favor a certain plant and be less strict because they want their cheap prices. However, it is the author's opinion that a corporation should not delegate all its auditing responsibilities to a third-party auditor. Corporate personnel need to make regular visits to show suppliers that they are serious about animal welfare.

On a farm, the farm's regular veterinarian or other professional can do the "internal" audits. Due to problems with conflict of interest, the entire responsibility for auditing should not be done by the farmer's local veterinarian. The veterinarian may be reluctant to fail a client farm because they would risk getting fired by the farmer. To avoid this potential problem of conflict of interest, the auditor should not be the farmer's regular veterinarian. The person doing the audit should be on someone else's payroll. They should be paid by either the government, a third-party auditing company, a meat buying corporation such as McDonald's, a livestock association, or work as field staff for a large, vertically integrated company. In many systems, farms and slaughter plants have to pay for audits. To avoid a conflict of interest, the farm or plant pays the auditing company and the auditing company pays the auditor's salary. A similar system is used for government inspection. The inspector's salary is paid by the government. In both of these systems, either the auditing company or the government makes inspector and auditor assignments. A farm or plant should never be allowed to choose their favourite inspector. To help prevent personality conflicts, a third-party auditing firm should rotate auditors to avoid having the same person always visiting the same farm or plant.

Each plant or farm is a separate unit

Big multinational companies own many plants and farms. Each plant or farm site should be treated as an independent unit, regardless of ownership. It either passes or it fails an audit. A big company may have bad operations in another country. If your job is to implement a welfare audit program in your country, it is best to concentrate on working with each individual plant or farm in your country. A plant or a farm is either taken off the approved supplier list or put back on the approved list. Meat buying companies that have made the greatest progress in improving animal welfare have enough plant or farm suppliers so that a few can be taken off the approved supplier list and they will still have enough product. Even if all the meat comes from two big corporations, a buyer still has sufficient economic power to bring about change if they buy from several different plants owned by each company. In government programs, there are different levels of penalties, ranging from fines to shutdown of the farm or plant.

Procedures for handling noncompliance

Most government and industry systems have a formal procedure that is followed when there is a non-compliance. For both minor and major noncompliance, a letter (email or text message) has to be written explaining the corrective action that will be taken to rectify the problem. After a specified period of time, the farm or plant will have to have a new audit or inspection. Throughout this chapter, the author has emphasized the need for clear guidelines and standards that can be applied consistently by different auditors or inspectors.

The penalty for being one or two points below the minimum passing score should be much less severe than the penalty for a serious violation such as dragging a non-ambulatory animal down the unloading ramp. The author has worked with many large meat buyers on decisions to remove a supplier from the approved list. Usually, one or two points below the minimum standard requires a corrective action letter and a reaudit. Buying from the plant would continue. Dragging the non-ambulatory animal is a serious act of abuse and the meat buyer would stop buying from the plant for a minimum of 30 days. If they had another serious violation, they would be suspended for a longer time. Unfortunately, there are certain people who should never be handling animals. Data from many audits have repeatedly shown that approximately 10% of people need to be removed.

In conclusion of this section, the auditing and inspection process should be very objective, but sometimes great wisdom is required to determine the appropriate punishment. From over 20 years of experience implementing auditing systems, the author has learned that suppliers with a poor, uncooperative attitude need harsher punishment than more cooperative suppliers. When the original McDonald's audits were started in 1999 and 2000, out of 75 pork and beef plants, three plant managers had to be fired before improvements in animal welfare occurred. To get these three plants to conform to the standard, all purchases from them were suspended until new managers were hired. With new management, one of the plants went from being one of the worst plants in the system to one of the best. For both food safety and animal welfare, the best places have managers who want to do the right thing.

Use of remotely viewed video cameras to audit animal welfare

Most major meat companies in the USA and abattoirs in many countries are successfully using video cameras that can be viewed from a remote location to audit animal welfare at the slaughter plant. The cameras solve the problem of people "acting good" when they know they are being watched. The cameras are used to assess stunning efficacy and animal handling. The cameras have brought about the biggest improvements in animal handling and reducing electric procedures. To be effective, the cameras must be monitored by auditors who are part of a structured program. Reliance on plant management is not effective because they stop looking at the cameras after the novelty wears off. Video cameras are also being used to improve interaction between people and cows on dairy farms. The cameras are being monitored to document positive interactions with cows and determine the efficacy of different stockperson training methods.

CONCLUSION

The implementation of an effective auditing and inspection program can improve animal welfare greatly. Clear guidelines will improve the consistency of judgement between different people. This will help prevent the problem of one inspector being super strict and another failing to improve the treatment of animals.

REFERENCES

Alsaaod, M., Fadul, M., and Steiner, A. (2019) Automatic lameness detection in cattle. *Veterinary Journal* 246, 35–44.

Andreasen, S.N., Sandoe, P. and Forkman, B. (2014) A comparison of the Welfare Quality® protocol for dairy cattle and the simpler and less time*consuming protocol developed by the Danish Cattle Federation. *Animal Welfare* 23, 81–94.

AVMA (2016) *Guidelines for the Humane Slaughter of Animals*, 2016 edn. American Veterinary Medical Association, Schaumberg, Illinois.

Barnett, J.L., Hemsworth, P.H., Cronin, G.M., Jongman, E.C. and Hutson, G.D. (2001) A review of the welfare issues for sows and piglets in relation to housing. *Australian Journal of Agricultural Research* 52, 1–28.

Beattie, V.E., O'Connell, N.E., Kilpatrick, D.J. and Moss, B.W. (2000) Influence of environmental enrichment on welfare related behavioral and physiological parameters in growing pigs. *Animal Science* 70, 443–450.

Beggs, D.S., Jongman, E.C., Hemsworth, P.H., and Fisher, A.D. (2019) Lame cows on Australian dairy farms: a comparison of farmer identified lameness and formal lameness scoring and position of lame cows within the milking order. *Journal of Dairy Science* 102, 1522–1529.

Benjamin, M. and Yiks, S. (2019) Precision livestock farming in swine welfare: a review for swine practitioners. *Animals* 9(4), 133. doi: 10.3390/ani9040133

Bilcík, B. and Keeling, L.J. (1999) Changes in feather condition in relation to feather pecking and aggressive behaviour in laying hens. *British Poultry Science* 40, 444–451.

Blomke, L., Volkmann, N. and Kemper, N. (2020) Evaluation of an automated assessment system for ear and tail lesions as animal welfare indicators in pigs at slaughter. *Meat Science* 159, 107934.

Boissy, A., Manteuffel, G., Jensen, M.B., Moe, R.O., Spruijt, B. et al. (2007) Assessment of positive emotions in animals to improve welfare. *Physiology and Behavior* 92, 375–397.

Bracke, M.B.M., Zonderland, J.J. and Bleumer, E.J. (2007) Expert consultation on weighing factors of criteria for assessing environmental enrichment materials for pigs. *Applied Animal Behaviour Science* 104, 14–23.

Brünger, J., Duppel, S., Koch, R. and Veit, C. (2019) Tailception: using neural networks for assessing tail lesions as pictures of pig carcasses. *Animal* 13(5), 1030–1036.

Dawkins, M.S., Donnelly, C.A. and Jones, T.A. (2004) Chicken welfare is influenced more by housing conditions than stocking density. *Nature* 427, 342–344.

D'Eath, R.B. (2012) Repeated locomotion scoring of a sow herd to measure lameness: consistency over time. The effect of sow characteristics and interobserver reliability. *Animal Welfare* 21, 219–231.

de Graaf, S., Ampe, B., Buijs, S., Andreason, S.N., Roches, A. et al. (2018) Sensitivity of the Integrated Welfare Quality Scores to changing values in individual dairy cattle welfare measures. *Animal Welfare* 27, 156–157.

de Jong, I.C., Hindle, V.A., Butterworth, A. and Engel, B. (2016) Simplifying the welfare quality assessment protocol for broiler chicken welfare. *Animal* 10, 117–127.

De Rosa, G., Tripaldi, C., Napolitano, F., Saltalamacchia, F., Grasso, F. et al. (2003) Repeatability of some animal related variables in dairy cows and buffaloes. *Animal Welfare* 12, 625–629.

de Vries, M., Bokkers, E.A.M., Van Schaik, G., Botreau, R., Engel, B., et al. (2013) Evaluating the results of the Welfare Quality multicriteria evaluation model for classification of dairy cattle welfare at the herd level. *Journal of Dairy Science* 96, 6264– 6273.

Doughty, A.K., Horton, B.J., Huven, N.T.D., Ballagh, L.R., Corkrey, R. and Hinch, G.N. (2018) The influence of lameness and individuality on movement patterns in sheep. *Behavioral Processes* 151, 34–38.

Edgar, J.L., Mullen, S.M., Pritchard, J.C., McFarlane, U.J.C. and Main, D.C. (2013) Towards a good life for farm animals: development of resource tier framework to achieve positive welfare for laying hens. *Animals* 3, 584–605.

Edge, M.K. and Barnett, J.L. (2008) Development and integration of animal welfare standards into company quality assurance programs in the Australian livestock (meat) processing industry. *Australian Journal of Experimental Agriculture* 48(7), 1009–1013.

Edmonson, A.J., Lean, I.J., Weaver, L.D., Farver, T. and Webster, G. (1989) A body condition scoring chart of Holstein dairy cows. *Journal of Dairy Science* 72, 68–76.

EdwardsCallaway, L.N., CalvoLorenzo, M.S., Scanga, J.A. and Grandin, T. (2017) Mobility scoring in finished cattle. *Veterinary Clinics of North America, Food Animal Practice* 33, 235–250.

EFSA (European Food Safety Authority) (2012) Statement on the use of Animal based measures to assess the welfare of animals. *EFSA Journal* 10(6), 2767. doi: 10.2903/j.efsa.2012.2767

EFSA (European Food Safety Authority) (2019) Slaughter of Animals: Poultry. *EFSA Journal* 17(11), 5849.

Engblom, L., Lundeheim, N., Dalin, A.M. and Anderson, K. (2007) Sow removal in Swedish commercial herds. *Livestock Science* 106, 76–86.

Ferguson, J.O., Galligan, D.T. and Thomsen, N. (1994) Principal descriptors of body condition score in Holstein cows. *Journal of Dairy Science* 77, 2695–2703.

Fulwider, W.K., Grandin, T., Garrick, D.J., Engle, T.E., Lamm, W.D. et al. (2007) Influence of free-stall base on tarsal joint lesions and hygiene in dairy cows. *Journal of Dairy Science* 90, 3559–3566.

Gaughan, J.B. and Mader, T.L. (2014) Body temperature and respiratory dynamics in unshaded beef cattle. *International Journal of Biometeorology* 58, 1443–1450.

Gaughan, J.B., Mader, T.L., Holt, S.M. and Lisle, A. (2008) A new heat load index for feedlot cattle. *Journal of Animal Science* 86, 226–234.

Gibbons, J., Haley, D.B., Cutler, J.H., Nash, C., Heyerhoff, J.Z. et al. (2014) Technical note: A comparison of 2 methods of assessing lameness prevalence in tie-stall herds. *Journal of Dairy Science* 97, 350–353.

Grandin, T. (1998a) Objective scoring of animal handling and stunning practices at slaughter plants. *Journal of the American Veterinary Medical Association* 212, 36–39.

Grandin, T. (1998b) The feasibility of using vocalization scoring as an indicator of poor welfare during slaughter. *Applied Animal Behaviour Science* 56, 121–128.

Grandin, T. (2001) Cattle vocalizations are associated with handling and equipment problems in beef slaughter plants. *Applied Animal Behaviour Science* 71, 191–201.

Grandin, T. (2005) Maintenance of good animal welfare standards in beef slaughter plants by use of auditing programs. *Journal of the American Veterinary Medical Association* 226, 370–373.

Grandin, T. (2010) Auditing animal welfare at slaughter plants. Meat Science 86, 56–65. Grandin, T. (ed.) (2014) *Livestock Handling and Transport*, 4th edn. CAB International, Wallingford, UK.

Grandin, T. (2017) On-farm conditions that compromise animal welfare that can be monitored at the slaughter plant. *Meat Science* 132, 52–58.

Grandin, T. (2019) *Recommended Animal Handling Guidelines and Audit Guide*, 2019 edn. American Meat Institute, Washington, DC.

Grandin, T., Curtis, S.E. and Taylor, I.A. (1987) Toys, mingling and driving reduce excitability in pigs. *Journal of Animal Science* 65 (Suppl. 1), 230 (Abstract).

Heath, C., Lin, Y.C., Mullan, S., Brown, W.J. and Main, D.C.J. (2014) Implementing Welfare Quality® in UK assurance schemes: evaluating the challenges. *Animal Welfare* 23, 95–107. Hemsworth, P.H., Barnett, J.L., Coleman, G.J. and Hansen, C. (1989) A study of the relationships between the attitudinal and behavioral profiles of stock persons and the level of fear of humans and reproductive performance of commercial pigs. *Applied Animal Behaviour Science* 23, 301–314.

Hewson, C.J. (2003) Can we access welfare? *Canadian Veterinary Journal* 44, 749–753.

Hoffman, A.C., Moore, D.A., Wenz, J.R. and Vanegas, J. (2013) Comparison of modeled sampling strategies for estimation of dairy herd lameness, prevalence and cow level variables associated with lameness. *Journal of Dairy Science* 96, 5746–5755.

Jones, E.K.M., Wathes, C.M. and Webster, A.J.F. (2005) Avoidance of atmospheric ammonia by domestic fowl and the effect of early experience. *Applied Animal Behaviour Science* 90, 293–308.

Kniese, C.I., Buttner, C., Bellenge, K., Grosse, E., Schrader, L. and Kricter, J. (2016) Test retest reliability of the Welfare Quality animal welfare assessment protocol for growing pigs. *Animal Welfare* 25, 447–459.

Knowles, T.G., Kestin, S.C., Haslam, S.M., Brown, S.N., Green, L.E. et al. (2008) Leg disorders in broiler chickens, prevalence, risk factors and prevention. *PLoS ONE* 3(2), e1545. doi: 10.1371/journal.pone.0001545

Kristensen, H.H. and Wathes, C.M. (2000) Ammonia and poultry: a review. *World Poultry Science Journal* 56, 235–243.

Kristensen, H.H., Burgess, L.R., Demmers, T.G.H. and Wathes, C.M. (2000) The preferences of laying hens for different concentrations of ammonia. *Applied Animal Behaviour Science* 68, 307–318.

LayWel (2006) Welfare implications of changes in production systems for laying hens. Available at: http://ec.europa.eu/food/animal/welfare/farm/laywel_final_report_en.pdf (accessed 14 August 2020).

Leary, N.W., Byrne, D.T., O'Connor, A.H. and Shalloo, L. (2020) Invited review: Cattle lameness detection with acetometers. *Journal of Dairy Science* 103(5), 3895–3911.

Losada-Espinosa, N., Villarrael, M., Maria, G.A. and Miranda de la Lama, G.C. (2017) Preslaughter cattle welfare indicators for use in commercial abattoirs with voluntary monitoring systems: a systematic review. *Meat Science* 138, 34–38.

Main, D.C.J., Mulan, S., Atkinson, C., Cooper, M., Wrathall, J.H.M., et al. (2014) Best practice framework for animal welfare certification schemes. *Trends in Food Science and Technology* 37, 127–136.

Maria, G.A., Villarroel, M., Chacon, G. and Gebresenbet, G. (2004) Scoring system for evaluating the stress to cattle of commercial loading and unloading. *Veterinary Record* 154, 818–821.

Mellor, D.J. (2017) Operational details of the Five Domains Model and its key applications to assessment and management of Animal Welfare. *Animals* 7(8), 60. doi: 10.3390/ani7080060

Meluzzi, A., Fabbri, C., Folegatti, E. and Sirri, F. (2008) Survey of chicken rearing conditions in Italy: effects of litter quality and stocking density on productivity, foot dermatitis and carcass injuries. *British Poultry Science* 49, 257–264.

Mollenhorst, H., Rodenburg, T.B., Bokkers, E.A.M., Koene, P. and de Boer, I.J.M. (2005) On farm assessment of laying hen welfare – a comparison of one environment based and two Animal based methods. *Applied Animal Behaviour Science* 90, 277–291.

Montoro, C., Miller-Cushon, E.K., DeVries, T.J. and Bach, A. (2013) Effect of physical form of forage on performance, feeding behavior, and digestibility of Holstein calves. *Journal of Dairy Science* 96, 1117–1124.

Mulling, C., Dopfer, D., Edwards, T., Larson, C., Tomlinson, D. et al. (2014) *Cattle Lameness, Identification, Prevention, and Control of Claw Lesions*, 2nd edn. Zinpro Corporation, Eden Prairie, Minnesota. To obtain, email: contact@zinpro.com.

Mullins, I.L., Truman, C.M., Campler, M.R., Bewly, J.P. and Costa, J.H.C. (2019) Validation of a commercial automated body condition scoring system on a commercial dairy farm. *Animals* 9(6), 287. doi: 10.3390/ani9060287

Munoz, C., Campbell, A., Hemsworth, P. and Doyle, R. (2018) Animal based measures to assess the welfare of extensively managed ewes. Animals 8(1), 2. doi: 10.3390/ani8010002 NAMI (2019) *Recommended Animal Handling Guidelines and Audit Guide*. North American Meat Institute, Washington, DC.

Nicole, R.A. (2014) A comparison of three animal welfare assessment programs on Canadian Swine Farms. Thesis, University of Guelph, Canada.

OIE (2019a) Chapter 7.9. Animal welfare and beef cattle production systems. In: *Terrestrial Animal Health Code*, 28th edn. World Organisation for Animal Health, Paris. Available at: oie.int/standardsetting/terrestrialcode/accessonline (accessed 16 August 2020).

OIE (2019b) Chapter 7.10. Animal welfare and broiler chicken production systems. In: *Terrestrial Animal Health Code*, 28th edn. World Organisation for Animal Health, Paris. Available at: oie.int/standardsetting/terrestrialcode/accessonline (accessed 16 August 2020).

OIE (2019c) Chapter 7.5. Slaughter of animals. In: *Terrestrial Animal Health Code*, 28th edn. World Organisation for Animal Health, Paris. Available at: oie.int/standard setting/terrestrialcode/accessonline (accessed 16 August 2020).

OIE (2019d) Chapter 7.3. Transport of animals by land. In: *Terrestrial Animal Health Code*, 28th edn. World Organisation for Animal Health, Paris. Available at: oie.int/standard setting/terrestrialcode/accessonline (accessed 16 August 2020).

OIE (2019e) Chapter 7.13. Animal welfare and pig production systems. In: *Terrestrial Animal Code*, 28th edn. World Organisation for Animal Health, Paris. Available at: oie.int/standardsetting/terrestrialcode/accessonline (accessed 16 August 2020).

Pfeifer, M., Eggemann, L., Kransmann, J., Schmitt, A.O. and Hessel, E.F. (2019) Inter and intra-observer reliablity of animal welfare indicators for on-farm self-assessment of fattening pigs. *The Animal Consortium* 13(8), 1712–1720. doi: 10.1017/S175173118003701

Phythian, C.J., Toft, N., Cripps, P.J., Michalopoulou, E., Winter, A.C. et al. (2013) Interobserver agreement diagnostic sensitivity and specificity of Animal based indicators of young lamb welfare. *Animal* 7, 1182–1190.

Phythian, C.J., Michalopoulou, E. and Duncan, J.S. (2019) Assessing the validity of animal-based indicators of sheep health and welfare: do observers agree? *Agriculture* 9(5), 88. doi: 10.3390/agriculture 9050088

Robichaud, M.V., Rushen, J., de Passille, A.M., Vasseur, E., Orsel, K., and Pellerin, D. (2019) Association between on-farm animal welfare indicators and productivity and profitability on Canadian dairies: 1. On free stall farms. *Journal of Dairy Science* 102, 4341–4351.

Rutherford, K.M., Donald, R.D., Lawrence, A.B. and Wemelsfelder, F. (2012) Qualitative behavioural assessment of emotionality in pigs. *Applied Animal Behaviour Science* 139, 218–224.

Saraiva, S., Saraiva, C. and Stillwell, G. (2016) Feather condition and clinical scores as indicators of poultry welfare at the slaughterhouse. *Research in Veterinary Science* 107, 75–79. doi: 10.1016/j.rvsc.2016.05.005

Schlageter-Tello, A., Hertem, T.V., Bokkers, E.A.M., Viazzi, S., Bahr, C. and Lokhorst, K. (2018) Performance of human observers and automatic 3-dimensional computer vision-based locomotion scoring method to detect lameness and hoof lesions in dairy cows. *Journal of Dairy Science* 101, 6322–6335.

Silvera, A.M., Knowles, T.G., Butterworth, A., Berckmans, D., Vranken, E. and Blokhuis, H.J. (2017) Lameness assessment with automatic monitoring of activity in commercial broiler flocks. *Poultry Science* 96, 2013-2017.

Smerchek, D. and Smith, Z. (2020) Effects of bedding application on feedlot receiving phase growth performance and estimated maintenance requirement in newly weaned beef steers. *American Society of Animal Science, Annual Meeting Abstract* 162.

Smulders, D., Verbeke, G., Mormede, P. and Geers, R. (2006) Validation of a behavioral observation tool to assess pig welfare. *Physiology and Behavior* 89, 438–447.

Stockman, C.A., McGilchrist, P., Collins, T., Barnes, A.L., Miller, D. et al. (2012) Qualitative behavioural assessment of Angus steers during preslaughter handling and relationships with temperament and physiological responses. *Applied Animal Behaviour Science* 142, 125–133.

Stull, C.L., Reed, B.A. and Berry, S.L. (2005) A comparison of three animal welfare assessment programs in California dairies. *Journal of Dairy Science* 88, 1595–1600.

Thomsen, P.T., Munksgaard, L. and Togersen, F.A. (2008) Evaluation of lameness scoring of dairy cows. *Journal of Dairy Science* 91, 119–126.

University of Wisconsin (2005) *Body Condition Score—What is Body Condition Score? What Does it Help Us Manage?* University of Wisconsin, Madison.

Van Os, J.M.C., Winckler, C., Treb, J., Matarazzo, S.W., Lehenbauer, T.W. et al. (2018) Reliability of sampling strategies for measuring dairy cattle welfare on commercial farms. *Journal of Dairy Science* 101, 1495–1504.

Vasseur, E., Gibbons, J., Rushen, J. and de Passille, A. (2013) Development and implementation of a training program to ensure high repeatability of body condition scoring of dairy cows. *Journal of Dairy Science* 96, 4725–4737.

Velarde, A. and Dalmau, A. (2012) Animal welfare assessment at slaughter in Europe: moving from inputs to outputs. *Meat Science* 92, 244–251.

Viera, A., Battini, M., Can, E. and Mattiello, S. (2018) Interobserver reliability of animal-based welfare indicators included in the Animal Welfare indicators welfare assessment protocol for dairy goats. *Animal* 9, 1942–1949. doi: 10.1017/S1751731117003597

von Borell, E., Bockisch, F.J., Büscher, W., Hoy, S., Krieter, J. et al. (2001) Critical control points for on-farm assessment of pig housing. *Livestock Production Science* 72, 177–184.

Wallgren, T., Larsen, A., Lundeheim, N., Westin, R., and Gunnarsson, S. (2019) Implication and impact of straw provision on behaviour, lesions, and pen hygiene on commercial farms rearing undocked pigs. *Applied Animal Behavior Science* 210, 26–37.

Webb, L.E., Bokkers, E.A.M., Heutinck, L.F.M., Engel, B., Buist, W.G. et al. (2013) Effects of roughage source, source amount and particle size on the behaviour and gastrointestinal health of veal calves. *Journal of Dairy Science* 96, 7765–7776.

Webster, A.B., Fairchild, B.D., Cummings, T.S. and Stayer, P.A. (2008) Validation of three-point gait scoring system for field assessment of walking ability of commercial broilers. *Journal of Applied Poultry Research* 17, 529–539.

Webster, J. (2005) The assessment and implementation of animal welfare: theory into practice. *Revue Scientifique et Technique (International Office of Epizootics)* 24, 723–734.

Welfare Quality Network (2009) Assessment protocols. Available at: www.welfarequality.net (accessed 12 August 2020).

Whay, H.R., Main, D.C.J., Green, L.E. and Webster, A.J.E. (2003) Assessment of the welfare of dairy cattle using animal-based measurements: direct observations and investigation of farm records. *Veterinary Record* 153, 197–202.

Whay, H.R., Leeb, C., Main, D.C.J., Green, L.E. and Webster, A.J.F. (2007) Preliminary assessment of finishing pig welfare using animal-based measurements. *Animal Welfare* 16, 209–211.

Wickham, S.L., Collins, T., Barnes, A.L., Miller, D.W., Beatty, D.T. et al. (2015) Validating the use of qualitative behavioural assessment as a measure of the welfare of sheep during transport. *Journal of Applied Animal Welfare Science* 18(3), 269–286.

Wildman, E.E., Jones, G.M., Wagner, P.E., Boman, R.L., Troutt, H.F. et al. (1982) A dairy cow body condition scoring system and its relationship to selected production characteristics. *Journal of Dairy Science* 65, 495–501.

Winckler, C. and Willen, S. (2001) The reliability and repeatability of a lameness scoring system for use as an indicator of welfare in dairy cattle. *Acta Agriculturae Scandinavica: Section A, Animal Science* 30 (Suppl. 1), 103–107.

Xiong, J. and Grankow, J. (2020) Winter effects on cattle performance. *Feedlot Magazine*, Dighton, Kansas.

Zinpro Corporation (no date) *Locomotion Scoring of Dairy Cattle.* Zinpro Corporation, Eden Prairie, Minnesota. Available at: zinpro.com/lameness/dairy/locomotionscoring (accessed 14 August 2020).

TWENTY-THREE

On-Farm Conditions That Compromise Animal Welfare That Can Be Monitored at the Slaughter Plant

T. Grandin (2017) On-Farm Conditions That Compromise Animal Welfare That Can Be Monitored at the Slaughter Plant, *Meat Science*, Vol. 132, pp. 52-58.

The use of objective numerical scoring is described for use in slaughter plants to document poor living conditions on the farm. Today some of the scoring methods outlined in this paper are being done by cameras linked to artificial intelligence programs. There are some welfare issues that are impossible to assess at slaughter such as the type of housing, environmental enrichments, and anesthesia for painful surgeries.

ABSTRACT

Handling and stunning at slaughter plants has greatly improved through the use of numerical scoring. The purpose of this paper is to encourage the use of numerical scoring systems at the slaughter plants to assess conditions that compromise welfare that occurred either during transport or on the farm. Some of the transport problems that can be assessed are bruises, death losses, and injured animals. Welfare issues that occurred on the farm that can be assessed at the abattoir are body condition, lameness, lesions, injuries, animal cleanliness and internal pathology. There are important welfare issues that cannot be assessed at slaughter. They are on-farm euthanasia methods, use of analgesics during surgeries, and the type of animal housing systems. Welfare evaluations at slaughter have the potential to greatly improve welfare.

INTRODUCTION

Handling and stunning of cattle, pigs, and sheep has improved in many countries. Audits and standards required by major buyers of meat have greatly improved conditions in the United States (Grandin, 2010, 2000a, 2005). The slaughter plants were evaluated with numerical scoring of stunning efficacy, slips and falls, vocalization, electric prod use and insensibility (Grandin, 2010, 1998). Numerical scoring for assessing stunning and handling has also been used by Welfare Quality Network (2009); (Velarde & Dalmau, 2012; Dalmau & Nande, 2016). In many cases, both stunning and handling was improved without major investments in equipment (Grandin, 2000a). In one study, a more highly skilled operator stunned a higher percentage of cattle accurately (Atkinson, Velarde & Algers, 2013). Poor maintenance was a major cause of ineffective captive bolt stunning (Grandin, 1998). Another method that has been used to improve both stunning and handling is video auditing where a third-party auditing firm monitors unloading, handling, and stunning with remotely viewed video cameras. The use of numerical scoring should be expanded to determine the percentages of animals that have welfare issues that occurred on the farm.

The purpose of this paper is to review animal welfare problems that have occurred either on the farm or during transport that can be easily assessed at the slaughter plant. It is much easier to monitor the large numbers of animals that arrive at a slaughter plant than to visit the many farms where they originate. Several research groups have already determined that many conditions that may compromise animal welfare can be easily assessed at the abattoir (Llonch, King, Clarke, Downes & Green, 2015; Harley, Moore, O'Connell, et al., 2012; Harley, Moore, Boyle, et al., 2012). There are two categories of animal welfare programs that can be assessed at a slaughter plant. They are: 1) Acute or traumatic conditions that recently occurred that are associated with loading on the farm or transport and 2) long-term chronic conditions. Chronic problems were present before animals were loaded for transport. Some examples of recently occurring conditions that occur during transport are bruises, dead animals (DOAs), fresh injuries, and non-ambulatory animals. Some examples of chronic long-term welfare problems that are not usually associated with transport are lameness (difficulty walking), shoulder sores on sows, swollen hocks on dairy cows, breast blisters on chickens, neglected injuries, necrotic prolapses or advanced cancer eye.

PRINCIPLES OF ASSESSMENT TOOL USE

There are many different assessment tools for evaluating lameness, lesions and other problems. For example, different tools for assessing lameness and hock lesions may have three to five categories (Welfare Quality Network, 2009; Grandin, 2015; Angell, Cripps, Grove-White & Duncan, 2015; Zinpro, 2016; Nalon, Conte, Maes, Tuyltens & DeVillers, 2013; Gibbons, Vasseur, Rushen & dePasille, 2012). When assessments are compared between different abattoirs it is important that they both used the same assessment tool. Photos of lesions that evaluators can hold during assessments may improve accuracy (Foddar, Green, Mason & Kaler, 2012). Training programs also help improve the repeatability of assessments (Gibbons et al., 2012). In the next section of this paper, transport and on-farm welfare problems that can be assessed at the abattoir will be reviewed.

ASSESSMENT OF ACUTE OR TRAUMATIC CONDITIONS THAT RECENTLY OCCURRED

Bruises

The first step is to start measuring bruises to determine where the baseline is. When a baseline is determined, it will make it possible to determine if bruise levels are increasing or decreasing. To improve the accuracy of bruise assessment, it is important to have the same person do the scoring. Strappini, Frankena, Metz, and Kemp (2012) found that intra-observer reliability was higher than interobserver reliability (Strappini, Frankena,

Metz, Gallo & Kemp, 2011; Strappini et al., 2012). Several bruise scoring systems are available for cattle (Strappini, Metz, Gallo & Kemp, 2009; Anderson & Horder, 1979; Chile, 1992; Chile, 2002; McKeith et al., 2015). In supply chains where loading on the farm and transport to the abattoir occurs within 12 h, it is very difficult to tell the age of a bruise (Strappini et al., 2009). Bruises that are more than 18 h old can be differentiated from fresh bruises, because they will have a yellowish color (Langlois, 2007). Animals that are marketed through a series of markets or dealers may have old bruises where it can be easily determined that the bruises did not occur at the plant. Old bruises often have yellowish mucous (Grandin, 2000b). In cattle, bruises can be separated into two categories: fresh bruises and old bruises that definitely occurred outside the abattoir. In the U.S. and Canada, cattle are held for only a short time in the lairage. In feedlot beef, the total time from loading at the feedlot until stunning is usually under 12 h. In many supply chains, the time between loading on the farm and stunning may be much larger. In systems where cows may be moving through markets or on trucks for many hours, old bruises that occurred outside the abattoir can be easily identified (Strappini et al., 2011). A histological test can also be used to determine if a bruise is over 24 h old (McCausland & Dougherty, 1978). In broiler chickens, bruises can also be separated into old and new categories. Bruises that are over 24 h old will have a green color (Northcutt & Rowland, 2000).

Bruises can occur after captive bolt stunning and prior to bleeding (Meischke & Horder, 1976). They will be a bright red. Horns are another cause of increased bruises (Ramsey, Meischke & Anderson, 1976; Shaw, Baxter & Ramsey, 1976). Tipping horns does not reduce bruises (Ramsey et al., 1976). Minka and Ayo (2007) found that in Western Africa, breeds with huge horns had more bruises than cattle with smaller horns. To determine the origin of fresh bruises, differences in the percentages of bruised carcasses between different farms and transporters has to be tabulated. If either a single farm or a single transporter has a significantly higher percentage of bruises, then it is likely that the bruises are not occurring in the abattoir. Bruises that occur in the slaughter plant will usually occur on animals from many different farms or transporters. These bruises are often on the same location of the carcass or they may mainly occur on very tall cattle that hit their backs on equipment.

In a poultry plant with poorly supervised shacklers, chickens from multiple farms had bruised legs (Grandin, 2015). These bruises were definitely occurring in the abattoir and were caused by handlers who squeezed the thighs too hard. Changing how people are paid may also reduce bruises. When producers and transporters have to pay for bruises, they will be greatly reduced (Grandin, 1981).

Injuries inflicted by people

Danish researchers have developed a scoring system for pigs to differentiate scratches and injuries that are likely to be inflicted by humans from scratches caused by pigs fighting (Nielsen et al., 2014). Other injuries inflicted by people that can be easily detected at the plant are shotgun shot, broken tails on cattle, and hide damage due to poking cattle with sticks with nails in them. Hide damage or broken tails that definitively occurred on the farm can be easily differentiated from more recent injuries that may have been inflicted by a transporter. Older injuries will be healed. A healed broken tail will have a permanent bend or kink.

Dead on arrival or non-ambulatory

Both DOAs and downed non-ambulatory animals can be associated with either poor conditions during transport or conditions on the farm. Overloading of trucks with either cattle or pigs may increase bruises, non-ambulatory and dead-on arrival. In an overstocked truck, a downed animal cannot get back up (Tarrant, Kenny & Harrington, 1988). Genetics is also a factor. Pigs that are either heterozygous or homozygous for the porcine stress halothane gene will have a higher percentage of pigs dead on arrival (Murray & Johnson, 1998; Holtcamp, 2000). A dose of 200 mg/animal/day of the beta agonist zilpaterol was associated with a higher percentage of feedlot death losses (Longeragen, Thomson & Scott, 2014). Pigs fed a high dose of ractopamine may have more non-ambulatory animals if they are handled roughly (Peterson et al., 2015). The author has observed that charging producers a fee for handling non-ambulatory pigs greatly reduced them. At one slaughter plant where the pigs were grown under a contract, the authors observed high percentages of non-ambulatory pigs. The percentage of downed non-ambulatory pigs was cut in half by making three changes in farm production practices. These were 1) a change in boar genetics to eliminate lameness caused by poor leg conformation, 2) reduced the dose of ractopamine, and 3) acclimation of the pigs to people walking through their pens during the finishing period.

EASE OF HANDLING

The importance of acclimation to handling

Some animals will move more easily through alleys and races than others. People who work in the lairage (yards) have informed the author that pigs or cattle from different producers are either difficult to move or easy to move. An animal's previous experiences with handling at the farm can affect its reaction to being handled in the future (Grandin, 1997; Grandin & Shivley, 2015). Objective numerical scoring can be used to determine which groups of animals can be moved more easily. Animals that are more difficult to move may be more likely to be abused. Some of the handling variables that can be compared between different producers animals are: vocalization due to electric prod use, balking, refusing to move, backing up, or turning back (Grandin, 1998; Welfare Quality Network, 2009; Edwards et al., 2010). Three studies have shown that pigs will move more easily if they have been acclimated to being handled (Abbott, Hunter, Guise & Penny, 1997; Geverink et al., 1998; Krebs & McGlone, 2009). Producers should walk quietly through the fattening pens throughout the feeding period to improve ease of movement at the abattoir (Grandin, 2015). This will train the pigs to quietly get up and move away from the person. Animals differentiate between a person in the alley and a person in their pen. Pigs may be more difficult to load and unload from a truck if their first experience with people in their pens occurs the day of loading. Another problem area is extensively raised cattle that have been handled exclusively by people on horseback. This is a common problem in the U.S., Canada, Australia, and South America. The author has observed that they can be dangerous to handle by a person on foot (Grandin, 2015). A person on a horse is perceived as safe and familiar. A person on foot is novel and frightening, which greatly enlarges the animal's flight zone. To improve both animal welfare and safety for employees at the abattoir, cattle should become accustomed to being moved in and out of pens by people on foot before they leave the ranch or feedlot of origin.

Beta-agonists and handling

Another factor that may affect ease of handling is high doses of beta agonists. In pigs, a high dose of ractopamine made them reluctant to move (Marchant-Forde, Lay, Pajor, Richert & Schinckel, 2003). A high dose of 7.5 mg/kg increased the incidence of non-ambulatory pigs after they were handled in an aggressive manner (Peterson et al., 2015). Two new studies clearly show that high doses were more likely to cause problems. Peterson et al. (2015) fed pigs ractopamine for 28 days at doses of 0 mg/kg, 5 mg/kg, and 7.5 mg/kg. Pigs fed the highest dose had more non-ambulatory pigs (Peterson et al., 2015). Noel et al. (2016) compared handling of pigs fed 10 mg/kg of ractopamine or 0 mg/kg. The pigs fed 0 mg/kg were able to walk a further distance before they became exhausted. In the Noel et al. (2016) study, ractopamine was fed for 32 days.

In cattle, observations by the author indicated high doses of Zilpaterol was associated with stiff muscles and reluctance to move (Grandin, 2015). This was most likely to occur during hot weather. A dose of 200 mg/animal/day of ractopamine for 28 days had a negligible effect on handling through a squeeze chute (Baszczak et al., 2006). Cattle handling observations were done during a cool day. To help prevent handling or welfare problems associated with beta-agonists, the author has four recommendations: 1) Use lower doses, 2) hot weather over 90 °F is more likely to cause problems, 3) reduce the number of days the beta-agonist is fed, 4) allow cattle to rest after physical exertion. In large feedlots cattle often have to walk over a kilometer from their home pens to the loading ramp. Feedlot managers have observed that bringing them up close to the loading ramp the day before transport will allow them to recover and prevent handling problems. Cattle on beta-agonists may require more time to recover from physical exertion.

EXTERNAL LESIONS AND DAMAGE ASSOCIATED WITH HOUSING CONDITIONS

Cattle

The percentage of dairy cows with injured hocks, swollen leg joints, or lame (difficulty walking) can be easily evaluated at the abattoir. Scoring systems for evaluating hock lesions can be found in Fulwider, Grandin, Garrick, Engle & Rollin, 2007; Welfare Quality Network, 2009; Gibbons et al., 2012; Von Keyserlingk, Barrientos, Ito, Galo & Weary, 2012; and Gibbons et al., 2012. An increased percentage of swollen hocks in dairy cows is associated with poor management of the bedding in free stalls (cubicle) housing (Fulwider et al., 2007). Dairies with dirty stalls had more hock injuries than dairies that had free stalls with deep clean bedding (Barrientos, Chapinal, Weary, Gallo & Vonkeyserlingk, 2013). A Dutch study showed the importance of a soft lying surface to prevent injuries (deVries et al., 2015). Stalls that are too short can also damage the hocks on the concrete curb, Beef cattle housed on concrete slats for 128 days had swollen leg joints (Wagner, 2016). When rubber mats were installed over the concrete, swollen leg joints were reduced. Cattle can also get damage to the top grain of the leather from muddy feedlots.

Pigs

Shoulder lesions (decubital ulcers) may occur in sows. The presence or absence of shoulder lesions cannot be used to determine if a producer is housing sows in stalls. The author has observed shoulder lesions in group housed sows. Shoulder lesions are genetically correlated with thin back fat (Lundeheim, Lundgren & Rydhmer, 2014). Sows with thin body condition were also more likely to have shoulder lesions (Lundeheim et al., 2014). One method to help prevent shoulder lesions is to feed sows more to increase their body condition. The author has observed that when farm managers started measuring the prevalence of shoulder lesions, they greatly reduced them. People manage the things they measure.

POULTRY

In broilers

In broilers, the three main lesions associated with housing problems that can be evaluated at the abattoir are: breast blisters, hock burn and foot pad lesions (Saraiva, Saraiva & Stillwell, 2016; Mayne, 2005; deJong, Gunnink & van Harn, 2014). Scoring systems are available for foot pad lesions (Dawkins, Donelly & Jones, 2004; Ekstrand, Algers & Swedberg, 1997), breast blisters (Allain et al., 2009), and hock burn (Allain et al., 2009). Saraiva et al. (2016) has further information on scoring systems. Kjaer, Su, Nielsen, and Sorensen (2016) indicated that slow growing birds had less hockburn. Research shows that footpad lesions cause pain in turkeys (Wyneken, Sinclair, Veldkamp, Vinco & Hocking, 2015). Wet litter increases footpad lesions in broiler chickens (deJong et al., 2014). There is some evidence that genetic factors may contribute to susceptibility to footpad dermatitis and hock burn (Kjaer et al., 2016). A study done by Jacob, Baracho, Naas, Salgado, and Souza (2016) illustrates the importance of using outcome measures instead of input engineering standards. In this study, broilers on reused sawdust litter had lower levels of footpad lesions compared to new sawdust litter (Jacob et al., 2016). Broilers can also be inspected for eye damage due to high ammonia levels in the building. Ammonia can definitively irritate the eyes and mucous membranes (Kristensen & Wathes, 2000; Miles, Miller, Branton, Maslin & Lott, 2006). Another condition in broilers that can be assessed in the slaughter plant is twisted legs (tibial dysplasia). Bone abnormalities may be associated with rapid growth (Shim, Karnauah, Mitchell, Anthony & Aggrey, 2011). Progressive breeders have worked to correct these problems. There are some fast growing broilers that have strong bones (Shim et al., 2011). If leg problems are observed, the use of gait scoring at the farm is strongly recommended. Information on gait scoring and assessment of poultry mobility can be found in Berg and Sanotra (2003); Kestin, Gordon, Su, and Sorensen (2001); Knowles et al. (2008).

Another issue that may need to be addressed is woody breast (muscle myopathies) which can be easily observed at the slaughter plant as white streaks in the breast meat.

Research is needed to determine possible welfare issues associated with muscle myopathy (Thaxton et al., 2016).

Layers

They can be assessed for foot damage and damage from feather pecking. Scoring systems for foot problems in layers can be found in Welfare Quality Network, 2009; Blanchford, Fulton & Mench, 2015.

COAT/FEATHER CONDITION

Poorly maintained housing, parasites or behavior problems can damage the feathers or coats of livestock and poultry. In dairy goats, animals with rough matted coats were more likely to have nutritional deficiencies and be skinny compared to goats with normal, smooth, shiny hair (Battini et al., 2015). In cattle, bald spots from heavy lice infections can be easily observed. Bald spots from lice should not be confused with normal shedding. The author has observed lice problems in some organic cattle raised according to U.S. standards. The producer did not want to lose his/her U.S. organic status by treating them. Other parasites that can damage the hide are cattle grubs that drill holes in the back of the hide. In confined sheep, wool pulling by other sheep sometimes becomes a problem (Huang & Takeda, 2016). Sheep that have wool pulled out by other sheep can be easily observed. For laying hens, there are good scoring systems available for assessing feather condition (Featherwel, in press; Laywel, 2006). Loss of feathers can be caused by several factors. They are genetic factors that influence feather pecking and housing conditions. Some genetic lines of layers are more prone to feather pecking than others (Morrissey, Brockhurst, Baker, Widowski & Sandilands, 2016). Providing hens housed in a cage free system with hay bales for foraging will help reduce feather pecking (Daigle, Rodenburg, Bolhuis, Swanson & Siegford, 2014).

LAMENESS

Lame animals that have difficulty walking can be easily assessed when they are unloaded from the trucks at the abattoir. A major problem is that many different scoring systems are used. Nalon et al. (2013) reviewed ten different systems for assessing lameness and claw lesions in sows. For all species, the most common lameness scoring systems have three to five categories. To facilitate comparison between slaughter plans, both producers and the meat industry should choose a scoring tool that everybody in their country or region will use. Some of the common lameness scoring tools that are readily available for cattle are the Zinpro five point scale (Zinpro, 2016), the Welfare Quality three point scale (Welfare Quality Network, 2009), the four-point scale in Grandin (2015), and a grainfed beef mobility scoring system (NAMI, 2015). Another problem in comparing data from different lameness scoring systems is the numbering scales. Some systems designate a normal animal as 0 and others designate the normal animal as 1. Grandin (2015) has an easy to use four-point scale for cattle, pigs, and sheep. This system can be easily changed to designating the normal animal as zero. The author has observed that training people to accurately score slight lameness score 2 on a five point scale is difficult. This is in agreement with D'Eath (2012). On a four-point scale, Angell et al. (2015) got good intra-observer reliability for sheep, but only fair to moderate interobserver reliability. Use of the same observer is strongly recommended. The Welfare Quality Network (2009) three-point lameness scoring system loses severity information but it would have better inter–observer reliability. Teaching people to differentiate between lame and normal cows requires less training than determining degrees of lameness (March, Brinkman & Winkler, 2007). On a five point scale, interobserver reliability is worst at the mild to moderate lameness scores of 2 and 3 (Schlageter-Tlelo et al., 2014). In a three-point scoring system, animals are scored as normal, lame, or non-ambulatory. Below is a scoring system developed by Grandin (2015) that can be easily used at the abattoir when animals are unloaded at a slaughter plant.

1. Normal.
2. Walks with an obvious limp (difficulty walking) but keeps up with the walking group of animals.
3. Walks with an obvious limp, and not able to keep up with the walking group.
4. Almost a downer, can barely walk.

Lameness is associated with many conditions. Dairy cows with swollen hocks are also more likely to be lame (Kester, Holzhauer & Frankena, 2014). Lameness and painful conditions can also be caused by hoof diseases such as digital dermatitis or hoof rot (Higginson-Cutler, Cramer, Walter, Millman & Kelton, 2013). The author has observed lameness associated with poor leg conformation in pigs. Leg conformation is influenced by genetics (Le et al., 2015). The animals either had a collapsed pasture or the leg and ankle were too straight. Many people in the dairy industry consider lameness in dairy cows as a major welfare problem (Ventura, von Keyserlingk & Weary, 2015). Lameness causes pain. When cows are given an analgesic, lameness is reduced (Flower et al., 2008). It is a major welfare concern because it may cause a painful condition for a long period of the animal's life. Producers who use good production practices can greatly reduce lameness. There is a big difference between the best and the worst dairies (Cook, Hess, Foy, Bennett & Bratzman, 2016; Bennett, Barker, Main, Whay & Leach, 2014; Von Keyserlingk et al., 2012). The state of Wisconsin has worked hard to reduce lameness and their average dairy cow lameness is 13% (Cook et al., 2016) and the national average is almost double (Von Keyserlingk et al., 2012; Bennett et al., 2014). Cook et al. (2016) found that the best high producing dairies had 2.8% lame cows and the worst one had 36%. Chapinal, Weary, Collings, and von Keyserlingk (2014) report that producers are motivated to reduce lameness when they receive reports which show how they rank compared to other producers. There are many researchers studying automated systems to assess lameness. Many of these systems would require that animals unloading at a slaughter plant would have to walk through in single file. This would be likely to slow down unloading and be difficult to implement.

DIRTY ANIMALS

Manure and dirt on both mammals and birds can be easily assessed. Welfare Quality Network (2009) has an easy to use three-point scoring system for poultry. Saraiva et al. (2016) used the following system, 0 = clean, 1 = soiling limited to breast area, 2 = very dirty, dirt caked or adhering to the feathers. In a large survey of fed cattle arriving at eight large abattoirs in the U.S., they were scored with a three-point system (McKeith et al., 2015). Forty-nine percent were completely clean, 37% had dirty legs, 24% dirty belly and legs (McKeith et al., 2015). When indoor housing is used, sufficient bedding should be provided to prevent soil from transferring onto the feathers or coats of the animals. The author has observed that in bedded pack indoor barns, one of the biggest problems is not supplying sufficient bedding to soak up the moisture and keep animals clean. Dirty dairy cows may have higher somatic cell counts (Reneau et al., 2005).

BODY CONDITION SCORE

Animals with a poor body condition score may be due to either a lack of feed or disease. Cattle raised under extensive conditions may become thin and then regain their body condition during the rainy season when the grass returns. At what point is a thin extensively raised cow a welfare problem? That may be subject for debate. For intensively raised animals, such as Holstein dairy cows there are many scoring tools that are available (Wildman et al., 1982; Ferguson, Galligan & Thomsen, 1994). In any particular country, it is recommended to use the assessment tool that the producers in your area are accustomed to using. Some assessments have too many categories and achieving interobserver reliability may be more difficult. In the U.S., five point scales are popular.

NEGLECTED HEALTH PROBLEMS OR INJURIES

A survey done in the U.S. on cull cows arriving at slaughter showed that a major problem was timely marketing (Roeber, Belk, Field, Scanga & Smith, 2001). Producers need to bring animals to the abattoir before they become weak and debilitated. Some examples of neglected problems that would cause animals to suffer are necrotic infected prolapses, necrotic advanced cancer eye, advanced hoof disease in livestock and large hernias (ruptures) in pigs. There is an assessment tool for hernias in pigs in Welfare Quality Network (2009). Pigs must be marketed before hernias interfere with walking or become damaged by scraping on the ground.

ABNORMAL BEHAVIOR

Abnormal repetitive behavior such as tongue rolling may be observed at the abattoir. It is most often seen in Jersey dairy cows. Wool biting and pulling can also be easily detected at the abattoir. Feeding practices at the farm may have an effect on the incidence of wool pulling (Huang & Takeda, 2016). The author has observed that there are also big differences in the percentage of pigs from different genotypes that will fight or mount each other in the lairage pens. Pigs from some farms may have increased percentage of pigs with bitten tails. There are genetic differences in the aggression levels between pigs and tail biting. Genomic testing indicates that selecting pigs for rapid growth and lean backfat unintentionally selected for pigs that are more likely to be active tail biters or receivers of tail biting (Brunberg, Jansen, Isaksson & Keeling, 2013; Brunberg, Jensen, Isakssen & Keeling, 2013). Within a population of the same pig breed, there are "neutral" animals that are less likely to tail bite or receive tail bites (Brunberg, Jansen, et al., 2013; Brunberg, Jensen, et al., 2013).

INTERNAL ORGAN INSPECTION

Inspection of the animal's internal organs can detect diseases and conditions that may occur on the farm (Krage-Rasmussen, Rousing, Sorensen & Houe, 2014; Harley, Moore, O'Connell, et al., 2012; Harley, Moore, Boyle, et al., 2012). Some of the conditions that can be detected in the internal organs are gastric ulcers in pigs, parasites, liver abscesses, and pneumonia. Holstein dairy steers fed grain are more prone to liver abscesses than beef breed cattle (Renhardt & Hubbert, 2015). Research will need to be conducted to determine the severity of liver abscesses or gastric ulcers that would be detrimental to welfare. Animals with heavy loads of parasites or signs of severe respiratory illness would also have compromised welfare.

INSPECTION FOR SIGNS OF PROCEDURES THAT ARE PROHIBITED

Many welfare guidelines prohibit docking of dairy cow's tails or mulesing in sheep. Animals that have had these procedures can be easily observed at the slaughter plant.

WELFARE CONDITIONS ON THE FARM THAT CANNOT BE ASSESSED AT THE ABATTOIR

- The use of pain relief for surgeries such as castration or dehorning Type of housing used—individual gestation stalls or group housing
- Euthanasia methods used on the farm. This is an area of great public concern due to release of undercover videos on the internet.
- Accommodating behavioral needs on the farm
- Abuse by people that does not cause an injury. Example: Dragging conscious sows or cows.
- The quality of the animal's life on the farm and its ability to experience positive emotions. Did it have a life worth living? David Mellor (2016) and other scientists maintain that good welfare goes beyond avoiding negative emotional states such as fear and pain. Boissey et al. (2007) and colleagues introduced the concept of assessing positive emotions.

OTHER CONSIDERATIONS

The public is highly concerned about on-farm euthanasia methods, the lack of pain relief for routine surgeries and restrictive housing systems. Evaluation of on-farm euthanasia methods is impossible at a slaughter plant. The need for surgeries can be eliminated or reduced by finishing (fattening) either intact males, use of immune-castration or use of polled cattle. Determination of the type of on-farm housing system is another area where assessment cannot be done at the abattoir.

CONCLUSIONS

Even though there are limitations, assessment of animal welfare indicators at the slaughter plant would greatly improve animal well-being. It has the potential to reduce chronic painful conditions that are caused by either poor management or damage to the animal from housing. Some of the conditions that could be reduced by assessment at the slaughter plant are: lameness, leg injuries, damage caused by poor bedding/litter materials, bruises, and neglected health problems.

REFERENCES

Abbott, T. A., Hunter, E. J., Guise, J. H., and Penny, P. H. C. (1997). The effect of experience of handling on a pig's willingness to move. *Applied Animal Behaviour Science*, 54, 371–375.

Allain, V., Mirabitlo, L., Arnould, C., Calas, M., LeBouquin, S., Lupo, C., and Michel, V. (2009). Skin lesions in broiler chickens measured at the slaughterhouse relationships between lesions and between their prevalence and rearing factors. *British Poultry Science*, 50, 407–417.

Anderson, B., and Horder, J. C. (1979). The Australian carcass bruise scoring system. *Queensland Agricultural Journal*, 105, 281–287.

Angell, J. W., Cripps, P. J., GroveWhite, D. H., and Duncan, J. S. (2015). A practical tool for locomotion scoring of sheep: Reliability when used by veterinary surgeons and sheep farmers. *Veterinary Record*, 176(20), 521 (doi:10.1136vr.102882).

Atkinson, S., Velarde, A., and Algers, B. (2013). Assessment of stun quality at commercial slaughter in cattle shot with captive bolt. *Animal Welfare*, 22, 473–481.

Barrientos, A. A., Chapinal, N., Weary, D. M., Gallo, E., and Vonkeyserlingk, A. G. (2013) Herd level risk factors for hock injuries in free stalled housed dairy cows in northeastern United States and Canada. *Journal of Dairy Science*, 96, 3758–3765.

Baszczak, J. A., Grandin, T., Gruber, S. L., Engle, T. E., Platter, W. J., Laudert, S. B., . Tatum, J. D. (2006). Effect of ractopamine supplementation on behavior of British continental and Brahman crossbred steers during routine handling. *Journal of Animal Science*, 84, 3410–3414.

Battini, M., Peric, T., Ajuda, I., Viera, A., Grosso, L., Barbiera, S., et al. (2015). Hair coat condition: A valid and reliable indicator for on-farm welfare assessment of adult dairy goats. *Small Ruminant Research*, 123, 197–203.

Bennett, R., Barker, Z. E., Main, D. C. L., Whay, H. R., and Leach, K. A. (2014). Investigating the value dairy farmers place on a reduction of lameness in their herds using a willingness to pay approach. *Veterinary Journal*, 199, 72–75.

Berg, C., and Sanotra, G. S. (2003). Can a modified latency to lie test be used to validate gait scoring results in commercial broiler flocks. *Animal Welfare*, 12, 655–659.

Blanchford, R. A., Fulton, R. M., and Mench, J. A. (2015). The utilization of welfare quality assessment for determining laying hen condition across three housing systems. *Poultry*, 95, 154–163.

Boissey, A., Manteuffel, G., Jensen, M. B., More, R. O., Spruitj, B., Keeling, L. J., et al. (2007). Assessment of positive emotions in animals to improve their welfare. *Physiology and Behavior*, 92, 375–397.

Brunberg, E., Jansen, P., Isaksson, A., and Keeling, L. J. (2013a). Brain gene expression differences are associated with abnormal tail biting behavior in pigs. *Genes, Brain, and Behavior*. http://dx.doi.org/10.1111/gbb.12002.

Brunberg, E., Jensen, P., Isakssen, A., and Keeling, L. J. (2013b). Behavioral and brain gene expresson profiling in pigs during tail biting outbreaks—Evidence of a tail biting resistant genotype. *PLoS One*. http://dx.doi.org/10.1371/journalpone0066513.

Chapinal, N., Weary, D. M., Collings, L., and von Keyserlingk, M. A. G. (2014). Lameness and hock injuries improve on-farms participating in an assessment program. *The Veterinary Journal*, 202, 646–648.

Chile Instituto Nacional de Normalizacion INN (2002). *Norma Chilena Oficial Nch 1306 of 2002 Canales de bovine, definicianes y tipificacion*.

Chile, M.d. A. (1992). Establece sistema obligatorio de clasificacion de Ganado tipificacion y nomenclatura de sus carne y regula el funcimamiento de matadevos trigori ficos y estableciem tos de la industrio de la carne. *Diario Official de la Republica Ley*, no. 19 (pp. 162–???).

Cook, N. B., Hess, J. P., Foy, T. B., Bennett, T. B., and Bratzman, R. L. (2016). Management characteristics, lameness, and body injuries of dairy cattle housed in high performance day herds in Wisconsin. *Journal of Dairy Science*, 99, 5879–5891.

Daigle, C. L., Rodenburg, T. B., Bolhuis, J. E., Swanson, J. C., and Siegford, J. M. (2014). Use of dynamic and rewarding environmental enrichment to alleviate feather pecking in noncage laying hens. *Applied Animal Behaviour Science*, 161, 75–85.

Dalmau, A., and Nande, A. (2016). Application of the welfare quality protocol in pig slaughter houses in five countries. *Livestock Science*, 193, 78–87.

Dawkins, M. S., Donelly, C. A., and Jones, T. A. (2004). Chicken welfare is influenced more by housing conditions than by stocking density. *Nature*, 427, 342–344.

D'Eath, R. B. (2012). Repcated locomotion scoring of a sow herd to measure lameness: Consistency over time the effect of sow characteristics and interobserver reliability. *Animal Welfare*, 21, 219–223.

Edwards, L. N., Grandin, T., Engle, T. E., Porter, S. P., Ritter, M. J., Sosnicki, A. A., and Anderson, D. B. (2010). Use of exsanguination blood lactate to assess the quality of preslaughter pig handling. *Meat Science*, 86, 384–390.

Ekstrand, C., Algers, B., and Swedberg, J. (1997). Rearing conditions and footpad dermatitis in Swedish broiler chickens. *Preventive Veterinary Medicine*, 31, 167–174.

Elanco (2009). *The 5 point body condition scoring system*. Greenfield, Indiana: Elanco Animal Health (Accessed December 25, 2016).

Featherwel (2016). *Why feather score?* University of Bristolwww.featherwel.org/inuurious (pecking/how to feather score, (Accessed December 26, 2016)).

Ferguson, J. O., Galligan, D. T., and Thomsen, N. (1994). Principle descriptors of body condition score in Holstein cows. *Journal of Dairy Science*, 77, 2695–2703.

Flower, F. C., Sedlbauer, M., Carter, E., von Keyslerlingk, M. A. G., Sanderson, D. J., and Weary, D. M. (2008). Analgesics improve the gait of lame dairy cattle. *Journal of Dairy Science*, 91, 3010–3014.

Foddar, A., Green, L. E., Mason, S. A., and Kaler, J. (2012). Evaluating observer agreement of scoring systems for foot integrity ad foot rot lesions in sheep. *BMC Veterinary Research*. http://dx.doi.org/10.1186/174661488865.

Fulwider, W., Grandin, T., Garrick, D. J., Engle, T. E., and Rollin, B. E. (2007). Influence of freestall base on tarsal joint lesions and hygiene in dairy cows. *Journal of Dairy Science*, 90, 3559–3566.

Geverink, N. A., Kappers, A., van de Burgwal, J. A., Lambooij, E., Blockhuis, H. J., and Wiegant, V. M. Effect of regular moving and handling on the behavioral and physiological responses of pigs to preslaughter treatment and consequences for subsequent meat quality. *Journal of Animal Science*, 76, 2080–2085.

Gibbons, J., Vasseur, E., Rushen, J., and dePasille, A. M. (2012). A training program to ensure high repeatability of injury scoring of dairy cows. *Animal Welfare*, 21, 379–388.

Grandin, T. (1981). Bruises on southwestern feedlot cattle. *Journal of Animal Science*, 53 (Suppl. 1), 213 (Abstract).

Grandin, T. (1997). Assessment of stress during handling and transport. *Journal of Animal Science*, 76, 249–257.

Grandin, T. (1998). Objective scoring of animal handling and stunning practices at slaughter plants. *Journal of the American Veterinary Medical Association*, 212, 36–39.

Grandin, T. (2000a). Effect of animal welfare audits of slaughter plants by a major fast food company on cattle handling and stunning practices. *Journal of the American Veterinary Medical Association*, 216, 848–851.

Grandin, T. (2000b). *Livestock Handling and Transport* (2nd ed.). Wallingford, Oxfordshire UK: CABI International.

Grandin, T. (2005). Maintenance of good animal welfare standards in beef slaughter plants by using auditing programs. *Journal of the American Veterinary Medical Association*, 226, 370–373.

Grandin, T. (2010). Auditing animal welfare at slaughter plants. Meat Science, 86, 56–65. Grandin, T. (2015). *Improving Animal Welfare: A Practical Approach* (2nd ed.). Wallingford, Oxfordshire UK: CABI International.

Grandin T. , 2015 2nd edition. *Improving Animal Welfare: A Practical Approach*, Wallingford, Oxfordshire UK CABI International.

Grandin, T., and Shivley, C. (2015). How farm animals react and perceive stressful situations such as handling, restraint, and transport. *Animals*, 5(4), 1233–1251. http://dx.doi.org/10.3390/ani5040409.

Harley, S., Moore, S., Boyle, L., O'Connell, N., and Hanlon, A. (2012a). Good animal welfare makes economic sense: Potential of pig abattoir meat inspection as a welfare surveillance tool. *Irish Veterinary Journal*. http://dx.doi.org/10.1186/20460481 6511.

Harley, S., Moore, S. J., O'Connell, N. E., Hanlon, A., Teixeira, D., and Bogle, L. (2012b). Evaluating the prevalence of tail biting and carcass condemnations in slaughter pigs in the Republic of North Ireland and potential abbatoir meat inspection as a welfare surveillance tool. *Veterinary Record*. http://dx.doi.org/10.1136/vr.100986.

Higginson Cutler, J. H., Cramer, G., Walter, J. J., Millman, S. T., and Kelton, D. F. (2013). Randomized clinical trial of tetracycline hydrochloride bandage and paste treatments for resolution of lesions and pain associated with digital dermatitis in cattle. *Journal of Dairy Science*, 96, 7550–7557.

Holtcamp, A. (2000). Gut edema: Clinical signs, diagnosis and control. *Proceedings of the American association of swine practitioners 11–14, March* (pp. 337–340). Indianapolis, Indiana.

Huang, C. Y., and Takeda, K. (2016). Influence of food type and its effect on suppressing wool biting behavior in confined sheep. *Animal Science Journal*. http://dx.doi.org/10. 1111/asj.12664.

Jacob, F. G., Baracho, M. S., Naas, I. A., Salgado, D. A., and Souza, R. (2016). Incidence of pododermatitis in broilers reared under two types of environments. *Revista Brasileira de Ciencia, Avicola*. http://dx.doi.org/10.1590/1806906120150047.

deJong, I. C., Gunnink, H., and van Harn, J. (2014). Wet litter not only induces footpad dermatitis but reduces overall welfare, technical performance, and carcass yield in broiler chickens. *Journal of Applied Poultry Research*, 23, 51–58.

Kester, E., Holzhauer, M., and Frankena, K. (2014). A descriptive review of the prevalence and risk of hock lesions in dairy cows. *The Veterinary Journal*, 202, 222–228.

Kestin, S. C., Gordon, S., Su, G., and Sorensen, P. (2001). Relationships in broiler chickens between lameness, live weight, growth rate, and age. *The Veterinary Record*, 148, 195–197.

Kjaer, J. B., Su, G., Nielsen, B. L., and Sorensen, P. (2016). Foot pad dermatitis and hock burn in broiler chickens degree of inheritance. *Poultry Science*, 85, 1342–1348.

Knowles, T. G., Kestin, S. C., Haslam, S. M., Green, L. E., Butterworth, A., et al. (2008). Leg disorders in broiler chickens prevalence risk factors and prevention. *PLoS One*, 3(2), e1545. http://dx.doi.org/10.1371/journalpone.0001543.

KrageRasmussen, K. M., Rousing, T., Sorensen, J. T., and Houe, H. (2014). Assessing animal welfare in sow herds using data on meat inspection, medication, and mortality. *Animal*, 9, 509–515.

Krebs, N., and McGlone, J. J. (2009). Effects of exposing pigs to moving and odors in a simulated slaughter chute. *Applied Animal Behaviour Science*, 116, 179–185.

Kristensen, H. H., and Wathes, C. M. (2000). Ammonia and poultry welfare: A review. *World's Poultry Science Journal*, 56, 235–245.

Langlois, N. E. I. (2007). The science behind the quest to determine the age of bruises: A review of English language literature. *Forensic Science, Medicine, and Pathology*, 3, 241–251.

Laywel (2006). Welfare implications of changes in production in production systems for laying hens. *Photographic scoring system* (Available at laywel.eu. (Accessed December 21, 2016)).

Le, T. H., Norberg, E., Hielsen, B., Madsen, P., Nilssen, K., and Lundelheim, N. (2015). Genetic correlation between leg conformation in young pigs, sow reproduction, and longevity in Danish pig populations. *Acta Agriculturae Scandinavica Section A Animal Science*, 65, 132–138.

Llonch, P., King, E. M., Clarke, K. A., Downes, J. M., and Green, L. E. (2015). A systematic review of animal based indicators of sheep welfare on farm, and market, and during transport and qualitative appraisal of their validity and feasibility for use in the abattoirs. *Veterinary Journal*, 206, 289–297.

Longeragen, C. H., Thomson, D. U., and Scott, H. M. (2014). Increased mortality in groups of cattle administered the badrenergic agonist ractopomine hydrochloride or Zilpaterol hydrocholoride. *PLoS One*. http://dx.doi.org/10.37/journalpone0091177.

Lundeheim, N., Lundgren, H., and Rydhmer, L. (2014). Shoulder ulcers in sows are genetically correlated to leanness of young pigs and to litter weight. *Acta Agriculturae Scandinavica Section A Animal Science*, 64.

March, S., Brinkman, J., and Winkler, C. (2007). Effect of training on the interobserver reliability of lameness coring in dairy cattle. *Animal Welfare*, 16, 131–133.

MarchantForde, J. N., Lay, D. C., Pajor, H. A., Richert, J. A., and Schinckel, A. P. (2003). The effects of ractopamine on the behavior and physiology of finishing pigs. *Journal of Animal Science*, 81, 416–422.

Mayne, R. K. (2005). A review of the aetiology and possible causative factors of foot pad dermatitis in growing turkeys and broilers. *World's Poultry Science Journal*, 61, 256–267.

McCausland, I. P., and Dougherty, R. (1978). Histological aging of bruises in lambs and calves. *Australian Veterinary Journal*, 54, 525–527.

McKeith, R. O., Gray, G. D., Hale, D. S., Karth, C. R., Griffin, D. B., Savell, J. W., et al. (2015). National beef quality audit 2011: Harvest floor assessments of targeted characteristics that effect quality and value of cattle carcasses and byproducts. *Journal of Animal Science*, 90, 5135–5142.

Meischke, H. R. C., and Horder, J. C. (1976). Knocking box effect on bruising in cattle. *Food Technology in Australia*, 28, 369–371.

Mellor, D. J. (2016). Updating animal welfare thinking: Moving beyond the "five freedoms" towards "a life worth living". *Animals*, 6(3), 21. http://dx.doi.org/10. 3390/ani6030021.

Miles, D. M., Miller, W. W., Branton, S. L., Maslin, W. R., and Lott, B. D. (2006). Occular responses to ammonia in broiler chickens. *American Association of Avian Pathologists*, 50, 45–49.

Minka, N. S., and Ayo, J. O. (2007). Effects of loading behavior and road transport stress on traumatic injuries in cattle transported by road during the hot, dry season. *Livestock Science*, 107, 91–95.

Morrissey, K. L. H., Brockhurst, S., Baker, L., Widowski, T., and Sandilands, V. (2016). Cannon beak treated hens be kept in commercial furnished cages: Exploring the effects of strain and extra environmental enrichment on behaviour, feather cover and mortality. *Animals*, 6(3), 17 (doi.3390/ani6030017).

Murray, A. C., and Johnson, C. P. (1998). Importance of the halothane gene on muscle quality and preslaughter death in western Canadian pigs. *Canadian Journal of Animal Science*, 78, 543–548.

Nalon, E., Conte, S., Maes, D., Tuyltens, F., and DeVillers, N. (2013). Assessment of lameness and claw lesions in sows. *Livestock Science*, 156, 10–23.

NAMI. *Mobility scoring of cattle, video.* (2015). handling.org (Accessed December 26, 2016).

Nielsen, S. S., et al. (2014). The apparent prevalence of skin lesions suspected to be human inflicted in Danish finishing pigs at slaughter. *Preventive Veterinary Medicine*, 117, 200–206.

Noel, J. A., Broxterman, R. W., McCoy, G. M., Craig, J. C., Phelps, K. J., Bunett, D. D., et al. (2016). Use of electromyography to detect muscle exhaustion in finishing barrows fed ractopomine HC1. *Journal of Animal Science*, 94, 2344–2356.

Northcutt, J. K., and Rowland, G. N. (2000). Relationship of broiler bruise age to appearance and tissue histological characteristics. *Journal of Applied Poultry Science Research*, 9, 13–20.

Peterson, C. M., Pilcher, C. M., Rothe, H. M., MarchantForde, J. M., Ritter, M. J., Darr, N., Ellis, W. (2015). Effect of feeding ractopamine hydrochloride on growth performance and responses to handling and transport in heavy weight pigs. *Journal of Animal Science*, 93, 1239–1249.

Ramsey, W. R., Meischke, H. R. C., and Anderson, B. (1976). The effect of tipping horns and interception of the journey on bruising cattle. *Australian Veterinary Journal*, 52, 285–286.

Reneau, J. K., Seykova, A. J., Heins, B. J., Endres, M. I., Farnsworth, R. J., and Bey, R. F. (2005). Association between hygiene scores and somatic cell scores in dairy cattle. *Journal of the American Veterinary Association*, 227, 1297–1301.

Renhardt, C. D., and Hubbert (2015). Control of liver abscesses in feedlot cattle: A review. *The Professional Animal Scientists*, 31, 101–108.

Roeber, D., Belk, K. E., Field, T. G., Scanga, J. L., and Smith, G. C. (2001). National market cow and bull beef quality audit, 1999. A survey of producer related defects in market cows and bulls. *Journal of Animal Science*, 79, 658–665.

Saraiva, S., Saraiva, C., and Stillwell, G. (2016). Feather condition and clinical scores as indicators of broiler welfare at the slaughterhouse. *Research in Veterinary Science* (In Press).

SchlageterTlelo, A., Bokkers, E. A. M., Grootkoerkamp, P. W. G., Van Hertern, T., Viazzi, S., Romanini, C. B., et al. (2014). Effect of merging levels of locomotion scores for dairycows on intra and interrates reliability and agreement. *Journal of Dairy Science*, 97, 5533–5542.

Shaw, F. D., Baxter, R. I., and Ramsey, W. R. (1976). The contribution of horned cattle to carcass bruising. *Veterinary Record*, 98, 255–257.

Shim, M. Y., Karnauah, A. B., Mitchell, A. D., Anthony, N. B., and Aggrey, S. E. (2011). The effects of growth rate on leg morphology and tibia breaking strength, mineral density, mineral content and bone ash in broilers. *Poultry Science*, 91, 1790–1795.

Strappini, A. C., Frankena, K., Metz, J. H. M., Gallo, C., and Kemp, B. (2011). Characteristics of bruises in carcasses of cows sourced from farms or from livestock markets. *Animal*. http://dx.doi.org/10.1017/51751731111001698.

Strappini, A.C., Frankena, K., Metz, J.H.M., and Kemp, B. (2012) Intra and interobserver reliability of a protocol for postmortem evaluation of bruises in Chilean beef carcasses.

Strappini, A. C., Metz, J. H. M., Gallo, C. B., and Kemp, B. (2009). Origin and assessment ofbruises in beef cattle at slaughter. *Animal*, 3, 728–736.

Swaby, H., and Gregory, N. G. (2012). A note on the frequency of gastric ulcers detected during postmortem examination at a pig abattoir. *Meat Science*, 90, 269–271.

Tarrant, P. V., Kenny, F. J., and Harrington, D. (1988). The effect of stocking density during 4 hour transport to slaughter on behavior, blood constituents and carcass bruising in Friesian steers. *Meat Science*, 24, 209–222.

Thaxton, Y. V., Christensen, K. D., Mench, J. A., Runley, E. R., Daughterty, C., and Feinberg, B. (2016). Symposium: Animal welfare challenges for today and tomorrow. *Poultry Science*, 95, 2198–2207.

Velarde, A., and Dalmau, A. (2012). Animal welfare assessments at slaughter: Moving from inputs to outputs. *Meat Science*, 92, 244–251.

Ventura, B. A., von Keyserlingk, M. A. G., and Weary, D. M. (2015). Animal welfare concerns and values of stakeholders within the dairy industry. *Journal of Agricultural and Environmental Ethics*, 28, 109–126.

Von Keyserlingk, M. A. G., Barrientos, A., Ito, K., Galo, E., and Weary, D. M. (2012). Benchmarking cow comfort on North American free stall dairies: Lameness leg injuries, lying time, facility design and management for high producing Holstein dairy cows. *Journal of Dairy Science*, 95, 7399–7408.

deVries, M., Bukkers, E. A. M., van Reenen, C. G., Engel, B., van Schaik, G., Dijkstra, T., and de Boev, I. J. M. (2015). Housing and management factors associated with indicators of dairy cow welfare. *Preventive Veterinary Medicine*, 110, 80–92.

Wagner, D. (2016). *Behavioral analysis and performance response of feedlot steers on concrete slots versus rubber slats* (abstract). Salt Lake City, Utah: American Society of Animal Science.

Welfare Quality Network. *Assessment protocol for cattle*. (2009). www. welfarequalitynetwork.net (accessed, December 11, 2016).

Wildman, E. E., Jones, G. M., Wagner, P. E., Boman, R. L., Troutt, H. F., et al. (1982). A dairy cow body condition scoring system and its relationship to selected production characteristics. *Journal of Dairy Science*, 65, 495–501.

Wyneken, C. W., Sinclair, A., Veldkamp, T., Vinco, L. J., and Hocking, P. M. (2015). Foot pad dermatitis and pain assessment in turkey poults using analgesia and objective gait analysis. *British Poultry Science*, 56, 522–530.

Zinpro. *First step dairy locomotion scoring videos*. (2016). www.zinpro.com/videolibrary/ dairylocomotionvideos#/videos/list (Accessed December 26, 2016).

TWENTY-FOUR

Crossing the Divide Between Academic Research and Practical Application of Ethology and Animal Behavior Information on Commercial Livestock and Poultry Farms

T. Grandin (2019) Crossing the Divide Between Academic Research and Practical Application of Ethology and Animal Behavior Information on Commercial Livestock and Poultry Farms, *Applied Animal Behavior Science*, 218:104828.

One of the important things covered in this paper is the need for more animal behavior courses in both veterinary and animal science curriculums. Other topics that are covered are preparing students for careers and making research results accessible to the public.

ABSTRACT

There are young managers in commercial animal agriculture in the United States, United Kingdom, Asia, and other countries, who are unaware of the scientific field of animal ethology. They may have an agricultural degree with no training in animal behavior. Some have no idea that scientists have already conducted many research studies on animal behavior. In this opinion paper, the author discusses ways to cross this divide. Basic animal behavior principles should be taught to both veterinary and animal science undergraduate students. The basic information that should be taught to undergraduates is: 1) Behavioral principles of livestock handling, 2) importance of good stockmanship to improve animal productivity, 3) principles of animal learning, 4) bull, ram and boar safety, 5) the importance of behavioral needs and environmental enrichment, 6) how to recognize abnormal behaviors, and 7) formation of dominance hierarchies. This material should be in introductory courses with practical explanations about why it is important. For example, a nutritionist needs to understand how dominance behavior may reduce access to feed. When I communicate directly with students, they are eager to learn about behavior. Students should also be taught to use the major academic databases. The second step is that researchers must communicate with producers in jargon-free language. The third step is training graduate students for management jobs on farms or research careers in industry. In the developed countries, there is a shortage of academic positions for new Ph.D.'s. Graduates in animal behavior subjects can have excellent careers outside of academia. Training in animal behavior may help them influence the policies of their employers to improve animal welfare. There are also factors in the future that may block free flow of scientific information. Unfortunately, some research results remain proprietary commercial industry information and they are not published in the scientific literature. To promote the spread of knowledge, academic researchers should avoid signing long-term nondisclosure agreements with industry. These agreements may block scientific publication. Everybody in the field of animal behavior needs to communicate outside their field and explain why behavior is important.

INTRODUCTION

In many fields of science, there is a lack of communication between academic research scientists and people who could benefit from the results of the research (Butler, 2008). During many discussions with pork farm managers in the United States and Asia, I was shocked to learn that many managers were not aware that the field of ethology existed. They became interested when I showed them research papers that may be helpful to them. In this keynote address, I will discuss practical ways to reach across the divide between research scientists and agricultural managers who are not scientists. Many of the statements in this paper are my opinions. They are based on over 40 years of work in both academia and consulting work with the livestock industries. I have worked with farms, ranches, and abattoirs in the U.S., Canada, Europe, Asia, and South America.

BOTH ANIMAL SCIENCE AND VETERINARY STUDENTS NEED BEHAVIOR INFORMATION IN THEIR CURRICULUM

Some students in the U.S. and other countries are able to complete either an Animal Science degree, or a Veterinary Medicine degree and receive little or no instruction about farm animal behavior. Some aca demic programs do a much better job of teaching the importance of behavior compared to others. Shivley et al. (2016) found that a quarter of U.S. accredited veterinary colleges had no animal behavior course. Even in the United Kingdom, there are gaps in animal behavior education. In 2016, I talked to four students at a U.K. bovine veterinary medicine meeting and they had received no training in the basics of bull behavior and human safety. They were all studying at the same veterinary college. In years of consulting work on animal behavior, I have learned that some people have a hard time understanding why animal behavior training is important. Another factor is that behavioral consultations with clients who own pets bring in less money than doing a surgical procedure. U.S. veterinary students are hungry for behavior information. I have been invited to many student veterinary meetings.

MINIMUM BEHAVIOR BASICS ALL STUDENTS WORKING WITH FARM ANIMAL SHOULD LEARN

1. **Livestock handling principles**

   They need to learn the use of behavioral principles to move cattle, pigs, and sheep in a calm low stress manner. Handlers also need to understand that livestock have wide angle vision, and they are likely to balk at distractions such as shadows, reflections, or changes in flooring (Prince, 1977; Grandin, 1996). Some of the basics which should be learned are flight zone and point of balance. To create a safe working environment, people handling livestock need training in animal behavior (Lindahl et al., 2013). Further information is available in Grandin (2014, 2017) and Kilgour and Dalton (1984).

2. **Safety with bulls, boars, and rams**

   Many people are not aware that the bull which is most likely to attack a person is often a hand reared "pet." He has not learned to interact with his own species. When he becomes sexually mature, he attacks people instead of other bulls. Research shows that bulls reared alone are more likely to attack people (Price and Wallach, 1990). This occurs because the bull lacked the opportunity to learn how to interact with other cattle. Students should be trained to recognized the broad side threat a bull displays before he attacks.

3. **Good stockmanship is important**

   Animals may become behaviorally agitated due to fear and show eye white (Sandem et al., 2006). Pigs, chickens, and dairy cows that fear people are less productive (Hemsworth et al., 1986, 2000; Rushen and de Passille, 2015). Good stock people are highly skilled and should receive both more pay and recognition (Daigle and Ridge, 2018). Studies show the value of good stockmanship and it can improve animal productivity (Waiblinger et al., 2006). In my consulting work, I have often emphasized the economic benefits of good stockmanship. Progressive managers will use this information and improve practices. Unfortunately, I have observed poor managers who are not willing to change old rough practices. On many very large farms I have observed problems with understaffing and overworking of stock people.

4. **Learn to recognize stereotypies**

   There is a severe lack of studies on producer knowledge of abnormal animal behavior that does not physically damage the animal. Damaging behavior, such as tail biting, is recognized by farmers. In both the U.K. and the U.S., I have recently talked to pork farm managers and bovine veterinarians who did not know that repetitive stereotypic behaviors were abnormal. This is why students need to be specifically taught about stereotypies and other abnormal behavior. A basic reference for introductory students is Mason (1991).

5. **Learn why behavioral needs are important**
   Students should be taught that highly motivated innate behavioral needs, such as secluded nest boxes for laying hens and substrates for pigs to root in are important (Cooper and Appleby, 1996; Studnitz et al., 2007). Some veterinarians and animal scientists with no background in animal behavior may not understand the importance of behavioral needs and environmental enrichment.

6. **Teach students how to use academic databases and understand data generated by precision farming methods**
   In my livestock handling class, I have students pick out an animal behavior subject that interests them. Their assignment is to find abstracts and papers in four different databases and write a short summary about the papers. The databases are Science Direct, and Web of Science. Today, approximately a third of the undergraduate students in my class had never previously used databases. When they get out in industry, database use is an important skill. Precision livestock management is becoming increasingly mainstream. Students will also need to learn how to interpret animal behavior data from spread sheets and other computerized records.

7. **Learn basic principles of animal learning**
   Students in academic programs need to understand the basic principles of operant conditioning and how it can be used to train all types of animals. In many countries outside of Europe, a student can get a degree in animal agriculture with no training in animal learning. They also need to learn that animals will approach when they are positively reinforced and avoid places and people where they have had an aversive experience.

8. **Why animals form dominance hierarchies**
   Explain the basics of the formation of behavioral hierarches. When animals form a stable dominance order, it reduces constant fighting over resources such as feed and water. On the other hand males may fight to gain access to mates.

9. **Review of basics undergraduate students need to learn**
   The seven areas listed above are elementary basics that every educated manager who works with farm animals should be trained in. Unfortunately, there are still many managers in the animal agriculture industry who do not know these simple basics. Animal handling is one of the brighter spots. In the last ten years, cattle and pig handling classes are now being sponsored by livestock associations and industry suppliers throughout the U.S. and Canada. The other issues, such as understanding abnormal behavior and behavioral needs, is less likely to be taught. In 2016, a practicing bovine veterinarian in the U.K. asked, "What are stereotypies?" His statement may not be representative of most veterinarians, but I was surprised at his lack of knowledge.

RESEARCHER COMMUNICATION WITH PRODUCERS AND SCIENTISTS IN OTHER FIELDS

Often the disconnect between academic researchers and nonscientists is due to the two groups thinking differently. In a long career as an industry consultant, I have observed that academics are often too timid to make a recommendation to a producer because they feel that more research is needed. On the other hand, producers will often make big mistakes and try something new with too little information. In 2017, three U.S. professional societies in animal science split a joint meeting of three animal research societies into three separate meetings. This was a step backward in scientific communication. The Animal Scientists, Dairy scientists, and Poultry scientists stopped meeting together. This hampers transfer of knowledge to the farmers because consultants who go to the meetings will receive less information. In my own consulting work with pigs and cattle, I have learned things in the poultry meeting that were relevant to livestock. Most people do not have the time or the money to attend three separate meetings. In conclusion of this section, Bush et al. (2018) discusses the need for public outreach of animal behavior research.

1. **Write in industry publications**
   Researchers should communicate research findings in an easy-to-understand format. Publish in blogs, websites, and livestock and poultry trade publications. Eliminate jargon and tell producers and industry people the basic findings of your results. Instead of stating that pigs had "agonistic behavior with conspecifics" state that the pigs had fights with their herd-mates. When you communicate outside your field you must simplify the information.

2. **Develop good public speaking and writing skills**
   To help prepare my students, I have often recommended that they take either a public speaking class or a writing class. In our animal science department, all graduate students have to present seminars to help improve public speaking skills. Another factor that may hamper career development is poor writing skills. There are some students who got through college without writing term papers. They never had a teacher correct their papers and edit their grammar. Nobody had taught them how to write clearly.

3. **Learn to talk to non-scientists**
   During training, students need actual experience presenting their data to non-scientists. A recent discussion with a graduate student indicated that she became frustrated when talking to nonscientists. This occurred during an internship where she worked with government legislators. She had to learn how to state data simply with no jargon. It was an important learning experience for her.

4. **Choose Topics that will be interesting to producers and non-scientists**
   In my own work, I have learned to choose animal behavior topics that the public or producers will find interesting. I chose studies that could either help them with

their own animals or stimulate their interest. Below I will give an example from my consulting work. When I consulted on pork farms, I often discussed research on how genetics has an effect on either tail biting or fighting when pigs are mixed. Tail biting and fighting are different traits with different motivations. Pork producers I have talked to are really interested in these topics. They were especially interested in a study where neutral pigs were discovered, that do not inflict tail bites on others or allow other pigs to bite their tails (Brunberg et al., 2013, 2016). Many U.S. producers who have successfully switched to group sow housing have made changes in sow genetics to reduce fighting. Aggressive behavior in sows is heritable (Lovendahl et al., 2005). When I visited these farms, the sows gently nibbled my clothes. An older genetic line of pigs that had some of the worst behavior problems in group housing would bite my boots so hard it hurt. These pigs also startled more easily.

Another important principle I teach producers is that first experiences with new people, places or equipment should be positive. When they introduce a sow or beef cow to something new, such as a feeder, a negative first experience must be avoided. Producers who have been successful with electronic sow feeders report that they need to find a patient person (Hog Whisperer) to train new gilts to use the feeder (Coleman, 2016). Another question I get asked all the time is "My cattle are gentle at home and they went crazy at the auction barn. Why?" Then I explain how a sudden new novel experience is more likely to be frightening to an animal with more flighty genetics. A review of studies that support this concept are in Grandin (1997). Producers are not interested in lectures on statistics or explanations of results loaded with statistical jargon.

5. **Emphasize economic benefits**

In my own work, I often used economic justification for the use of good stockmanship and low stress handling methods. The two benefits that I communicated to producers were improved productivity and reduction of accidents during animal handling. To help convince producers, I showed them the work of Hemsworth et al. (1986, 2000) and Rushen and de Passille (2015). When working with clients in the meat packing industry, I sold design projects by emphasize reduction of bruises and reducing the number of people required to move the animals. Another thing I learned from working with many clients is that selling them new equipment was much easier than implementation of long-lasting improvements in management. Animal handling practices often reverted back to old rough methods unless the plant manager was committed to maintaining the improvements.

6. **Interdisciplinary research moves science forward**
   Doing research either across disciplines or between academia and commercial agriculture is often difficult. Commercial agriculture may not like the results of a study and stop funding it. A recent editorial in *Nature* (2017) discussed problems with communicating with other scientists in disciplines outside of their area. They may have different ways of evaluating the data. Collaboration between different disciplines can lead to advancements in science.

PREPARING BEHAVIOR STUDENTS FOR THE FUTURE

In many scientific fields in the U.S., there are fewer academic positions for the student who is trained to become a professor and support their research with grants. For animal behavior specialists with advanced degrees, U.S. university positions are available. Some students avoided them because the pressure to bring in grant money is too great. In the field of engineering, only 13% of new Ph.D.'s can fill an academic position (Larsen et al., 2014). There are similar but less severe shortages of tenure track positions in the biomedical sciences. There is a U.S. tenure track position for one out of six Ph.D. students (Ghaffarzadegan et al., 2014). Nelson (2017) states that in the field of chemistry, too many students are being trained for academic research jobs that do not exist. In chemistry, there is increasing emphasis on training students for careers in the industry. Nelson (2017) maintains that students need leadership skills and problem-solving ability. There may be a lack of university research positions for Ph.D.'s in the U.S. but this is not the case for countries that are developing (Santos et al., 2016). During the last five years, I have lectured about animal behavior to students and agricultural managers in Brazil, Chile, Uruguay, and China and other countries that are developing their livestock industry. The students are eager to learn about behavior. In Brazil, I told students in a recently opened veterinary college that "the future is right here." Many U.S. academic departments have reached out to foreign students because their governments pay their tuition. In my program, I had a visiting student from Brazil who participated in a cattle handling study. She went back to Brazil and published research on importance of good stockmanship (Peira Lima et al., 2018). To assist in communicating the importance of animal behavior, I have reached out to South American authors to write book chapters (Grandin, 2014). It was worth the effort to spend hours on a plane correcting their English. These authors had excellent information.

Careers for graduate students

There are important differences between U.S. and Canadian veterinary training and other countries. In North America a veterinary degree DVM requires a total of eight years of academic study. If the student gets either a master's or a Ph.D., additional years are required. In Europe and many other countries, a veterinary degree requires only five to six years of study. In the U.S. and Canada, many students who get Ph.D.'s in animal behavior subject's start their Ph.D. program after they have completed a four-year undergraduate animal science curriculum. There is little available information on where students who receive either a master's or Ph.D. degree in farm animal behavior are employed. A summary of the career paths of some of my graduate students may provide insights. From 2012 until 2017, six graduate students in my program have received either a masters or Ph.D. in animal behavior related subjects. Three of these students have gone into livestock industry careers. They are working for commercial companies in either farm animal welfare auditing or as a welfare officer for a pork company. Two other recent former students are working with laboratory animals and a third one is a government scientist doing research in epidemiology and disease. Four students who graduated before 2012 are either university professors in behavior, or they received a veterinary degree after they completed their masters. One veterinarian is now a government meat inspector and the other is in a private practice. The six most recent graduates avoided academic careers that required supporting their work with grants. There were several job offers available but they depended too heavily on getting grants. In the U.S., winning grants in behavior is getting more and more difficult. To fund my programs, I have had the good fortune of being able to use money from books, consulting, and speaking engagements to support my program for behavior graduate students. Today I have several students who are doing more meat science than behavior. These projects will probably direct them into careers in the meat industry. They are researching stunning methods and bruising because the industry supported this type of research. Even though these students may not enter a behavior career, their training in animal behavior will probably be beneficial to animal welfare. When they rise to corporate level management positions their animal behavior training is likely to influence their decisions.

PROPRIETARY RESEARCH IS NOT PUBLISHED IN OPEN SCIENTIFIC LITERATURE

In the poultry and pork industry there is more and more commercial research where the results never get published. During the last 10 years from 2009 to 2019, I have observed that both the chicken and the pork industry made real progress in correcting some severe animal welfare problems related to genetics. Unfortunately, neither the public nor the academic community knows about these improvements because the companies who implemented the improvements never wrote about them in either academic journals or industry trade press. To improve communication, I have written about some of the improvements, and I have had discussions with industry leaders on the importance of communicating their successes.

In my book, Animals in Translation, I wrote about "rapist roosters" that severely injured broiler breeder hens during breeding. In 2017, I visited a U.S. broiler breeder farm and the problem with the roosters had been corrected. Unfortunately, I was not able to find any published scientific literature. I have also observed that the same poultry breeding company had corrected lameness problems in heavy broiler chickens. They bred chickens with thicker legs.

I was pleased to see these improvements, but the problem is that the public does not know about them. The poultry industry has done a poor job communicating with the public about improvements that they have made. Fifteen years ago, indiscriminate selection for pork carcass traits resulted in leg conformation defects such as "post legged" or collapsed ankles. I observed market swine arriving at the abattoir where 50% of the pigs were lame. Today, in 2018, the pigs are heavier, weighing up to 132 kg and less than 5% were lame. To achieve this, the company changed genetics to improve the structural soundness of the feet and legs. Years ago, I introduced the concept of lameness scoring to a major breeding to help bring about improvements.

NONDISCLOSURE AGREEMENTS BLOCK COMMUNICATION

Researchers in academic positions should avoid signing long-term nondisclosure agreements. If you have to sign one, amend it so it does not apply to the rest of your life. Limit it to one to five years. A student graduating today with either an M.S. or a Ph.D. in behavior, should also avoid signing long-term noncompete clauses. These can really retard a person's professional advancement. If you are forced to sign a non-compete contract, limit it to one or two years.

During a long career where I consulted for many different corporations, I have observed that sometimes the places that are most secretive are obsolete. The people at highly secretive corporations become so inside the box that they do not realize that they have become obsolete. Corporations that communicate between each other are more likely to have leading edge technology.

CONCLUSION

Students studying ethology need to be able to communicate to both the agricultural industry and the public. Good communication skills are essential. Students who are studying for general veterinary or animal science degrees should also receive training in livestock handling and ethology. This will enable them to transfer behavior knowledge to both producers and the public. Another important skill is learning how to use academic databases. To facilitate communication of valuable information, ethologists should avoid signing long-term nondisclosure or non-complete agreements with industry.

REFERENCES

Brunberg, E., Jensen, P., Isaksson, A., Keeling, L.J., 2013. Behavioral and brain gene expression profiling in pigs during tail biting outbreaks, evidence for a tail biting resistant phenotype. *PLoS One* June 18;8(6):e 66513.

Brunberg, E.L., Rodenburg, T.B., Rhahmer, L., Kjaer, J.B., Jensen, P., Keeling, L.J.H., 2016. Omnivores going astray: a review of new synthesis of abnormal behavior in pigs and laying hens. *Front. Vet. Sci.* 3, 57.

Bush, J.M., Jung, H., Connell, J.P., Freeberg, T.M., 2018. Duty now for the future: a call for public outreach by animal behavior researchers. *Anim. Behav.* 139, 161–169.

Butler, D., 2008. Transitional research: crossing the valley of death. Nature 453, 840–842. Coleman, L., 2016. Achieving high productivity in group housed sows. *Advances in Pork Production Proceeding of the 2016 Banff Pork Seminar* 95–101.

Cooper, J.J., Appleby, M.C., 1996. Demand for next boxes in laying hens. *Behav. Proc.* 36, 171–182.

Daigle, C.L., Ridge, E.E., 2018. Investing in stock-peopleis an investment in animal wel fare and agricultural sustainability. *Anim. Front.* 8, 53–59.

Ghaffarzadegan, N., Hawley, J., Larson, R., 2014. A note on PhD. Population growth in biomedical sciences. *Syst. Res. Behav. Sci.* 32, 402–405.

Grandin, T., 1996. Factors that impede animal movement at slaughter plants. *J. Amer. Vet. Med. Assoc.* 209, 757–759.

Grandin, T., 1997. Assessment of stress during handling and transport. *J. Anim. Sci.* 75, 249–257.

Grandin, T., 2006. Animals in Translation. Houghton Mifflin, Harcourt, New York, NY. Grandin, T., 2014. *Livestock Handling and Transport*, 4th ed. CAB International, Wallingford, Oxfordshire, UK.

Grandin, T., 2017. *Temple Grandin's Guide to Working With Farm Animals.* Storey Publishing, North Adams, MA, USA.

Hemsworth, P.H., Barnett, J.L., Hansen, G., 1986. The influence of handling by humans on the behavior, reproduction and corticosteroids ofmale and female pigs. *Appl. Anim. Behav. Sci.* 15, 303–314.

Hemsworth, P.H., Coleman, G.J., Barnett, J.L., Borg, S., 2000. Relationships between human animal interactions and productivity of commercial dairy cows. *J. Dairy Sci.* 78, 2821–2831.

Kilgour, R., Dalton, C., 1984. *Livestock Behavior: A Practical Guide.* Granada Publishing, St. Albans, UK.

Larsen, L.C., Ghaffarzadegan, N., Yue, Y., 2014. Too many PH.D. Graduates or too few academic job openings. The basic number Ro in Academia. *Syst. Res. Behav. Sci.* 31, 745–750.

Lindahl, C., Hagevoort, G.R., Kolstrup, C.L., Doughrate, D., Pinzke, S., 2013. Occupational health and safety aspects of animal handling in dairy production. *J. Agromedicine* 18, 274–283.

Lovendahl, R., Damgaard, L.H., Nielson, B.L., Thodberg, K., Su, G., Rydhmer, L., 2005. Aggressive behavior in sows at mixing and maternal behavior are heritable geneti cally correlated traits. *Livest. Prod. Sci.* 93, 73–85.

Mason, G.J., 1991. Stereotypes: a critical review. Anim. Behav. 41, 1015–1037. Nature, 2017. Get past glib, editorial. *Nature* 552, 148.

Nelson, D.J., 2017. Improving the Employment of Chemical Professionals Vol. 18 *CEN*, American Chemical Society.org December 2017.

Peira Lima, M.L., Negrao, J.A., Para de Paz, C.C., Grandin, T., 2018. Minor corral changes and adoption of good handling practices can improve the behavioral and reduce cortisol release in Nellore cows. *Trop. Anim. Hlth. Prod.* 50, 525–530.

Price, E.O., Wallach, S.J.R., 1990. Physical isolation of hand reared bulls Hereford bulls increases their aggressiveness towards humans. *Appl. Anim. Behav.* Sci. 27, 263–267.

Prince, J.H., 1977. *The Eye and Vision.* Swenson, M.J. (editor) Dukes Physiology of Domestic Animals. Cornell University Press, New York, pp. 696–712.

Rushen, J., de Passille, A.M., 2015. The importance of Good stockmanship and its benefits to animals. In: Grandin, T. (Ed.), *Improving Animal Welfare: A Practical Approach*, 2nd edition. CABI Publishing, Wallingford, Oxfordshire, UK, pp. 125–138.

Sandem, A.I., Janczak, A.M., Salte, R., Brastead, B.O., 2006. The use of diazepam as a pharmacological validation of eye white and an indicator of the emotional state of dairy cows. *Appl. Anim. Behav. Sci.* 96, 177–183.

Santos, J.M., Horta, H., Manuel, Heitor, 2016. Too many Ph.D.'s? An invalid argument in countries developing their scientific and academic systems. The case for Portugal. Technol. *Forecast. Soc. Change* 113, 352–362.

Shivley, C., Garry, F.B., Kogan, L.R., Grandin, T., 2016. Survey of animal welfare, animal behavior, and animal ethics courses in the curricula of AVMA Councilaccredited veterinary colleges and schools. *J. Am. Vet. Med. Assoc.* 248, 1166–1170.

Studnitz, M., Jensen, M.B., Pedersen, L.J., 2007. Why do pigs root and what will they root: a review of exploratory behavior of pigs in relation to environmental enrich ment. *Appl. Anim. Behav. Sci.* 107, 183–197.

Waiblinger, S., Bovin, X., Pedersen, V., Tosi, M.W., Janczak, A.M., et al., 2006. Assessing the human animal relationship in farmed species: a critical review. *Appl. Anim.*

TWENTY-FIVE

Evaluating the Reaction to a Rotated Complex Object in the American Quarter Horse

Corgan, M.E., Grandin, T and Matlock, S. (2021) Evaluating the Reaction to a Rotated Complex Object in the American Quarter Horse (Equus caballus) *Animals*, 11:1383, MDPI, Basal, Switzerland.

I really like this study because it is the first scientific study that shows that animal memories are highly specific. When a colorful children's plastic playset was rotated, the horses thought it was something new. Animal memories are sensory-based and not word-based. This study may also explain why a horse may spook when it is in a familiar place. Megan Corgan was one of my master's students.

SIMPLE SUMMARY

Horses are prey animals and exhibit behaviors that help them adapt and survive in their environment. These reactions are often referred to as spooking and they have the potential to be dangerous to the horse, handler and rider. Spooking consists of avoidance reactions that include suddenly moving away or running away from the perceived danger. It is dangerous for both riders and horses when a horse suddenly startles. Sometimes horses do this in familiar environments because familiar objects may look different when rotated. The purpose of this study was to determine whether horses that had been habituated to a complex object (children's playset) would react to the object as novel when rotated 90 degrees. Twenty young horses were led past the playset 15 times by a handler. Next, the rotated group was led past the rotated playset 15 times. Each time the horse was led by the object was counted as a pass. An increasing reactivity scale was used to quantify behavioral responses. Being aware of potential reactions to changes in the orientation of previously familiar objects can help keep the handler safer.

ABSTRACT

It is dangerous for both riders and horses when a horse suddenly startles. Sometimes horses do this in familiar environments because familiar objects may look different when rotated. The purpose of this study was to determine whether horses that had been habituated to a complex object (children's playset) would react to the object as novel when rotated 90 degrees. Twenty young horses were led past the playset 15 times by a handler. Next, the rotated group was led past the rotated playset 15 times. Each time the horse was led by the object was a pass. The behavioral responses observed and analyzed were ears focused on the object, nostril flares, neck raising, snort, avoid by stopping, avoid by moving feet sideways, and avoid by flight. An increasing reactivity scale was used to quantify behavioral responses. A two-sample *T-test* was performed on the reactivity scores comparing the first pass by the novel object to the first pass by the rotated object. The horses in the rotated group reacted to the rotated orientation similarly to the first exposure ($p = 0.001$, $\alpha < 0.05$). Being aware of potential reactions to changes in previously familiar environments can help keep the handler safer.

INTRODUCTION

Spooking was associated with 27% of horse accidents.[1] Researchers have described "spooking" as "horse reacted in fear of something, unseating the rider."[1] Another study showed that injuries to riders were more likely to occur "when the horse behaved in an unexpected manner."[2] Some of the more severe injuries occur when a rider falls of a horse. Data collected from hospital emergency rooms indicated that falling off a horse resulted in more injuries than injuries that occurred when the rider was not mounted.[3] Another study showed that "danger to the rider increases as the horse's speed increases."[4] When riders fall of a horse, their head is more likely to be injured compared to other parts of the body.[5] Riders have reported that sometimes they do not know what caused their horse to spook. Experienced horse trainers know that it is important to habituate young horses to new things in their environment. One study showed that gradually desensitizing horses to stimuli such as a moving white nylon bag had fewer flight responses than horses where it was suddenly introduced.[6] Several studies have also indicated that a sudden stimuli such as an umbrella that opens suddenly can startle a horse.[7–11] A sudden novel stimulus such as a waving flag or a bell will raise a horse's heart rate.[12] Researchers have also reported the intensity of a horse's reaction was influenced by horse genetics.[13] Sudden novel stimuli will startle many different species of animals.[14] The dangerous and usually unexpected "spook" reaction is likely to be a startle response.

A review of the horse behavior literature showed that there are different ways that horses are introduced to new objects. They consist of voluntary approach in an open field, led towards the object, or ridden toward the object.[15-17] Open field tests are usually conducted in an arena. In these experiments a free moving horse is allowed to voluntarily approach novel objects.[15] In many of these studies the horse learns to associate a particular novel object with a food reward.[18,19] These studies are extremely useful for the study of horse cognition and learning. However, they do not provide information on the conditions that may cause a dangerous "spook" reaction that could seriously injure a rider. One study showed that horses that were more willing to investigate a novel object were able to learn more quickly.[20] There may be differences in a horse's reaction to a novel object when the way it is presented is changed. Research has shown that a horse's reaction to a novel object is different during a voluntary approach compared to being led or ridden.[17] The authors found that when including a rider in temperament tests and performance tests, results were more reliable and repeatable. When positive reinforcement such as food is used, horses are able to choose the correct object and be presented with a reward. This type of learning and test does not often include a handler and is referred to as discrimination.[18]

The purpose of this research was to conduct a study to identify a possible situation where a large complex novel object may have the potential to cause a serious horse accident. The object we chose was a child's colored plastic playset with a small slide and swing. It is a large object and large objects may be more frightening stimulus compared to smaller objects.[21] The playset is an object that a horse could possibly see during riding near residences. It is known that horses can learn to discriminate between different colors and shapes.[18,22] In our experiment, no attempt was made to control for color or shape, but a playset was chosen that had approximately the same dimensions when it was rotated.

The purpose of this preliminary study was to determine if a habituated horse may react differently when the playset was rotated. Both its shape and color would change. A nonvoluntary approach was used. All horses were led at a walk for both habituation and the test after the playset was rotated. For safety reasons, we conducted all experiments at a walk and measured behavioral reactions used by Leiner and Fendt.[7] Their study may provide insights into a situation that might trigger dangerous "spooking." If the rotated playset is associated with increased mild behavioral reactions when a horse is being led towards it, there is the possibility that it might trigger a more dangerous sudden reaction if a horse and rider was trotting towards it. A high-speed study would be too dangerous and would be unethical. This study may help identify one cause of spooking when a rider is not able to identify an obvious possible trigger. It will also provide a starting point for further studies into horse perception.

MATERIALS AND METHODS

The sample population consisted of twenty-two 2 and 3yearold American Quarter Horse horses (15 fillies and seven gelded colts) in a university horse training program. The horses had 4 months of handling training at the time of this study, all trained at the same place. The horses were taught using methods of pressure and release training to halter, lead, lunge, and acclimate to being groomed and handled. Of the twenty-two horses, one posed a safety concern for the research handlers by its continued attempt to pull away and was excluded from the study. Another horse was removed from the study on day 4 for soundness issues. Twenty horses continued through the entire study (*n* = 20). All horses were housed at the Colorado State University Equine Teaching and Research Center (CSU ETRC) in outdoor pens with ad libitum water and access to shelter. Horses were fed a mix of grass and alfalfa hay once per day on a feed bunk.

The test environment was an alley (4.57 meters wide) in an indoor horse barn in front of empty stalls with the doors closed. The barn had concrete flooring and electric lights above the alleyway. The horses were led in through the entrance, walked down the alleyway, past the novel object and led out of the test area through the exit (Figure 1). Two Hero 5 video cameras (GoPro, San Mateo, CA, USA) were placed in the test environment for later observation of behavioral responses.

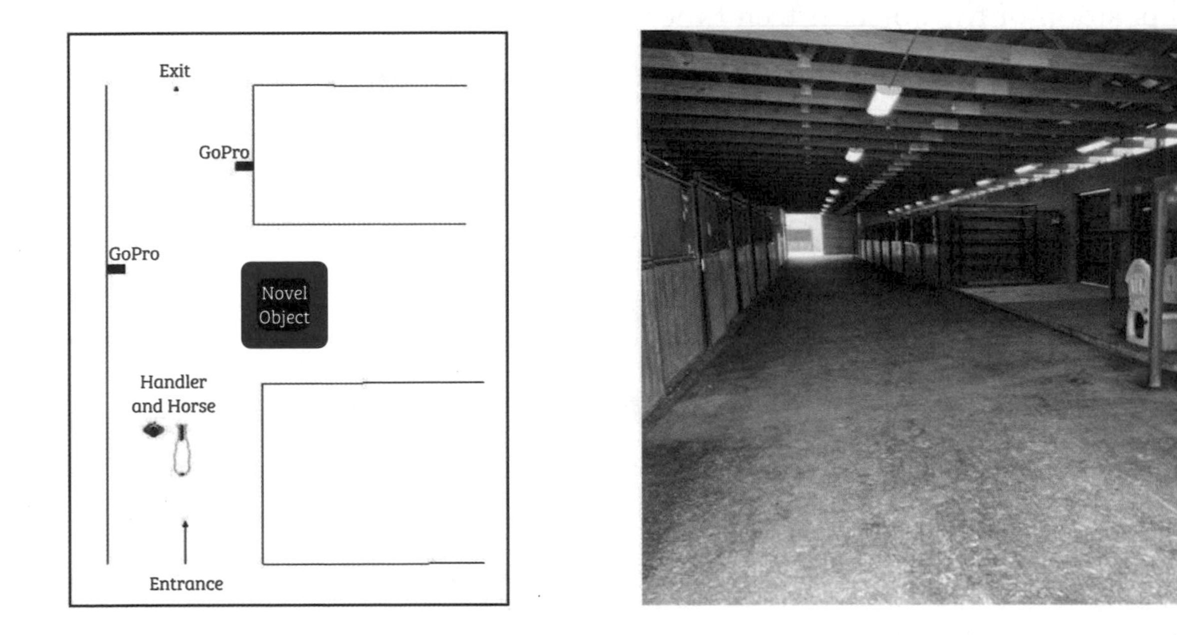

Fig. 1.
The test area consisted of GoPros (rotation days.) and the novel object placed during habituation to the novel object

The novel object was a children's plastic playset (Little Tikes Hide and Seek Climber and Swing Brown and Tan) (Table 1). The object was 134.62 l × 132.08 w × 104.14 h cm. This object was used because, in both orientations, its outer dimensions are similar. The playset had a different shape when rotated ninety degrees. Rope halters with 2 m lead ropes were used to lead the young horses past the playset.

On day 1–3 of the study (habituation to the test area) (Table 1), the horses were led through the test area five passes each day without the novel object to habituate the horses to the test area. Each time a horse was led past the object was counted as a pass. The horses were given 15 total passes through the test area based on Christensen et al.[15] The authors found that horses needed 4–13 exposures before meeting habituation criterion.

On day 4, the novel object was placed in the test area in the original position. Days 4–6 of the study (habituation to the novel object) consisted of the same procedure for the first three days with the novel object in its original position (Table 1). Each horse passed the original position of the object fifteen times over three days.

To assess the effect of object rotation on the horses' behavior, the horses were randomized into a control group and a rotated group. On days 7–9 (test days), the control group had five passes each day through the test area, with the novel object in the original position (Table 1). The control group passed the original position of the object fifteen times over the three days. The rotated group was led through the test environment for five passes each day with the novel object rotated 90 degrees clockwise (Table 1). The rotated group passed the rotated position of the object fifteen times over the three days.

Two handlers were used and both led the horses at a slow walk. Each handler had an equal number of horses randomly assigned from both the control and rotated group. Each horse was led at a walk through the test area by the same handler for the entire study. The handlers wore the same clothes every day (black overalls, jacket, hat and black boots). The handlers were instructed to walk the horse with a lead rope through the center of the alley (1 m away from the object), and move with the horse, only stopping or turning when the horse stopped or turned towards the object. If the horse stopped, the handler waited 3 seconds before gently encouraging the horse forward by walking forward and slightly pulling on the lead rope. To facilitate habituation, if a horse stopped when it was either approaching or passing the novel object, it was allowed to stop for a period of 3 seconds. If the horse did not stop, the handler continued to lead it past the novel object.

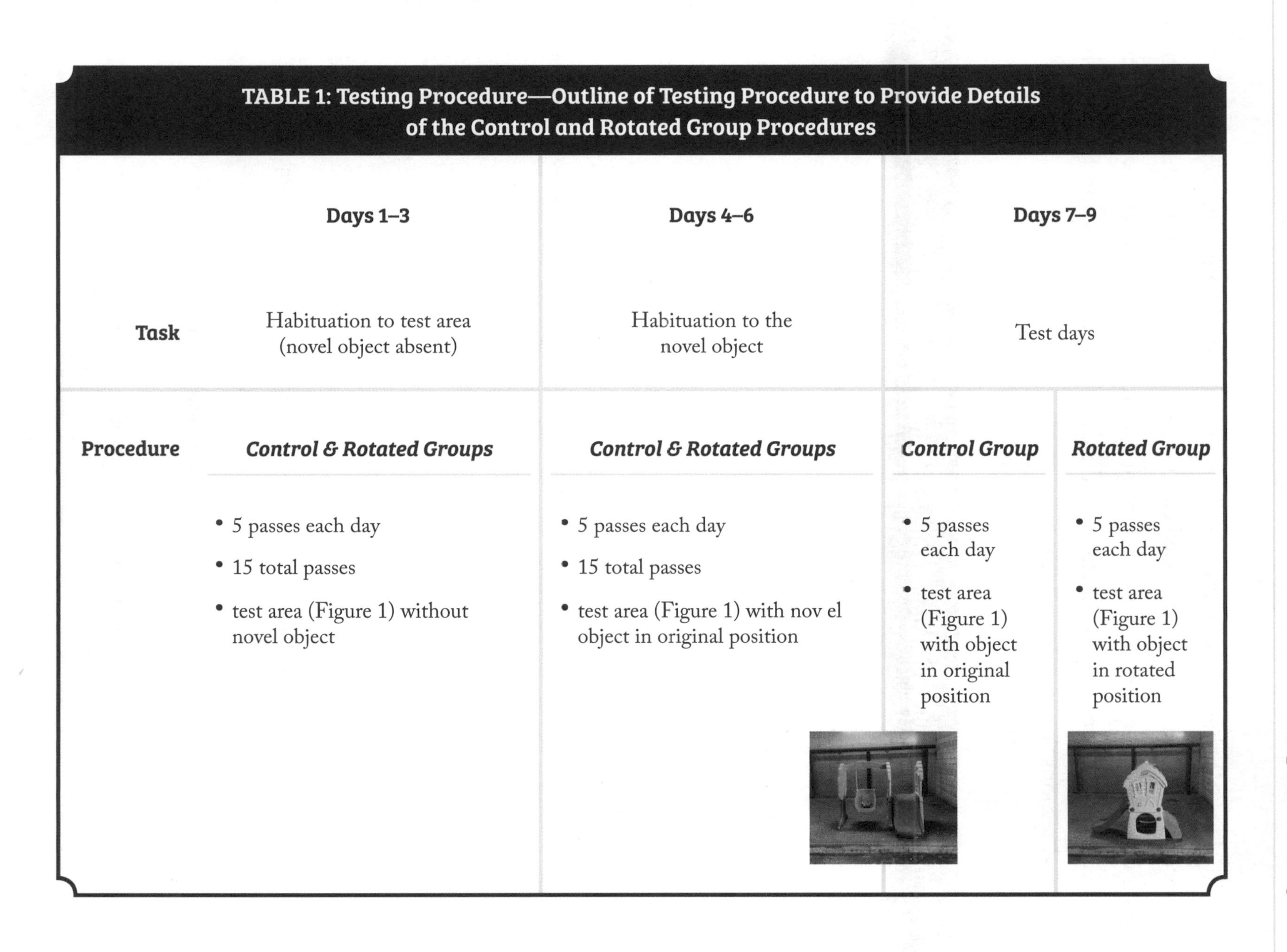

TABLE 1: Testing Procedure—Outline of Testing Procedure to Provide Details of the Control and Rotated Group Procedures

	Days 1–3	Days 4–6	Days 7–9	
Task	Habituation to test area (novel object absent)	Habituation to the novel object	Test days	
Procedure	***Control & Rotated Groups***	***Control & Rotated Groups***	***Control Group***	***Rotated Group***
	• 5 passes each day • 15 total passes • test area (Figure 1) without novel object	• 5 passes each day • 15 total passes • test area (Figure 1) with nov el object in original position	• 5 passes each day • test area (Figure 1) with object in original position	• 5 passes each day • test area (Figure 1) with object in rotated position

The horses' behavioral responses to the object exposure were analyzed based on the video recordings. One observer recorded eight different behavioral signs during each pass on each day. The behavioral responses recorded were ears focused on the object, nostril flares, neck raising, snorting, avoid stop, avoid side, avoid back, and avoid flight (Table 2). Behavioral responses were adapted from [7].

TABLE 2: Behavioral Responses and Definitions Used for Behavioral Analysis

Behavioral Responses	Definition
ears focused on the object	ears are pointed toward the novel object
nostril flares	nostrils overly expanded (nose elongation)
neck raising	neck raised above normal headset and/or neck muscles tense
snorting	"short powerful exhale"[23]
avoid stop	avoiding the object by stopping, feet stop moving
avoid side	avoiding the object by evasive steps to the side, away from the object
avoid back	avoiding the object by evasive steps backwards, backing up
avoid flight	avoiding the object by jumping away in a sudden movement, feet moving faster a walk

A reaction scale was created from the behaviors observed on a scale from 0–3 (Table 3). This reaction scale was adapted from Christensen et al.[15] to evaluate reactivity based on behaviors observed in this study. The reactivity scores increase with a bigger response to the object, score 3 showing the biggest response.

TABLE 3: Reaction Scale Used to Quantify Behavioral Responses (Adapted from Christensen et al.[15])

Score 0–3	Behavioral Responses Observed
0	No behavioral signs observed
1	Ears focused, nostril flares, and/or neck raising
2	Snorting and/or avoid stop
3	Avoid side, avoid back, avoid flight

To assess whether the horses' behavior was affected by the object rotation we compared difference in the reaction score per individual horse using a two-sample *T-test* ($\alpha < 0.05$) (R with R Studio, PBC, Boston, MA, USA). Christensen et al. and Tidyverse (www.tidyverse.org, accessed on 28 March 2021) was used for the figures.[15, 24, 25] This test was done for each pass 1–15 comparing the corresponding passes from the original position to the rotated position.

RESULTS

The control and rotated group showed significant differences between the change in reaction score from the first pass by the novel object to the first pass on the Test days (*T-test p-value* = 0.001) (Figure 2). Horses that reacted to the novel object in the Rotated group, reacted similarly on the first pass by the rotated position of the object as they did on the initial pass by the novel object.

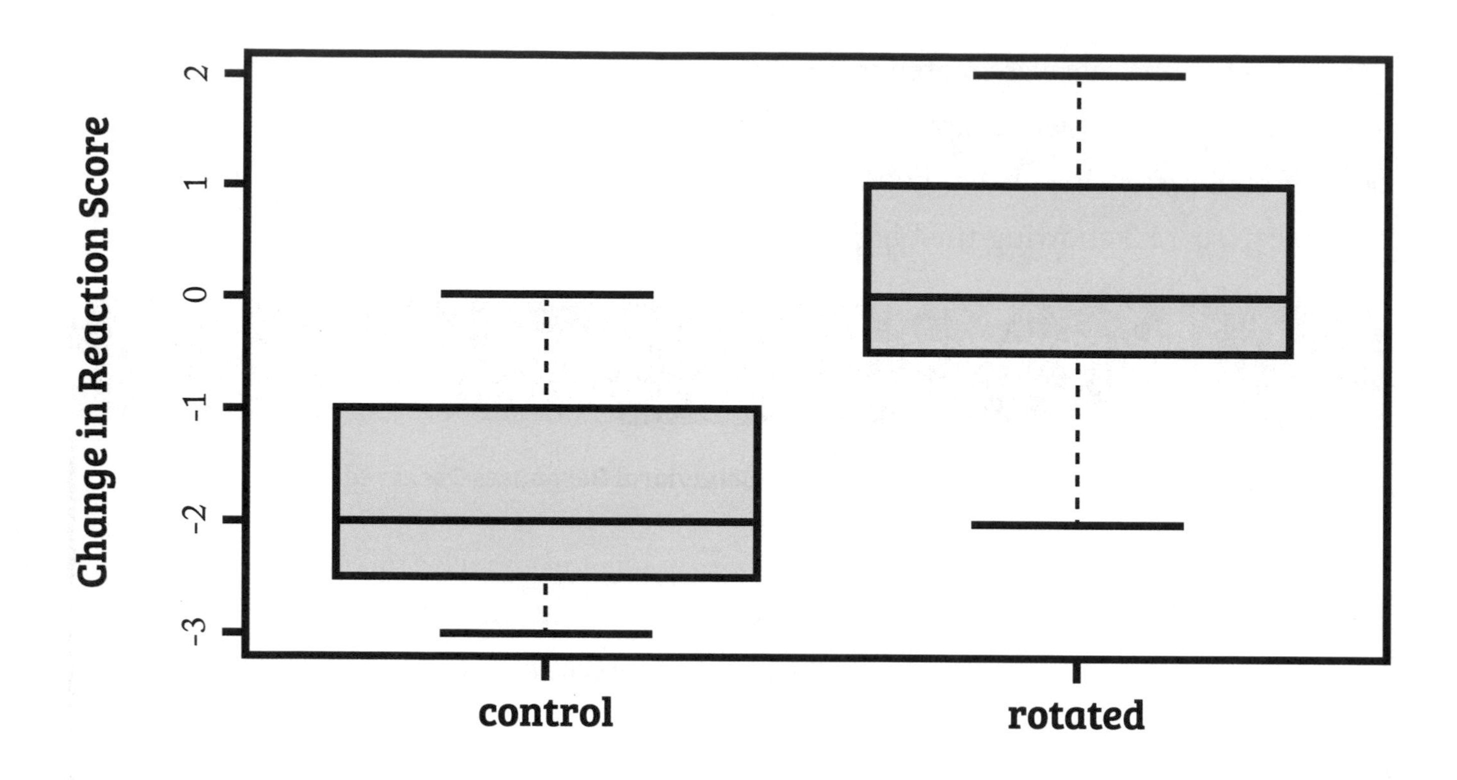

Fig. 2.
Boxplot of differences in Reaction Score for pass 1 by the novel object to pass 1 by the rotated object. The Control group mean (bold line) showed a decrease in reaction. The Rotated group mean showed no change in reaction. There was a significant difference between the means of the two groups.

Passes 1–4 after rotation in the rotated group showed a significant difference between the two groups change in reaction (*p-values* = 0.001, 0.010, 0.004, 0.001). After pass 4 by the rotated object, there was little significant difference between the rotated and control groups (*p-values* > 0.05). As noted in Table 4, some later passes also showed significant differences in the change in reaction between the two groups (Passes 1–4, 8, 9, 12: *p-value* > 0.05). Figures 2 and 3 show the significance in the change in reactions for the rotated group when the horses were exposed to the rotated object.

TABLE 4: Values for Differences in Reaction Score for Corresponding Passes 1–15 by the Novel Object to Rotated Object

Pass #	Control			Rotated			*p-value*
	Mean	Min.	Max.	Mean	Min.	Max.	
1	–1.75	–3	0	0.083	–2	2	0.001
2	–0.875	–2	0	0.25	–1	2	0.010
3	–0.875	–2	0	0.167	–1	1	0.004
4	–1	–2	0	0.333	–1	1	0.001
5	–0.375	–2	1	0.083	–2	0	0.312
6	–0.375	–1	0	0.25	–1	1	0.719
7	–0.5	–2	0	0.167	–1	1	0.062
8	–1	–2	0	0.333	–1	2	0.005
9	–0.875	–2	0	0.583	–2	2	0.002
10	–0.125	–1	1	0.167	–1	2	0.537
11	0	0	0	0.167	–2	2	0.656
12	0.125	–1	1	0.333	–1	0	0.010
13	–0.125	–1	1	0.333	–2	1	0.226
14	–0.25	–1	2	0.083	–2	2	0.554
15	0	–1	1	0.167	–2	2	0.700

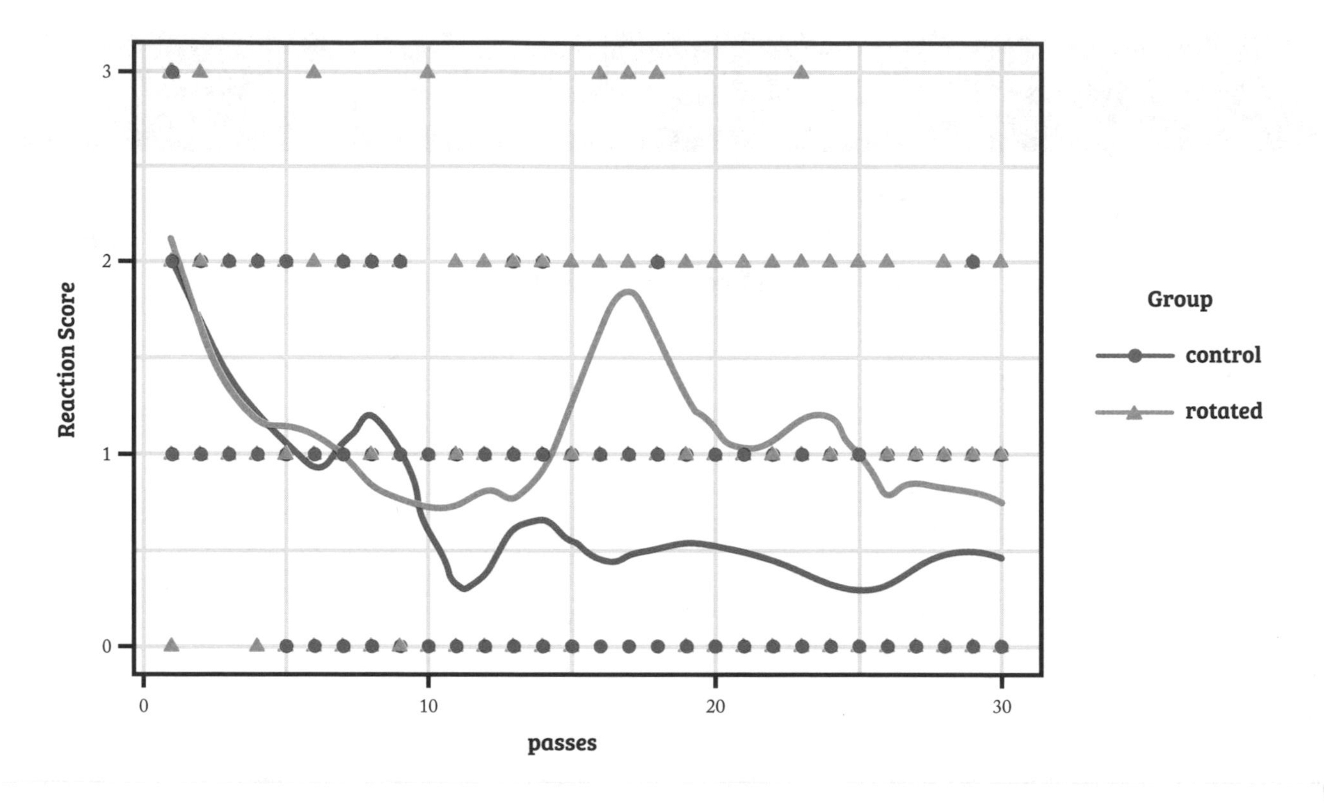

Fig. 3.
Graph of reaction scores from pass 1–30 for the control and rotated groups. The triangles indicate individual horses' reactivity scores. The curves depict mean reaction score by group over passes.

DISCUSSION

When a previously familiar complex novel object is rotated, the rotated object may cause reactions similar to the initial exposure to the novel object. This confirms what handlers and riders have described anecdotally. Understanding horses' reaction to a rotated object is important for the safety of riders and handlers. If handlers expect horses not to react to subtle changes in a familiar environment, they are less prepared for a horse spooking which could lead to an accident. Additionally, studies have shown that investigative behavior is correlated with learning.[20] Allowing a horse to investigate and become familiar with all orientations of an object can help to avoid spooking. Future studies are needed to evaluate if allowing a horse to fully investigate a novel object will help with habituation and decrease spooking.

As shown in Figure 3, there was a decline in the horses' reactions with each successive pass by the rotated object. Table 4 shows the significant decrease in the horses' reaction to the object with each successive pass, for the first four passes. After pass 4, the changes in reactions between the rotated and nonrotated groups seem to be less consistent. This inconsistency in changes in reactions between the two groups shows the unpredictable nature of the horse.[26] Even subtle changes to a familiar object can cause horses to react again. These subtle changes can cause the horse to need more exposure until they are habituated or until no reactions are shown again. Handlers need to be aware of this for safety of themselves and the horses.

This study shows that despite findings from previous research,[19] horses may not recognize different orientations of previously familiar objects, when being led by a handler. While assumptions cannot be made about the horse's recognition of the rotated object from the present study, there is an obvious reaction to the rotated object. This reaction is important to note and important for anyone handling horses to be aware of. There are possible differences in personality and reactivity by breed.[27] The traits that showed the most variability between breeds were excitability and anxiousness. This is thought to be because of the purpose each breed has been bred for over time. For example, Thoroughbreds tend to show a bigger flight response because they are bred to race and are expected to leave the gate quickly.[28] Age has also been shown to have an effect on reactivity. In a study comparing horses ranging in age from two months to two years, younger horses were more reactive.[29]

Training methodologies are worth further exploration when researching equine perception of novel objects. Humans can have an impact on how the horse reacts and behaves.[16] There may be a difference between a voluntary approach, as compared to being led by a handler. Marsboll and Christensen[30] and Hartmann et al.[31] found that a familiar handler can have a calming effect on the horses response to a novel object as well as a change in fear response. The present study used unknown handlers to evaluate how horses will react to a rotated object with a handler. Christensen et al.[20] showed that fearfulness was not correlated with learning, whereas investigative behavior was correlated with learning. While their study did not use handlers, the present study allowed some investigative behavior and kept the handler involved, creating a more practical scenario. The present study did not use food or positive reinforcement when evaluating recognition or reactivity, as compared to [19]. Using food as a reinforcement in training is similar to using latency to eat in research. It is important to note that most trainers do not use food as reinforcement in their training.

The purpose of this study was to evaluate how horses being led and habituated to a previously familiar complex object would react after it was rotated ninety degrees. This study showed that horses' reaction to a rotated orientation of a familiar complex object was similar to its reaction when it first saw it.

CONCLUSIONS

Horses may have a greater reaction to new orientations of previously familiar objects. This may cause accidents that lead to injury of the horse or handler. If handlers and riders can be prepared for how a horse may react, they may be able to help reduce risk by adjusting training methods to allow for investigation of all sides of an object. Additionally, while horses show a decrease in reactions to novel objects and novel orientations of familiar objects, there is the possibility for reactions during habituation. Further research needs to be conducted to evaluate how different methods of handling and training affect the horses' reaction to changes in their environment.

RESEARCH ARTICLE INFORMATION

Author Contributions: Conceptualization, M.E.C., T.G., and S.M.; methodology, M.E.C. and T.G.; formal analysis, M.E.C.; investigation, M.E.C. and S.M.; resources, M.E.C. and T.G.; data curation, M.E.C.; writing—original draft preparation, M.E.C.; writing—review and editing, M.E.C., T.G., and S.M.; visualization, M.E.C., T.G., and S.M.; project administration, M.E.C.; funding acquisition, T.G. All authors have read and agreed to the published version of the manuscript.

Data Availability Statement: The data presented in this study are openly available in FigShare at 10.6084/m9.figshare.14256512.v1.

Acknowledgments: The authors would like to thank John Snyder and the Legends of Ranching Program for use of the horses in the study and the Equine Teaching and Research Center for use of the facility. The authors would also like to thank Ben Sharp for assistance in statistical analysis.

REFERENCES

1. Camargo, F.; Gombeski, W.R.; Barger, P.; Jehlik, C.; Wiemers, H.; Mead, J.; Lawyer, A.; Gonzalez Redondo, P. Horse-related injuries: Causes, preventability, and where educational efforts should be focused. Cogent Food Ag. 2018, 4, 14322168.
2. Meredith, L.; Ekman, R.; Brolin, K. Epidemiology of Equestrian Accidents: A Literature Review. Internet J. Allied Health Sci. Pract. 2018, 17, 9.
3. Acton, A.S.; Gaw, C.E.; Chounthirath, T.; Smith, G.A. Nonfatal horse-related injuries treated in emergency departments in the United States, 1990–2017. Am. J. Emerg. Med. 2020, 38, 1062–1068.
4. Hawson, L.A.; McLean, A.N.; McGreevy, P.D. The roles of equine ethology and applied learning theory in horse related human injuries. J. Vet. Behav. 2010, 5, 324–338.
5. Krüger, L.; Hohberg, M.; Lehmann, W.; Dresing, K. Assesing the risk for major injuries in equestrian sports. BMJ Open Sport Exerc. Med. 2010, 4, 1–9.
6. Christensen, J.W.; Rundgren, M.; Olsson, K. Training methods for horses: Habituation to a frightening stimulus. Eq. Vet. J. 2010, 38, 439–443.
7. Leiner, L.; Fendt, M. Behavioural fear and heart rate responses of horses after exposure to novel objects: Effects of habituation. Appl. Anim. Behav. Sci. 2011, 131, 104–109.
8. Dai, F.; Heinzl, N.H.C.E.U.I.; Costa, E.D.; Canali, E.; Minero, M. Validation of a fear test in sport horses using infrared thermogra phy. J. Vet. Behav. 2015, 10, 128–138.
9. Gorecka-Bruzda, A.; Jastrzebska, E.; Sosnowska, Z.; Jaworksi, Z.; Jezierski, T.; Chruszczewski, M.H. Reactivity to humans and fearfulness tests: Field validation in Polish Cold Blood Horses. Appl. Anim. Behav. Sci. 2011, 133, 207–215.
10. Scopa, C.; Palagi, E.; Sighieri, C.; Baragli, P. Physiological outcomes of calming behaviors support the resilience hypothesis in horses. Sci. Rep. 2018, 8, 1–9.
11. Lansade, L.; Bouissou, M.F.; Erhard, H.W. Fearfulness in horses: A temerament trait stable across time and situation. Appl. Anim. Behav. Sci. 2008, 115, 182–200.
12. Noble, G.K.; Blackshaw, K.L.; Cowling, A.; Harris, P.A.; Sillence, M.N. An objective measure of reactive behavior in horses. Appl. Anim. Behav. Sci. 2013, 144, 121–129.
13. Borstel, U.K.v.; Duncan, I.J.H.; Lundin, M.C.; Keeling, L.J. Fear reactions in trained and untrained horses from dressage and showjumping breeding lines. Appl. Anim. Behav. Sci. 2010, 125, 124–131.
14. King, T.; Hemsworth, P.H.; Coleman, G.J. Fear of novel and startling stimuli in domestic dogs. Appl. Anim. Behav. Sci. 2003, 82, 45–64.
15. Christensen, J.W.;malmkvist, J.; Nielsen, B.L.; Keeling, L.J. Effects of a calm companion on fear reactions in naive test horses. Equine Vet. J. 2008, 40, 46–50.
16. Visser, E.K.; van Reenen, C.G.; van der Werf, J.T.; Schilder, M.B.H.; Knaap, J.H.; Barneveld, A.; Blokuis, H.J. Heart rate and heart rate variability during a novel object test and a handling test in young horses. Physio. Behav. 2002, 76, 289–296.
17. Borstel, U.K.v.; Pirsich, W.; Gauly, M.; Bruns, E. Repeatability and reliability of scores from ridden temperament tests conducted during performance tests. Appl. Anim. Behav. Sci. 2012, 139, 251–263.
18. Hanggi, E.B. Discrimination learning based on relative size concepts in horses (Equus caballus). Appl. Anim. Behav. Sci. 2003, 83, 201–213.
19. Hanggii, E.B. Rotated object recognition in four domestic horses (Equus caballus). J. Equine Vet. Sci. 2010, 30, 175–186.

20. Christensen, J.W.; Ahrendt, L.P.; Malmkvist, J.; Nicol, C. Exploratory behaviour towards novel objects is associated with enhanced learning in young horses. Sci. Rep. 2021, 14, 1428.

21. Bulens, A.; Sterken, H.; Van Beirendonk, S.; Van Thielen, J.; Driessen, B. The use of different objects during a novel object test in stabled horses. J. Vet. Behav. 2015, 10, 54–58.

22. Hall, C.A.; Cassaday, H.J.; Vincent, C.J.; Derrington, A.M. Cone excitation ratios correlate with color discrimination performance in the horse (Equus caballus). J. Comp. Psych. 2006, 120, 438–448.

23. McDonnell, S.M. The Equid Ethogram: A Practical Field Guide to Horse Behavior; Eclipse Press: Lexington, KY, USA, 2003; p. 99.

24. R Core Team. R: A Language Environment for Statistical Computing R Function for Statistical Computing; R Foundation for Statistical Computing: Vienna, Austria. (accessed on 24 March 2021).

25. www.tidyverse.org (accessed on 24 March 2021).

26. Thompson, K.; McGreevy, P.; McManus, P. A Critical Review of Horse-Related Risk: A Research Agenda for Safer Mounts, Riders and Equestrian Cultures. Animals 2015, 5, 561–575.

27. Lloyd, A.S.; Martin, J.E.; Bornett-Gauci, H.L.I.; Wilkinson, R.G. Horse personality: Variation between breed. Appl. Anim. Behav. Sci. 2008, 112, 369–383.

28. McGreevy, P.D.; Thomson, P.C. Differences in motor laterality between breeds of performance horses. Appl. Anim. Behav. Sci. 2006, 99, 183–190.

29. Caviello, R.F.; Titto, E.A.L.; Infante, P.; Lemedos-Santos, T.M.d.C.; Neto, M.C.; Pereira, A.M.F.; Titto, C.G. Proposal and Validation of a Scale of Composite Measure Reactivity Score to Characterize the Reactivity in Horses During Handling. J. Eq. Vet. Sci. 2016, 47, 62–70.

30. Marsboll, A.F.; Christensen, J.W. Effects of handling on fear reactions in young Icelandic horses. Equine. Vet. J. 2014, 47, 615–619.

31. Hartmann, E.; Rehn, T.; Christensen, J.W.; Nielsen, P.P.; McGreevy, P. From the Horse's Perspective: Investigating Attachment Behaviour and the Effect of Training Method on Fear Reactions and Ease of Handing—A Pilot Study. Animals 2021, 11, 457.

TWENTY-SIX

Cattle and Pigs That Are Easy to Move and Handle Will Have Less Preslaughter Stress

T. Grandin (2021) Cattle and Pigs That Are Easy to Move and Handle Will Have Less Preslaughter Stress, *Foods*, 10:2583.

When I see a severe animal welfare problem at a slaughter plant that is inspected and audited by both federal USDA inspectors and audits by customers, it is most likely to be caused by on-farm conditions. Lameness is a big issue in old dairy cows and some feedlot cattle. It is caused by a variety of factors, and it makes handling and transport of animals more stressful. Genetic selection for greater production for meat or milk is one factor. Poor management is another factor. This paper outlines on-farm factors that may seriously compromise animal welfare during handling and transport.

ABSTRACT

Previous research has clearly shown that short-term stresses during the last few minutes before stunning can result in Pale Soft Exudative (PSE) pork in pigs or increased toughness in beef. Electric prods and other aversive handling methods during the last five minutes are associated with poorer meat quality. Handlers are more likely to use aversive methods if livestock constantly stop and are difficult to move into the stun box. Factors both inside and outside the slaughter plant contribute to handling problems. Some in-plant factors are lighting, shadows, seeing motion up ahead, or air movement. Nonslip flooring is also very important for low-stress handling. During the last ten years, there have been increasing problems with on-farm factors that may make animals more difficult to move at the abattoir. Cattle or pigs that are lame or stiff will be more difficult to move and handle. Some of the factors associated with lame cattle are either poor design or lack of adequate bedding in dairy cubicles (free stalls) and housing beef cattle for long periods on concrete floors. Poor leg conformation in both cattle and pigs may also be associated with animals that are reluctant to move. Indiscriminate breeding selection for meat production traits may be related to some of the leg conformation problems. Other on-farm factors that may contribute to handling problems at the abattoir are high doses of beta-agonists or cattle and pigs that have had little contact with people.

INTRODUCTION

Many research studies have previously shown that short-term stresses during the last few minutes before stunning may result in both poorer meat quality and severely compromised animal welfare. In pigs, multiple shocks from electric prods within five minutes before stunning resulted in higher lactate levels and more Pale Soft Exudative meat (PSE).[1-3] Shocking pigs multiple times with an electric prod greatly increased lactate and glucose levels compared to low-stress handling and no use of electric prods. In cattle, short-term stresses shortly before stunning such as the use of electric prods or agitated behavior were associated with increased toughness in the meat.[4, 5] The purpose of this commentary is to discuss the author's observations of the increase in on-farm factors that may make animals more difficult to move at the abattoir.

When cattle or pigs are difficult to move and constantly keep stopping, handlers are more likely to use aversive methods to drive them such as electric prods or tail twisting.[6] Pigs that move easily also require fewer touches, slaps, or pushes. In this article, the author discusses some of the factors that are associated with animals that are difficult to move at the slaughter plant. Correcting these problems will help improve both welfare and meat quality.

There are factors both inside and outside the abattoir that can have an effect on the ease of animal movement. Major factors inside the plant are distractions such as sharp shadows on the floor, reflections on shiny metal, or a noisy vehicle near the lairage (stockyards) that may cause animals to stop.[6-9] Illuminating a dark restrainer entrance facilitated the movement of cattle and it reduced vocalization associated with electric prod use.[7] Other factors that can slow down animal movement are air blowing out through the stun box entrance towards approaching cattle or layout mistakes in the design of facilities.[9] Abattoir yards and lairages should also have nonslip flooring in pens, alleys, and stun boxes. If animals slip and fall on a slick floor, they are more likely to become stressed. It is also essential to have well-trained stock people who understand and use behavioral principles of moving livestock.[9, 10] Training of stock people will improve livestock handling and reduce aversive methods of driving livestock.[11]

The main emphasis of this article is to discuss on-farm factors such as housing problems, growth promotants, or over-selection for production traits that may be associated with lame cattle and pigs that are reluctant to move. It contains both scientific studies and observations from the author's experiences with handling livestock. Since the early 1970s, the author has consulted on improving livestock handing in abattoirs in the U.S., Europe, South America, Australia, and many other countries. High preslaughter standards for animal welfare are difficult to maintain if animals are lame, stiff, or have reduced mobility. These problems must be corrected at the farm of origin. The author has observed that handling problems at the abattoir are increasingly associated with breeding, feeding, or housing practices on the farm.

ON-FARM FACTORS ASSOCIATED WITH HANDLING PROBLEMS AT THE ABATTOIR

Within the last ten years, the author has observed that problems with cattle and pigs that are less willing to move have increased. Recently, a lairage manager at a large beef abattoir told the author that cattle from certain feedlots would immediately lie down after arrival. They were so reluctant to move that he had to get them up a few minutes before they went to the stunner. There are a variety of conditions that may have contributed to these handling problems. Over the years, the author has observed that these handling problems have slowly become worse. Issues with poor mobility of pigs and cattle have increased slowly and people did not notice it. The author calls this "bad becoming normal."[12]

In the U.S., the percentage of lame grain-fed cattle has increased. In 2020, only 74.5% of grain-fed beef cattle were free of lameness.[13] These data were collected during the months of July to October on 16,262 fed feedlot cattle that arrived at a large abattoir located in the Central Plains of the U.S. This area is in the heart of the U.S. feedlot industry. In the previous years of 2016–2019, 96.19% to 89.32% of the fed feedlot cattle that arrived at an abattoir were free of lameness.[13] There are also some dairies with high percentages of lame cows. Lame livestock that are reluctant to move may be more likely to have stressful aversive handling methods used on them at the slaughter plant. A recent Brazilian survey of 50 dairies showed that 41% of the cows were lame.[14] The percentage of lame cows varied from 13.8% in the best dairy to 64.5% in the worst dairy.[14] Recent studies conducted in the UK and Canada indicated that 31.8% of the dairy cows in the UK were lame[15] and 21% of the dairy cows kept in cubicles were lame. Research also clearly shows that dairy producers will often greatly underestimate the percentage of lame dairy cows.[17] When lameness is actually measured, they will discover that the percentage of lame dairy cows may be double their estimate.

POOR STRUCTURAL LEG CONFIRMATION

The author has observed that grain-fed market cattle or pigs, which are indiscriminately bred for growth, are more likely to be lame and reluctant to move. At one large abattoir, 50% of the incoming market weight pigs were lame. Approximately half of the pigs had poor leg conformation and exhibited traits such as legs too straight (post-legged), collapsed ankles, or the feet were rotated. The problem probably starts with the sow herd. Breeding stock with poor leg conformation had a higher rate of culling due to lameness.[18] Pork and beef producer organizations have now recognized the problem and they have distributed leg conformation charts for producers to use when they select breeding stock.[19, 20] The American Angus Association has an EPD for leg conformation.[21] It was created in response to producer reports that leg conformation was worsening in Angus cattle selected for rapid weight gain and large muscles. In Thailand, the author observed severe lameness issues in pigs that had been selected to have small feet.

DEFICIENCIES IN HOUSING ASSOCIATED WITH LAMENESS OR SWOLLEN JOINTS

Poorly designed housing or lack of management of housing is associated with injuries to legs that may lead to lameness in both dairy cows and beef cattle. Housing fattening beef cattle on concrete for long periods of time can lead to swollen joints.[22] In dairy cows, free stalls (cubicles) that are either too small or poorly bedded are associated with more leg problems and swollen joints.[23, 24] Farms with better bedding management had fewer cows with swollen joints.[23] Other factors that were associated with increased lameness were slippery floors and poor body condition.[24] In one survey, 40% of the skinny dairy cows with a body condition score under 2.5 were lame.[24] Improvements in flooring and maintaining cow body condition may also help reduce lameness. This shows the importance of good management on the farm for reducing lameness. Charolais bulls housed on a concrete slatted floor had significantly more lameness than bulls housed on deep litter.[25] The author has observed that lameness in cattle housed on concrete can be reduced by shortening the period of time they are kept on concrete. Covering the slats with rubber mats may also help reduce lameness. More research is needed to determine guidelines for the maximum length of time that fattening cattle should be housed on concrete floors.

EXCESSIVE USE OF GROWTH PROMOTANTS AND HANDLING PROBLEMS

Research has clearly shown that pigs fed high doses of beta-agonists are more likely to become fatigued and non-ambulatory.[26] Non-ambulatory pigs increase labor requirements at the abattoir and their welfare is severely compromised. Ractopamine and zilpaterol are feed additives that are used to increase the amount of muscle.[27] High doses of ractopamine were associated with greater difficulty in handling pigs.[28, 29] Hot weather is also more likely to increase death losses in cattle fed ractopamine.[30, 31] Handling problems observed by both the author and reports in the scientific literature both indicate that high doses, combined with hot weather over 32°C, caused the most problems.[8, 29] Pigs fed beta-agonists must be handled in a low-stress manner to prevent downed non-ambulatory animals.[32] Researchers at a Colorado feedlot also reported that feeding zilpaterol predisposed grain-fed cattle to heart problems.[33] Behavioral observations of cattle indicate that they may also have muscle stiffness. Feeding zilpaterol at the recommended label dose to grain-fed cattle resulted in 31% of the animals lying in an abnormal posture.[34] It is likely that the abnormal lying posture is related to attempts to reduce muscular discomfort. In one case, a group of cattle fed a high carbohydrate potato byproduct diet combined with high doses of zilpaterol presented some cases where the outer hoof sloughed off.[35, 36] Some specialists who are concerned about both meat quality and animal welfare may ask if transportation practices contribute to this problem. It is likely that transportation is not the main cause of this problem. Both the location of the abattoirs and the feedlots that supply the cattle had not changed. Both before and after the appearance of the handling problems, the cattle were transported the same distances from most of the same feedlots. The use of beta-agonists is banned in Europe and China.[37] They are legal in the U.S., Canada, Brazil, and many other countries. It is important for readers in countries where beta-agonists are legal to be aware of possible handling and welfare problems. These problems are more likely to occur when higher doses are used.

THE CONCEPT OF BIOLOGICAL SYSTEM OVERLOAD

The author has been in the livestock industry for many years. In the 1970s through the 1990s, most welfare and handling problems in an abattoir were due to either poorly designed facilities, lack of equipment maintenance, or rough abusive handling by people.[9] Today, in a well-managed U.S. slaughter plant, handling problems with grain-fed cattle or pigs are more likely to be associated with on-farm factors. The problem may also be due to pushing the livestock to gain weight fast. This is accomplished by both genetic selection for production traits and feeding practices. The animal's biology is pushed to the point where it starts to break down. Cardiac problems in cattle used to occur only at high altitudes. Researchers have found that they are now occurring at lower altitudes.[38] Heart problems in cattle associated with high altitude are heritable.[39, 40] It is possible that heart problems are related to a greater emphasis on breeding cattle for large amounts of muscle mass. In 2015, veterinarians described a condition in cattle called fatigued cattle syndrome. This is similar to problems with weak fatigued pigs. There are four factors that may have led to the relatively recent observations of more problems with both cattle and pigs that are reluctant to move:

1. Cattle fed to heavier weights at a younger age and more cattle fattened for highly marbled USDA prime beef[41]
2. Indiscriminate breeding and selection for growth and muscle mass in both species[42]
3. Feeding high-grain diets to cattle and a lack of roughages[25]
4. High doses of beta-agonists fed to both cattle and pigs[28, 29]

It is the author's opinion that pushing the animal's biology until it starts to break down may be one of the most serious animal welfare problems.[43] These problems will also contribute to meat quality problems. The author has tracked non-ambulatory pigs through to the meat cutting floor and they had high levels of PSE. Increasing muscle growth with beta-agonists also resulted in increased beef toughness.[44] Producers should strive for optimum performance and not maximum growth and muscle. Producing animals that convert feed more efficiently into muscle is good from a sustainability standpoint because they eat less feed. There is a point where it is both not sustainable and bad for animal welfare. An animal that dies shortly before it is time to slaughter it wastes all the feed it has eaten.

ON-FARM BEHAVIORAL AND MANAGEMENT FACTORS

The discussions in the previous sections of this paper covered physical problems that made animals more difficult to handle. In this section, factors that are purely behavioral will be covered. The author has observed that pigs and cattle will move more easily during handling at the abattoir if they have become accustomed to people walking through them. An animal's experiences on the farm will affect its behavior during handling in the future. When people regularly walk through the finishing pens several times each week, pigs will move more easily at the slaughter plant. Finishing-market-weight pigs that have been moved several times on the farm will be easier to move and drive through alleys in the future.[45-48] Pigs differentiate between a person walking in the aisle and a person walking through their pens. From the author's experience on farms, pigs will be easier to handle if they become accustomed to quietly moving away when a person walks through their pens. The author has observed that cattle that have been extensively raised are sometimes difficult and dangerous to handle at the slaughter plant. Cattle can tell the difference between a person walking on the ground and a person riding a horse. Extensively raised cattle that have been handled with horses may have a greatly increased flight zone when they first encounter a person walking on the ground. The horse and rider were perceived as familiar and safe, and the person walking is new and novel.[49] Handling at the abattoir will be safer and cattle will be less stressed if they become accustomed to moving in and out of pens by people on foot before they arrive at a slaughter plant.

TWO OBSERVATIONAL CASE HISTORIES WHERE ON-FARM PRACTICES HAD SIGNIFICANT EFFECTS ON EASE OF HANDLING

The author consulted with a large pork plant that had severe problems with downed non-ambulatory finisher pigs and pigs that were difficult to move. To handle all the fatigued non-ambulatory pigs required five or six full-time people to stun the pigs in the lairage and transport the stunned animals to the bleeding area. After three changes were made on the farm, the numbers of downed non-ambulatory pigs dropped to the point where only one halftime person was required to handle downers. The three things the farms changed were (1) eliminated breeding to a boar line that had poor leg conformation, (2) reduced or eliminated ractopamine use, and (3) started a program that required producers to walk through the finishing pens every day. This trained the pigs to quietly get up and walk away from the person walking amongst them. These observations clearly showed how on-farm factors can have detrimental effects on both handling practices and animal welfare.

There has been much discussion about designing and building better vehicles to reduce stress on pigs during loading and transport. In Europe, many new vehicles have power lifts and movable decks to eliminate ramps for loading and unloading pigs. In the eastern U.S., the height of almost all trucks is restricted to 13 feet 6 in. (4.11 m) due to low bridges.[50] The vehicle will be too tall if two decks of cattle are positioned above the wheels on a level floor. Two decks of cattle are accommodated in a compartment between the axles.[51] This design has internal ramps to load and unload the animals on and off the two decks. These trailers work well for cattle, but some pigs have difficulty negotiating the ramps. This has resulted in a straight trailer design for pigs that have two decks located over the top of the wheels. Use of these trailers is limited to pigs, sheep, or other small animals. Many independent truckers in the U.S. who own their own vehicles need to have the flexibility of transporting both cattle and pigs in the same trailer. These owner–operators will usually use a cattle trailer that has internal ramps to transport pigs. Some people who are concerned about animal welfare believe that this trailer design should not be used for pigs.

In the spring of 2021, the author visited an abattoir that processed pigs that lived outdoors. All of the pigs arrived in cattle trailers that had the internal ramps. The author watched four trailers unload at this abattoir. The pigs moved easily up the ramp from the belly compartment and down the ramp from the top compartment. There was zero use of electric prods and none of the animals fell down during unloading. For these pigs, the cattle trailers with the internal ramps were satisfactory. People who are concerned about animal welfare or meat quality need to think about how to improve handling. The question is: Do you improve the pig, or should you improve the design of the vehicle? This recent experience reinforced my opinion that many of the intensively raised pigs have become very difficult to handle. It is the author's opinion that the pig needs to be improved by changing breeding, feeding, and production practices. The author is not suggesting raising all finishing pigs outside. What is being suggested is that the emphasis needs to be on improving the pigs, so they are stronger and more willing to move. One study showed that exercising pigs improved ease of handling.[52] Possible ways of doing this are genetic selection and regular moving of pigs on the farm. During the springtime observation of many pigs in the stockyard (lairage), there was only one group that had poor leg conformation. The author warned the company that they need to keep working with producers to breed pigs that have good leg conformation.

CONCLUSIONS

Previous research clearly shows that to preserve meat quality and maintain good animal welfare, cattle and pigs should move easily with a minimum use of aversive driving methods such as electric prods, tail twisting, or hitting. The meat industry needs to address increasing problems with on-farm factors that may make cattle and pigs more difficult to move at the abattoir. Lame animals that have difficulty walking are more difficult to handle. Some of the on-farm factors that may contribute to these problems are poor leg conformation in both pigs and cattle, housing finishing cattle for long periods on concrete, poor management of dairy cow cubicles, or feeding high doses of beta-agonists. To improve both animal welfare and meat quality, producers need to correct these problems.

REFERENCES

1. Edwards L.N., Grandin T., Engle T.E., Porter S.P., Ritter M.J., Sosnicki A.A., Anderson D.B. Using exsanguination blood lactate to assess the quality of preslaughter pig handling. Meat Sci. 2010; 86:384–390. doi: 10.1016/j.meatsci.2010.05.022.
2. Hambrecht E., Eissenn J.J., Deklein W.J.H., Duevo B.J., Smits C.H.M., Verstegen M.W.A., den Hartog L.A. Rapid chilling cannot prevent inferior pork quality caused by high preslaughter stress. J. Anim. Sci. 2004; 82:551–556. doi: 10.2527/2004.822551x.
3. Hambrecht E., Eissen J.J., Newman D.J., Verstegen M.W., Hartog L.A. Preslaughter handling affects pork quality and glycolytic potential of two muscles differing in fibertype composition. J. Anim. Sci. 2005; 83:900–907. doi: 10.2527/2005.834900x.
4. Warner R.D., Ferguson D.W., Cotrell J.J., Knee B.W. Acute stress induced by the preslaughter use of electric prodders causes tougher beef meat. Aust. J. Exp. Agric. 2007; 47:782–788. doi: 10.1071/EA05155.
5. Gruber S.L., Tatum J.D., Engle T.E., Chapman P.L., Belk K.E., Smith G.C. Relationships between behavioral and physiological symptoms of preslaughter stress on beef longissimus muscle tenderness. J. Anim. Sci. 2010; 88:1148–1159. doi: 10.2527/jas.20092183.
6. Willson D., Baier F., Grandin T. An observational field study on the effects of changes in shadow contrasts and noise on cattle movement in a small abattoir. Meat Sci. 2021;179 doi: 10.1016/j.meatsci.2021.108539.
7. Grandin T. Cattle vocalizations are associated with handling and equipment problems at beef slaughter plants. Appl. Anim. Behav. Sci. 2001; 71:191–201. doi: 10.1016/S01681591(00)001799.
8. Grandin T. Factors that impede animal movement at slaughter plants. J. Am. Vet. Med. Assoc. 1996; 209:757–759.
9. Grandin T., Cockram M. The Slaughter of Farmed Animals: Practical Ways of Enhancing Animal Welfare. CABI International; Wallingford, Oxfordshire, UK: 2020.

10. Yost J.K., Yates J.W., Davis M.P., Wilson M.E. The Stockman's scorecard: Quantitative evaluation of cattle stockmanship. Trans. Anim. Sci. 2020;4: txaa175. doi: 10.1093/tas/txaa175.

11. Ceballos M.C., Sant'Anna A.C., Boivin X., de Olivera Costa F., Monique V.D.L., de L. Carvahal M.V., Paranhos da Costa M.J.R. Import of good practices of handling training on beef cattle welfare and stock people attitudes and behaviors. Livest. Sci. 2018; 216:24–31. doi: 10.1016/j.livsci.2018.06.019.

12. Grandin T. Improving Animal Welfare: A Practical Approach. CABI International; Wallingford, Oxfordshire, UK: 2021.

13. Mijares S., Calvo Lorenzo M., Betts N., Alexander L., Edwards Callaway L.N. Characterization of fed cattle mobility during the COVID-19 pandemic. Animals. 2021; 11:1749. doi: 10.3390/ani11061749.

14. Bran J.A., Costa J.H.C., von Keyserlingk M.A.G., Hotzel M.J. Factors associated with lameness prevalence in lactating cows housed in free stall and compost pack bedded pack dairy farms in Southern Brazil. Prev. Vet. Med. 2019; 172:15. doi: 10.1016/j.prevetmed.2019.104773.

15. Griffith B.E., White D.G., Oikunomous G. A cross-sectional study into the prevalence of dairy cattle lameness and associated herd level risk factors in England and Wales. Font. Vet. Sci. 2018; 5:65. doi: 10.3389/fvets.2018.00065.

16. Jewell M.T., Cameron M., Spears J., McKenna S.L., Cockram M.S., Sanchez J., Keefe G.P. Prevalence of lameness and associated risk factors on dairy farms in the Maritime Provinces of Canada. J. Dairy Sci. 2019; 102:3392–3405. doi: 10.3168/jds.201815349.

17. Fabian J., Laven R.A., Whay H.R. The prevalence of lameness on New Zealand dairy farms: A comparison of farmer estimate and locomotion scoring. Vet. J. 2014; 201:31–38. doi: 10.1016/j.tvjl.2014.05.011.

18. de Seville X.F., Fabrege E., Tibau J., Casellas J. Effect of leg conformation on the survivability of Durue, Landrace, and Large White Sows. J. Anim. Sci. 2008; 86:2391–2400.

19. Moeller S.J., Stalder K.J. Genetic Aspects of Female Longevity. Pork Information Gateway. 2006. (accessed on 4 September 2021).

20. Bertram J. Selecting Bulls for Structural Soundness in Beef Cattle. The Beef Site. 2005. (accessed on 4 September 2021).

21. Retallick K. New Foot Structure EDD's: Claw and Foot Angle. Angus Journal. 2019. (accessed on 22 August 2021).

22. Dewell R.D., Dewell G.A., Euken R.M., Sadler L.J., Wang C., Carmichael B.A. Association of floor type with health and wellbeing and performance parameters of beef cattle fed in indoor confinement facilities during finishing phase. Bov. Pract. 2018; 52:16–25. doi: 10.21423/bovine.

23. Fulwider W.K., Grandin T., Garrick D.J., Engle T.E., Lamm W.D., Dalsted N.L., Rollin B.E. Influence of free stall base on tarsal joint lesions and hygiene in dairy cows. J. Dairy Sci. 2007; 80:3559–3566. doi: 10.3168/jds.2006793.

24. Solano L., Barkema H.W., Pajor E.A., Masson S., LeBlance S.J., Heyerhoff J.C.Z., Nash G.G.R., Haley D.B., Vasseur E., Pellerin D., et al. Prevalence of lameness and associated risk factors in Canadian Holstein Friesian cows housed in free stall barns. J. Dairy Sci. 2015; 98:6978–6991. doi: 10.3168/jds.20159652.

25. Magrin L., Gottardo F., Contiero B., Brscic M., Cozzi G. Time of occurrence and prevalence of severe lameness in fattening Charolais bulls: Impact of type of floor and space allowance with type of floor. Livest. Sci. 2019; 221:86–88. doi: 10.1016/j.livsci.2019.01.021.

26. Peterson C.M., Pilcher C.M., Rothe H.M., Marchant-Forde J.N., Ritter M.J., Carr S.N., Puls C.L., Ellis M. Effect of feeding ractopamine hydrochloride on the growth performance and response to handling and transport in heavy weight pigs. J. Anim. Sci. 2015; 93:1239–1249. doi: 10.2527/jas.20148303.

27. Scanlin S.M., Platter W.J., Gomez R.A., McKeith F.K., Killefer J. Comparative effects of ractopamine hydrochloride and zilpaterol hydrochloride on growth performance, carcass traits, and longissimus tenderness of finishing steers. J. Anim. Sci. 2009; 88:1823–1829.

28. MarchantForde J.N., Lay D.C., Pajor E.A., Richert B.T., Schinckel A.P. The effects of ractopamine on the behavior and physiology of finishing pigs. J. Anim. Sci. 2003; 81:416–422. doi: 10.2527/2003.812416x.

29. Ritter M.J., Johnson A.K., Benjamin M.E., Carr S.N., Ellis M., Faucitano L., Grandin T., SalakJohnson J.L., Thomson D.U., Goldhawk G., et al. Review of ractopamine hydrochloride (Paylean) on welfare indicators for market weight pigs. Trans. Anim. Sci. 2017; 1:533–558. doi: 10.2527/tas2017.0060.

30. Longeragan G.H., Thomson D.U., Scott H.M. Increased mortality in groups of cattle administered Badrenergic agonists ractopamine hydrochloride and zilpaterol hydrochloride. PLoS ONE. 2014; 9:3. doi: 10.1371/journal.pone.0091177.

31. Montgomery J.L., Krehict C.R., Cranston J.J., Yates D.A., Hutchinson J.P., Yates D.A., Galyean M.L. Effects of dietary zilpaterol hydrochloride on feedlot performance and carcass characteristics of beef steers fed without monensin and tylosin. J. Anim. Sci. 2009; 87:1013–1023. doi: 10.2527/jas.20081169.

32. James B.W., Tokach M.D., Goodband R.D., Nelssen J.L., Dritz S.S., Owen K.Q., Woodworth J.C., Sulabo R.D. Effect of dietary Larnitine and ractopamine HCL on the metabolic response to handling in finishing pigs. J. Anim. Sci. 2013; 91:4426–4439. doi: 10.2527/jas.20114411.

33. Neary J.M., Gary F.B., Gould D.H., Holt T.N., Brown R.D. The beta agonist zilpaterol hydrochloride may predispose feedlot cattle to cardiac remodeling and dysfunction. (accessed on 23 October 2021); F1000 Res. 2018 7:399. doi: 10.12688/f1000research.14313.1.

34. Tucker C.B., CalvoLorenzo M.S., Mitloehner F.M. Effects of growth promoting technology on feedlot cattle behavior 21 days before slaughter. Appl. Anim. Behav. Sci. 2015; 162:1–8.

35. Huffstutter P.J., Polansek T. Special Report: Left hooves, dead cattle before Merck halted Zilmax sales. Reuters. 2013. (accessed on 5 September 2021).

36. Thomsen D.U., Loneragan G.H., Henningson J.N., Ensley S., Baw B. Description of novel fatigue syndrome of finished feedlot cattle following transportation. J. Am. Vet. Med. Assoc. 2015; 247:66–72. doi: 10.2460/javma.247.1.66.

37. Davis N.E., Belk K. Managing meat exports considering production technology challenges. Anim. Front. 2019; 8:23–29. doi: 10.1093/af/vfy007.

38. Moxley R.A., Smith D.R., Grotelveschen D.M., Edwards T., Steffen D.J. Investigation of congestive heart failure in beef cattle in a feedlot at moderate altitude in western Nebraska. J. Vet. Diagn. Inv. 2019; 31:509–522. doi: 10.1177/1040638719855108.

39. Will D.H., Hick J.L., Card C.S., Alexander A.F. Inherited susceptibility of cattle to high altitude pulmonary hypertension. J. Appl. Physiol. 1975; 38:491–494. doi: 10.1152/jappl.1975.38.3.491.

40. Newman J.H., Holt T.N., Cogan J.D., Womack B., Phillips J.A., Li C., Kendall Z., Stenmark K.R., Thomas M.G., Brown D., et al. Increased prevalence of EPAS variant in cattle with high altitude pulmonary hypertension. Nat. Comm. 2015:6.

41. Thorton P.K. Livestock production: Recent trends, future prospects. Philos. Trans. R. Soc. B. Biol. Sci. 2010; 365:2853–2867. doi: 10.1098/rstb.2010.0134.

42. Peel D.S. Carcass Weights Keep Increasing. Agri View. 2021. (accessed on 27 August 2021).

43. Grandin T., Whiting M. Are We Pushing Animals to Their Biological Limits? Welfare and Ethical Implications. CABI International; Wallingford, Oxfordshire, UK: 2018.

44. Moholisa E., Hugo A., Strydom P.E., Heerden T.V. The effects of animal age feeding regime and dietary beta-agonist on tenderness of three beef muscles. J. Sci. Food. Agric. 2016; 97:2375–2381. doi: 10.1002/jsfa.8049.

45. Abbot T.A., Hunter E.J., Gruise J.H., Penny R.H.C. The effect of experience of handling on a pig's willingness to move. Appl. Anim. Behav. Sci. 1997; 54:371–375. doi: 10.1016/S01681591(97)000452.

46. Geverink N.A., Kappers A., van de Burgwal E., Lambooij E., Blokhuis J.H., Wiegant V.M. Effects of regular moving and handling on the behavioral and physiological responses of pigs to preslaughter treatment and consequences for meat quality. J. Anim. Sci. 1998; 76:2080–2085. doi: 10.2527/1998.7682080x.

47. Lewis G.R.G., Hulbert C.E., McGlone J.J. Novelty causes elevated heart rate and immune changes in pigs exposed to handling alleys and ramp. Livest. Sci. 2008; 116:338–341. doi: 10.1016/j.livsci.2008.02.014.

48. Krebs N., McGlone J.J. Effects of exposing pigs to moving and odors in a simulated slaughter chute. Appl. Anim. Behav. Sci. 2009; 116:179–185. doi: 10.1016/j.applanim.2008.10.007.

49. Grandin T., editor. Livestock Handling and Transport. CABI Publishing; Wallingford, Oxfordshire, UK: 2019. The effects of both genetics and previous experiences on livestock behavior during handling; pp. 80–109.

50. Trucking Legal Height Map. (accessed on 5 September 2021).

51. Schwartzkoph-Genswein K., Grandin T. Livestock Handling and Transport. 5th ed. CABI International; Wallingford, Oxfordshire, UK: 2019. Cattle transport in North America; pp. 153–183.

52. Goumon S., Bergeron R., Faucitano L., Crowe T., Connor M.L., Gonyou H.W. Effect of previous ramp exposure and regular handling on heart rate, ease of handling and behaviour of near market weight pigs during simulated loading. Can. J. Anim. Sci. 2013; 93:461–470. doi: 10.4141/cjas2013166.

TWENTY-SEVEN

Grazing Cattle, Sheep, and Goats Are Important Parts of a Sustainable Agricultural Future

T. Grandin (2022) Grazing Cattle, Sheep, and Goats are Important Parts of a Sustainable Agricultural Future, *Animals* 12:2092, MDPI, Basel, Switzerland.

I wrote this paper because the cattle industry has been attacked by many environmental activists. I reviewed numerous research that shows that grazing done right is sustainable and can improve soil. About 20 percent of habitable land can only be used for grazing. It is too arid for crops.

SIMPLE SUMMARY

Increasing attacks on animal agriculture have forced many people to question the use of animals for food. Grazing livestock are part of a sustainable agricultural future. Vast amounts of land all around the world can only be used for grazing. It is either too arid or the terrain is too rough for growing crops. Rotating cattle, sheep, or other livestock between different pastures can improve both soil health and plant biodiversity. This is a sustainable use of land that cannot be cropped. On cropland, the rotation of conventional crops, such as corn or soybeans, with livestock grazing on a forage crop can improve both soil health and reduce the need for artificial fertilizer. Successful grazing programs must be adapted to local conditions. When grazing is performed correctly, it will improve the land.

ABSTRACT

Many people believe that animal agriculture should be phased out and replaced with vegetarian substitutes. The livestock industry has also been attacked because it uses vast amounts of land. People forget that grazing cattle or sheep can be raised on land that is either too arid or too rough for raising crops. At least 20% of the habitable land on Earth is not suitable for crops. Rotational grazing systems can be used to improve both soil health and vegetation diversity on arid land. Grazing livestock are also being successfully used to graze cover crops on prime farmland. Soil health is improved when grazing on a cover crop is rotated with conventional cash crops, such as corn or soybeans. It also reduces the need for buying fertilizer. Grazing animals, such as cattle, sheep, goats, or bison, should be used as part of a sustainable system that will improve the land, help sequester carbon, and reduce animal welfare issues.

INTRODUCTION

This paper contains a combination of observations I have made during a fifty-year career in the cattle industry, combined with references from scientific research. I have visited ranching operations in many different countries during my work as a livestock handling consultant. Recently, I taught a seminar to a group of animal science students at the University of Nebraska. We entered into an intense discussion about the future sustainability of animal agriculture. I told the students that I have been in the cattle industry for 50 years. I have worked hard to improve livestock handling, design equipment, and implement animal welfare auditing programs.[1,2] My work in animal agriculture has improved livestock handling. Unfortunately, it has been attacked by some people in the animal rights movement. They believe that using animals for food should be abolished. The animal advocacy movement has shifted toward veganism.[3] Another issue that concerns many people is that scientists are also learning that cattle, pigs, octopuses, and other animals are sentient beings.[4, 5] This knowledge has also increased ethical concerns about how people use animals. I have always believed that cattle and other animals are conscious. On many ranches, I have observed good stock people having positive interactions with their cattle. At ranches in Texas and Australia, I have watched Brahman cows approach people to be petted and scratched. The animal was motivated to initiate human contact and not food. During a long career, I have observed increasing societal concerns on how animals we use for food are treated.[6] Many people currently believe that animal agriculture should be phased out and replaced with vegetarian substitutes or meat cells grown in bioreactors. There are now many scientific papers about cultured meat.[7-9]

Another concern is the effect of livestock on the environment. Cattle definitely emit methane, but there are other sources such as wetlands, leaking oil field equipment, and landfills.[10] Wetlands may emit one-third of atmospheric emissions of methane.[11] Before Europeans came to North America, large herds of bison emitted large amounts of methane. Bison methane emissions in the past would have been equal to 86% of farmed ruminants in the United States.[12]

I have visited ranches and observed sheep and cattle grazing in every state and province in the U.S. and Canada. In Central and South America, I have been to Mexico, Brazil, Argentina, Uruguay, and Chile. In Europe, I have visited cattle pasture operations in the United Kingdom, Ireland, France, Denmark, and Portugal. Other places I have visited are the Australian Outback and the green hills of New Zealand. I have learned from my extensive travels that there are vast amounts of land that can only be used for grazing. The first type of land is too arid and hilly to raise crops. Some examples are the high desert in Arizona, Sand Hills in Nebraska, Texas Panhandle, and the Australian Outback. The second type of grazing land has higher moisture but it is also too hilly and rough for growing crops. Some examples in the U.S. would be the hill country in eastern Kentucky and southern Missouri. The green, steep hills of New Zealand are another place where grazing is the only option. During my travels, I have seen the very best and the very worst grazing operations. I have talked to many family ranchers who care deeply about the land. They are proud of being good stewards of both the land and their animals.

All the issues that have just been discussed have motivated me to think deeply about the use of animals for food. Ruminant animals are the only way to produce food on these lands. On a well-managed grazing operation, cattle and sheep welfare would be better compared to either indoor confinement or a muddy feedlot. When I started learning more about plants and agronomy, I discovered that grazing is part of the natural grassland system. My conversation with the students shifted to their own state of Nebraska. Nebraska is an excellent starting point for discussion due to its diverse land types. The western sandhills are too dry and hilly to be cropped. The eastern part of the state is prime farmland that is being used to grow cash crops, such as corn (maize) and soy.

INSIGHT FROM LEARNING THAT ANIMALS ARE PART OF THE LAND

A few years ago, I attended a departmental seminar at my university. A visiting agronomist explained how grazing bison created the best U.S. cropland in Iowa and Illinois. This was an important insight for me, and I learned about how grazing animals were a natural part of the land. The next section of this paper discusses scientific literature that supports the concept of the use of grazing to improve the land. Richard Teague from Texas A&M University explains that grasslands evolved with grazing animals. There were intense periods of grazing followed by a long period of rest, which allowed the pasture fully recover.[13] There is evidence that grazing can also be used to improve soil health and sequester carbon.[13-16] Another sustainability benefit of well-managed grazing is increased plant biodiversity.[15-17] There is also an abundance of scientific evidence that shows that rotating crops, such as corn or soy, with livestock grazed on cover crops can improve soil health,[18] sequester carbon,[14, 19] and increase the abundance of pollinating insects.[20] High intensity agriculture with low crop diversity will decrease insect biodiversity.[21] A long-term study conducted in Ontario Canada showed that soybean monocultures were detrimental to soil health.[2]

I told the students that they should work on developing systems to integrate grazing animals, such as cattle, sheep, goats, or bison, into both pasture and crop-based systems. Nebraska is a unique state that contains both prime cropland in the eastern half for non-irrigated dryland cash crop farming and western sandhills land. The sand hills are the largest field of dunes in the Western Hemisphere.[23] It is stabilized by surface vegetation that can be grazed by livestock. They are unique land that is not suitable for crops. It is too hilly and underground water supplies are limited. At present, the sand hills are used for grazing livestock. There are also many other parts of the world where the land is either too arid or lacks a sufficient underground water supply to raise crops.[24] Estimates show that at least 20% of the habitable land on Earth would only be suitable for grazing.[25] I informed the students that we needed to continue to grow food on the sand hills with grazing livestock. They also needed to work with the farmers on the prime farmland in eastern Nebraska to integrate grazed cover crops with standard cash crops, such as corn (maize) or soy.

BASIC PRINCIPLES OF ROTATIONAL GRAZING

There are some basic principles of grazing, but I always tell producers to obtain good local advice. A method that works well and benefits the soil in one part of the world may not work in another. Grazing in different regions often needs to be managed differently.[26] The basic principle of intense rotational grazing is to mimic the way herds of wild herbivores, such as bison or wildebeest, grazed the land.[27, 28] A herd of ruminants would migrate in a tight bunch. The animals grazed a patch of land and then moved on. They would "mow" the vegetation and then not return until next year. This would provide an entire year for the plants to regenerate. One of the early rotational grazing mistakes was not giving the pasture sufficient time to recover until it was grazed again.[29, 30] Modern electric fencing that can run on solar power has made it much easier and more economical to create small paddocks. The animals can be bunched relatively densely to force them to eat all types of vegetation before being moved. This prevents the problem of animals eating only the highly palatable plants and leaving the less palatable plants behind. Grazing specialist Fred Provenza calls this "eat the best and leave the rest."[31] There is a basic principle of plant physiology that many people forget. When a pasture regrows, the green leaves recover before the roots.[29, 30] The recovery time has to be sufficiently long to allow the roots to fully recover.

FOUR BASIC TYPES OF GRAZING

In the scientific literature, there are many different terms being used to describe grazing systems. The four basic types are (1) continuous grazing on a single pasture; (2) conventional rotation between two to four pastures, which is the most common system in low rainfall areas; (3) multiple paddocks where the animals are moved every four to eleven days; and (4) mob grazing, also called adaptive multi paddock.[13, 16] In mob grazing, the animals are stocked very densely, and they may need to be moved either once or twice a day. In many parts of the U.S. with higher rainfall levels, the most common system is using paddocks where the livestock are moved every few days. This is especially true in the wetter parts of the U.S.

SOIL HEALTH BENEFITS OF ROTATIONAL GRAZING

The full soil health benefits of rotational grazing cannot be achieved quickly.[32] Three to five years are required to start seeing soil health benefits.[33] In northern Spain, which averages 1000 mm (40 in) of annual rain, the intensive rotational grazing of sheep resulted in higher forage production and increased carbon storage.[34] I traveled to parts of the world where intensive rotational grazing is really successful. These areas were in New Zealand, the United Kingdom, southern U.S., and Uruguay. Another issue that was reported is reduced livestock productivity.[35] This is most likely to occur when rotational grazing is first started. When the animals become accustomed to rotational grazing, this may be less likely to be an issue. Intensive rotational grazing is more likely to be successful on pastures with higher rainfall.[36] On more arid land, high, intensive, rotational mob grazing with frequent moves is more likely to reduce weight gain.[37, 38] Research is needed to determine if low-stress cattle-handling methods would reduce this problem.

HIGH-PLAINS UNITED STATES LIVESTOCK GRAZING

The high plains area of the United States is in the midsection of the country, and it includes Colorado, Kansas, Montana, North Dakota, and South Dakota. I have driven in this part of the country many times during both the summer and winter. The eastern parts of the high plains have sufficient rainfall for dryland farming and the more arid west plains are used for cattle ranching.

Grazing cover crops in Nebraska and other high plains states

The research clearly shows that rotating cash crops, such as corn (maize) or soybeans, with a cover crop improves soil health.[39] To get the field ready for the next cash crop, the cover crop is often terminated with a herbicide, such as glyphosate.[40] From a sustainability point of view, this is a poor practice. There are many new research studies in Nebraska on grazing cover crops. Grazing cover crops is a win–win situation. It pays for both the cover crop,[41, 42] and provides food for people from the livestock. Grazing also greatly reduces the use of herbicides and chemical fertilizer. The cattle or other livestock are used to terminate the cover crop instead of using herbicides.[40]

Farmers are often concerned that grazing a cover crop and having animals walking on the field could cause soil compaction and hurt the yields of their crops. A three-year study conducted in Nebraska showed that if stocking density is kept low, grazing a rye cover crop would have no effect on corn yield.[43] Other studies have also shown that grazing a cover crop had no effect on grain, corn, or soybean yields.[44, 45]

To obtain the maximum benefits, cattle are often grazed on both the cover crop and the stubble that is left over from the harvested corn (maize) crop.[44] On a recent drive through Nebraska, I observed many fields where cattle were grazing corn (maize) stubble. One study in Nebraska showed that an oat–rapeseed mixture cover crop was good for growing cattle before they are put into feedyards.[46] This is another win–win situation because growing young cattle on cover crops reduces the amount of grain that is fed to them. This is due to the animal being larger and heavier before it is placed in a feedlot.

Rotational grazing on the Nebraska Sandhills and eastern Colorado

I have driven through the sand hills many times and there is no way that most of this land can be used for growing crops. It consists of steep sandy hills with small areas of flatland. On this land, Nebraska ranchers often used a four-pasture rotational grazing system. This had the added benefit of increasing beneficial dung beetles.[47] The plant diversity index was worse in the no grazing and continuously grazed pastures, compared to pastures that were rotated.[47]

High intensity mob grazing with high stocking rates works really well in areas with high rainfall.[34] In arid areas with lower rainfall, two long-term studies showed that it caused decreased cattle weight gain in Eastern Colorado and the Nebraska Sandhills.[38, 48] In the Nebraska study,[48] fencing methods and the use of small numbers of yearling steers in three different rotational pasture treatments may have influenced the results. When the steers were in the high-density mob condition, a portable electric fence was used to corral 36 steers and move them twice each day. In two other more lightly stocked rotational treatments, 8 to 9 yearlings were moved every ten days and permanent electric fencing was used. The differences in handling fencing and very small groups may have been one cause of the poor mob grazing results. When either a permanent electric fence is used, or a fence is moved gradually across a field, the cattle will always know where it is located. I observed that when small numbers of cattle were held in small electrically fenced pens, the cattle were more likely to accidentally contact the fence. On a regular commercial operation, the groups of cattle would be larger. This makes it easier for the cattle to avoid being shocked.

Another study conducted in Colorado compared very high stocking density rotational mob grazing with continuous grazing. The very high stocking density was detrimental to weight gains.[38] The next step of the research was to experiment with lower stocking densities and longer periods between rotations. A rancher in Colorado reported that he had good cattle performance and improved plant heterogeneity after three years. He moved his cattle every three to seven days, depending on grass growth. For many people, mob grazing where animals have to be moved at least once a day is too difficult. It is my opinion that more research is needed on rotational systems where the animals are moved every few days.

Grazing the eastern wetter parts of the U.S.

In other parts of the U.S. where it is warmer and wetter, a system of ten rotational grazing pastures worked well.[49] In a six-year trial on a native tallgrass prairie, the rotational system was compared to continuous grazing. Cattle movement to the next pasture was based on the animals eating 50% of the grass instead of a set number of days. In southern Missouri, I observed that ranchers are improving plant diversity and improving soil health with six to eleven paddocks. The cattle were moved when 50% of the grass was consumed. Moves during the growing season were every five to eleven days. Intensive rotational grazing with more frequent moves was beneficial in the warmer wetter parts of the U.S., such as Georgia and Mississippi.[50, 51]

In many parts of central and southern U.S., cattle graze fescue grass.[52] One of the problems with fescue is that most of this grass is infected with fungal endophyte, which increases the plant's tolerance to environmental stresses.[52] To reduce toxicity to cattle, ranchers often seed clover in their fescue pastures.[52] Endophytic-infected fescue can impair the animal's ability to thermal regulate.[53] Other methods to reduce fescue toxicity is breeding cattle to tolerate fescue.[54] Another study showed that the microbiota in the animal's gastrointestinal tract had an effect on fescue toxicity.[54] Ranchers have had to learn how to adapt to endophytes infected fescue because fescue grass that lacks the endophyte is less tolerant of environmental stressors.[55]

Increasing rancher interest in grazing

All over the U.S., there is increasing interest in using grazing to help sequester carbon and reduce the amount of grain fed to cattle.[56, 57] There is also research that shows that targeted grazing can be used to improve rangeland.[58] When this method is used, livestock are brought in to manage a specific area by having animals eat plants that they want to eliminate. In some parts of the U.S., such as West Texas, the ground had to be converted to grazing because there was no longer sufficient groundwater to irrigate with center pivot sprinklers. While flying over this area in a commercial airliner, I observed that many of the center pivots were turned off. There were brown circles next to circles green with growth. In the more arid parts of the world, grazing is the only method for raising food on the land.

I emphasized that good managers could use cattle, sheep, and goats to improve the land. I traveled to all the areas of the U.S. where the land is grazed. During these travels, I observed both the best rotational grazing where the land was improved and the worst overgrazed fields where the land was degraded. I has also had the opportunity to observe land that was not grazed adjacent to arid land that was managed by good, conventional rotational grazing. The grazed land was in much better condition. The un-grazed pasture was full of dead grass surrounded by bare dirt. The grazed pasture had a much more even growth of grass. Ranchers are adopting rotational grazing that ranges from intense mob grazing to conventional pasture rotations. A recent survey showed that 60% of U.S. ranchers use some form of rotational grazing.[59] Recent discussions in 2022 with ranchers in southern Missouri indicated that about half the ranchers in the area used some form of rotational grazing.

SOUTH AMERICAN SUSTAINABLE GRAZING SYSTEMS

I had the opportunity to view well-managed grazing lands in Brazil, Argentina, Chile, and Uruguay. Uruguay is raising 90% of its beef on pasture.[60] In the more temperate parts of South America, there has been extensive research on integrating crops and livestock. Researchers have learned that soy monocultures are detrimental because key nutrients, such as potassium, are depleted.[61] In the temperate southern parts of Brazil, Holstein heifers are grazed on an integrated soy and ryegrass system. Many South American researchers are working to integrate livestock and crops.[62-64] Another truly sustainable system that utilizes livestock is Silvo-pasture systems. They are used in tropical and sub-tropical areas. They consist of both pasture and edible leaves or scrubs.[65-67] Silvo-pastures can be used to improve the soil and rehabilitate degraded pastures.

EUROPEAN GRAZING SYSTEMS

In both western and eastern Europe, there is adequate rainfall and both rotational grazing and integrating crops with livestock has been successful.[68-71] In Germany, grazing of sheep and goats has been used to rehabilitate land that has been overgrazed by cattle.[72] These animals eat weedy, woody plants that cattle avoid. The research conducted in Hungary clearly observed that "one size fits all" strategies cannot be recommended.[73] They used a combination of sheep and cattle grazing. In Ireland, cattle are raised on pasture and housed in buildings in the winter because the fields are too muddy. The use of mixed grazing with sheep and cattle resulted in more diversity in both the vegetation and soil microbes.[68] The diversity of arthropods and birds was also improved with low intensity grazing of mixed species.[74] I recently went on a tour of Southern Portugal. Cattle and sheep were being grazed among ancient olive trees. I believe that this practice was totally sustainable because the trees were not being damaged by the cattle and the ground had forage growing. Unfortunately, a growing export market for olive oil motivated producers to plant intensive irrigated monocultures of olive trees.

CHINA AND NORTHERN STEPPES LAND GRAZING

Many northern areas of China are more arid than Europe, the eastern U.S., or South America. China has large parts of the country that are grasslands.[75] Unfortunately, 90% became degraded.[76] This was due to overstocking the pastures, and not providing them with sufficient rest to recover. When pasture is severely damaged, excluding livestock increases soil organic carbon.[77,78] There is a point where exclusion provides no additional benefit, and the introduction of well-managed rotational grazing improves organic matter.[79] Grazing does not need to be eliminated but reducing "grazing pressure conserves diversity."[79] Moderate grazing can improve species richness.[80] The key is just the right amount of moderate grazing.[81,82]

An important pillar of sustainability is the local people should be able to have a viable livelihood. Traditional herding practices may mimic rotational grazing. A four-season nomadic herding system and a four-season rotational grazing were both less detrimental than "settle" continuous grazing.[83] Converting grassland to crops caused losses of nitrogen and soil organic carbon.[79] During my consulting business, I have been on extensive road trips. In many parts of China, I observed diverse agriculture with many different crops grown in strips on the same field. Few cattle or other grazing animals were visible. An area for possible future innovation would be incorporating grazing animals into these multiple crop systems.

ARID, HOT AUSTRALIAN OUTBACK

A trip to the Australian Outback in 2018 made me think deeply about using livestock to produce food. I left Darwin and flew for two hours south in a small plane. I observed an arid vastness with no signs of houses, electrical wires, or roads. As I looked out over this vast land, I thought that people need to be able to raise food on this land. The plane flew almost directly over a single road and then, all of a sudden, a cattle station appeared.

This station grazed cattle that were a mixture of European and *Bos indicus* breeds. Grazing cattle, sheep, or goats is the only way that food can be grown on this vast land. There is only enough water to water grazing livestock. Research is needed on the best way to manage this land. There were numerous studies done in other parts of the world. Outback land may be too arid and sparse to make intensive rotational grazing work. A nine-year study near Darwin showed that an intensive rotational system lowered weight gain.[37] In a study in the wetter, semiarid parts of southeastern Australia, rotational grazing improved plant biodiversity.[36]

GRAZING IN AFRICA

In South Africa, the integration of crops and grazing has similar results to other countries. Grazing livestock reduced the use of chemical herbicides.[84] In Africa, the vast semiarid savannah pastures are inhabited by vast variety of grazing wild ungulates. The overstocking of pasture with cattle in these areas is detrimental to wild herbivores. However, lower stocking rates with cattle may represent a win–win for wild herbivore conversation and the individual performance of livestock.[85] Further research and innovation from local ranchers are required to develop sustainable grazing that can coexist with wildlife.

CHOOSING THE RIGHT LIVESTOCK FOR EXTENSIVE GRAZING FOR BOTH PRODUCTIVITY AND ANIMAL WELFARE

I observed that the animal that performs best in an extensive grazing environment is not the same animal that performs well being fed grain in a feedlot. When I visited the hot Australian outback, I quickly learned that purebred British breeds could not survive there. Ranchers told me that they tried using Angus sires and they all died. This was a serious welfare problem. For heat tolerance, it is essential to have cattle with high percentages of *Bos indicus* breeding. These cattle can have good welfare if they eat sufficient forage to maintain body condition. U.S. ranchers learned that when cattle are selected solely for meat production, the result is often a cow that is too large and requires too many calories to be successful on extensive pasture. John Scasta, from the University of Wyoming, stated that the cattle industry encouraged ranchers to select large cattle that may not be suitable for harsh rangeland environments.[86] Ranchers in Nebraska learned that moderate sized, early maturing genetics cows are the best. Many ranchers told me that it was too expensive and difficult to feed larger cows hay in the wintertime. The original Angus and Hereford cattle from the United Kingdom were originally bred to fatten on grass. In the southern U.K., lush grass is plentiful. Producers who raise grassfed beef in the U.S. often use cattle that have not been bred for maximum meat production. I observed that the Hereford and Angus breeds have undergone extensive selection since the 1970s. Over the years, these breeds have been selected to fatten on grain. They were selectively bred to reduce the amount of external fat. Breeds, such as Murray Grey cattle, which have not been intensively selected for meat production, are popular in grass feeding programs.[87] In the U.S., in southern Missouri where it is colder and wetter, I observed that smaller, moderate frame sized cattle are being used at present.

OTHER WAYS TO USE LIVESTOCK TO IMPROVE SUSTAINABILITY

Effects of fire

Another use of grazing animals that needs further research is fire suppression. In Colorado, over 1000 homes were destroyed in a disastrous fire.[88] The fire started in a large, open, grassy area near the houses called Marshall Mesa Trailhead.[89] The grazing of this land near the houses might have prevented this fire. There are already some producers who are using grazing animals to reduce fire risk. Fire and planned burning is also needed to increase biodiversity on Savannah land.[90] I observed arid land in Arizona that has been ruined by invasive juniper bushes. This is not natural. University of Nebraska, Range Ecologist Dirac Twinwell is concerned that juniper bushes are taking over pastures.[91]

Grazing under solar panels

Large amounts of land are now covered with solar panels. Innovative producers are now grazing sheep on the land that is underneath the panels.[92, 93] This is another win–win practice. The sheep control weeds and mow the vegetation under the panels. Grazing sheep enables food production on land covered with solar panels. To graze cattle under solar panels requires the use of stronger supporting posts and the installation of the panels higher off the ground.

Combine the best of organic and conventional agriculture

To achieve the most sustainable system may require the use of many of the principles of organic agriculture.[94] A combination of organic and chemical fertilizer increased crop yield.[94] Both the scientific research and a discussion with farmers indicated that a total ban on synthetic fertilizer or crop chemicals may not be the best approach. One farmer told me: "It is a mistake to be too pure." The use of rotational livestock grazing on cover crops makes it possible to greatly reduce chemical use. I predict that the most sensible approach is the hybrid approach that combines the best principles of both organic and conventional agriculture. Merging a mostly organic approach with some use of industrial methods may work best.[95,96] Prices for conventional fertilizer have skyrocketed.[97] This is an additional motivator to integrate livestock and crops.

ANIMAL WELFARE ISSUES IN GRAZING SYSTEMS

When grazing is performed properly, the animals can have a high standard of welfare. Welfare scientists agree that preventing suffering is not sufficient. The animals should also have a "life worth living" and opportunities for positive experiences.[98,99] Good stockmanship is essential when livestock have to be continuously moved. Over the course of my long career, I have observed that cattle handling has greatly improved. Many people are now teaching low-stress methods. When handling is performed correctly, cattle become calmer when pastures are switched.[100] In very dry, extensive areas, such as the outback in Australia and desert grazing in Arizona, I observed that body condition must be closely monitored. In the wetter areas of the world, such as the green, lush pastures in Ireland, the cattle graze in the summer and they are kept in buildings in the winter because the fields are too muddy. The welfare of cattle in some of these barns may be compromised. I observed very filthy, dirty cattle arriving at slaughter plants in the U.K. and Ireland. The cattle originated from these barns. The successful use of Holstein heifers on a corn (maize) rye system in Brazil could serve as an example that could improve Holstein dairy cattle welfare. This would allow Holstein steers to stay on pasture and not enter a feed-yard at a young age.

LOCAL GRAZING MAY REDUCE SUPPLY CHAIN FRAGILITY

Local grazing operations may help provide food when there are supply chain disruptions. Everybody has now become more aware about the fragility of food supply chains. COVID-19 disrupted supply chains.[101] This has forced many people to become aware about how food enters their local supermarket. This has made more people become interested in local sources of food.[102] Before the COVID-19 pandemic, many people never thought about how both goods and food moved across their country or around the world. The public has now learned that big is fragile.[103] When COVID-19 sickened the workers in two large U.S. pork processing plants, over 300,000 pigs had to be destroyed on the farm. This caused both a severe animal welfare issue and a huge waste of food.

The COVID-19 pandemic has motivated many U.S. cattle producers to start regional beef slaughter facilities.[105-107] These smaller plants will not be able to compete with the low-cost production of large processors. They will have to enter premium niche markets, such as grassfed, family farm, organic, or locally raised, and charge more for their meat. Ranchers also need to receive good local advice on the use of both cover crops and pasture rotation. The most effective systems for both the pasture, rotation, and integration of crops and livestock needs to be adapted to local conditions. A system that works well in one part of the U.S. may not work in another part. One size does not fit all. I observed some early attempts at intense rotational grazing in arid Arizona in the 1970s that were not successful. The pastures were stocked too heavily and not given sufficient time to recover.

CONCLUSIONS

There are vast areas of land in the world that can only be used for grazing. Grazing makes it possible to produce food on land that cannot be cropped. Rotating grazing cattle or other livestock on a cover crop rotated with cash crops, such as corn (maize) or soy, will improve both soil health and reduce dependence on expensive, fragile fertilizer supply chains. The welfare of these animals is likely to be better compared to livestock housed indoors. There are many grazing alternatives that will be sustainable if the system is not overloaded. I observed that it was difficult for people to determine what was optimal production versus maximum production. Both the scientific research and practical applications indicate that both overgrazing and no grazing is detrimental to the land. There needs to be just the right amount of grazing. Well-managed grazing systems can be truly sustainable and improve soil health, help sequester carbon, and maintain plant biodiversity. The grazing animals are part of the cycle of life and the natural grass ecosystem. They are a natural part of the land.

REFERENCES

1. Grandin, T. Auditing animal welfare at slaughter plants. Meat Sci. 2010, 86, 56–65.
2. Grandin, T.; Cockram, M. The Slaughter of Farmed Animals: Practical Ways of Enhancing Animal Welfare; CABI: Wallingford, UK, 2020.
3. Pendergrast, N. The vegan shift in the Australian animal movement. Int. J. Soc. Soc. Policy 2020, 41, 407–423.
4. Birch, J.; Burn, C.; Schnell, A.; Browning, H.; Crump, A. Review of Evidence of Sentience in Cephalopod Mollusks and Decapod Crustaceans. 2021. Lse.ac.uk/business/consulting/asset/documents/SentienceincephalopodmollusksandDecapodcrustaceansFinalReportNovember2021.pdf (accessed on 21 June 2022).
5. de Waal, F.B.; Andrews, K. The question of animal emotions. Science 2022, 6587, 1351–1352.
6. Hampton, J.O.; Jones, B.; McGreevy, P.D. Social licenses and animal welfare: Developments from the past decade in Australia. Animals 2020, 19, 2237.
7. Rischer, H.; Szilvay, G.R.; OksmanCaldentey, K.M. Cellular agricultureindustrial biotechnology for food and materials. Curr. Opin. Biotechnol. 2020, 61, 128–134.
8. Sergelidas, D. Lab grown meat: The future sustainable alternative to meat or a novel functional food. Biomed. J. Sci. Tech. Res. 2019, 17, 12440–12444.
9. Lee, S.Y.; Kang, H.J.; Lee, D.A.; Kang, J.H.; Ramanz, S.; Park, S.; Huv, S.J. Principle protocols for processing cultured meat. J. Anim. Sci. Tech. 2021, 63, 673–680.
10. Vafi, K.; Rafiq, T.; Biraud, S.; Thorpe, A.; Duren, R.; Hopkins, F.M. Methane super emitters in California oil fields. Res. Sq. 2021, 1–31.
11. Tiwari, S.; Singh, C.; Singh, J.S. Wetlands: A major natural source responsible for methane emission. In Restoration of Wetland Ecosystem, a Trajectory Towards a Sustainable Environment; Upadhyay, A.K., Ed.; Springer: Singapore, 2020.
12. Hristov, A.N. Historic preEuropean settlement and presentday contribution of wild ruminants to enteric methane emissions in the United States. J. Anim. Sci. 2012, 90, 1371–1375.
13. Teague, R.; Kreuter, U. Managing grazing to restore soil health, ecosystem function, and ecosystem services. Front. Sustain. Food Sci. 2020, 157.
14. Byrnes, R.; Eastburn, D.J.; Tate, K.W.; Roche, L.M. A global met analysis of grazing impacts on soil health indicators. J. Environ. Qual. 2018, 47, 758–765.
15. Apfelbum, S.L.; Thompson, R.; Wang, F.; Mosier, S.; Teague, R.; Byck, P. Vegetation water in filtration in soil carbon response to adaptive multi paddock and conventional grazing and southeastern USA ranches. J. Environ. Manag. 2022, 308, 114576.
16. Teague, W.R.; Apfelbaum, S.; Lai, R.; Kneotes, U.P.; Rowntree, J.; Davies, C.A.; Conser, R.; Rasmussen, M.; Hatfield, J.; Wang, T.; et al. The role of ruminants in reducing agriculture's carbon footprint in North America. J. Soil Water Conserv. 2016, 71, 156–164.
17. Barry, S.; Huntsinger, L. Rangeland sharing, livestock grazing's role in conservation of imperiled species. Animals 2021, 13, 4466.
18. Menefree, D.S.; Collins, H.; Smith, D.; Haney, R.L.; Fay, P.; Polley, W. Cropping management in a livestockpasturecrop rotation integration modifies microbial communities, activity, and soil health scare. J. Environ. Qual. 2021.
19. Becker, A.E.; Horowitz, L.S.; Ruark, M.D.; Jackson, R. Surface soil carbon stocks are greater under well-managed grazed pasture than row crops. Soil Water Manag. Conserve 2022, 86, 758–768.
20. LeeMader, E.; Stine, A.; Fowler, J.; Hopwood, J.; Vaughan, M. Cover Cropping for Pollinators and Beneficial Insects, SARE (Sustainable Agriculture Research Education USDA) 2014. Save.org/wpcontact/uploads/covercroppingforpollinatorsandbeneficialinsects.pdf (accessed on 6 July 2022).
21. Outhanwaite, C.L.; McCann, P.; Newbold, T. Agriculture and climate change are reshaping insect biodiversity worldwide. Nature 2022, 605, 97–102.
22. PerezGusman, L.; Phillips, L.A.; Seuradge, B.J.; Agomoh, I.; Drury, C.F.; Acosta-Martinez, V. An evaluation of biological soil health indicators in four long-term agro ecosystems in Canada agro systems. Agrosyst. Geosci. Environ. 2021, 4, e20164.
23. Ahlbrandt, T.S.; Fryberger, S.G. Eolian deposit in the Nebraska sandhills, Geologic and Paleoecologic studies in Nebraska sandhills. In Geological Survey Professional Paper; 1120A.B.C.; United States Printing Office: Washington, DC, USA, 1980.
24. Nunez, C. Grassland, like the little Missouri National Grassland in the United States, fill a niche between forests and deserts often bordering on the two. In National Geographic; National Geographic Society: Washington, DC, USA, 2020.
25. Seo, N.S. Sublime grasslands: A story of the Pampas, Prairie, Steppe, and Savannas, Where Animals Graze. In Climate Change and Economics; Palgrave Macmillan: London, UK, 2021.
26. McSherry, M.E.; Ritchie, M.E. Effects of grazing on grassland carbon: A global review. Global Chang. Biol. 2013, 23, 585–594.
27. Frank, D.A.; McNaughton, S.J.; Tracey, B.F. The ecology of earth's grazing ecosystems. BioScience 1998, 48, 513–521.
28. Bailey, D.W.; Woodward, O.K.; Gross, J.E.; Laca, E.A.; Rittenhouse, L.R. Mechanisms that result in large herbivore grazing distribution patterns. J. Range Manag. 1997, 49, 386–400.

29. Machado, L.O.C.P.; Sco, H.L.S.; Daros, R.R.; Enriquez, D.; Wendling, A.V.; Pinheiro, L.C. Voisin rotational grazing as a sustainable alternative for livestock production. Animals 2021, 11, 3434.

30. Aljoe, H. The keys to successful regenerative grazing management. In Progressive Cattle; Progressive Publishing: Jerome, ID, USA, 2022; pp. 33–35.

31. Provenza, F.D.; Villalba, J.J.; Dziba, L.E.; Atwood, S.B.; Banner, R.E. Linking herbivore experience, varied diets, and plant biochemical diversity. Small Rumin. Res. 2003, 49, 257–274.

32. Rui, Y.; Jackson, R.D.; Cotrufo, M.F.; Ruak, M.D. Persistent soil carbon enhanced in mollisols by well-managed grassland but not annual grain or dairy forage cropping systems. Proc. Natl. Acad. Sci. USA 2022, 119, e2118931119.

33. Mosier, S.; Apfelbaum, S.; Byck, P.; Cotrufo, F.M.F. Adaptive multi paddock grazing enhances nitrogen stock and stabilization throughout south-eastern grazing lands. J. Environ. Manag. 2021, 288, 112409.

34. de Otalora, X.; Epelde, L.; Arranz, J.; Garbisu, C.; Ruiz, R.; Mandaluniz, N. Regenerative rotational grazing management of dairy sheep increases springtime grass production and topsoil carbon storage. Ecol. Indictors 2020, 125, 107484.

35. DiVirgilio, A.; Lambertucci, S.A.; Morales, J.M. Sustainable grazing management of rangelands: Over a century of searching for a silver bullet. Agric. Ecosyst. Environ. 2019, 283, 106561.

36. McDonald, S.E.; Reid, N.; Smith, R.; Waters, C.M.; Hunter, J.; Rader, R. Rotational grazing management achieves similar plant diversity outcome in areas managed for conservation in a semiarid rangeland. Rangel. J. 2019, 41, 135–143.

37. Schatz, T.; Ffloukes, D.; Shotton, P.; Hearnden, M. Effect of high intensity rotational grazing on the growth of cattle grazing buffalo pasture in the northern territory and on carbon sequestration. Anim. Prod. Sci. 2020, 60, 1814–1821.

38. Augustine, D.J.; Derner, J.D.; FernandezGimenez, M.E.; Porensky, L.M.; Wilmer, H.; Briske, D.D.; FernandezGimenez, M.E.; Parensky, L.M.; Wilmer, H.; Briske, D.D.; et al. Adaptive multi paddock rotational grazing management a ranch scale assessment of effects on vegetation and livestock performance on semiarid rangeland. Rangel. Ecol. Manag. 2020, 3, 796–810.

39. Drewnoski, M.; Parsons, J.; Blaner, H.; Redfeam, D.; Hales, K.; MacDonald, J. Forages and pastures symposium—Cover crops in livestock production whole system approach: Can cover crops pull double duty: Conservation and profitable forage production in midwestern United States? J. Anim. Sci. 2018, 96, 3503–3512.

40. Wyffels, S.A.; Bourgault, M.; Dafoe, J.M.; Lamb, P.F.; Boss, D.L. Introducing cover crops as a fallow replacement in Northern Great Plains 1. Evaluation of cover crop mixes as forage source for grazing cattle. Renew. Agric. Food Syst. 2021, 37, 292–302.

41. Farney, J.K.; Sassenrath, G.F.; Davis, C.; Presley, D.A. Growth forage quality and economics of cover crop mixes for grazing. In Kansas Agricultural Experiment Station Reports; Kansas State University: Manhattan, KS, USA, 2018.

42. Rai, T.; Tieya, T.; Kumer, S.; Sexton, P. The medium-term impacts of integrated crop performance. Agronomy J. 2021, 113, 5207–5221.

43. BlancoCanqui, H.; Drewnoski, M.F.; MacDonald, J.C.; Redfearn, D.D.; Parsons, J.; Lesoing, G.W.; Williams, T. Does cover crop grazing damage soils and reduce crop yields. Agrosyst. Geosci. Environ. 2020, 3, e20102.

44. Kuhn, A. Livestock Grazing Impacts on Crop and Soil Responses for Two Cropping Systems. Master's Thesis, University of Nebraska, Lincoln, NE, USA, 2021.

45. Conway, A.C.; Bondurant, R.G.; Hilscher, F.H.; Parsons, J.; Redifearn, D.; Drewnoski, M.F. Impact of grazing spring rye on subsequent crop yields and profitability. In Nebraska Beef Report; University of Nebraska: Lincoln, NE, USA, 2019; pp. 47–48.

46. Riley, H.E.; Hales, K.E.; Shackelford, S.D.; Freetly, H.C.; Drewnoski, M.E. Effect of Rapeseed Inclusion in Late Summer Planted Oats Pasture on Growing Performance of Beef Steers. Nebraska Beef Cattle Reports. 2019. (accessed on 15 June 2022).

47. Wagner, P.M.; Abagandura, G.O.; Mamo, M.; Weissingling, T.; Wingeyer, A.; Bradshaw, J.D. Abundance and diversity of dung beetles (Coleoptera scarab aeoida) as affected by grazing management in Nebraska Sandhill Ecosystem. Environ. Entomol. 2021, 50, 222–231.

48. Andradi, B.Q.; Shropshire, A.; Johnson, J.J.; Redden, M.D.; Semerad, T.; Soper, J.; Beckman, B.; Mibby, B.; Eskridge, K.M.; Velesky, J.D.; et al. Vegetation and animal performance responses to stocking density grazing systems in Nebraska Sandhills Meadows. Rangel. Ecol. Manag. 2022, 82, 86–96.

49. Steiner, J.L.; Starks, P.J.; Neel, J.P.S.; Northrop, B.; Turner, K.E.; Gowda, P.; Calemon, S.; Brown, M. Managing tallgrass prairies for productivity and ecological function: A long-term grazing experiment in the Southern Great Plains. USA, Grasslands Manage. Sustain. Agrosyst. 2019, 11, 699.

50. Machmuller, M.G.; Kramer, M.G.; Cyle, T.K.; Hill, N.; Hancock, D.; Thompson, A. Emerging land use practices rapidly increase soil organic matter. Nat. Commun. 2015, 6, 6995.

51. Wang, F.; Apfelbaun, S.I.; Thompson, R.L.; Teague, R.; Byke, P. Effect of adaptive multiple paddock and continuum grazing on fine scale spatial patterns: Vegetation species and biomass in commercial ranches. Landsc. Ecol. 2021, 36, 2725–2741.

52. Aiken, G.E.; Henning, J.C.; Rayburn, E. Chapter 9—Management strategies for pasture, beef cattle, and marketing of stocker-feeder calves in the Upper South I64 Corridor. In Management Strategies for Sustainable Cattle Production in Southern Pastures; Rouquette, M., Aiken, G.E., Eds.; Academic Press: Cambridge, MA, USA, 2020; pp. 227–264.

53. Mote, R.S.; Hill, N.S.; Skarlupka, J.H.; Tran, V.T.; Walker, D.I.; Turner, Z.B.; Sanders, Z.P.; Jones, D.P.; Suen, G.; Filipov, N.M. Toxic fall fescue grazing increases susceptibility of Angus steer fecal microbiota and plasma/urine metabolome to environmental effects. Sci. Rep. 2020, 10, 2497.

54. Poole, D.H.; Mayberry, K.J.; Newsome, M.; Poole, R.K.; Gallious, J.M.; Khanal, P.; Poore, M.H.; Serao, N.V.L. Evaluation of resistance to fescue to toxicosis in purebred Angus cattle utilizing animal performance and cytokine response. Toxins 2020, 12, 796.

55. Kaester, L.R.; Poole, D.H.; Serao, N.V.L.; Schmitz-Esser, S. Beef cattle that respond differently to fescue toxicosis have distinct gastrointestinal tract microbiota. PLoS ONE 2020, 15, e0229192.

56. Jackson, R.D. Grazed perennial grasslands can match current beef production while contributing to climate mitigation and adaptation. Agric. Environ. Lett. 2022, 7, e20059.

57. Aide, M.; Graden, I.; Murray, S.; Schabbing, C.; Scott, S.; Siemers, S.; Svenson, S.; Weather, J. Optimizing beef cow grazing across Missouri with an emphasis on protecting ecosystem services. Land 2021, 10, 1076.

58. Bailey, D.W.; Moslely, J.C.; Estell, R.E.; Cibils, A.F.; Horney, M.; Hendrickson, J.R.; Walker, J.W.; Launchbaugh, K.L.; Burritt, E.A. Synthesis paper targeted livestock as a prescription for healthy rangelands. Rangel. Ecol. Manag. 2019, 72, 865–877.

59. Wang, T.; Jin, H.; Kreuter, U.; Feng, H.; Hennessy, D.A.; Teague, R.; Cho, Y. Challenges of rotational grazing practices. Views from nonadopters across the Great Plains. USA. J. Environ. Manag. 2020, 256, 109941.

60. Koop, F. Uruguay Plans to Boost Beef Production and Lessen Its Climate Footprint. Dialogo Chino 2021. (accessed on 10 August 2022).

61. Alves, L.A.; de Oliveira Denardin, L.G.; Martins, A.P.; Anghinoni, I.; de Faccio Carvalho, P.C.; Tiecher, T. Soil acidification and P, K, Ca, and Mg budget is affected by sheep grazing and crop rotation in a long-term integrated crop livestock system in southern Brazil. Geoderma 2019, 351, 197–208.

62. de Faccio, P.C.; PontesPrater, A.; Szymcak, L.S.; Filho, W.S.; Moojen, F.G.; Lemain, G. Reconnecting grazing livestock to crop landscapes: Reversing specialization trends to restore landscape multifunctionality. Front. Sustain. Food Syst. 2021, 5, 750765.

63. Alves, L.A.; Denardin, L.G.O.; Martins, A.; Bayer, C.; Veloso, M.G.; Bremm, C.; Carvalho, P.C.R.; Machado, D.R.; Tiecher, T. The effect of crop rotation and sheep grazing management on plant production and soil C and N stock in a long-term integrated crop livestock system in southern Brazil. Soil Tillage Res. 2020, 203, 104678.

64. Schuster, M.Z.; Harrison, S.K.; de Moraes, A.; Sulk, R.M.; Carvalho, P.C.F.; Lang, C.R.; Anghinoni, L.; Lustosa, S.B.C.; Gastal, F. Effects of crop rotation and sheep grazing management on seedbank and emerged weed flora under no tillage integrated crop livestock system. J. Agric. Sci. 2018, 156, 810–820.

65. Broom, D.M. A method for assessing sustainability with beef production as an example. Biol. Rev. 2021, 96, 1836–1853.

66. Polania-Hincapiem, K.L.; Olaya-Montes, A.; Cherubin, M.R.; Herrera-Valencia, W.; OrtizMorea, F.A.; SilvaOlaya, A.M. Soil physical quality responses to silvo-pastoral implementation in Columbia Amazon. Geoderma 2021, 136, 114900.

67. Kumar, R.V.; Roy, A.K.; Kumar, S.; Gautam, K.; Singh, A.K.; Ghosh, A.; Singh, H.V.; Koli, P. Silvopastural systems for restoration of degraded lands in a semiarid region of India. Land Degrad. Dev. 2022, in press.

68. Fraser, M.D.; Garcia, R.R. Mixed species grazing management to improve sustainability and biodiversity, The Contribution of Animals to Human Welfare. Rev. Sci. Tech. 2018, 37, 247–252.

69. Soussana, J.F.; Lemaire, G. Coupling carbon and nitrogen cycles for environmentally sustainable intensification of grasslands and crop production systems. Agric. Ecosyst. Environ. 2014, 190, 9–17.

70. Bonaudo, T.; Bandahan, A.B.; Sabatier, R.; Ryschawny, J.; Bellon, S. Agroecological principles for the redesign of integrated crop livestock systems. Eur. J. Agron. 2014, 57, 43–51.

71. Veysset, P.; Lherm, M.; Bebin, D.; Rouiene, M. Mixed crop livestock farming systems: A sustainable way to produce beef? Commercial farms results in questions and perspectives. Animals 2014, 8, 1218–1228.

72. Benthien, O.; Braun, M.; Rieman, J.C.; Stolter, C. long-term effect of sheep and goat grazing on plant diversity in a semi dry natural grassland habitat. Hellyon 2018, 4, e00556.

73. Toth, E.; Deak, B.; Valko, O.; Keleman, A.; Miglecz, T.; Tothmeresz, B.; Tarok, P. Livestock type is more crucial than grazing intensity: Traditional cattle and sheep grazing in short grass steppes. Land Degrad. Dev. 2016, 29, 231–239.

74. Evans, D.M.; Redpath, S.M.; Evans, S.A.; Elston, D.A.; Gardner, C.J.; Dennis, P.; Pakeman, R.J. Low intensity mixed livestock grazing improves breeding abundance of common insectivorous passerine. Biol. Lett. 2016, 2, 636–638.

75. Zhang, Y.J.; Zhang, X.Q.; Wang, X.Y.; Liu, N.; Kan, H.M. Establishing the carrying capacity of grasslands of China: A Review. Rangel. J. 2014, 36, RJ13033.

76. Kemp, D.; Han, G.; Hou, F.; Hou, K.; Li, Z.; Sun, Y.; Wang, Z.; Wu, J.; Zhang, X.; Zhang, Y.; et al. Sustainable management of Chinese grassland issues and knowledge. Front. Agric. Sci. Eng. 2018, 5, 9–23.

77. Dong, L.; Martinson, V.; Wu, Y.; Zheng, Y.; Liang, C.; Liu, Z.; Mulder, J. Effect of grazing exclusion and rotational grazing on labile soil organic carbon in North China. Soil Sci. 2021, 72, 372–384.

78. Wang, L.; Gan, Y.; Wiesmeier, M.; Zhao, G.; Zhang, R.; Han, G.; Siddique, K.H.M.; Hou, F. Grazing exclusion—An effective approach for naturally restoring degraded grasslands in Northern China. Land Degrade Dev. 2018, 29, 4439–4455.

79. Li, C.; Dong, G.; Sui, B.; Wang, H.; Zhao, L. Effects of grassland conversion in the Chinese Chernozem region in soil carbon, nitrogen, and phosphorus. Sustainability 2021, 13, 2554.

80. Zhang, R.; Wang, J.; Ni, U.S. Towards a sustainable grazing management based on biodiversity and ecosystem multifunctionality in dryland. Curr. Opin. Environ. Sustain. 2001, 48, 36–43.

81. Munkhzul, O.; Oyundelger, K.; Narantuga, N.; Tuushintogtokh, I.; Oyuntsetseg, B.; Jaschke, Y. Grazing effects on Mongolian steppe vegetation: A systemic review of local literature. Front. Ecol. Eval. 2021, 9, 703220.

82. Dai, L. Moderate grazing promotes the root biomass in kobresia meadow on northern Qinhai-Tibet Plateau. Ecol. Evol. 2019, 9, 9395–9406.

83. Bao, X.; Yi, J.; Liu, S.; Gaowa, J.; Wureqimuge; Jigejidesuran; Budebateer; Wang, P.; Lian, Y. Effects of different grazing on the typical steppe vegetation characteristics on the Mongolian plateau. Nomadic Peoples 2020, 14, 53–66.

84. MacLaren, C.; Storkey, J.; Strauss, J.; Swanepool, P.; Dehnen-Schmitz, K. Livestock on diverse cropping systems improve weed management and sustainable yields whilst reducing inputs. J. Appl. Ecol. 2019, 56, 144–156.

85. Wells, H.B.M.; Crego, R.D.; Ekadeli, J.; Namoni, M.; Kimuyo, D.M.; Odadi, W.O.; Porenskym, L.M.; Dougill, A.J.; Stringer, L.C.; Young, T.P. Less is More: Lowering cattle stocking rates enhances wild herbivore habitual use and cattle foraging efficiency. Front. Ecol. 2022, 11, 825689.

86. Scasta, J.D.; Lalman, D.L.; Henderson, L. Drought mitigation for grazing operations: Matching the animal to the environment. Rangelands 2016, 38, 204–210.

87. Thomas, H.S. Murray Grey cattle fit the grassfed niche. In The Stockman Grass Farmer; Mississippi Valley Publishing: Ridgeland, MS, USA, 2022; pp. 7–9.

88. Case, A. Updated Numbers Show 1,084 Homes Destroyed in Marshall Fire. 9 News 2022. (accessed on 18 June 2022).

89. Carrasco, A. Marshall Mesa Trailhead Reopens 4 Months after Wildfire. 2022. (accessed on 10 August 2022).

90. Morris, C.D.; Everson, C.S.; Everson, T.M.; Gordijn, P.J. Frequent burning maintained a stable grassland over four decades in Drakensberg, South Africa. Afr. J. Range Forage Sci. 2021, 1, 39–52.

91. Smith, T. RAP reveals grassland stories. In Angus Beef Bulletin; Angus Association: Saint Joseph, MO, USA, 2021; pp. 84–85.

92. Alyssa, A.C. Lamb Growth and Pasture Production in Agrivoltaic Production System. Honors Thesis, Oregon State University, Corvallis, OR, USA, 2020. (accessed on 10 August 2022).

93. Kochencloerfer, N.; Thonney, M.L. Grazing Sheep on Solar Sites in New York, Opportunities and Challenges; Cornell University: Ithaca, NY, USA, 2021. (accessed on 18 June 2022).

94. Li, X.; Li, B.; Chen, L.; Liang, J.; Huang, R.; Tang, X.; Zhang, X.; Wang, C. Partial substitution of chemical fertilizer with organic fertilizer over seven years increases yields and restores soil bacterial community diversity in wheat-rice rotation. Eur. J. Agron. 2022, 133, 126445.

95. Heyman, S. The Planter of Modern Life: Louis Broomfield and the Seeds of Modern Food Production; W.W. Norton: New York, NY, USA, 2020.

96. Montgomery, D. The Novelist Who Loved the Soil. Nature 2020, 580, 319–320.

97. Whitlock, J. New Report Estimates Fertilizer Prices to Increase by 80%. TexasFarmBureau.org, 2022. (accessed on 18 June 2022).

98. Mellor, D.J.; Beausoleil, N.J.; Littlewood, K.E.; McLean, A.N.; McCueevy, P.D.; Jones, E.; Wilkins, C. The 2020 Five Domains Model: Including Human-Animal Interactions in Assessments of Animal Welfare. Animals 2020, 10, 1870.

99. Grandin, T. Improving Animal Welfare: A Practical Approach. In CABI Publishing; Wallingford: Oxford, UK, 2020.

100. Cebellos, M.; Gois, K.C.R.; Sant Anna, A.C.; de Costa, M.J.R.P. Frequent handling of grazing beef cattle maintained under rotational stocking method improves temperament over time. Anim. Prod. Sci. 2016, 58, 307–313.

101. Szegedi, K. COVID-19 Has Broken the Global Food Supply Chain. So Now What? Deloitte Consumer Business. 2021. (accessed on 19 June 2022).

102. DuPuis, M.E.; Ransom, E.; Worosz, M.R. Food supply chain shocks and the pivot toward local: Lessons from the global pandemic. Front. Sustain. Food Syst. 2022, 6, 836574.

103. Grandin, T. Temple Grandin: Big Meat Supply Chains are Fragile. Forbes 2020. (accessed on 20 June 2022).

104. Grandin, T. Methods to prevent future severe animal welfare problems caused by COVID-19 in the Pork Industry. Animals 2021, 11, 830.

105. Thomas, P. Cattle ranchers take aim at meatpackers' dominance. Wall Str. J. 2022. (accessed on 19 June 2022).

106. Eller, D. Lowan Company Plans to Build a $325 Million Beef Processing Plant in Council Bluffs. Iowa, Des Moines Regist. 2021. (accessed on 10 August 2022).

107. Kovner, G. Bay Area Ranchers Open Their Own Mobile Meat Processing Plant, Filling a Key Gap for Local Industry. Press Democrat, 2022. Pressdemocrat.com/article/news/bayarearanchersopentheirownmobilemeatprocessingplantfillingkey/ (accessed on 19 June 2022).

TWENTY-EIGHT

Practical Application of the Five Domains Animal Welfare Framework for Food Supply Animal Chain Managers

T. Grandin (2022) Practical Application of the Five Domains Animal Welfare Framework for Food Supply Animal Chain Managers, *Animals*, 12:2831. MDPI, Basel, Switzerland.

The Five Domains Animal Welfare model is increasingly being adopted. Preventing suffering is not sufficient. The animals must also have opportunities for positive emotional experiences. Many papers have been published on the theoretical basis of the Five Domains System. My paper shows how to incorporate easy-to-use, numerical based welfare scoring tools into the Five Domains.

SIMPLE SUMMARY

The Five Domains model is being increasingly used as a framework for assessing animal welfare on farms. This commentary is focused on the practical application of the Five Domains by supply chain managers who buy food animal products and often work in global supply chains. Assessments used in commercial supply chains need to be simpler than assessment tools used in scientific research. There needs to be very clear guidance on conditions that should result in a failed audit. Welfare auditors can be easily trained to assess Animal based outcome measures such as body condition score, foot pad lesions on poultry or lameness. A farm would also have to have the type of housing that is specified in the buyer's welfare guidelines. Easy to evaluate animal welfare indicators should be included in each of the four domains of nutrition, environment, health, and behavioral interaction.

ABSTRACT

The author has worked as a consultant with global commercial supply managers for over 20 years. The focus of this commentary will be practical application of The Five Domains Model in commercial systems. Commercial buyers of meat need simple easy-to-use guidelines. They have to use auditors that can be trained in a workshop that lasts for only a few days. Auditing of slaughter plants by major buyers has resulted in great improvements. Supply chain managers need clear guidance on conditions that would result in a failed audit. Animal based outcome measures that can be easily assessed should be emphasized in commercial systems. Some examples of these key animal welfare indicators are: percentage of animals stunned effectively with a single application of the stunner, percentage of lame animals, foot pad lesions on poultry, and body condition scoring. A farm that supplies a buyer must also comply with housing specifications. The farm either has the specified housing or does not have it. It will be removed from the approved supplier list if housing does not comply. These types of easy to assess indicators can be easily evaluated within the four domains of nutrition, environment, health and behavioral interactions. The Five Domains Framework can also be used in a program for continuous improvement of animal welfare.

INTRODUCTION

Both animal welfare researchers and commercial buyers of animal products are moving towards adoption of the Five Domains Model for their animal welfare programs.[1-3] The author has worked extensively as a consultant with commercial supply managers on both training of animal welfare auditors and development of their auditing programs.[4,5] Most older animal welfare assessments, such as the Five Freedoms, emphasized the importance of preventing suffering.[6] The Five Domains model states that preventing suffering is not sufficient.[1] The animals must also have opportunities for positive emotional (affective) experiences. Research clearly shows that animals have affective experiences that are both positive and negative.[7,8] In the Five Domains model, there are four domains where animal welfare indicators can be assessed. They are: Nutrition, Environment, Health and Behavioral Interactions.[2] These four domains have either a positive or negative effect on the fifth affective domain that cannot be directly measured. The four domains are similar to the four parts of the European Welfare Quality Protocols of Good Feeding, Good Housing, Good Health, and Appropriate Behavior.[9] It was developed by European welfare specialists. This similarity will make it easier for supply chain managers to incorporate the Five Domains if they are already familiar with Welfare Quality. The first three domains are very similar. The main difference is in the fourth domain of Behavioral Interactions. Welfare Quality includes positive emotional states in the fourth principle. In the Five Domains, positive emotions are removed from this section, and they become part of the fifth domain. It contains both positive and negative affective emotional states that cannot be directly measured. Before discussing specific ways to incorporate the Five Domains, the author is first going to discuss how commercial supply chains operate.

COMMERCIAL SUPPLY CHAIN MANAGERS NEED SIMPLE EASY-TO-USE GUIDELINES

During the author's many years of consulting work with many commercial buyers, supply managers, and producers, she has learned that if an assessment guideline becomes too complicated, they cannot effectively implement them. A welfare audit conducted by either a commercial auditing firm or a corporate buyer has to be able to be conducted in a single day on each farm. Assessments and audits that are used commercially must be simpler than measurements and assessments used in research. The commercial reality is, that a typical welfare auditor is trained in a two- or three-day workshop.[10] At the end of the training, they have to take an exam. To become fully qualified, they also have to conduct two or three shadow audits with an experienced auditor. There is also a requirement to attend either online or in person livestock meetings to fulfill requirements for continuing education. This is a very short period of training, compared to the studying that is required to become a veterinarian, or scientific researcher. Many corporations also have advisory boards or welfare officers that have advanced degrees. These people provide advice on a corporation's animal welfare standards. The author has observed that an effective supplier auditing program for animal welfare has three components.[11] They are: (1) independent third-party audits conducted by an auditing company, (2) Audits by corporate supply chain buyers, and (3) Internal self-audits by the farmer or slaughter plant.[11] Further observations by the author indicate that all three of these components are essential to help ensure that animal welfare guidelines are being followed.

A commercial supply chain manager also needs clearly written guidance on both poor practices and housing that would result in a failed audit. This is essential from a legal standpoint. When a supplier has to be "delisted" and removed from a company's approved supplier list, the reason for failure has to be clear. The author observed a bad situation where a company that was delisted sued a third-party independent auditing company. The delisted supplier argued that the guideline was vague and did not clearly specify the reason for being delisted.

Wording in a guideline must not be vague.[11] For example, wording such as provide sufficient space or handle animals calmly is too vague. There is no easy objective way to train an auditor to assess this. Two examples of clear guidelines are either a numerical space requirement for a housing, or a statement that all the animals must have sufficient space to all be able to lie down at the same time without resting on top of another animal. Photographs that show a correctly stocked pen and an overstocked pen are another easy method to provide guidance.[12] Another example of vague unclear guidance is the term "unnecessary suffering". This statement appears in both the United Kingdom and Irish animal welfare acts.[13,14] This will not be effective in a commercial system because there are too many different ways that it can be interpreted.

Non-compliances that would result in an automatic failed welfare audit

Many existing welfare guidance documents that are used commercially or by a government have criteria for severe animal welfare problems. Most of these conditions or abusive acts by people would result in being delisted in a commercial system or major regulatory penalties in a government system. From a supply chain manager's viewpoint, it is essential that the wording is very clear.

Acts of abuse or neglect—automatic failure

Managers of supply chains need to enforce severe penalties on suppliers who allow acts of abuse to occur. Photos of people abusing animals or neglected health problems can go viral online. This may be really costly for a food company.

Below is a list of Acts of Abuse that would result in an automatic failed audit.

During a slaughter audit, cutting or dismembering an animal that is showing signs of returning to consciousness.[15, 16]

Allowing conscious poultry or pigs to enter the scaulder.

Acts of abuse during handling such as dragging conscious non-ambulatory animals,[16] beating animals,[15-17] breaking tails,[17, 18] poking sensitive areas of the animal such as rectum, eyes, mouth, ears, or udder,[15] or lifting sheep or goats by the wool or horns.[19]

Severe neglected health problems such as necrotic prolapses, necrotic ocular neoplasia that has invaded the face or deep infected cuts.

Key welfare indicators

Key welfare indicators are Animal based measurements that seriously compromise animal welfare. These indicators are scored and tabulated as the percentage of the animals that have a condition that is a serious welfare problem. Other names for these indicators are core criteria,[15] critical control points, or critical non-compliances. They identify the most important welfare problems. One example that has been determined by experts is lameness (difficulty walking). In the broiler poultry industry, the author is currently working with a buyer who is in the process of adopting three major key welfare indicators for broiler chickens. Two of them are foot pad lesions and hock burn and they will be used in a global supply chain. These two indicators are associated with poor housing, and they can be easily measured at slaughter. The original Welfare Quality protocols were too time consuming for use in many commercial systems. Shorter versions that use the key indicator concept are being developed for indoor beef cattle and have already been implemented for dairy cattle.[20, 21]

Animal housing specifications

If a commercial guideline prohibits certain types of housing such as sow gestation stalls or small battery cages for laying hens, the audit is failed if the prohibited housing is being used. The guideline should clearly state the types of housing that are definitely not permitted. The housing guidelines must also clearly state things that are required. Some examples may be pasture access for dairy cows, nest boxes for laying hens or straw bedding for pigs. Both the supply chain manager and the auditor need clear guidance on both prohibited forms of housing and required housing features. The guideline should not be too prescriptive on permitted housing, because producers need to be free to innovate.

Problems with a single combined welfare score

The Welfare Quality system[9] attempted to convert a large number of welfare measurements into a single score. One of the problems with a single score is that a serious welfare problem may be concealed.[22] In one study, this enabled a dairy with 47% lame cows to pass a welfare audit because they had high scores on other welfare indicators such as access to clean water.[23] Other researchers have also found problems with converting animal welfare data into a single score.[24] The author has observed that supply chain managers like numbers. Many managers want to convert the data from each animal welfare audit into a single score for each farm or slaughter plant. One way to successfully do this, is to have two different percentage levels for the most important Animal based measures. There would be a high automatic fail level and a lower level that would be a substantial number of points off. This could be done with measures for lameness, body condition score, animal cleanliness and other key welfare indicators. There are certain egregious abusive acts which should always result in an automatic failure. The number of points off for different key welfare indicators must be calculated so that a combined aggregate score does not conceal a serious welfare problem.

Emphasize conditions an auditor can directly observe

In the author's work as a welfare consultant to commercial supply managers, she recommends putting the most emphasis on conditions that can be directly observed. A farm either has the specified housing or does not have it. Animal based key welfare indicators can be easily observed. There are many scoring tools available for different species. Some of the examples of readily available scoring tools are feather condition in laying hens,[25] lameness in cattle,[26, 27] hock swelling in dairy cows,[9, 28] and gait score in broiler chickens. In systems where the farms are integrated with a slaughter facility, many on-farm measurements can be easily done at the slaughter plant.[30] A common mistake in developing an effective commercial supply chain welfare auditing program is an over reliance on data from records. The author has been a welfare auditor for many years, and she has observed many falsified records. Unfortunately, falsified records are very common.

The OIE and other welfare assessments also put a lot of emphasis on having animal-based outcome measures[18] or measurables. For example, instead of specifying exactly how to design a dairy cow cubicle (free stall), the outcome of either poorly designed or poorly managed cubicles is assessed. Problems with cubicle design or management would result in a greater percentage of lame cows or cows with swollen hocks.[28] A recent easy-to-use dairy farm welfare assessment has been developed in the Netherlands.[21] It contains specifications for free stall dimensions. This resource measurement will work well in the Netherlands, but it may not be effective in a worldwide supply system with many types of dairies. In these situations, a greater reliance on outcome measures may be required.

INCORPORATING THE FIVE DOMAINS INTO EXISTING COMMERCIAL WELFARE AUDITING PROGRAMS

For supply managers who are already using Welfare Quality assessments, the Five Domains will be easy to implement. As discussed in the Introduction, the first four domains are very similar to Welfare Quality.[9] Welfare Quality has four welfare areas of good feeding, good housing, good health, and appropriate behavior. The four domains listed above all influence the Fifth Affective Domain, which cannot be directly measured.[1,2]

Animal-based outcome measures for the four domains that can be directly assessed

The First Domain: Nutrition

Body condition scoring is one method for assessing good nutrition. For example, there are many different body condition scoring tools for assessing dairy cows.[9,31-33] All of these tools are slightly different, and a supply chain manager has to specify which specific tool is used in their system. For beef cows living on arid pastures, it is essential to assess body condition to insure they are eating sufficient feed. Body condition of breeding sows, ewes, and all farm animals needs to be assessed to insure that they are getting sufficient feed.[34-37] To help improve interobserver reliability, visual scoring charts should be made available to welfare auditors. These photographic visual aids will make it easy for auditors to identify noncompliant skinny animals. Auditors should be encouraged to have these charts on either their phone or laminated cards. They should refer to them often. Nutrition is not the only cause of poor body condition. Parasites or disease may also reduce body condition.[38, 39] The Welfare Quality system puts a lot of emphasis on both clean water and animal cleanliness. Their protocols have good photographic tools for scoring water trough cleanliness.[9]

The Second Domain: Environment

The second domain covers both the environment in a handling facility and the environment in housing. Problems with animal handling may be associated with deficiencies in the environment such as slick floors[40] or design mistakes in the handling facility.[40] Two important measurables that can be scored numerically during animal handling are slipping and falling.[15, 18, 19, 41, 42] Animals may become injured or stressed if they fall. Ease of animal movement through a handling facility, on a farm or at a slaughter plant can be assessed by counting the numbers of animals, turning back, stopping, balking or refusing to move forward.[9, 19, 43]

If the housing is too hot, outcome measurables such as panting in cattle and sheep can be used as indicators of heat stress.[19, 44, 45] Cold stress in piglets can be assessed by observing huddling or piling.[46] Another important measure is cleanliness of the hide or feathers of both livestock and poultry. Animals that are lying in wet manure may have poor welfare and lack a positive affective state. Some examples of hygiene scoring tools can be found in [9,11]. In both cattle and poultry, a poor environment may also be associated with damage to either the skin or feathers. In broiler chickens, poor conditions of the litter can be assessed by measuring the percentage of birds with breast blisters and foot pad lesions.[48] Groups of broilers with foot pad and lung lesions had more deads on arrival at the abattoir.[49] In laying hens, feather damage associated with housing can be assessed.[25, 50] Another important variable which should be evaluated is atmospheric ammonia levels. High ammonia levels are detrimental to welfare.[51]

In every supply chain there will be requirements and specifications for the types of housing that are permitted. Europe has already banned sow gestation stalls[52] but in other countries, they are still allowed. Many food companies have specifications that require group sow housing for gestating sows. In pork supply chains where group housing is required, there needs to be clear guidance on whether individual gestation stalls can be used for a short period of time for breeding and pregnancy conformation. Both the supply chain managers and the auditors need clear guidance on types of housing systems that are not allowed. It is important to avoid being too prescriptive on housing design because producers need to have the freedom to innovate. For laying hens, there are many types of systems that are now available to replace small battery cages.[53] When evaluating mortality data from different types of hen housing systems, the effect of a producer's experience with a new system has to be taken into account. When a producer gains more experience, mortality often decreases.[53]

In high animal welfare product lines, there also needs to be clear guidance on both bedding and environmental enrichment requirements. For example, the author has visited both excellent and really dirty straw bedded systems for pigs and cattle. In the dirty facilities, the producer did not use sufficient straw to keep the animals clean. Scoring animal cleanliness would have quickly detected this problem. Guidance is also required for environmental enrichment. It is beyond the scope of this paper to provide a review of all of the research on environmental enrichment. Below are some examples of typical environmental enrichments. For pigs, a variety of objects that they can chew are now commercially available.[54] Pigs prefer environmental enrichment devices that are chewable.[54] A ball on a chain is not effective.[55] The author has observed broiler chickens actively using ramps where they can either climb on top or hide under them. Another enrichment device that broilers will actually use are peck stones. For cattle or sheep, pasture access is required in many high welfare programs. One study showed that pasture may be a highly rewarding environment.[56] Two of the most innovative enrichments are laser beams for broilers[57] and motorized grooming brushes for dairy cows. Research clearly shows that cows are highly motivated to use the brushes.[58] The motorized grooming brush would be an example of an animal experiencing a positive affective state.[58]

The Third Domain: Health

Good health is essential to have good welfare, but health alone is not sufficient. An animal can be healthy and free of disease and still engage in abnormal repetitive stereotypic behavior.[59] On each farm, data must be collected on both mortality and morbidity for all species of animals. Lameness (difficulty walking) has been placed in the health domain because it may be associated with either disease[60] or deficiencies in the environment.[28] Many welfare specialists consider lameness to be one of the most serious welfare issues.[61-63] There are many scoring tools available that have been previously discussed. It is important for a supply chain manager to use the same scoring tool throughout their supply chain. Some tools use a "0" for the best rating and others use a "1" for best rating. Some of the popular lameness scoring tools for cattle are in [9, 27, 63, 64]. Information on scoring lameness in sheep can be easily found.[65] Leg problems are really common in broiler chickens. Information on leg problems and lameness in broiler chickens can be found in [29].

Any condition that causes pain is included in the Health Domain,[1,2] such as broken bones and bruises during handling and transport. Painful procedures such as dehorning and castration are also included in the Health Domain.[1,2] Animals can definitely experience and feel pain.[66] This would create a negative affective state in the Fifth Domain. Providing analgesics for pain relief definitely reduced indicators of stress such as high cortisol levels in the blood.[66] Research also showed that beak trimming in chickens may cause long-term pain.[67] To assess bruising, there are many scoring tools that are available.[68] The acts of abuse that were described earlier in this paper, would be included in the Health Domain because they cause pain. The effectiveness of stunning methods at slaughter and euthanasia on the farm would also be in the Health Domain. Poor stunning or poor euthanasia causes pain. There are many guidelines for assessing the effectiveness of stunning at slaughter.[4, 15, 69] Some of the problems that can be easily monitored at the slaughter plant are foot pad lesions in poultry, lameness in livestock and dirty animals.[30, 70] Monitoring of compliance with housing requirements and the use of analgesics after surgery cannot be assessed at a slaughter plant. Most of the currently available scoring tools assess conditions that would cause pain or discomfort. There is a need to also have easy-to-use assessments for positive experiences.

The Fourth Domain: Behavioral Interactions

Mellor has split the Behavioral Interaction domain into three parts.[1] They are:

1. Animal interactions with the environment
2. Animal interactions with other animals
3. Animal interactions with people[2]

Some examples of outcome measurables for animal interactions with the environment may be either repetitive stereotypic behavior (negative)[71] or motivation to use enrichment devices (positive).[72] Some of the most highly motivated behavioral needs are: nest boxes for laying hens,[73] materials for pigs to chew or root,[54] motorized brushes for cows,[58] and devices for broilers to hide under, climb on, or peck.[74] Supply managers will need to rely on evidence from research studies to determine which enrichments that farms in their supply chain should be required to have.

For the second category on interactions with other animals, there have been many studies on damaging behaviors inflicted by pen-mates or flock-mates. Damage can result from feather pecking in hens,[75] wounds from fighting in pigs,[76] and tail biting in pigs.[77] A recent study showed that sows from one genetic line can be selected to be less aggressive and have fewer injuries when they are mixed.[76] Unfortunately, the piglets from the less aggressive genetic line had lower survivability. Breeding to reduce aggression in group housed sows reduced piglet numbers. It is possible that this issue may not occur in all sow genetic lines. Breeders are challenged to think about difficult tradeoffs between animal welfare and economics. All of the above would be examples of behaviors that would lead to a negative affective state. There is also need to assess positive behaviors between animals such as grooming each other.[78]

For assessing the third section on interactions with humans, there are many handling scoring systems for assessing handling.[9, 15, 69, 79, 80] Numerical scoring of animal handling is already being used by many companies such as McDonald's[5] and commercial third-party auditing companies. The following variables are scored in livestock handling systems that are used for loading trucks, vaccination or moving animals to the stunner at a slaughter plant. They are easy to measure indicators of poor handling practices that compromise animal welfare. Some examples of common measurements are the percentage of animals moved with electric prod (goad)[30, 42, 43, 80] and vocalization during handling and restraint of cattle.[30, 43, 79, 81, 82] Vocalization during handling in cattle is associated with high cortisol levels.[81, 82] Squealing in pigs is also used to assess handling practices.[9, 15] In pigs, squealing is associated with physiological measures of stress.[83] Another measure that has been used in both cattle and sheep is speed of exiting after an animal is released from restraint.[84-87] High speeds are associated with greater stress. The percentage of animals turning back during handling or falling is also part of many assessments of animal handling.[9,15,69] These two variables have already been discussed in the Environment Section. Good stockmanship practices such as moving small groups of cattle and not yelling resulted in improved scores on handling measures.[88]

All of the above handling measurements assess negative interactions between stock people and animals. There is also evidence that both positive attitudes by handlers and positive interactions between people and animals improved both productivity and welfare. Dairy cows with low somatic cell counts were more willing to approach people. There is a need to develop simple measures for positive interaction between people and animals. Most of the measures which are currently being used to assess welfare in commercial supply chains are used to prevent suffering.

SUMMARY OF KEY WELFARE INDICATORS FOR USE WITH THE FIVE DOMAINS

Tables 1–5 contain key welfare indicators for cattle, swine, broiler chickens, and laying hens. References to both scientific studies and easy to use scoring tools are included on the tables. When welfare auditors have to be trained in a very short time, the use of videos and pictorial scoring aids is really useful. Recently one of the author's former graduate students was hired as a welfare auditor. He is now traveling all over the U.S. and assessing slaughter plants, dairies, laying hens, broiler chickens, and pig farms. To help him evaluate body condition score, feather condition, lameness, and other key welfare indicators, the author helped him download some of the most useful online scoring tools. References for these scoring tools are in Tables 1–5. Information that is available on open access is given priority on the tables. Corporate managers often stop at paywalls. Their reading of scientific literature is often limited to open access materials. In many countries, managers cannot afford to pay fees to download papers. In the future, new welfare scoring tools will be developed. The author encourages open access publishing of scoring tools used to assess animal welfare.

TABLE 1: Guide to Assessment Tools for Key Welfare Indicators for Beef Cattle

Domain	Parameter Assessed	
First: Nutrition	Body condition of breeding cows	96
	Water trough cleanliness	97
Second: Environment	Slipping and falling during handling	15,18,79,98
	Turning back balking during handling	97
	Heat stress—Open mouth breathing	44,45
	Hygiene Scoring—Dirty hide	97
	Housing Requirements and Specifications	
Third: Health	Lameness—All types of cattle	20,26,64,99
	Bruise scoring	68
	Effectiveness of stunning at slaughter	15,69
	Swollen hocks	97
Fourth: Behavior	Electric prod use during handling	15,69,79,98
	Vocalization during handling	15,69,79,98
	Acts of abuse	15,16,17,18
	Animal refusing to move forward during handling	43,97

TABLE 2: Key Welfare Indicators for Dairy Cows		
Domain	**Parameter Assessed**	
First: Nutrition	Body condition of dairy cows	21,33,100,101
	Water trough cleanliness	21,97
Second: Environment	Slipping and falling during handling	15,16,17,18
	Hygiene scoring—Dirty hide	21,100
	Swollen hocks	21,28,102
	Housing Requirements and Specifications	
Third: Health	Lameness	21,27,103
	Bruises at slaughter	68
	Effectiveness of stunning at slaughter	15,69
Fourth: Behavior	Electric prod use during handling	15,69,79,98
	Vocalization during handling	15,69,79,98
	Flight distance from people	21
	Acts of abuse	15,16,17,18
	Environmental enrichment requirements	58

TABLE 3: Key Welfare Indicators for Pigs

Domain	Parameter Assessed	
First: Nutrition	Body condition of breeding sows	34
	Water trough cleanliness	104
Second: Environment	Slipping and falling during handling	15,18
	Shoulder lesions	105
	Swollen joints and hoof damage	106
	Huddling—Cold stress piglets	10,46,105
	Housing requirements and specifications	
Third: Health	Lameness sows and finishing pigs	106,107
	Skin damage	70,108
	Effectiveness of stunning at slaughter	15,69
	Hernias	104
	Tail damage	70,77
	Housing requirements and specifications	
Fourth: Behavior	Electric prod using during handling	15
	Vocalization scoring during handling	15
	Lesions from fighting	105
	Stereotypic behavior	59
	Acts of abuse	15,16,17,18
	Environmental enrichment	54,55

TABLE 4: Key Welfare Indicators for Laying Hens		
Domain	**Parameter Assessed**	
First: Nutrition	Water	
Second: Environment	Feather condition scoring	25
	Foot pad lesions	109
	Wounds	25
	Foot pads	25
	Housing requirements and specifications	
Third: Health	Beak abnormalities	109
	Wounds	25
Fourth: Behavior	Feather pecking	25

TABLE 5: Key Welfare Indicators for Broiler Chickens		
Domain	**Parameter Assessed**	
First: Nutrition	Water	109
Second: Environment	Breast blisters	110
	Hock burn	110
	Foot pad lesions	48,110
	Cleanliness of plumage	47,110
	Space—Day of catch: Birds can move away 1 m	
	Housing requirements and specifications	
Third: Health	3-point gait scoring	29,61,109,111
	Bruise scoring	
	Effectiveness of stunning	
Fourth: Behavior	Environmental enrichments	57
	Acts of abuse	
	Broken wings due to poor handling	11

In the future, many of the key welfare indicators will be able to be evaluated at slaughter with camera systems and artificial intelligence programs. These systems are already being developed for scoring foot pad lesions in broiler chickens.[92, 93] In pigs, there are systems for assessing tail damage in pigs[94] and body condition in cattle.[95]

THE IMPORTANCE OF COMMITMENT BY BOTH CORPORATE SUPPLY CHAIN MANAGERS AND UPPER MANAGEMENT

During the last twenty-five years, the author has worked with both training animal welfare auditors and has served on the welfare panels of many large corporations, some of which include McDonald's Corporation, Wendy International, Tyson, Costco Foods, Maple Leaf Foods, and others. The most effective programs have managers who are committed to improving animal welfare. The author has observed either an improvement in a company's program, or a decline with a change in top management. When the corporate welfare auditing programs started in the late 90s, the author took executives from McDonald's and other companies on their first tours of farms and slaughterhouses. During these initial tours, the author observed that animal welfare changed from being an abstract issue that was delegated to the legal or public relations department, to a real issue that needed to be addressed. When one executive observed debilitated, emaciated dairy cows going into their product, they were appalled. Seeing bad animal welfare motivated them to start auditing programs and implement improvements. A retired executive from McDonald's, has written a book about his experiences with starting the McDonald's slaughter plant audits.[112] Recently I was with a top animal welfare executive from another company. We were visiting a beef slaughter plant that he purchased from. He became upset when he saw skinny emaciated organic cull dairy cows. This motivated him to make improvements.

To be effective, corporate buyers have to get out of the office and see what is occurring on farms and in slaughterhouses. Reading third party auditor reports in the office is not sufficient. Over the years, the author has observed that the best programs have senior executives who actually visit their supplier's facilities. The author has to disclose that she is currently a paid consultant for McDonald's Corporation and Costco Corporation. For over twenty years, she has continued to observe a reoccurring pattern. Big corporations have constant shifts in top management. She has observed that the effectiveness of their animal welfare program may vary depending on the objectives of top managers. The greatest improvements occur when motivated supply chain managers are allowed to get out on the farms and make changes. Most supply chain managers and animal welfare officers the author has worked with want to make improvements. Recently the author talked to a supply chain person who felt that upper management held them back. That person was told to wait until upper management changed, and then they may have a window of opportunity to make significant improvements. Supply chain managers and animal welfare officers who work for large corporations are in a position to greatly improve animal welfare on farms and in slaughterhouses. The information in this paper will help them to implement effective programs.

CONCLUSIONS

The Five Domains Framework can be easily incorporated into many existing animal welfare programs conducted by commercial supply chain managers. They can also be incorporated into programs of continuing improvement. Many supply chain managers can relate to The Five Domains because it goes beyond the prevention of suffering.

REFERENCES

1. Mellor, D.J. Updating animal welfare thinking: Moving beyond the "Five Freedoms" towards a life worth living. Animals 2015, 6, 21.
2. Mellor, D.J.; Beausoleil, N.J.; Littlewood, K.E.; McLean, A.N.; McGueevy, P.D.; James, B.; Wilkins, C. The 2020 Five Domains Model: Including Human-Animal Interactions in Assessments of Animal Welfare. Animals 2020, 10, 1870.
3. Tyson Foods. https://www.tysonfoods.com/news/newsreleases/2021/7/tysonfoodsintegratingfive domainsanimalwelfareframeworkacross (accessed on 21 August 2022).
4. Grandin, T. Effect of animal welfare audits of slaughter plants by a major fast-food company on cattle handling and stunning practices. J. Amer. Vet. Med. Assoc. 2000, 216, 848–851.
5. Grandin, T. Maintenance of good animal welfare standards in beef slaughter plants by use of auditing programs. J. Amer. Vet. Assoc. 2005, 226, 370–373.
6. FAWC (Farm Animal Welfare Council). FAWC Updates—Five Free. Vet. Res. 1992, 131, 357.
7. Birch, J.; Schell, A.K.; Clayton, N.S. Dimensions of animal consciousness. Trends Cogn. Sci. 2020, 24, 789–901.
8. Panksepp, J. The basic emotional circuits of mammalian brains: Do animals have affective lives? Neurosci. Biobehav. Rev. 2011, 35, 1791–1804.
9. Welfarequalitynetwork.net. Assessment Protocols, Welfare Quality. 2018. http://www.welfarequalitynetwork. net/enus/reports/assessmentprotocols (accessed on 4 October 2022).
10. PAACO (Professional Animal Auditor Certification Organization). Kearney, Missouri, USA. 2022. https://www.animalauditor.org (accessed on 27 August 2022).

11. Grandin, T. (Ed.) Implementing Effective Animal Based Measurements for Assessing Animal Welfare on Farms and Slaughter Plants. In Improving Animal Welfare, 3rd ed.; CABI Publishing: Wallingford, Oxfordshire, UK, 2021; pp. 60–83.

12. Kline, H.C.; Edwards Callaway, L.N.; Grandin, T. Short Communication: Field observation, pen stacking capacities for overnight lairage of finished steers and heifers in a commercial slaughter facility. Appl. Anim. Sci. 2019, 35, 130–133.

13. UK Animal Welfare Act. UK Public General Ach, Preventing Harm, Section 4, Legislation. 2006. https://www.legislation.gov.uk/ukpga/2006/45/section/4/enacted (accessed on 4 July 2022).

14. Irish Animal Health Welfare Act. 2013. https://www.irishstatutebook.ie/el:/2013/act/15/enacted/en/print. html (accessed on 16 October 2022).

15. NAMI—Recommended Animal Handling Guidelines and Audit Guide, Revision 2, North American Meat Institute. Washington, DC, USA, 2021. https://www.animalhandling.org/producers/guidelines_audits (accessed on 21 August 2022).

16. FSIS/USDA. Humane Handling and Slaughter of Livestock, Revision 3, FSIS Directive 6900.2, 24 September 2020, Food Safety and Inspection Service, United States Dept. of Agriculture, Washington, DC, USA, FSIS.USD. https://www.gov/policy/fsisdirectives/6900.2 (accessed on 6 July 2022).

17. Gov.UK. Guidance at Farm Shows and Markets: Welfare Regulations, Dept. for the Environment and Rural Affairs (DEFRA). 2019. https://www.gov.uk/guidance/farmedanimalwelfareatshowsandmarkets (accessed on 6 July 2022).

18. OIE (World Animal Health Organization). Chapter 7.3 Transport of Animals by Land, Terrestrial, Animal Health Code. 2021. https://www.woah.org/whatwedpstandards/codesandmanuals/terrestrialcodeonlineaccess/?id= 16a8L=18htmfile=chapiter_aw_land_transphtm (accessed on 16 October 2022).

19. EFSAWelfare of Sheep and Goats at Slaughter, EFSA European Food Safety Authority. 2021. https://efsa. onlinelibrary.wiley.com/doi.2903/j.efsa.2021.6882 (accessed on 16 October 2022).

20. Lorenzi, V.; Sagoifo Rossi, C.A.; Compiani, R.; Grossi, S.; Bolzoni, L.; Marza, F.; Clemente, G.A.; Fusi, F.; Bertocchi, L. Using expert elicitation for ranking hazards, promoters and animal-based measures for on-farm welfare assessment of indoor reared beef cattle: An Italian experience. Vet. Res. Commun. 2022.

21. Van Eerdenbury, F.J.C.M.; DiGianto, A.M.; Hulsen, J.; Snel, B.; Stegman, A.J. A new practical animal welfare assessment for dairy farms. Animals 2021, 11, 881.

22. Sandoe, P.; Corr, S.A.; Lund, T.B.; Forkman, B. Aggregating animal welfare indicators: Can it be done in a transparent and ethnically robust way? Anim. Welf. 2019, 28, 67–76.

23. Devries, M.; Bokker, E.A.M.; van Schaik, G.; Bottreau, R.; Engel, B. Evaluating the results of Welfare Quality multicriteria evaluation model for classification of dairy cow welfare at the herd level. J. Dairy Sci. 2013, 96, 6264–6273.

24. De Graaf, S.; Ampe, B.; Buijs, S.; Andreason, S.N.; de Boyer Des Roches, A.; van Eerdenburg, F.J.C.M.; Haskell, M.J.; Kirchner, M.K.; Mounier, L.; Radeski, M.; et al. Sensitivity of integrated Welfare Quality scores to changing values in individual dairy cattle welfare measures. Anim. Welf. 2018, 27, 156–157.

25. European Research Program. Laywel Welfare Implications of Changes in Production Systems for Laying Hens. https://www.laywel.eu (accessed on 21 August 2022).

26. EdwardsCallaway, L.N.; CalvoLorenzo, M.S.; Scanga, J.A.; Grandin, T. Mobility scoring in finished cattle. Vet. Clin. N. Amer. Food. Anim. Pract. 2017, 33, 235–250.

27. Zinpro Locomotion Scoring. Dairy Australia, Zinpro Corporation, Eden Prairie, Minnesota. https://www. youtube.com/watch?v=OPHOM-WSK518 (accessed on 6 July 2022).

28. Fulwider, W.K.; Grandin, T.; Garrick, D.J.; Engle, T.L.; Lamm, W.D. Influence of free stall base on tarsal joint lesions and hygiene in dairy cows. J. Dairy Sci. 2007, 90, 3559–3566.

29. Knowles, T.G.; Kestin, S.C.; Haslam, S.M.; Brown, S.N.; Green, L.E.; Butterworth, A.; Pope, S.J.; Pfeiffer, D.; Nicol, C.J. Leg disorders in broiler chickens, prevalence risk factors, and prevention. PLoS ONE 2008, 3, e1545.

30. Grandin, T. On-farm conditions that compromise animal welfare that can be monitored at the slaughter plant. Meat Sci. 2017, 132, 52–58.

31. Wildman, E.E.; Jones, G.M.; Wagner, P.E.; Boman, R.L.; Troutt, H.F., Jr.; Lesch, T.N. Body condition scoring system and its relationship to selected production characteristics. J. Dairy Sci. 1982, 65, 495–501.

32. Ferguson, J.D.; Galligan, D.T.; Thomsen, N. Principle descriptions of body condition score in Holstein cows. J. Dairy Sci. 1994, 77, 2695–2703.

33. Mullins, I.J.; Truma, C.M.; Campler, M.R.; Bewly, J.P.; Costa, J.H.C. Validation of a commercial body condition scoring system on a commercial dairy farm. Animals 2019, 9, 287.

34. Coffey, R.D.; Parker, G.R.; Laurent, K.M. Assessing Sow Body Condition, Cooperative Extension Service, University of Kentucky. 1999. https://www.ASC158ww2.ca.uky.edu/agcomm/pubs/asc158/asc158.pdf (accessed on 10 July 2022).

35. Lionch, P.; King, E.M.; Clarke, K.A.; Down, J.M.; Green, L.E. A systematic review of animal-based indicators of sheep welfare on farm and market an during transport and qualitative appraisal of their validity and feasibility for use in abattoirs. Vet. J. 2015, 206, 289–297.

36. Thompson, J.; Meyer, H. Body Condition Scoring of Sheep, Oregon State University Extension, Corvallis, Oregon, US EC1433. 1994. https://www.agsci.Oregonstate.edu/sites/agscid7/files/ec1433.pdf (accessed on 21 August 2022).

37. Farm Advisory Service, Scottish Government Technical note TN702, Body Condition Scoring of Mature Sheep, National Advice Hub. 2018. fas.scot/downloads/tn/702bodyconditionscoringmaturesheep/ (accessed on 21 August 2022).

38. Esteves, A.; VieiraPinto, M.; Quintas, H.; Orge, L.; Gama, A.; Alves, A.; Seeixas, F.; Pirex, I.; de Lordes Pinto, M.; Mendonca, A.P. Seraphic at abattoir: Monitoring, control and differential diagnosis of wasting conditions at meat inspection. Animals 2021, 11, 3028.

39. Cornelius, M.P.; Jacobson, C.; Besier, R.B. Body condition score as a selection tool for targeted selective treatment based nematode control strategies in Merino ewes. Vet. Parasitol. 2014, 206, 173–181.

40. Grandin, T. (Ed.) How to Improve Livestock Handling and Reduce Stress. In Improving Animal Welfare: A Practical Approach, 3rd ed.; CABI Publishing: Wallingford, Oxfordshire, UK, 2021; pp. 84–112.

41. Edge, M.K.; Barnett, J.L. Development and integration of animal welfare standards into company quality assurance programs in Australian livestock (meat) processing industry. Aust. J. Exper. Agric. 2008, 48, 1009–1013.

42. Simon, G.E.; Hoar, B.R.; Tucker, C.B. Assessing cowcalf welfare, Part 1: Benchmarking beef cow health and behavior during handling and management, facilities and producer perspective. J. Anim Sci. 2016, 94, 3476–3487.

43. Hultgren, J.; Segekrist, K.A.; Berg, C.; Karlsson, A.H.; Alberts, B.O. Animal handling and stress related behavior at mobile slaughter of cattle. Prev. Vet. Med. 2020, 177, 104959.

44. Gaughan, J.R.; Mader, T.L. Body temperature and respiratory dynamics in unshaded beef cattle. Int. J. Biometeor. 2014, 58, 1443–1450.

45. Gaughan, J.B.; Mader, T.L.; Hult, S.M.; Lisle, A. A new heat load index for feedlot cattle. J. Anim. Sci. 2008, 86, 226–234.

46. VillanuevaGarcía, D.; MotaRojas, D.; MartínezBurnes, J.; OlmosHernández, A.; MoraMedina, P.; Salmer-n, C.; G-mez, J.; Boscato, L.; GutiérrezPérez, O.; Cruz, V.; et al. Hypothermia in newly born piglets: Mechanisms of thermoregulation and pathophysiology of death. J. Anim. Behav. Biometeorol. 2020, 9, 1–10.

47. Saraiva, S.; Saraiva, C.; Stilwell, G. Feather conditions and clinical scores as indicators of broilers welfare at the slaughterhouse. Res. Vet. Sci. 2016, 107, 75–99.

48. Heitmann, S.; Stracke, J.; Petersen, H.; Spindler, B.; Kemper, N. First approach validating a scoring system for foot pad dermatitis in broiler chickens developed for application in practice. Prev. Vet. Med. 2018, 154, 63–70.

49. Kettelsen, K.E.; Moe, R.O.; Hoel, K.; Kolbjornsen, O.; Nafstad, O.; Granquist, E.G. Comparison of flock characteristics, journey, duration and pathology between flocks with a normal and a high percentage of broilers, dead on arrival at abattoirs. Animals 2017, 12, 2301–2308.

50. Mollenhorst, H.; Rodenburg, T.B.; Borkkers, E.A.M.; Koene, P.; DeBoer, I.J.M. On-farm assessment of laying hen welfare: A comparison of one environment based and two animal-based methods. Appl. Anim. Behav. Sci. 2005, 90, 277–291.

51. Naseem, S.; King, A.T. Ammonia production in poultry houses can affect health of humans, birds, and the environment: Techniques for its reduction during poultry production. Environ. Sci. Pollut. Res. Int. 2018, 16, 15269–15293.

52. European Parliament—Parliamentary Questions, Implementation of Ban on Individual Sow Stalls in Force Since 1 January 2013 in Accordance with Directive 20087/120/EC on Production of Pigs, European Parliament. 2013. https://www.europarl.euro.eu/doceo/document/E72013000321_EN.html?redirect (accessed on 10 July 2022).

53. SchuckPaim, C.; NegroCalduch, E.; Alonso, W.J. Laying hen mortality in different indoor housing systems: A metaanalysis of data from commercial farms in 16 countries. Sci. Rep. 2021, 11, 3052.

54. Vanderweerd, H.; Ison, S. Providing effective environmental enrichment for pigs: How far have we come? Animals 2019, 9, 254.

55. Bracke, M.B.M.; Koene, P. Expert opinion on metal chains and other indestructible objects as proper enrichment for intensively farmed pigs. PLoS ONE 2019, 14, e0212610.

56. Crump, A.; Jenkins, K.; Bethell, E.J.; Ferris, C.P.; Kabboush, H.; Weller, J.; Arnott, G.I. Optimism and pasture access in dairy cows. Sci. Rep. 2021, 1, 4882.

57. Meyer, M.M.; Johnson, A.K.; Bobeck, E.A. A novel environmental enrichment device increased physical activity and walking distance in broilers. Poult. Sci. 2020, 99, 48–60.

58. McConnachiz, E.; Maricke, A.; Alexander, C.S.; Thompson, J.; Weary, D.M.; Gaworski, M.A.; VonKeyserlingk, A.G. Cows are highly motivated to access grooming substrate. Biol. Lett. 2018, 14, 20180303.

59. Radkowska, I.; Dogyn, D.; Fie, K. Stereotypic behavior in cattle, pigs, and horses. Anim. Sci. Pap. Rep. 2020, 38, 303–319.

60. Trott, D.J.; Moeller, M.R.; Zuerner, R.L.; Goff, J.P.; Waters, W.R.; Alt, D.P.; Walker, R.L.; Wannamuehler, M.J. Characterization of Treponema phagedenislikespirochates isolated from papillamatous digital dermatitis in dairy cattle. J. Clin. Microbiol. 2003, 41, 2522–2529.

61. Webster, A.B.; Fairchild, B.D.; Cummings, T.S.; Stayer, P.A. Validation of a threepoint gait scoring system for field assessment of walking ability of commercial broilers. J. Appl. Poult. Res. 2008, 17, 529–539.

62. Whay, H.R.; Main, D.C.L.; Green, L.F.; Webster, A.J.E. Assessment of the welfare of dairy cattle using animal based measurements: Direct observation and investigation of farm records. Vet. Rec. 2003, 153, 197–202.

63. Grandquist, E.G.; Vasdal, G.; DeJong, I.C.; More, R.O. Lameness and its relationship to health and production measures in broilers. Animals 2019, 13, 2365–2372.

64. Daly, R. Lameness in Cattle: Causes associated with infections, South Dakota State University Extension 2021. South Dakota State University: Brookings, SD, USA. https://www.extension.sdstate.edu/lamenesscattlecausesassociated infections (accessed on 16 October 2022).

65. Munoz, C.; Campbell, A.; Hemsworth, P.; Doyle, R. Animal based measures to assess the welfare of extensively raised ewes. Animals 2018, 8, 2.

66. Colpaert, F.C.; Tarayre, J.P.; Alliaga, M.; Slot, L.A.B.; Attal, N. Opiate selfadministration as a measure of chronic pain in arthritic rats. Pain 2001, 91, 33–45.

67. Gentle, M.J.; Waddington, D.; Hunter, L.N.; Jones, R.B. Behavioral evidence of persistent pain following partial beak amputation in chickens. Appl. Anim. Behav. Sci. 1990, 27, 149–157.

68. EdwardsCallaway, L.C.; Kline, H.C. The basics of bruising in cattle. What, When and How. In The Slaughter of Farmed Animals; Grandin, T., Cockram, M., Eds.; CABI Publishing: Wallingford, Oxfordshire, UK, 2020; pp. 78–89.

69. AVMAAVMA. Guidelines for the Humane Slaughter of Animals, 2016 Edition; American Veterinary Medical Association: Schaumberg, IL, USA; https://www.avma.org/site/default/files/resources/Humaneslaughterguidelines.pdf (accessed on 16 October 2022).

70. DeLuca, S.; Zanardim, E.; Alberali, G.L.; Ianiera, A.; Ghidina, S. Abattoir based measures to assess swine welfare: Analysis of method adopted in European slaughterhouses. Animals 2020, 11, 226.

71. Garner, J.P. Stereotypies and other abnormal repetitive behaviors: Potential impact on validity, reliability, and replicability on scientific outcome. ILAR J. 2005, 46, 106–117.

72. Follensbee, M.E.; Duncan, I.J.H.; Widowski, T.M. Quantifying nesting modification of domestic hens. J. Anim. Sci. 1992, 70, 1994.

73. WoodGush, D.G.M.; Gilbert, A.B. Observations of laying behavior in hens in battery cages. Brit. Poult. Sci. 1969, 10, 29–36.

74. Riber, A.B.; van de Weerd, H.A.; DeJong, I.C.; Steenfeldt, S. Review of environmental enrichment for broiler chickens. Poult. Sci. 2018, 97, 378–396.

75. VanStaaverson, N.; Ellis, J.; Baes, C.F.; HarlanderMatauschek, A. A metaanalysis on the effect of environmental enrichment on feather pecking and feather damage in laying hens. Poult. Sci. 2021, 100, 397–411.

76. Brajon, S.; AhloyDallaire, J.; Devillers, N.; Guay, F. The role of genetic selection on agonistic behavior and welfare of gestating sows housed in large semistatis groups. Animals 2020, 10, 2299.

77. Keeling, L.J.; Wallenbeck, A.; Holmgren, N. Scoring tail damage in pigs: An evaluation based on recordings in Swedish slaughterhouse. Aeta Vet. Scand. 2012, 54, 32.

78. De Fresion, I.; Peralta, J.M.; Strappin, A.C.; Monti, G. Understanding allogrooming through a dynamic social network approach: An example in a group of dairy cows. Front. Vet. Sci. 2020, 7, 535.

79. Woiwode, R.; Grandin, T.; Birch, B.; Patterson, J. Compliance of large feed yards in Northern Great Plains with Beef Quality Assurance Feed-yard Assessment. Prof. Anim. Sci. 2016, 32, 750–757.

80. Maria, G.A.; Villarroel, M.; Chacon, G.; Debresenbet, G. Scoring system for evaluating the stress to cattle during commercial loading and unloading. Vet. Rec. 2004, 154, 818–821.

81. Dunn, C.S. Stress reactions in cattle undergoing ritual slaughter using two methods of restraint. Vet. Rec. 1990, 126, 522–525.

82. Hemsworth, P.H.; Rice, M.; Karlan, M.G.; Calleja, L.; Barnett, J.L. Human animal interactions at abattoirs relationships between handling and animal stress in sheep and cattle. Appl. Anim. Behav. Sci. 2022, 135, 24–33.

83. Warris, P.D.; Brown, S.N.; Adams, S.J.M.; Carlett, I.K. Relationship between subjective and objective assessment of stress at slaughter and meat quality in pigs. Meat Sci. 1994, 38, 329–340.

84. Curley, K.O.; Paschal, J.C.; Welsh, T.H.; Randel, R.D. Technical Note. Exit velocity as a measure of cattle temperament is repeatable and associated with serum cortisol in Brahman bulls. J. Anim. Sci. 2006, 84, 3100–3103.

85. Vetters, M.D.D.; Engle, T.E.; Ahola, J.K.; Grandin, T. Comparison of flight speed and exit score measurements and temperament in beef cattle. J. Anim. Sci. 2013, 91, 374–381.

86. Parham, J.T.; Blevins, S.R.; Turner, A.E.; Wahlberg, M.I.; Swecker, W.S.; Lewis, R.M. Subjective methods of quantifying temperament in heifers as indicative of physiological stress. Appl. Anim. Behav. Sci. 2021, 234, 105197.

87. Brown, D.J.; Fogerty, N.M.; Iker, C.L.; Ferguson, D.M.; Blanche, D.; Gaunt, G.M. Genetic evaluation of maternal behavior and temperament in Australian sheep. Anim. Prod. 2015, 56, 767–774.

88. Yost, J.K.; Yates, J.W.; Davis, M.P.; Wilson, M.E. The stockmanship Scorecard: Quantitative evaluation of beef cattle stockmanship. Trans Anim. Sci. 2020, 4, txaa175.

89. Hemsworth, P.H.; Coleman, G.J.; Barnett, J.L.; Borg, S.; Dowling, S. The effects of cognitive behavioral intervention on the attitude and behavior of stockpersons and the behavior and productivity of commercial dairy cows. J. Anim. Sci. 2002, 80, 68–78.

90. Fukasawa, M.; Kawahata, M.; Higashiyama, Y.; Komatsu, T. Relationships between the stockperson's attitude and dairy productivity in Japan. J. Anim. Sci. 2017, 88, 394–400.

91. Kauppinen, T.; Vesala, K.M.; Valros, A. Farmer attitudes towards improvement of animal welfare and piglet production. Livestock Sci. 2012, 143, 142–150.

92. Kaewtapee, C.; Thepparaki, S.; Rakangthong, C.; Burichasak, C.; Supratak, A. Objective scoring of foot pad dermatitis in broiler chickens using image segmentation and a deep learning approach: Camera base system. Br. Poult. Sci. 2022, 3, 427–433.

93. Louton, H.; Bergmann, S.; Pillar, A.; Erhand, M.; Stracke, J.; Spindler, B.; Schmidt, P.; SchultzLandwehr, J.; Schwarzer, A. Automatic system for monitoring foot pad dermatitis in broiler. Agriculture 2022, 12, 221.

94. Wang, S.; Jiang, H.; Qiao, Y.; Jiang, S.; Lin, H.; Sun, Q. The research progress of vision based artificial intelligence in smart pig farming. Sensors 2022, 22, 6541.

95. Albornoz, R.I.; Giri, K.; Hannah, M.C.; Wales, W.J. An improved approach to automated assessment of body condition score in dairy cows using a three-dimensional camera system. Animals 2022, 12, 72.

96. Lahman, D.; Selk, G.; Stein, D. Body Condition Scoring of Cows; Oklahoma State University Extension: Stillwater, OK, USA, 2017.

97. Welfarequalitynetwork.com. Protocol for cattle. 2009. https://www.welfarequality.network.net/media/1088/cattle_protocol_without_vealcalves.pdf (accessed on 27 September 2022).

98. Calabasa, E.; Clowser, M.; Weller, Z.D.; Bigler, L.; Fulton, J.; EdwardsCallaway, L.N. Benchmarking animal handling outcomes on cowcalf operation sand identifying associated factors. Trans. Anim. Sci. 2022, 6, txac106.

99. Meat News Network Mobility Scoring of Cattle; North American Meat Institute: Washington, DC, USA, (VIDEO); https://www.youtube.com/watch?v=feGv03gRack (accessed on 27 September 2022).

100. Cook, N.B. Hygiene Scoring Card; University of Wisconsin: Madison, WI, USA, 2020; https://www.vetmed.wisc. edu/fapm/wpcontent/uploads/2020/01/hygiene.pdf (accessed on 4 October 2022).

101. Ferguson, J.D.; Azzaro, G.; Licita, G. Body condition assessment using digital images. J. Dairy Sci. 2006, 89, 3833–3841.

102. McGill. MacDonald Campus Farm Complex, Injury Scoring for Dairy Cows, Standard Operating Procedure #DC303. 2018. https://www.mcgill.ca/research/files/research/dc303_injury_scoring_for_dairycows.pdf (accessed on 27 September 2022).

103. Dairy Australia. Lameness Scoring Video; Dairy Australia: Southbank, VIC, Australia, 2016.

104. Welfarequalitynetwork.net. Welfare Quality Assessment for Pigs. 2009. https://www.welfarequalitynetwork. net/media/1018/pigprotocol.pdf (accessed on 27 September 2022).

105. National Pork Board. Common Swine Industry Audit; National Pork Board: Des Moines, IA, USA, 2022; Porkcdn.com/com/sites/porkcheckoff/CSIA/2022.CSIA.01.03.22.pdf (accessed on 27 September 2022).

106. Linden, J. The Problem of Lameness on IRISH Pig Farms. 2013. https://www.thepigsite.com (accessed on 1 October 2022).

107. Zinpro. What Does Swine Locomotion Scoring of 0 Look Like? 2018. https://www.youtube.co/watch?v=_laying_hen_protocol_20_defdecember_2019_pdf (accessed on 4 October 2022).

108. Meyer, D.; Hewicker-Trautwein, M.; Hartmann, M.; Beilage, E.G. Scoring shoulder ulcers in breeding sows in a distinction between substantial and insubstantial animal welfare related lesions possible in clinical examination. Porc. Health Mgt. 2019, 5, 1–9.

109. Welfarequalitynetwork.net. Welfare Quality Assessment Protocol for Laying Hens. 2019. https://www.welfarequalitynetwork,net/media/1294/wq_laying_hen_protocol_20_def_december_2019_pdf (accessed on 4 October 2022).

110. Welfarequalitynetwork.net, Welfare Quality Assessment Protocol for Poultry. 2009. https://www.welfarequalitynetwork.net/media/1293/poultry_protocol_watermark_6_2_2020.pdf (accessed on 4 October 2022).

111. Kittelson, K.E.; David, B.; More, R.O.; Poulson, H.D.; Young, J.F.; Granquist, E.G. Associations among gait score, production data, abattoir registrations, and postmortem tibia measurements in broiler chickens. Poult. Sci. 2017, 96, 1033–1040.

112. Langert, B. The Battle to Do Good: Inside the McDonald's Sustainability Journey; Emerald Publishing: West Yorkshire, UK, 2019.